HANDBOOK OF FIBER OPTIC
DATA COMMUNICATION

HANDBOOK OF
FIBER OPTIC
DATA COMMUNICATION

Second Edition

CASIMER DeCUSATIS

Editor
IBM Corporation
Poughkeepsie, New York

ACADEMIC PRESS

An Elsevier Science Imprint

San Diego   London   Boston
New York   Sydney   Tokyo   Toronto

ACADEMIC PRESS
An Elsevier Science Imprint
525 B Street, Suite 1900, San Diego, CA 92101-4495, USA
http://academicpress.com

ACADEMIC PRESS LIMITED
An Elsevier Science Imprint
Harcourt Place, 32 Jamestown Road, London NW1 7BY, UK
http://academicpress.com

Library of Congress Catalog Card Number: 2001095438
International Standard Book Number: 0-12-207891-8

Printed in China
02 03 04 05 RDC 9 8 7 6 5 4 3 2 1

*To the people who give meaning to my life
and taught me to look for wonder in the world:
my wife, Carolyn, my daughters, Anne and Rebecca, my parents,
my godmother, Isabel, and her mother, Mrs. Crease. — CD*

Contents

Chapter 8   Planning and Building the Optical Link   292

R. T. Hudson, D. R. King, T. R. Rhyne, and T. A. Torchia

Chapter 9   Testing Fiber Optic Local Area Networks (LANs)   333

Jim Hayes and Greg LeCheminant

Part 3   The Applications

Chapter 10   Introduction to Industry Standards   367

Schelto Van Doorn

Chapter **19**   Alignment Metrology and Techniques   699

Darrin P. Clement, Ronald C. Lasky, and Daniel Baldwin

Contributors

Numbers in parentheses indicate the pages on which the authors' contributions begin.

Daniel Baldwin (699), Manufacturing Research Center, School of Mechanical Engineering, Georgia Institute of Technology, 813 Ferst Drive Northwest, Atlanta, Georgia 30332

Carl Beckmann (435), Thayer School of Engineering, 8000 Cummings Hall, Dartmouth College, Hanover, New Hampshire 03755

Alan F. Benner (379, 464), IBM Corporation, 2455 South Road, MS P967, Poughkeepsie, New York 12601

Darrin P. Clement (699), Maponics, 468 Garey Road, East Thetford, Vermont 05043

John D. Crow (379), IBM Corporation, Thomas J. Watson Research Center, Yorktown Heights, New York 10598

Carolyn J. Sher DeCusatis (89), formerly from Lighting Research Center, Rensselaer Polytechnic Institute, Troy, New York 12180

Casimer DeCusatis (217), IBM Corporation, 2455 South Road, MS P343, Poughkeepsie, New York 12601

Jim Hayes (333), Fotec, Incorporated, 151 Mystic Avenue, Suite 7, Medford, Massachusetts 02155

R. T. Hudson (292), Siecor Corporation, 800 17th Street Northwest, Hickory, North Carolina 28601

Ching-Long (John) Jiang (89), Amp Incorporated, Lytel Division, MS 300-001, 61 Chubb Way, Post Office Box 1300, Somerville, New Jersey 08876

Wenbin Jiang (41, 603), Phoenix Applied Research Center, Motorola, Incorporated, 2100 East Elliot Road, MS EL703, Tempe, Arizona 85284

Richard Johnson (159), Infineon Technologies North America Corporation, 1730 North First Street, San Jose, California 95112

D. R. King (292), Siecor Corporation, 800 17th Street Northwest, Hickory, North Carolina 28601

Michael Langenwalter (159), Infineon Technologies AG, Fiber Optics Division, Wernerwerkdamm 16, D-13629, Berlin, Germany

Ronald C. Lasky (699), Consultant, 26 Howe Street, Medway, Massachusetts 02053

Michael S. Lebby (41, 603), Phoenix Applied Research Center, Motorola, Incorporated, 2100 East Elliot Road, MS EL703, Tempe, Arizona 85284

Greg LeCheminant (333), Hewlett-Packard, Santa Rosa Systems Division, Santa Rosa, California 95403

Eric Maass (127), Motorola, Incorporated, 2100 East Elliot Road, Tempe, Arizona 85284

Ulf L. Österberg (3), Thayer School of Engineering, Dartmouth College, Hanover, New Hampshire 03755

Glenn Raskin (675), Motorola, Incorporated, 2501 South Price Road, M/D G651, Chandler, Arizona 85248

T. R. Rhyne (292), Siecor Corporation, 800 17th Street Northwest, Hickory, North Carolina 28601

Dave Siljenberg (655), IBM Corporation, 3605 Highway 52 North, Rochester, Minnesota 55901

Herwig Stange (143), Infineon Technologies, Fiber Optics, 13623, Berlin, Germany

Daniel J. Stigliani, Jr. (506), IBM Corporation, 2455 South Road, Poughkeepsie, New York 12601

Ray D. Sundstrom (127), Motorola, Incorporated, 2501 South Price Road, M/D G235, Chandler, Arizona 85248

Rakesh Thapar (564), Marconi, Warrendale, Pennsylvania, 15086

T. A. Torchia (292), Siecor Corporation, 800 17th Street Northwest, Hickory, North Carolina 28601

Schelto Van Doorn (367), Siemens Corporation, Santa Clara, California 95054

Preface to the Second Edition

SONET[1] on the Lambdas[2]
(by C. DeCusatis, with sincere apologies to Milton[3])

When I consider how the light is bent
By fibers glassy in this Web World Wide,
Tera- and Peta-, the bits fly by
Are they from Snell and Maxwell sent
Or through more base physics, which the Maker presents
(lambdas of God?) or might He come to chide
"Doth God require more bandwidth, light denied?"
Consultants may ask; but Engineers to prevent
that murmur, soon reply "The Fortune e-500 do not need
mere light alone, nor its interconnect; who requests
this data, if not clients surfing the Web?" Their state
is processing, a billion MIPS or CPU cycles at giga-speed.
Without fiber optic links that never rest,
The servers also only stand and wait.

When the first edition of this Handbook was published, the world of optical data communication appeared to be already well established, and its technology was a major driving force in the development of computers (both servers and clients), storage, and real-world applications in business, education, transportation, and many other sectors of the economy. Even so, we recognized at the time that the revolution in optical communications was only just beginning. Today, at the dawn of a new millennium, there are tremendous opportunities for growth in this field, and we have seen the pace of new technology accelerate even further in recent years. These many new developments, coupled with the success of the first edition, led to the conclusion that the time was right to update and expand on this Handbook.

[1] Synchronous Optical Network
[2] The Greek symbol "lambda" or λ is commonly used in reference to an optical wavelength
[3] The original author of the classic sonnet "On his blindness"

The intervening period has seen many significant changes, which required updating the original chapters; new types of lasers and photodetectors, a new physical layer interface for ESCON, new packaging technology for parallel optics, and many more revisions too numerous to mention have been included. Many new chapters have also been added to address the rapidly accelerating rate of change which has characterized wavelength multiplexing, optically clustered servers, small form factor connectors and transceivers, special types of fiber optic cable, and other areas which would have been classified as emerging technologies only a short while ago. Open standards, which to a great extent have created the Internet and the Web (remember TCP/IP?) also continue to evolve, and new standards are emerging to deal with the requirements of the next-generation intelligent optical infrastructure; some of these standards, such as MPLS and Infiniband, have been added to this edition. There are also new chapters on the history of communications technology (with apologies to those who have noted that it remains difficult to determine exactly who invented the first one of anything, and that the history of science is filled with tales of misplaced credit). There are also new predictions of the future, as envisioned by some of the leading commercial technology forecasts. Since the first edition, DeCusatis' Law has continued to hold; available data suggests that high-end network bandwidth will continue to grow at this pace for the foreseeable future. Indeed, the proliferation of new optical interfaces has led the editor to coin the Law of Cable Growth, also known as the First Corollary to DeCusatis' Law: the development of each new type of optical connector interface will proliferate 2 orders of magnitude in new cable assembly types. For example, if you happen to be a supplier of fiber optic cables, each new connector interface needs to be offered with 3 different fiber types (single mode, 50 micron multimode, and 62.5 micron multimode), 2 or 3 different cable jacket types (riser, LSZH, and plenum), 5 or 6 standard lengths for common applications, and hybrid cables or adapter kits that are backward-compatible with all the previous types of optical connectors. Thus, one new optical connector easily drives a hundred new cable assemblies! We've attempted to create a guide to the increasing complexity of optical cabling, and many other areas as well. Jargon and buzzwords in the industry have also continued to grow; when the first edition was published, not too long ago, nobody knew what a SAN or NAS was, let alone the difference between them. Today, a resurgence of interest in Storage Area Networks and Network Attached Storage has made these and many other acronyms commonplace, but our readers have expressed their ongoing frustration at repeatedly encountering AUA (another useless acronym). In an effort to help, we've included a new glossary and table of acronyms in this edition, as well as an expanded index.

Today's business initiatives are inexorably linked to network bandwidth; after all, your business can only grow as fast as information can be exchanged and acted upon. Perhaps this is why leading analysts tell us worldwide bandwidth demand grew roughly 200% from 1998 to 2000, and will increase more than 400% before 2002. The statistics of bandwidth growth are truly astonishing. It has been estimated that the entire Library of Congress can be stored in about 17 TeraBytes of disk space; this is also the current size of the package tracking database at United Parcel Services. In July 2000, the communication platform at the Web search engine Yahoo! delivered 4.4 billion messages and averaged 680 million page views per day. In the Information Economy, the Fortune 1000 companies are expected to add over 150 TeraBytes of storage capacity by 2003. Simply put, the demand for bandwidth exceeds the supply. And this is just the beginning; even today, less than 25% of these companies have electronic transaction capabilities over the Web, and even less are enabled for true e-business applications. And a recent University of California study, having estimated that the entire human race has accumulated about one exabyte (1018) of information to date, goes on to predict that the second exabyte will be generated within the next 3 years. In this environment, we are beginning to see the promise of all-optical networking emerge — application-neutral, distance-independent, infinitely scalable, user-centric networks that catalyze real-time global computing, advanced streaming multimedia, distance learning, telemedicine, and a host of other applications. We hope that those who build and use these networks will benefit in some measure from this book.

An undertaking such as this would not be possible without the concerted efforts of many contributing authors and a supportive staff at the publisher, to all of whom I extend my deepest gratitude. The following associate editors contributed to the first edition of the Handbook of Fiber Optic Data Communication: Eric Maass, Darrin Clement, and Ronald Lasky. As always, this book is dedicated to my parents, who first helped me see the wonder in the world; to the memory of my godmother Isabel; and to my wife, Carolyn, and daughters Anne and Rebecca, without whom this work would not have been possible.

Dr. Casimer DeCusatis, Editor
Poughkeepsie, New York

Part 1 | The Technology

Chapter 1 | Optical Fiber, Cable, and Connectors

Ulf L. Österberg

Thayer School of Engineering, Dartmouth College, Hanover,
New Hampshire 03755

1.1. Light Propagation

1.1.1. RAYS AND ELECTROMAGNETIC MODE THEORY

Light is most accurately described as a vectorial electromagnetic wave. Fortunately, this complex description of light is often not necessary for a satisfactory treatment of many important engineering applications.

In the case of optical fibers used for tele- and data communication it is sufficient to use a scalar wave approximation to describe light propagation in single-mode fibers and a ray approximation for light propagation in multimode fibers.

For the ray approximation to be valid the diameter of the light beam has to be much larger than the wavelength. In the wave picture we will assume a harmonically time-varying wave propagating in the z direction with phase constant β. The electric field can be expressed as

$$E = E_o(x, y) \cos(\omega t - \beta z). \tag{1.1}$$

This is more conveniently expressed in the phasor formalism as

$$E = E_o(x, y)e^{j(\omega t - \beta z)}, \tag{1.2}$$

where the real part of the right-hand side is assumed.

HANDBOOK OF FIBER OPTIC
DATA COMMUNICATION
$35.00

A wave's propagation in a medium is governed by the wave equation. For the particular wave in Eq. (1.2) the wave equation for the electric z component is

$$\nabla_t^2 E_z(x, y) + \beta_t^2 E_z(x, y) = 0, \qquad (1.3)$$

where we have introduced

$$\nabla_t^2 = \frac{\partial^2}{\partial x^2} + \frac{\partial^2}{\partial y^2} \qquad \text{[Transverse Laplacian]},$$

$$\beta_t^2 = k^2 n^2 - \beta^2 \qquad \text{[Transverse phase constant]},$$

$$k = \frac{2\pi}{\lambda} \qquad \text{[Free space wave vector]},$$

$$n(x, y) \qquad \text{[Refractive index]}.$$

The variable kn corresponds to the phase constant for a plane wave propagating in a medium with refractive index n. There is an equivalent wave equation to Eq. (1.3) for the H_z component. We have to solve only the wave equation for the longitudinal components E_z and H_z. The reason for this is that E_x and E_y can both be calculated from E_z and H_z using Maxwell's equations.

1.1.2.   SINGLE-MODE FIBER

In an infinitely large isotropic and homogeneous medium a light wave can propagate as a plane wave and the phase constant for the plane wave can take on any value, limited only by the available frequencies of the light itself. When light is confined to a specific region in space, boundary conditions imposed on the light will restrict the phase constant β to a limited set of values. Each possible phase constant β represents a mode. In other words, when light is confined it can propagate only in a limited number of ways.

For an engineer it is important to find out how many modes can propagate in the fiber, what their phase constants are, and their spatial transverse profile. To do this we have to solve Eq. (1.3) for a typical fiber geometry (Fig. 1.1). Because of the inherent cylindrical geometry of an optical fiber, Eq. (1.3) is transformed into cylindrical coordinates and the modes of spatial dependence are described with the coordinates r, ϕ, and z. Because the solution is dependent on the specific refractive index profile, it has to be specified. In Fig. 1.2 the most common refractive index profiles are shown. For step-index profile in Fig. 1.2c, a complete analytical set of solutions can

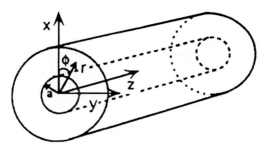

Fig. 1.1 Typical fiber geometry. Reprinted from Ref. [1], p. 12, courtesy of Academic Press.

be given [3]. These solutions can be grouped into three different types of modes: TE, TM, and hybrid modes, of which the hybrid modes are further separated into EH and HE modes. It turns out that for typical fibers used in tele- and data communication the refractive index difference between core and cladding, $n_1 - n_2$, is so small ($\sim$0.002–0.008) that most of the TE, TM, and hybrid modes are degenerate and it is sufficient to use a single notation for all these modes—the LP notation. An LP mode is referred to as $LP_{\ell m}$, where the ℓ and m subscripts are related to the number of radial and azimuthal zeros of a particular mode. The fundamental mode, and the only one propagating in a single-mode fiber, is the LP_{01} mode. This mode is shown in Fig. 1.3. To quickly figure out if a particular LP mode will propagate, it is very useful to define two dimensionless parameters, V and b.

$$V = ka\sqrt{n_1^2 - n_2^2} \approx \frac{2\pi}{\lambda} \cdot a \cdot n_1\sqrt{2\Delta}, \qquad (1.4)$$

where a is the core radius, λ is the wavelength of light, and $\Delta \approx \frac{n_1 - n_2}{n_1}$. The V number is sometimes called the normalized frequency.

The normalized propagation constant b is defined as

$$b = \frac{(\beta^2/k^2) - n_2^2}{n_1^2 - n_2^2}, \qquad (1.5)$$

where b is the phase constant of the particular LP mode, k is the propagation constant in vacuum, and n_1 and n_2 are the core and cladding refractive indexes, respectively.

Equation (1.5) is very cumbersome to use because b has to be calculated from Eq. (1.3). For LP modes Gloge *et al.* [4] have shown that to a very good accuracy the following formulas can be used to calculate b for different

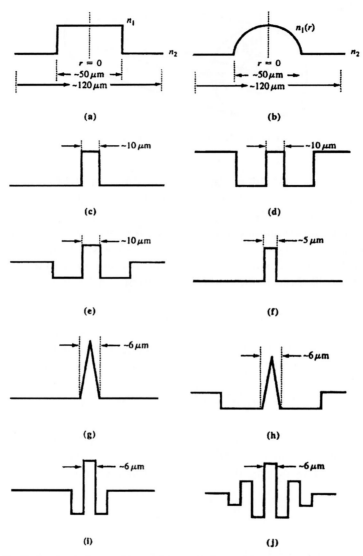

Fig. 1.2 Refractive index profiles of (a) step-index multimode fibers, (b) graded-index multimode fibers, (c) match-cladding single-mode fibers, (d, e) depressed-cladding, single-mode fibers, (f–h) dispersion-shifted fibers, and (i, j) dispersion-flattened fibers. Reprinted from Ref. [2], p. 125, courtesy of Irwin.

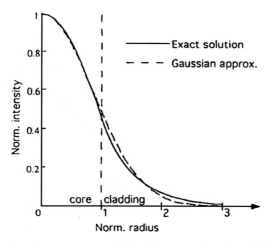

Fig. 1.3 Comparison between exact and Gaussian approximation of the radial mode field distribution for the lowest order mode. Reprinted from Ref. [1], p. 14, courtesy of Academic Press.

$LP_{\ell m}$ modes:

$$LP_{01}: b_{01} = 1 - \left[\frac{1 + \sqrt{2}}{1 + (4 + V^4)^{\frac{1}{4}}} \right]^2 \qquad (1.6)$$

$$LP_{\substack{\ell m \\ \ell m \neq 01}} : b_{\ell m} = 1 - \frac{U_c^2}{V^2} \cdot exp\left[\frac{2}{S} \left(\arcsin\left(\frac{S}{U_c}\right) - \arcsin\left(\frac{S}{V}\right)\right) \right]$$

$$S = \sqrt{U_c^2 - \ell^2 - 1} \qquad (1.7)$$

$$U_c = A - \frac{B - 1}{8A} - \frac{4(B - 1)(7B - 31)}{3(8A)^3}$$

$$A = \pi \cdot \left[m + \frac{1}{2}(\ell - 1) - \frac{1}{4} \right]; \qquad B = 4(\ell - 1)^2.$$

The graphs in Fig. 1.4 were generated using Eqs. (1.6) and (1.7). The normalized propagation constant b can vary only between 0 and 1 for guided modes; this corresponds to

$$n_2 k < \beta < n_1 k. \qquad (1.8)$$

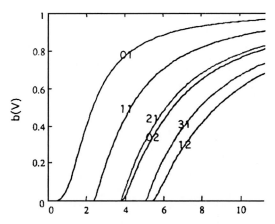

Fig. 1.4   Cutoff frequencies for the lowest order LP modes. Reprinted from Ref. [1], p. 15, courtesy of Academic Press.

The wavelength for which b is zero is called the cutoff wavelength, i.e.,

$$b_{\ell m}(V_{\text{co}}) = 0 \Rightarrow \lambda_{\text{co}} = \frac{2\pi}{V_{\text{co}}} \cdot a \cdot n_1 \sqrt{2\Delta}. \tag{1.9}$$

Therefore, for wavelengths longer than the cutoff wavelength the mode cannot propagate in the optical fiber.

Cutoff values for the V number for a few LP modes are given in Table 1.1. The fundamental mode can, to better than 96% accuracy, be described using a Gaussian function

$$E(r) = E_o \exp\left[-\left(\frac{r}{w_g}\right)^2\right], \tag{1.10}$$

where E_o is the amplitude and $2w_g$ is the mode field diameter (MFD) (Fig. 1.3). The meaning of the MFD is shown in Fig. 1.5. The MFD for the fundamental mode is larger than the geometrical diameter in a single-mode (SM) fiber and much smaller than the geometrical diameter in a multimode (MM) fiber. The optimum MFD is given by the following formula [7]:

$$\frac{w_g}{a} = 0.65 + 1.619 \cdot V^{-3/2} + 2.87 \cdot V^{-6}, \tag{1.11}$$

where a is the core radius. Equation (1.11) is valid for wavelengths between $0.8\,\lambda_{\text{co}}$ and $2\,\lambda_{\text{co}}$.

Table 1.1   **Cutoff Frequencies of Various LP$_{\ell m}$ Modes in a Step Index Fiber**[a]

$\ell = 0$ modes	$J_1(V_c) = 0$	$\ell = 1$ modes	$J_0(V_c) = 0$
Mode	V_c	Mode	V_c
LP$_{01}$	0	LP$_{11}$	2.4048
LP$_{02}$	3.8317	LP$_{12}$	5.5201
LP$_{03}$	7.0156	LP$_{13}$	8.6537
LP$_{04}$	10.1735	LP$_{14}$	11.7915
$\ell = 2$ modes	$J_1(V_c) = 0; V_c \neq 0$	$\ell = 3$ modes	$J_2(V_c) = 0; V_c \neq 0$
LP$_{21}$	3.8317	LP$_{31}$	5.1356
LP$_{22}$	7.0156	LP$_{32}$	8.4172
LP$_{23}$	10.1735	LP$_{33}$	11.6198
LP$_{24}$	13.3237	LP$_{34}$	14.7960

[a] Reprinted from Ref. [5], p. 380, courtesy of Cambridge University Press.

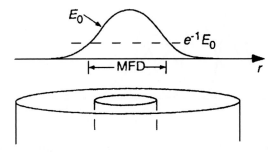

Fig. 1.5   The electric field of the HE$_{11}$ mode is transverse and approximately Gaussian. The mode field diameter is determined by the points where the power is down by e^{-2} or where the amplitude is down by e^{-1}. The MFD is not necessarily the same dimension as the core. Reprinted from Ref. [6], p. 144, courtesy of Irwin.

If the radial distribution for higher order modes is needed, it is necessary to use the Bessel functions [3]. In Fig. 1.6 the radial intensity distribution is shown for five LP modes in a fiber with $V = 8$. Recommended specifications for a single-mode fiber are summarized in Table 1.2.

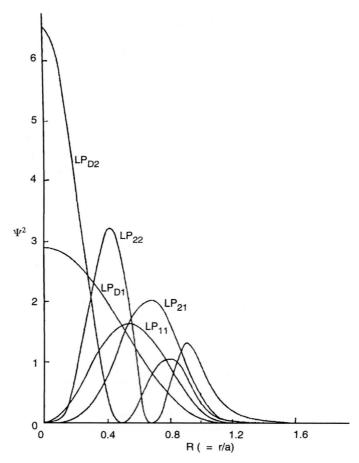

Fig. 1.6 Radial intensity distributions (normalized to the same power) of some low-order modes in a step index fiber for $V = 8$. Notice that the higher order modes have a greater fraction of power in the cladding. Reprinted from Ref. [5], p. 382, courtesy of Cambridge University Press.

1.1.3.   MULTIMODE FIBER

The previous discussion has in principle been for a step-index MM fiber. Because of the severe differences in propagation time between different modes in a step-index fiber, these are not commonly used in practice. Instead, a graded refractive index core is used for a MM fiber (Fig. 1.7).

Table 1.2 **CCITT Recommendation G.652^a**

Parameters	Specifications
Cladding diameter	125 μm
Mode field diameter	9–10 μm
Cutoff wavelength λ_{co}	1100–1280 nm
1550 nm bend loss	$\leq$1 dB for 100 turns of 7.5 cm diameter
Dispersion	$\leq$3.5 ps/nm · km between 1285 and 1330 nm
	$\leq$6 ps/nm · km between 1270 and 1340 nm
	$\leq$20 ps/nm · km at 1550 nm
Dispersion slope	$\leq$0.095 ps/nm^2 · km

^{a} Reprinted from Ref. [2], p. 126, courtesy of Irwin.

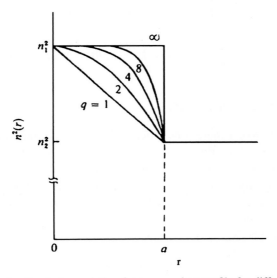

Fig. 1.7 The refractive index variation for a power law profile for different values of q. Reprinted from Ref. [5], p. 396, courtesy of Cambridge University Press.

The various graded-index profiles are generated by

$$n^2(r) = \begin{cases} n_1^2\left(1 - 2\Delta\left(\frac{r}{a}\right)^q\right); & 0 \leq r \leq a \\ n_1^2(1 - 2\Delta) = n_2^2; & r \geq a, \end{cases} \tag{1.12}$$

for different q's, q is called the profile exponent. The optimum profile is the one that gives the minimum dispersion; this occurs for q at slightly less than 2.

The total number of modes that can propagate in a MM fiber is given by

$$N = \frac{1}{2}\frac{q}{q+2}V^2. \tag{1.13}$$

For a step-index fiber $q = \infty$ and $N = \frac{V^2}{2}$. Equation (1.13) is only valid for large V numbers.

One can calculate the different ray paths that are possible in a MM fiber using a geometrical optics approach [8]. A more accurate view of the different modes can be obtained by solving Eq. (1.3) using the so-called WKB approximation [3]. Using this analysis the phase constants for the different modes can be shown to obey the following relationship:

$$\beta_m = nk\sqrt{1 - 2\Delta\left(\frac{m}{n}\right)^{q/q+2}}, \quad m = 1, 2, \ldots, N. \tag{1.14}$$

1.1.4. OPTICAL COUPLING

A first approach to estimate how much light can be coupled into an optical fiber is to use the ray picture. In this picture the light is confined within the core if it undergoes total internal reflection at the core–cladding boundary. This will occur only for light entering the fiber within an acceptance cone defined by the angle θ (Fig. 1.8). Rather than stating the angle θ for an optical fiber, it is the convention to give $\sin\theta$, which is called the numerical aperture (NA). The NA is defined as

$$NA = n \cdot \sin\theta, \tag{1.15}$$

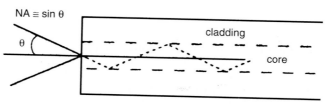

Fig. 1.8 Acceptance angle for an optical fiber. Reprinted from Ref. [1], p. 10, courtesy of Academic Press.

where n is the refractive index of the medium the light is coming from. In the case of coupling into an optical fiber, light is usually coming from air and subsequently $n \approx 1$. A more useful formula for the NA can be obtained if we use the dimensionless parameter Δ,

$$\text{NA} \approx n_1 \sqrt{2\Delta}. \tag{1.16}$$

For an incoherent light source such as a light emitting diode (LED) one can show that the total power accepted by the fiber is given by [9]

$$P \approx \pi \cdot B \cdot A_{\text{fiber}} \cdot (\text{NA})^2, \tag{1.17}$$

where B is the LED's radiance (units for radiance is watts per area and steradian).

It is more common to give a coupling efficiency; thus, giving the total power accepted by the fiber, the efficiency is defined as [10]

$$\eta = \frac{P_{\ell m}}{P_{in}}, \tag{1.18}$$

where P_{in} is the power launched into the fiber and $P_{\ell m}$ the power accepted by the $LP_{\ell m}$ mode. For link budget analyses it is more convenient to deal with coupling losses in units of decibels α:

$$\alpha = 10 \cdot \log \eta. \tag{1.19}$$

A more general definition of the coupling efficiency η for an optical fiber that obeys the weekly guiding approximation can be given as [10]

$$\eta = \left| \frac{n_2}{Z_o} \iint_{A^\infty} E_i \cdot E_{\ell m}^* dA \right|^2, \tag{1.20}$$

where n_2 is the refractive index of the cladding, z_o is the free space wave impedance, and E_i and $E_{\ell m}$ are the electric field amplitudes for the incoming light and for the light propagating in mode $LP_{\ell m}$ in the fiber, respectively.

Coherent light from a laser can often be approximated with a Gaussian beam; furthermore, if we restrict ourselves to a SM fiber, so that the LP_{01} mode can also be approximated as a Gaussian field, it is possible to calculate η analytically [7, 11]:

$$\eta = \left(\frac{4D}{B} \right) e^{-AC/B}, \tag{1.21}$$

where

$$A = \frac{(k's_1)^2}{2}$$

$$B = G^2 + (D+1)^2$$

$$C = (D+1)F^2 + 2DFG\sin\theta + D(G^2 + D + 1)\sin^2\theta$$

$$D = \left(\frac{s_2}{s_1}\right)^2$$

$$F = \frac{2\Delta}{k's_1^2}$$

$$G = \frac{2\Delta Z}{k's_1^2}$$

$$k' = \frac{2\pi n_o}{\lambda}$$

where ΔZ is the separation distance between source and fiber, Δ is the lateral displacement, θ is the angular displacement, and s_1 and s_2 are the mode field radii or spot sizes of the source field and fiber field, respectively. The refractive index of the medium between source and fiber is denoted n_o.

Equation (1.21) takes into account four different coupling cases at once (Fig. 1.9). If only one of these different coupling cases is present at a time Eq. (1.21) can be simplified to:

Case 1.1.   Spot-size mismatch $s_1 \neq s_2$:

$$\eta = \left(\frac{2s_1 s_2}{s_1^2 + s_2^2}\right)^2. \tag{1.22}$$

Case 1.2.   Transverse offset Δ:

$$\eta = \exp\left(-\left(\frac{\Delta}{s_2}\right)^2\right). \tag{1.23}$$

Case 1.3.   Longitudinal offset ΔZ:

$$\eta = \frac{1}{1 + \left(\frac{\Delta Z}{2Z_0}\right)^2}, \tag{1.24}$$

where Z_0 is the Rayleigh range.

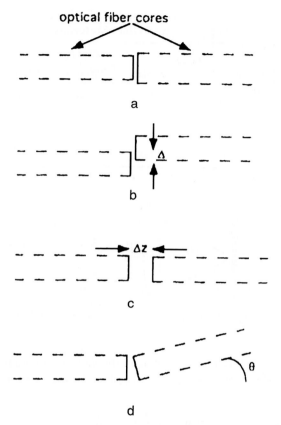

Fig. 1.9 Different coupling cases. Reprinted from Ref. [1], p. 18, courtesy of Academic Press.

Case 1.4.   Angular misalignment θ:

$$\eta = \exp\left(-\left(\frac{\theta}{\theta_0}\right)^2\right). \tag{1.25}$$

If a lens is used in between the emitter and fiber, some modifications to the previous formulas have to be done. What the lens can do for us is to match the output angle of the emitter to the acceptance angle of the receiving fiber. If properly done, the power coupled into the fiber is multiplied with the lens magnification factor M: $M = \frac{d_{rec}}{d_{em}}$, (see Fig. 1.10). All the preceding formulas need to be corrected for reflection losses. If the

Table 1.3   Scattering Loss for Several Representative Glass Materials[a]

Material	dB/km		
	633 nm	800 nm	1060 nm
Fused silica	4.8	1.9	0.6
Soda lime	8.5	3.3	1.1
Borosilicate crown	7.7	3.0	1.0
Lead silicate	47.5	18.6	6.0

[a] Reprinted from Ref. [1], p. 21, courtesy of Academic Press.

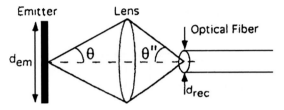

Fig. 1.10   Lens for improving coupling efficiency. Reprinted from Ref. [1], p. 20, courtesy of Academic Press.

refractive index of the medium between the source and the fiber is denoted n_o, the coupled power into the fiber is reduced with a factor R,

$$R = \left(\frac{n_1 - n_o}{n_1 + n_o} \right)^2. \tag{1.26}$$

1.2.   Optical Fiber Characterization

1.2.1.   OPTICAL FIBER MATERIALS

The material is primarily chosen to provide the minimum attenuation. Table 1.3 shows order of magnitude attenuation at three different wavelengths for four common glass types.

For tele- and data communication fibers, fused silica glass is the preferred material. To provide guiding of the light, the core of the fiber is doped with a few molar percentage of a substance that increases the refractive index. It is also possible to dope the cladding such that its refractive index becomes lower than the pure silica glass index in the core (Fig. 1.11).

Table 1.4   **Factors Affecting Attenuation**[a]

Intrinsic loss mechanisms

    Tail of infrared absorption by Si–O coupling

    Tail of ultraviolet absorption due to electron transitions in defects

    Rayleigh scattering due to spatial fluctuations of the refractive index

Absorption by impurities

    Absorption by molecular vibration of OH

    Absorption by transition metals

Structural imperfections

    Geometrical nonuniformity at core–cladding boundary

    Imperfection at connection or splicing between fibers

[a] From Ref. [3].

Fig. 1.11   Influence of different dopant mole percentages on refractive index of silica glass. Reprinted from Ref. [1], p. 22, courtesy of Academic Press.

1.2.2.   ATTENUATION

Attenuation is a very important factor in designing effective long-distance fiber optic networks. Consequently, the fabrication methods have improved dramatically during the past 30 years so that attenuation is measured in a few tenths of a dB/km. The dB is defined in Eq. (1.19). The various factors affecting the attenuation, in the 0.8- to 1.6-μm wavelength region, are listed in Table 1.4. Figure 1.12 shows schematically how some of the factors

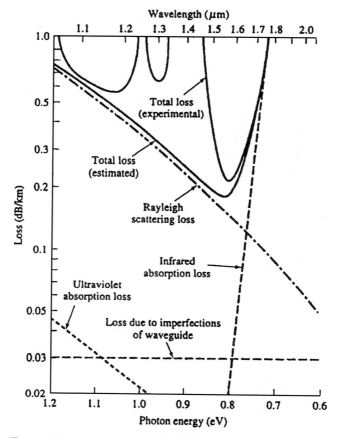

Fig. 1.12 Transmission loss in silica-based fibers. Reprinted from Ref. [12], p. 474, courtesy of Irwin.

contributing to the overall light attenuation vary with the wavelength in the near-infrared wavelength region. Typical total losses for an optical fiber, at the three different transmission windows at 800, 1300, and 1550 nm, are ≈ 2 or 3, ≤ 0.5, and ≤ 0.2 dB/km, respectively. These numbers are for SM fibers; from Fig. 1.13 it can be seen that MM fibers have slightly higher losses. The losses dealt with to date have been due to either intrinsic properties of the glass or extrinsic properties (such as OH and transition metal contents) that come from the particular fabrication method used. In addition, there are bending losses. If the fiber has been improperly cabled or installed these

Fig. 1.13 Wavelength dependence of fiber attenuation. Reprinted from Ref. [14], p. 8. Copyright © 1989 by Hewlett Packard Company. Reproduced with permission.

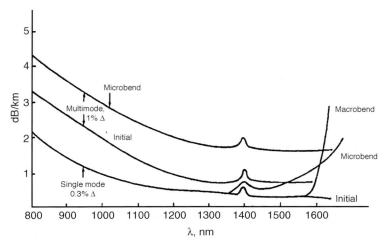

Fig. 1.14 Bend-induced losses of optical fibers. Reprinted from Ref. [15], p. 1.33, courtesy of McGraw-Hill.

bending losses can be substantial. Bending losses are divided into micro- and macro-bending losses. Micro-bending losses are due to nanometer-size deviations in the fiber. Macro-bending losses are due to visible bends in the fiber. Figure 1.14 shows qualitatively how micro- and macrobends contribute to the overall loss in a SM and MM fiber.

1.2.3. DISPERSION

Dispersion is due to the fact that different wavelengths experience different propagation constants, $\beta(\lambda)$, and therefore travel with different velocities causing a longer temporal pulse at the end of the fiber. Dispersion does not alter the wavelength (frequency) content of the light pulse. From a communications point of view dispersion is a very important factor because it directly affects the bit rate. The following are three major components contributing to the dispersion:

- Material
- Waveguide
- Intermodal

The material and waveguide dispersion are often referred to as chromatic dispersion, measured in units ps/nm · km, which is defined as

$$D_{\text{chr}} = -\frac{1}{L}\frac{dt_g}{d\lambda}, \tag{1.27}$$

where L is the fiber length and t_g is the time required to propagate a distance L. The subscript g refers to group velocity. When a pulse propagates in a dispersive medium it propagates with the group velocity,

$$V_g = \frac{d\omega}{d\beta}. \tag{1.28}$$

The phase travels with the phase velocity given by

$$V_p = \frac{\omega}{\beta} = \frac{c}{n}. \tag{1.29}$$

The dispersion properties are completely determined by the group velocity.

Material dispersion originates from the fact that the refractive index is a function of wavelength, $n(\lambda)$:

$$D_{\text{mat}} = -\frac{\lambda}{c}\frac{d^2 n_1}{d\lambda^2}. \tag{1.30}$$

The refractive index can be calculated from the Sellmeier equation:

$$n_1^2(\lambda) = 1 + \sum_{i=1}^{3} \frac{a_i \lambda^2}{\lambda^2 - b_i}, \tag{1.31}$$

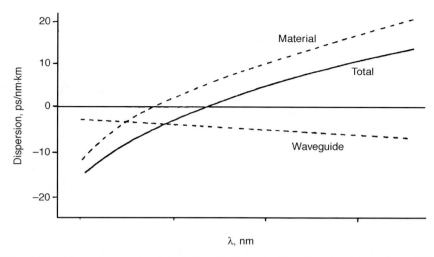

Fig. 1.15 Dispersion vs wavelength for single-mode fiber. Reprinted from Ref. [15], p. 1.38, courtesy of McGraw-Hill.

where the constants a_i and b_i may vary with the particular fiber composition. These constants have been accurately measured and tabulated for different Ge, B and P concentrations [16].

Waveguide dispersion is due to the different propagation characteristics of the light in the core and cladding. Keep in mind that a large portion of the light (30–40%) travels in the cladding for the LP_{01} mode around the cutoff frequency for the next higher order mode. The waveguide dispersion can be approximated, using the normalized propagation constant b, as

$$D_{wg} \simeq -\frac{\lambda}{c} \cdot n_1 \cdot \Delta \cdot \frac{d^2 b}{d\lambda^2}. \tag{1.32}$$

The total chromatic dispersion as well as its constituents are plotted in Fig. 1.15. As shown in Fig. 1.2, an optical fiber can have many different refractive index profiles. How the refractive index profile can affect the dispersion properties of the fiber is shown in Fig. 1.16.

Intermodal dispersion arises from the different travel times for the different modes in the fiber. For modal dispersion the units are nsec/km and D_{mod} is defined as follows [2]:

$$D_{mod} = t_g^{max} - t_g^{min}, \tag{1.33}$$

Fig. 1.16 Spectral disposition for various single-mode fiber profiles. Reprinted from Ref. [15], p. 1.38, courtesy of McGraw-Hill.

where t_g^{max} and t_g^{min} are the propagation times for the modes traveling the longest and shortest distances, respectively. Using the group index N_1 [3],

$$N_1 = n_1 + k\frac{dn_1}{dk}, \tag{1.34}$$

where the term $k\frac{dn_1}{dk}$ is the material dispersion for a plane wave traveling in the core region; we can approximately write the modal dispersion in a step-index fiber as follows [2]:

$$D_{\text{mod}} \simeq -\frac{N_1}{c}\Delta, \tag{1.35}$$

and for a graded-index fiber, with $q = 2(1 - \Delta)$, as

$$D_{\text{mod}} \simeq -\frac{N_1}{c}\frac{\Delta^2}{8}. \tag{1.36}$$

Note that for most fibers $N_1 \simeq n_1$ is an excellent approximation [3].

The total fiber disposition in a MM fiber, units nesc/km, can be shown to be the following [2]:

$$D_{\text{tot}}^2 = D_{\text{chr}}^2 \cdot \Delta\lambda^2 + D_{\text{mod}}^2, \tag{1.37}$$

where $\Delta\lambda$ is the spectral bandwidth of the light source.

Knowledge of the total dispersion, D_{tot}, can be transmitted into bandwidth (BW), units MHz · km, as follows [17]:

$$\text{BW} \simeq \frac{350}{D_{\text{tot}}}. \tag{1.38}$$

1.2.4. MECHANICAL PROPERTIES

Optical fibers have outstanding signal characteristics such as low attenuation and small dispersion. For optical fibers to be practically useful they must also have good mechanical properties to withstand environmental strains. In Table 1.5, physical properties for different materials used in communication cables are compared.

Silica glass has a very high threshold for breaking when there are no flaws, either on the surface or inside the glass. The tensile strength has been estimated to be 20 GPa [19]. Because optical glass fibers have a large surface-to-volume ratio they are very prone to failure from surface flaws that propagate into the glass. The actual stress for which the fiber breaks as a function of the size of the surface flaw is shown in Fig. 1.17.

Table 1.5   **Physical Properties of Materials**[a]

Property	Silica glass	Copper	Aluminum	Steel
Chemical sign	SiO_2	Cu	Al	Fe
Specific gravity (g/cm^3)	2.20	8.9	2.70	7.9
Tensile strength (kg/mm^2)	500	25	10	120
Young's modulus (kg/mm^2)	7200	12,000	6300	20,000
Elongation (%)	2–8	20–30	7–20	5–15
Coefficient of thermal expansion	5×10^{-7}	1.7×10^{-6}	2.3×10^{-5}	1×10^{-5}
Specific heat (cal/°C · g)	0.20	0.09	0.5	0.1

[a] Reprinted from Ref. [18], p. 53, courtesy of Dekker.

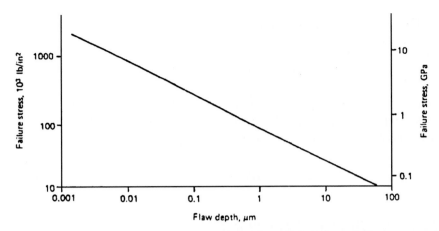

Fig. 1.17 Failure stress vs flaw size. Reprinted from Ref. [15], p. 1.40, courtesy of McGraw-Hill.

When fibers are tested for their mechanical reliability, it is common to use short fibers for destructively testing the fibers' breaking threshold for static and dynamic loads. Additionally, a screen test is performed on long samples of the fibers to find the effective strength [19].

1.2.5. MEASUREMENT TECHNIQUES

In this section I will briefly introduce one measurement method each for measuring attenuation, mode field diameter, cutoff wavelength, and chromatic dispersion.

Attenuation (the Cutback Method)

The transmission is measured as a function of wavelength for the full-length fiber, $P_L(\lambda)$. The fiber is then cut approximately 1 or 2 m away from the input end and a new transmission measurement is performed, $P_r(\lambda)$ (Fig. 1.18). The attenuation constant α is then calculated from as follows [20]:

$$\alpha = 10 \cdot \log \frac{P_L(\lambda)}{P_r(\lambda)}. \tag{1.39}$$

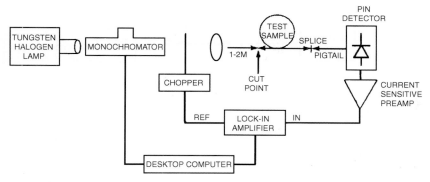

Fig. 1.18 Experimental setup for attenuation measurements by the cutback technique. This same apparatus can be used to measure cutoff wavelength. Reprinted from Ref. [20], p. 336, courtesy of Academic Press.

Fig. 1.19 Offset technique for mode field measurement—a block diagram; CMS, cladding mode stripper. Reprinted from Ref. [18], p. 426, courtesy of Marcel Dekker Inc.

Mode Field Diameter (the Near-Field Method)

A small area detector is used to scan the magnified image of the fiber end (Fig. 1.19). The experimental data are fitted to Eq. (1.10), from which the mode-field diameter can be calculated.

Cutoff Wavelength (the Single-Bend Attenuation Method)

In this method the power transmission is measured as a function of wavelength for two different fiber bending situations. In measurement 1 a prescribed bend diameter of 28 cm is used (CCITT and EIA standards). In the

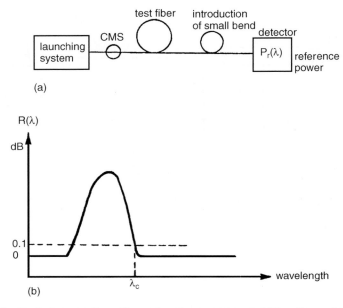

Fig. 1.20 Typical setup of cutoff wavelength measurement: (a) bending and (b) typical measurement of graph of the cutoff wavelength, single-mode reference (bending method). Reprinted from Ref. [18], (a) p. 430, and (b) p. 431, courtesy of Marcel Dekker Inc.

second measurement an additional loop with diameter 4–6 cm is used to, as much as possible, attenuate the higher order modes. The cutoff is then defined as the wavelength for which the normalized power becomes linear to within 0.1 dB (Fig. 1.20).

Chromatic Dispersion (Interferometric Method)

This method measures the phase delay for different wavelengths. The source is an incoherent high-pressure arc lamp and a typical delay curve is shown in Fig. 1.21.

1.3. Cable Designs

Because an optical fiber is very brittle and susceptible to chemical degradation it is important to protect the optical performance. Two common types of fiber optic cable construction are shown in Fig. 1.22. Common for all cable types is that the individual fibers are coated with some kind of organic

Fig. 1.21 Optical arrangement of dispersion measured by the interferrometric technique and the delay curve. Reprinted from Ref. [20], p. 351, courtesy of Academic Press.

material, examples of which include the following [16]:

- Polydimethyl siloxane
- Silicone oils
- Extrudates
- Acrylates

Fig. 1.22 Common single-fiber cable types. Reprinted from Ref. [1], p. 36, courtesy of Academic Press.

Fig. 1.23 Fiber optic cable-corrugated armor. Reprinted from Ref. [21], p. 247, courtesy of Academic Press.

For ease of manufacturing and to prevent the glass surface from being scratched the coating is applied during the drawing process. Often, the first coating is not sufficient to protect the fiber and additional coatings are used, which are commonly referred to as buffering. The buffered fiber is then embedded in various strength members and additional protective jackets. These additional layers have as their primary goals to protect the fibers from water, rodents, and temperature fluctuations and to reduce stress in the fiber (Fig. 1.23).

1.4. Connectors

1.4.1. CONNECTOR TYPES

It is important to be able to easily reconfigure communication links for cost and performance optimization, to replace link subunits, and for relocating equipment [1]. To do this, connectors are a key component. There are two types of fiber optic connections:

- Fiber to transceiver
- Fiber to fiber

An example of a laser transceiver module is shown in Fig. 1.24. The fiber-to-fiber connectors are divided into simplex (one fiber to one fiber) and duplex (two fibers to two fibers) connectors (Fig. 1.25). The two basic connector designs are butt joints and expanded beam joints (Figs. 1.26a and 1.26b).

1.4.2. MECHANICAL PROPERTIES

Connectors are further differentiated depending on the particular mechanical method used to join the connectors.

FC connectors use threaded fasteners (Fig. 1.27). SC connectors use a spring latch (Fig. 1.28). These connectors can come as both simplex and duplex.

Variations on the SC connector, developed for specific standardized local area networks, are FDDI, ESCON, and FCS connectors (Figs. 1.29a–c). A common connector for MM data communication networks is the SMA connector (Fig. 1.30).

1.4.3. OPTICAL PROPERTIES

There is obviously some additional loss produced into a communications link with the incorporation of a connector (Fig. 1.31). As can be seen from Fig. 1.31 there is both a local and a distributed loss resulting from the connector joint. Contributing connector loss factors are shown in Table 1.6. The local loss factors can be calculated using the formulas in Section 1.1.4. The distributed losses are due to changes in the model launching conditions [15].

Table 1.6   **Connector Loss Factors**[a]

Extrinsic	*Intrinsic*
Lateral offset	Mode field diameter mismatch
Axial gap	Index profile mismatch
Tilt	Core eccentricity
End-face quality	Fiber and ferrule hole clearance
Reflections	Ferrule hole eccentricity

[a] Reprinted from Ref. [1], p. 55, courtesy of Academic Press.

Fig. 1.24   Common solution today; Laser TO can → optical subassembly (OSA) → module. There are three levels of packaging and several active alignments. Reprinted from Ref. [22], p. 134, courtesy of Academic Press.

ST SERIES CONNECTOR

Fig. 1.25 Straight tip ferrule single fiber connector. Reprinted from Ref. [23], p. 303, courtesy of Academic Press.

Fig. 1.26 (a) Example of two popular alignment mechanisms used in fiber optic connectors. (b) Schematic representation of a typical expanded-beam type connector. Reprinted from Ref. [23], p. 316 (a) and p. 318 (b), courtesy of Academic Press.

Fig. 1.27 FC connectors and adaptors. Reprinted from Ref. [24], p. 16, courtesy of Molex Fiber Optics, Inc.

Connector losses can be as high as 0.5 dB (this corresponds to an additional km of fiber at 1.3 μm). With improved manufacturing technology, losses between 0.1 and 0.2 dB are obtained for new connectors. Loss data for a FDDI connector are shown in Fig. 1.32.

1.5. Optical Fiber Bragg Gratings

An optical glass fiber has many advantages compared to other materials such as a copper cable. One of these advantages is its inertness to external changes such as heat, pressure, electric and magnetic fields, toxicity, etc. However, the glass in an optical fiber can be altered by intense laser light [25, 26]. This fact has led to the manufacturing of so-called fiber Bragg gratings (FBG). These in-core fiber gratings have revolutionized optical fiber communications systems, [27, 28]. The FBGs can be used for dispersion compensation [29], filter applications [30], mode conversion [31], gain equalization [32], and wavelength multiplexing [33]. FBGs were first discovered in 1978 by Hill *et al.* [25]. These early gratings were created by the interference between forward and backward traveling waves in the fiber. The backward traveling wave comes from Fresnel reflection at the output end. At the crests of this interference pattern, where the light is the strongest, interactions between the light and Ge defects in the glass alter

Fig. 1.28 (A, B) Spring latches. Reprinted from Ref. [24], p. 24 (A), and p. 34 (B), courtesy of Molex Fiber Optics, Inc.

the refractive index $\Delta n \approx 10^{-3}$–10^{-5}. The periodicity of these gratings is given by

$$\Lambda = \frac{\lambda}{2n_{eff}} \qquad (1.40)$$

where n_{eff} is the effective index of the fundamental mode. Evident from equation 1.40 is the fact that for a given laser only one periodicity can be produced. Not very practical! Because the interaction between the light and the defects is optimum for a small wavelength region in the UV one cannot just change laser to obtain a new grating periodicity. It was therefore a major breakthrough when Meltz *et al.* [34] invented a new holographic

34   Ulf L. Österberg

Fig. 1.29 Duplex connectors: (a) ESCON (R), (b) FDDI, (c) FCS duplex (fiber channel standard compliant). Reprinted from Ref. [1], p. 50, courtesy of Academic Press.

Fig. 1.30 SMA connectors and adapters. Reprinted from Ref. [24], p. 42, courtesy of Molex Fiber Optics, Inc.

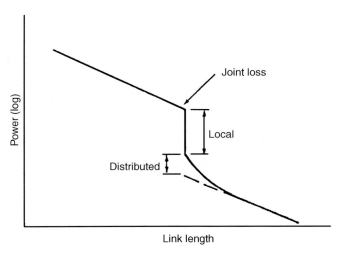

Fig. 1.31 Local and distributed (transient) losses and their launch dependence. Reprinted from Ref. [15], p. 4.40, courtesy of McGraw-Hill.

Fig. 1.32 Insertion loss performance for FDDI connectors. Reprinted from Ref. [24], p. 46, courtesy of Molex Fiber Optics, Inc.

method with which gratings of an (almost) arbitrary periodicity can be written into the fiber. This opened up the possibility for making gratings at wavelengths important for tele- and data communications. Further improvements have led to the fabrication of gratings using phase masks [35], a method that allows gratings to be manufactured while the fiber is being drawn.

An optical fiber grating reflects wavelengths at and close to the Bragg wavelength ($\lambda_B = 2n_{eff} \cdot \Lambda$). A typical reflection spectrum is shown in Fig. 1.33. The equations used to calculate the spectrum in Fig. 1.33 are found in [26]. Reflection takes place because light is partially reflected at each refractive index crest and maximum reflection occurs when each partial reflection is in phase with its neighbors, Fig. 1.34.

Let's look at two specific applications of these Bragg gratings. The first one is dispersion compensation. As mentioned in Section 1.2.3, the refractive index is a function of wavelength in the glass, so a light pulse will always broaden in time as it propagates through the fiber. This broadening becomes a big problem in high bandwidth time-multiplexed systems because the pulses, if left alone, will eventually merge together and destroy any hope of detecting individual pulses. By using an FBG in which the periodicity varies linearly along the fiber, the grating is said to be chirped, one can delay (in time) different wavelengths different amounts and thereby sharpen the pulse in time. Dispersion compensating FBGs of up to 1500 ps/nm in the wavelength region around 1550 nm are commercially available.

The second application is concerned with add/drop multiplexer/demultiplexers (mux/demux). One example of such a device is shown in Fig. 1.35 [33]. The purpose of such a device is to insert or remove a particular wavelength in a wavelength-division system. For example, to remove one

Fig. 1.33 Reflectivity spectra for two different values of the induced refractive index change in the core.

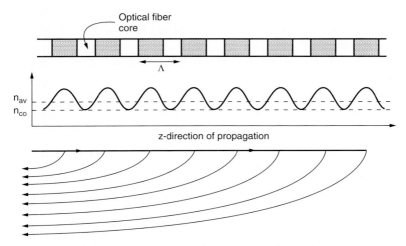

Fig. 1.34 Buildup of back-reflected light in an optical fiber grating.

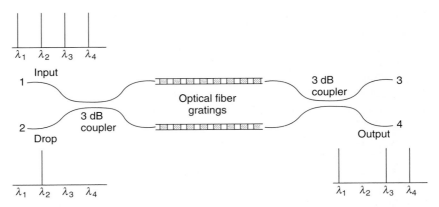

Fig. 1.35 Example of a mux/demux device utilizing a FGB in a Mach-Zehnder interferometer.

wavelength out of a stream of wavelengths one can use the combination of a Mach-Zehnder interferometer in conjunction with FBGs. Assuming one would want to remove wavelength λ_2, it is necessary to have two identical FBGs, one in each of the two interferometer arms (Fig. 1.35) and both optimized for λ_2 (i.e., only λ_2 satisfies the Bragg condition in Eq. 1.40). At the input the 4 wavelengths are divided so that 50% of each wavelength goes into each arm. Due to the FBGs λ_2 will be reflected from each arm and the other three wavelengths will be transmitted. The two streams of transmitted light are subsequently combined in the right coupler and interfering in such a way that all the transmitted light will exit through port 4. Similarly, the two portions of λ_2 will be combined and interfere in the left coupler to emerge at port 2.

Due to the inherent symmetry of these balanced Mach-Zehnder interferometers they can simultaneously perform both multiplexing and demultiplexing. A commercial mux/demux device is capable of adding or dropping wavelengths that are <1 nm apart (<100 GHz @ 1550 nm) with less than −30 dB crosstalk between different wavelengths.

References

1. Webb, J. R., and U. L. Österberg. 1995. In *Optoelectronics for Datacommunication.* San Diego: Academic Press.
2. Liu, M. M. K. 1996. Fiber, cable and coupling. In *Principles and applications of optical communications,* eds. R. C. Lasky, U. L. Österberg, and D. P. Stigliani. Chicago: Irwin.

3. Okoshi, T. 1982. *Optical fibers.* New York: Academic Press.
4. Marcuse, D., D. Gloge, and E. A. J. Marcatiti. 1979. *Guiding properties of fibers. Optical fiber telecommunications.* Orlando: Academic Press.
5. Ghatak, A., and K. Thyagarajan. 1989. *Optical electronics.* Cambridge, UK: Cambridge University Press.
6. Pollock, C. R. 1995. *Fundamentals of optoelectronics.* Chicago: Irwin.
7. Marcuse, D. 1977. *Loss analysis of single-mode finger splices. Bell System Tech. J.* 56: 703–718.
8. Cheo, P. K. 1990. *Fiber optics and optoelectronics.* 2nd. ed. Englewood Cliffs, N.J.: Prentice-Hall.
9. Powers, J. P. 1993. *An introduction to fiber optic systems.* Homewood, Ill.: Aksen.
10. Neumann, E. G. 1988. *Single-mode fibers: Fundamentals.* Berlin: Springer-Verlag.
11. Nemoto, S., and T. Makimoto. 1979. Analysis of splice loss in single-mode fibers using a Gaussian field approximation. *Opt. Quantum Electron.* 11: 447–457.
12. Maurer, R. D. 1973. Glass fibers for optical communication. *Proc. IEEE* 61: 452–462.
13. Chen, C. L. 1996. *Principles and applications of optical communications.* Chicago: Irwin.
14. Hentschel, C. 1989. *Fiber optics handbook.* Boeblingen, Germany: Hewlett Packard.
15. Allard, F. C. 1990. *Fiber optics handbook. For engineers and scientists.* New York: McGraw-Hill.
16. Shibata, N., and T. Edahiro. 1982. *Trans. Inst. Electron. Commun. Eng. Jpn.* 65: 166–172.
17. Hecht, J. 1987. *Understanding fiber optics.* Indianapolis, Ind.: Sams.
18. Murata, H. 1996. *Handbook of optical fibers and cables.* New York: Dekker.
19. Kalish, D., Key, P. L., Kurkjian, C. R., Tanyal, B. K., and Wang, T. T. 1979. Fiber characterization — mechanical. In *Optical fiber telecommunications,* eds. S. E. Miller and C. G. Chynoweth. Boston: Academic Press.
20. Philen, D. L., and W. T. Anderson. 1988. Optical fiber transmission evaluation. In *Optical fiber telecommunications II,* eds. S. E. Miller and I. P. Kaminow. Boston: Academic Press.
21. Gartside, C. H., P. D. Patel, and M. R. Santana. 1988. Optical fiber cables. In *Optical fiber telecommunications II,* eds. S. E. Miller and I. P. Kaminow. Boston: Academic Press.
22. Radcliffe, J., C. Paddock, and R. Lasky. 1995. Integrated circuits, transceiver modules and packaging. In *Optoelectronics for datacommunication,* eds. R. C. Lasky, U. L. Österberg, and D. P. Stigliani. San Diego: Academic Press.
23. Young, W. C., and D. R. Frey. 1988. Fiber connectors. In *Optical fiber telecommunications II,* eds. S. E. Miller and I. P. Kaminow. Boston: Academic Press.

24. Molex Fiber Optics, Inc. 1996. *Fiber optic product catalog,* No. 1096.

25. Hill, K. O., Y. Fujii, and D. C. Johnson, *et al.* 1978. "Photosensitivity in Optical Fiber Waveguides: Application to Reflection Filter Fabrication," *Applied Physics Letters* **32**(10): 647–649.

26. Bennion, I., J. A. R. Williams, and L. Zhang, *et al.* 1996. "Tutorial Review, UV-Written In-Fiber Bragg Gratings," *Optical and Quantum Electronics* 28: 93–135.

27. Hill, K. O., D. C. Johnson, and F. Bilodeau, *et al.* 1987. "Narrow-Bandwidth Optical Waveguide Transmission Filters: A New Design Concept and Applications to Optical Fiber Communications," *Electronics Letters* **23**(9): 465–466.

28. Bilodeau, F., K. O. Hill, and B. Malo, *et al.* 1994. "High-Return-Loss Narrowband All-Fiber Bandpass Bragg Transmission Filter," *IEEE Photonics Technology Letters* **6**(1): 80–82.

29. Hill, K. O., F. Bilodeau, and B. Malo, *et al.* 1994. "Chirped In-Fiber Bragg Grating for Compensation of Optical-Fiber Dispersion," *Optics Letters* **19**(17): 1314–1316.

30. Matsuhara, M., and K. O. Hill. 1974. "Optical-Waveguide Band-Rejection Filters: Design," *Applied Optics* **13**(12): 2886–2888.

31. Hill, K. O., B. Malo, and K. A. Vineberg, *et al.* 1990. "Efficient Mode Conversion in Telecommunication Fiber Using Externally Written Gratings," *Electronics Letters* **26**(16): 1270–1272.

32. Vengsarkar, A. M., J. R. Pedrazzani, and J. B. Judkins, *et al.* 1996. "Long-Period Fiber-Grating-Based Gain Equalizers," *Optics Letters* **21**(5): 336–338.

33. Bilodeau, F., D. C. Johnson, and S. Thériault, *et al.* 1995. "An All-Fiber Dense-Wavelength-Division Multiplexer/Demultiplexer Using Photoimprinted Bragg Gratings," *IEEE Photonics Technology Letters* **7**(4): 388–390.

34. Meltz, G., W. W. Morey, and W. H. Glenn. 1989. "Formation of Bragg Gratings in Optical Fibers by a Transverse Holographic Method," *Optics Letters* **14**(15): 823–825.

35. Hill, K. O., B. Malo, and F. Bilodeau, *et al.* 1993. "Bragg Gratings Fabricated in Monomode Photosensitive Optical Fiber by UV Exposure Through a Phase Mask," *Applied Physics Letters* **62**(10): 1035–1037.

Chapter 2 | Optical Sources: Light-Emitting Diodes and Laser Technology

Wenbin Jiang
Michael S. Lebby

Phoenix Applied Research Center, Motorola, Incorporated,
Tempe, Arizona 85284

2.1. Introduction

In this chapter, we will present some fundamentals of optical transmitter design for fiber optic data communications. Specifically, we will consider the operating principles behind the three most common types of optical sources used for data communications:

- Light-emitting diodes (LEDs) — edge emitting (E-LEDs) and surface emitting
- Edge-emitting semiconductor laser diodes
- Vertical cavity surface-emitting lasers (VCSELs)

As discussed in Chapter 1, the sources for fiber optic communications most often operate at wavelengths near 1300 or 1550 nm, and VCSELS are currently limited to operation near 780–980 nm. A detailed discussion of laser and LED sources could easily fill a book by itself; for our purposes, we will present some of the fundamentals of this technology and refer the interested reader to some of the many good treatments in the literature [1–3]. Fabrication and manufacturing of these devices, with particular emphasis on VCSELs, will be discussed further in Chapter 15.

HANDBOOK OF FIBER OPTIC
DATA COMMUNICATION
$35.00

2.2. Technology Fundamentals

An optical source for fiber optic communications must have properties that are slightly different from those required for other applications. Specifically, it must exhibit a high radiance over a narrow band of wavelengths that coincide with the transmission window of the fiber, typically 0.8–1.55 μm. The emissive area should be no greater than the core diameter of the fiber, and the distribution of incident radiation should match the numerical aperture of the fiber (the acceptance cone). In addition, the source should be easily modulated at high frequencies (up to several gigahertz), have a modest cost compared with other components of the datacom system, and be at least as reliable as other large system computer components (several hundred thousand power-on hours with minimal variation in optical output power). Semiconductor sources satisfy these requirements by emission of light at a pn junction through the process of injection luminescence. In this chapter, we will begin with a brief review of semiconductor properties, then describe the fundamentals of LEDs, edge-emitting laser diodes, and VCSELs.

For the range of wavelengths commonly used in data communications, the efficiency of both sources and detectors is determined by the bandgap energy. It is often desirable for the bandgap of the detectors to be slightly less than that of the source materials. To begin, then, we will consider some fundamentals of semiconductor physics (more information on this topic is provided in Chapter 3). Intrinsic semiconductor materials are characterized by an electrical conductivity much lower than that of pure metals, which increases rapidly with temperature. As described in Chapter 3, the Fermi level lies between the conduction and valence bands of the material. These materials may be doped with various impurities as either n type or p type so that the Fermi level moves either closer to the conduction or valence bands. Common room-temperature semiconductors include germanium and silicon, from group IV(b) of the periodic table, as well as several binary compounds from groups III(b) and V(b), such as gallium arsenide (GaAs) and indium phosphide(InP).

Silicon or germanium can be made n type by doping with donor impurities, such as P, As, or other group V elements, whereas they can be made p type by doping with acceptor impurities such as B, Ga, or other group III elements. These dopant concentrations are fairly small, approximately 1e-14 to 1e-21 per cubic centimeter in Si — or approximately from 1 dopant atom per billion silicon atoms to approximately 1 per thousand silicon.

Typically, the Fermi level is uniform throughout the doped material; when an abrupt junction between p-type and n-type materials is formed, the conduction, valence, and Fermi bands bend to accommodate the difference. Electrons tend to accumulate on the n side of the junction, and holes develop on the p side. A thin depletion region is formed at the junction itself — the depletion region is effectively depleted of both hole and electron carriers.

An external voltage applied across the junction will have the effect of raising or lowering the potential barrier between the two materials, depending on its polarity. When the p side is connected to a positive voltage, we say that the junction is forward biased; the opposite condition is known as reverse bias. In a forward-biased junction, excess carriers are injected into either side of the depletion region, where they are the minority carriers. These excess minority carriers tend to diffuse away from either side of the depletion layer. Thickness of the depletion layer decreases under forward bias conditions. There is thus a balance of carrier flow; injection of carriers over the depletion region (which gives rise to an excess carrier concentration) is balanced by both diffusion away from the junction and carrier recombination.

Recombination is the process in which the electrons and holes combine. Several different kinds of recombination are possible, although we are only concerned with the radiative recombination processes that release photons; these are direct band-to-band transitions across the bandgap. Direct bandgap semiconductor materials (such as GaAs and other III–V compounds) have an energy–k relationship such that the conduction band minimum and the valence band maximum occur at the same value of k, as shown in Fig. 2.1. With direct bandgap materials, the holes and electrons can recombine directly; by contrast, in an indirect bandgap material such as silicon, the conduction band minimum does not correspond to the valence band maximum, and it is more complicated to recombine while satisfying the requirements of both conservation of energy and conservation of momentum. The nonradiative transitions, in which the energy is lost to thermal energy within the crystal, involve intermediate energy levels that trap carriers deep within the crystal lattice.

In direct bandgap materials, the radiative recombination process is proportional to the excess minority carrier concentration and gives rise to the creation of radiation from the junction. The process is known as injection luminescence and is the fundamental mechanism for the creation of optical radiation from semiconductor optical sources. We will not discuss this effect in great detail, other than to note that the operation of such sources

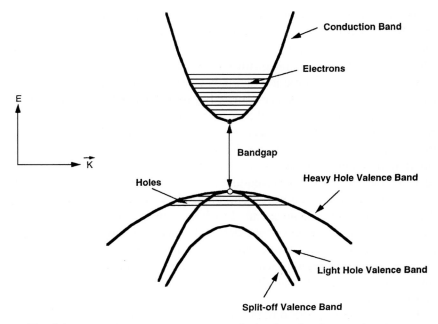

Fig. 2.1 Energy band structure (energy vs k) for direct bandgap semiconductor.

is also influenced by the minority carrier lifetime, diffusion length near the junction, injection efficiency, and width of the depletion layer. Equivalent circuits for the pn junction (discussed in Chapter 3) allow the modeling of junction capacitance and series resistance; these factors affect how rapidly the optical source can be modulated or, conversely, how fast a semiconductor photodiode can respond to incident radiation.

The recombination of electrons and holes in a forward-biased pn junction can produce optical radiation at wavelengths of 0.8–1.7 μm, suitable for data communications. If recombination takes place in several stages, it is also possible for more than one photon of longer wavelength to be emitted. When only a single photon of energy (E) is produced, the wavelength may be determined from Planck's Law:

$$\lambda = hc/E = 1.24/E\,(\mu\text{m/eV}),\tag{2.1}$$

where h is Planck's constant and c is the speed of light. The probability that an electron of energy E_2 will recombine with a hole of energy E_1 is proportional to the concentration of electrons at E_2 and the concentration of holes at E_1. The spectrum of the radiated energy may be determined by integrating the product of these carrier concentrations over all values of E_1 or E_2, subject to the constraint that the difference between E_1 and E_2 is the desired photon energy. Using a simple model for the behavior of carriers in a semiconductor, it is possible to derive the emission spectra due to injection luminescence; an LED will give rise to an approximately Gaussian spectra with a half-width of between 30 and 150 nm or more at room temperature.

It is desirable to design devices with high internal quantum efficiency, defined as the ratio of the rate of photon generation to the rate at which carriers are injected across the junction. A related parameter is the external quantum efficiency, defined as the ratio of the number of emitted photons to the number of carriers crossing the junction. External quantum efficiency is smaller than internal quantum efficiency for several reasons. For instance, some light will be reabsorbed before it can reach the emitting surface. Only light emitted toward the semiconductor–air interface is useful for coupling into the fiber, and a small percentage of this light is reflected back from the surface. Furthermore, only light reaching the surface at less than the critical angle will be coupled into an adjacent optical fiber. For a semiconductor material with an internal quantum efficiency of η_i, an injection current of I will give rise to an optical power, P, of

$$P = \eta_i \cdot (I/q) \cdot E, \qquad (2.2)$$

where q is the charge on an electron. It is also possible to derive expressions for the behavior of these sources at high modulation frequencies; it can be shown that they exhibit a high frequency cutoff at modulation frequencies greater than

$$f = 1/\tau_p, \qquad (2.3)$$

where τ_p is the mean lifetime of the excess minority carriers; it gives rise to a characteristic diffusion length, L,

$$L = \sqrt{\mu \cdot k \cdot T \cdot \tau_p/q} = \sqrt{(D \cdot \tau_p)}, \qquad (2.4)$$

where μ is the electron mobility, k is Boltzmann constant, and the quantity D is known as the electron diffusion coefficient. From these expressions, we can determine that modulation frequencies above $f = 20\text{--}25$ MHz are difficult to obtain; however, bandwidths as high as several hundred MHz can be obtained from the double-heterostructure diodes that we will discuss in this chapter.

Fig. 2.2   Cross-section view of an LED.

The structure of a typical surface-emitting LED is shown in Fig. 2.2. Note that for some structures, part of the substrate may be etched away to minimize the distance between the active area and the emitting surface. A thin layer of insulating oxide may be used to separate the positive contact from the *p* layer, except in the region of the active layer where the current flow is concentrated. The LED junction temperature is a critical parameter; the wavelength distribution of the LED changes with increasing temperature, internal quantum efficiency falls off, and the lifetime of the device is significantly decreased. In general, for GaAs and GaAlAs devices, the peak junction temperature should be kept below 50 or 100°C by the use of heat sinks or other devices. If the active layer is kept close to the heat sink, the thermal inpedance is small, and higher current densities may be used without an excessive rise in temperature.

In general, LEDs are diffuse (Lambertian) sources; for a typical GaAs LED forward biased at 100 mA and consuming 150 mW of power, the optical output is only on the order of 50 μW. To compensate for this rather low power, various optical coupling and packaging schemes are employed; these will be discussed in other chapters. One alternative is the edge-emitting LED, which offers small emissive area and higher radiance. The edge-emitting LED takes advantage of a slightly different approach — the double-heterojunction diode. Heterojunctions are produced at the boundary between two different types of materials whose physical lattice sizes match but that have different bandgap energies and other properties. Without going into details about the physics involved, we note that such structures offer higher injection efficiency, improved carrier confinement (and hence higher probability for radiative transitions to occur), and higher modulation bandwidths. The double heterostructure is simply a layered approach that further helps to confine excess minority carriers that are injected over the forward-biased pn junction. For such devices, the modulation bandwidth increases as the square root of the current density and inversely as the square root of the active layer thickness (note that the modulation current must be a small fraction of the DC bias current, or nonlinear effects will occur that limit device performance). With current densities of 50 A/mm^2, devices such as this routinely offer 1 or 2 mW output power from a 50-μm aperture. These can also be made to emit at the longer wavelengths — in the 1.3- to 1.55-μm range.

The light output of an LED is based on spontaneous emission and tends to be incoherent, with typical spectral widths on the order of 100–150 nm. The two types of LEDs, surface-emitting LEDs and ELEDs, are characterized by high divergence of the output optical beams (120° or more), slow rise times (>1 ns) that generally limit the data communication applications of LEDs to less than 200 MHz, low temperature dependence, insensitivity to optical reflections, and output powers on the order of 0.1–3 mW.

By contrast, laser sources tend to exhibit much higher output powers (3–100 mW or more), narrower spectral widths (<10 nm), smaller beam divergence (5–10°), faster modulation rates (hundreds of MHz to several GHz), and higher sensitivity to both temperature fluctuations and optical reflections back into the laser cavity. This is because they are based on stimulated emission of coherent radiation — indeed, the word "laser" is an acronym for light amplification by stimulated emission of radiation. The radiative recombination process responsible for injection luminescence in LEDs is the result of spontaneous bandgap emissions. Transitions may also be stimulated by the presence of radiation of the proper wavelength; all photons produced by stimulated emission have the same frequency and are in phase with the stimulating emission, making this a source of coherent radiation. This is the type commonly produced by semiconductor edge-emitting lasers and vertical cavity lasers. Like all lasers, the semiconductor laser diode requires three essential operating elements:

1. A means for optical feedback, usually provided by a cleaved facet or multilayer Fresnel reflector
2. A gain medium, consisting of a properly doped semiconductor material
3. A pumping source, provided by current applied to the diode

For a semiconductor material with gain g and loss α, placed in a cavity of length L between two partially reflective mirrors of reflectivity R_1 and R_2, the minimum requirement for lasing action to occur is that the light intensity after one complete trip through the cavity must at least be equal to its starting intensity. This is called the threshold condition, given by the following relation:

$$R_1 R_2 \exp(2(g - \alpha)L) = 1. \tag{2.5}$$

Thus, the laser diode begins to operate when the internal gain exceeds a threshold value. Two key parameters associated with the laser are its efficiency of converting electrical current into laser light, known as the slope efficiency dP/dI, and the amount of current required before the laser begins stimulated emission, known as the threshold current I_{th}. This information is summarized in the power vs current (P vs I) characteristic curve of a laser diode, as shown in Fig. 2.3. Applying current, I, to a laser device initially gives rise to spontaneous emission of light, as in an LED; as the current is increased, the cavity and mirror losses are overcome, and the laser passes the threshold and begins to emit coherent light. The optical power emitted

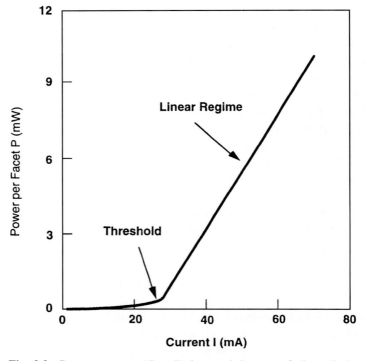

Fig. 2.3   Power vs current (P vs I) characteristic curve of a laser diode.

above threshold, P, is given by

$$P = (I - I_{th}) \, dP/dI. \tag{2.6}$$

Threshold current is found by extrapolating the linear region of the P/I curve. Low threshold currents and high efficiencies are desirable in laser diodes. The slope efficiency, dP/dI, is related to the external differential quantum efficiency, η, by

$$\eta = (\lambda q)/(hc)(dP/dI), \tag{2.7}$$

where λ is the wavelength, q is the charge of an electron, h is the Plank constant, and c is the speed of light. The threshold current is a strong function of temperature, T, according to the relation

$$I_{th}(T) = I_{th}(T_1) \exp[(T - T_1)/T_0], \tag{2.8}$$

where T_0 is the laser's characteristic temperature, and T_1 is room temperature (300 K). A large T_0 is desirable because it translates into a reduced sensitivity of threshold current to operating temperature. Usually, it is only possible to measure the case temperature of a packaged laser diode. The junction temperature is always higher than the case temperature in typical applications. The junction temperature can be estimated by taking into account the device thermal resistance, the input electrical power, and the output optical power.

Increasing the diode current above threshold allows the gain coefficient to stabilize at a value near the threshold point, just large enough to overcome losses in the media:

$$g = \alpha - \frac{1}{2L} \ln(R_1 R_2). \tag{2.9}$$

Because of the high gain coefficients that can be obtained, the cavity of a double heterostructure laser can be made much smaller than other types of lasers; 1 mm or less is typical. For lasers biased above threshold, the slope of the power vs current characteristic curve in the spontaneous emission region corresponds approximately to the external quantum efficiency. The slope in the lasing region is related to the differential quantum efficiency, η_d, by

$$\eta_d = \frac{q}{E} \, dP/dI. \tag{2.10}$$

In practice, this ideal characteristic curve may be less well behaved; considerable research has been devoted to eliminating undesirable effects, such as "kinked" curves that correspond to unwanted fluctuations in the laser power. This has led to the development of many features, such as stripe geometry and buried heterostructure lasers, that are beyond the scope of this discussion. Compared with LED emissions, laser light has a narrower bandwidth (<10 nm, making it less susceptible to certain kinds of noise in the fiber link), it is more directional (so that the external quantum efficiency may be improved), and its modulation bandwidth is greater (up to several Ghz). Laser sources at short wavelengths (780–850 nm) are typically used with multimode fiber, whereas longer wavelengths (1.3 or 1.55 μm) are used with single-mode fiber.

2.3.   Device Structure — LED

In this and the following sections, device structures for LEDs, edge-emitting lasers, and VCSELSs will be introduced, with particular emphasis on the laser technologies. In Chapter 15, the manufacturing issues for each will be introduced, again with particular emphasis on the laser technologies.

LEDs use either direct bandgap or indirect bandgap materials for spontaneous light emission. When holes, in the valence band of Fig. 2.1, combine with electrons in the conduction band, photons of light with energy consistent with the bandgap are emitted. The material selection is dependent on the desired wavelength of light to be emitted. For example, whereas GaAs emits light at approximately 880 nm, GaP emits light at approximately 550 nm and InGaP emits light at approximately 670 nm. The light emission begins as soon as the LED is forward biased: Unlike the laser diodes discussed in the following section, there is no threshold current.

A surface-emitting LED consists of a pn junction, with a controlled concentration profile between the p-typed and n-typed material. Figure 2.2 shows the cross section of an LED. The substrated in this example is GaAs; other III–V compounds, such as GaP, may be appropriate depending on the desired wavelength of light to be emitted. The substrate layer and the graded layer above are doped n type, and they are also doped to provide the bandgap for the desired wavelength of emitted light. A constant layer caps the graded layer and is capped in turn by a dielectric material such as silicon nitride. The p region of the pn junction is formed by the diffusion of Zn doping into an opening in the dielectric. The pn junction is contacted on the top by a metal such as gold, AuBe, or aluminum contacting the p-type, diffused Zn-doped region. The n-type region is contacted on the back side, with a metal such as AuSn or AuGe in contact with the n-doped substrate.

Device characteristic curves for LEDs include the forward current vs applied voltage curve and the light emission or luminous intensity vs forward current curve. Other useful characteristics in optoelectronic applications include the spectral distribution and the luminous intensity versus temperature and angle, the latter due to the need to couple into a medium such as optical fiber. An example of each is shown in Fig. 2.4. The spectral

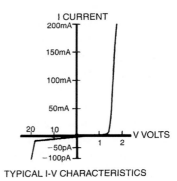

Fig. 2.4 Characteristics of an LED: spectral distribution, luminous intensity versus temperature and angle, and I vs V characteristics.

distribution of an LED is wider than that of a laser, incurring more chromatic dispersion. The lower frequency with which an LED can be switched, the wider angle of emission, and the larger spectral distribution are all disadvantages of LEDs compared to lasers. The primary advantage of LEDs over lasers for data communication is the low cost due to the significantly simpler processing.

2.4. Device Structure — Lasers

Edge-emitting semiconductor lasers have been around for more than 30 years. Semiconductor lasers were first reported in 1962 [4–7]. Initial devices were based on forward-biased GaAs pn junctions. Optical gain was provided by electron–hole recombination in the depletion region, and optical feedback was provided by polished facets perpendicular to the junction plane. This type of homojunction design meant that the carrier confinement of those lasers was poor, and the high laser threshold prohibited laser operation at room temperature. The concept of using wider bandgap material as one or both cladding layers to improve the laser carrier confinement and thus to reduce the leakage current was first proposed in 1963 [8]. Optical mode confinement was also expected to improve because a larger refractive index of the center active layer would provide a waveguide effect. Seven years later, a continuous wave (CW) GaAs/AlGaAs double-heterojunction (DH) semiconductor laser operating at room temperature was demonstrated using a liquid-phase epitaxial (LPE) growth technique [9, 10]. Commercial application of edge-emitting semiconductor lasers has since become practical. Today, the worldwide semiconductor laser annual sales revenue has exceeded $400 million [11].

In the late 1970s, Iga *et al.* [12] proposed a semiconductor laser oscillating perpendicular to the device surface plane to overcome the difficulties facing edge-emitting semiconductor lasers that oscillate in parallel to the device surface plane. This type of laser is termed VCSEL, pronounced "VIXEL." VCSELs have demonstrated many advantages over edge-emitting semiconductor lasers. First, the monolithic fabrication process and wafer-scale probe testing as per the silicon semiconductor industry substantially reduces the manufacturing cost because only known good devices are kept for further packaging [13, 14]. Second, a densely packed two-dimensional (2-D) laser array can be fabricated because the device occupies no larger of an area than a commonly used electronic device [15]. This is very important for applications in optoelectronic integrated circuits. Third, the microcavity length allows inherently single longitudinal cavity mode operation due to its large mode spacing. Temperature-insensitive devices can therefore be fabricated with an offset between the wavelength of the cavity mode and the active gain peak [16, 17]. Finally, the device can be designed with a low numerical aperture and a circular output beam to match the optical mode of an optical fiber, thereby permitting efficient coupling without additional optics [14, 18].

A conventional edge-emitting semiconductor laser utilizes its cleaved facets as laser cavity reflectors because the length of the active layer is usually several hundred micrometers. The long active length provides enough optical gain to overcome the cavity reflector loss even though the reflectivity of the facets is only $\sim$30%. In comparison, a VCSEL needs both of its surfaces to be highly reflective to reduce the cavity mirror loss because its active layer is less than 1 μm thick. The first VCSEL was demonstrated with GaInAsP/InP in 1979, which operated pulsed at 77°K with annealed Au at both sides as reflectors [12]. A room-temperature pulsed-operating VCSEL was demonstrated with a GaAs active region in 1984 [19]. Room-temperature CW-operating GaAs VCSELs were achieved by improving both the mirror reflectivity and the current confinement [20].

Currently, an output power of more than 100 mW has been obtained from an InGaAs/GaAs VCSEL with GaAs/AlAs monolithic diffractive Bragg reflectors (DBR) [21, 22]. VCSELs with lasing threshold of sub-100 μA [23, 24] or wall-plug efficiency of more than 50% have been reported with lateral oxidized–Al confinement blocks [25, 26]. Room-temperature CW InGaAsP/InP VCSELs have met some difficulties primary due to a low index difference between GaInAsP and InP, which causes difficulty in preparing a highly reflective monolithic DBR [27]. Nevertheless, CW InGaAsP VCSELs at 1.5 μm have been reported using GaAs/AlAs DBR mirrors [28, 29].

Two-dimensional (2-D) arrayed VCSELs can find important applications in stacked planar optics, such as the simultaneous alignment of a tremendous number of optical components used in parallel multiplexing lightwave systems and parallel optical logic systems, free space optical interconnects, etc. High-power lasers can also be made with phase-locked 2-D arrayed VCSELs. Some 2-D arrayed devices have been demonstrated [15, 30, 31] and efforts have been made in coherently coupling these arrayed lasers [32, 33].

2.4.1. EDGE-EMITTING LASERS

Two types of lasers have been extensively studied. They include near-infrared $Al_yGa_{1-y}As/Al_xGa_{1-x}As$ $(x > y \geq 0)$ and long wavelength $In_xGa_{1-x}As_yP_{1-y}/InP$ DH edge-emitting semiconductor lasers. The epitaxial structures for both types of lasers are similar. They are usually n–p–p type, n–i–p type, or n–n–p type. LPE used to be the dominant epitaxial growth technique during the 1970s and early 1980s for high-quality semiconductor laser material growth, but it has gradually been taken over predominantly by metal organic chemical vapor deposition (MOCVD) techniques. Molecular beam epitaxy (MBE) is also a growth technique that has been used to demonstrate semiconductor lasers in research and development environments with a limited commercial success. A $GaAs/Al_xGa_{1-x}As$ DH laser is shown in Fig. 2.5 consisting of a multiple of compound semiconductor layers grown on a n^+-type GaAs substrate. The p-type active layer of the GaAs semiconductor laser, in which stimulated emission is amplified, is made of GaAs doped by Be, C, or Zn at 1×10^{17} cm^{-3}. The active layer thickness ranges from 50 to 2000 Å. On top of the active layer is a p-type $Al_xGa_{1-x}As$ cladding layer doped to 1×10^{18} cm^{-3} using Be, C, or Zn. Below the active layer is a n-type $Al_xGa_{1-x}As$ cladding layer doped to 1×10^{18} cm^{-3} using Si, Sn, or Te.

Fig. 2.5 Schematic diagram of an ridge waveguide GaAs/AlGaAs double-heterojunction (DH) laser.

Cladding layer thickness usually ranges from 1500 Å to 1 μm. Within each $Al_xGa_{1-x}As$ layer, x is the value of aluminum mole fraction. When x is 0.3, for example, the energy bandgap of $Al_xGa_{1-x}As$ is approximately 1.8 eV — 0.4 eV wider than that of GaAs active layer. When the pn junction is forward biased, electrons in the n-type cladding region are injected into the p-type active region. With a p-type semiconductor of wide bandgap on the other side of the pn junction, the injected minority carriers are mostly confined within the p-type active region. This carrier confinement allows population inversion to occur and optical gain to increase efficiently. In addition, the refractive index of GaAs is higher than that of $Al_xGa_{1-x}As$, and this acts like a waveguide to confine the majority of generated light within the GaAs active layer. The light that is not confined and penetrates into the $Al_xGa_{1-x}As$ cladding layers will not be absorbed by the cladding materials because of the wider bandgap and will therefore benefit laser action.

Although an epitaxy-ready GaAs substrate can be of good quality, there may exist some level of surface defects due to either grown-in crystal defects or mechanical polishing. Microscopic substrate surface flatness and native oxide on the substrate surface are also of great concern because they add a degree of difficulty to the epitaxial growth. To ensure a high-quality epitaxial crystal structure, the substrate usually first goes through the cleaning procedure, and a GaAs buffer layer with the same type of doping as that in the substrate is then deposited before the growth of any DH structure. There are also reports that high-quality epitaxial structures can be grown without buffer layers if substrates are thoroughly cleaned before the growth [34]. For the device structure shown in Fig. 2.5, the GaAs buffer layer is doped with a n-type dopant such as Si, Sn, or Te, typically having a concentration on the order of 1×10^{18} cm^{-3}. Its thickness ranges from 100 nm to several micrometers. The GaAs cap layer on top of the DH structure is very heavily doped with a p-type dopant such as Be, C, or Zn, typically at a concentration above 1×10^{19} cm^{-3}. This permits a low-resistivity metal Ohmic contact to be used for electrical conduction. Although a higher impurity doping concentration helps reduce the device series resistance, there is a limit at which the impurities can be incorporated into the host crystal structure. The adverse effect of overdoping is to form impurity clusters that behave as nonradiative recombination centers in the active region and the cladding layers, introducing internal optical loss or so-called free carrier absorption. The net effect is that the laser threshold current will increase. Therefore, the doping level at each layer of the device structure should be carefully adjusted to achieve the optimum designed laser performance.

The double-heterostructure semiconductor laser represents the single largest constituent of today's total semiconductor laser production because of its application in compact disk (CD) players and CD data storage. The current CD laser market volume is greater than 80 million units per year [11]. The high volume of this market has driven the unit cost of a packaged laser to less than $1, with a large volume pricing of approximately $1. This has allowed businesses in the fiber optics market to use the low-cost CD laser in a historically high-cost environment. The CD laser has been used very successfully for some short-distance optical data links and has decreased laser wavelength specification from 850 nm for GaAs DH laser to 780 nm for CD laser [35, 36]. Because a CD laser operates at a wavelength of 780 nm, its active layer is made of AlGaAs with an Al mole fraction of approximately 15%. For example, a CD laser may have an epitaxial structure consisting of 0.1-μm-thick undoped $Al_{0.15}Ga_{0.85}As$ active layer sandwiched between a n-type $Al_{0.6}Ga_{0.4}As$ cladding layer and a p-type $Al_{0.6}Ga_{0.4}As$ cladding layer grown on a n-type GaAs substrate, as shown in Fig. 2.6 [37]. The device fabrication is similar to any other type of DH lasers. Typically, a CD laser has a threshold of 20–50 mA at room temperature and is operated at an output power of approximately 3–5 mW, as shown in Fig. 2.7. The laser wavelength is usually between 770 and 795 nm. The low-threshold CD laser is mostly used for portable consumer electronics powered by batteries. When used in a CD player, a CD laser is

p - metal contact

p^+ - GaAs cap

p - $Al_{0.6}Ga_{0.4}As$ cladding

$Al_{0.15}Ga_{0.85}As$ active (0.1 μm)

n - $Al_{0.6}Ga_{0.4}As$ cladding

n - GaAs buffer

n^+ - GaAs substrate

n - metal contact

Fig. 2.6   Schematic diagram of an epitaxial structure for an edge-emitting semiconductor laser at 780 nm.

Fig. 2.7 (Left) Output power vs current of a CD laser, and (right) the correspondent laser spectrum.

usually designed to operate at multimode or self-pulsation mode in GHz range in order to reduce the feedback noise due to light reflection from a disk [37]. This type of CD laser, however, cannot be used for multi-Gb data communications.

When the active thickness of a DH laser is reduced to become comparable to the de Broglie wavelength [38], quantum mechanical effect starts to occur and the layer becomes a quantum well (QW). These QWs have been specifically used to design a new class of single quantum well (SQW) lasers. Two or more QWs can be placed between the two cladding layers to form a multiquantum well (MQW) laser. The layers separating the wells in the MQW laser are called barrier layers. Compared to a SQW laser, the MQW laser has a larger optical mode confinement factor, resulting in lower threshold carrier density and lower threshold current density. In comparison with a DH laser, the QW laser has a smaller active volume, a lower lasing threshold, and a higher differential gain, leading to increased relaxation oscillation frequency and reduced relative intensity noise. Quantum well lasers with small signal modulation frequencies above 20 GHz have been demonstrated [39, 40]. High-speed semiconductor lasers are important for large-bandwidth optical communications, as can be seen by the rapid deployment of optical fibers for transoceanic telecommunication cables and networking backbones throughout the Untied States and the world.

Shown in Fig. 2.8 is an energy band diagram of a GaAs/AlGaAs graded-index (GRIN) SQW laser structure [41]. The epitaxial layers are grown by

GRIN region

$Al_{0.6}Ga_{0.4}As$ ◄──────► $Al_{0.18}Ga_{0.82}As$

(0.2 µm)

conduction band

n-cladding   SQW   p-cladding

$Al_{0.6}Ga_{0.4}As$ GaAs $Al_{0.6}Ga_{0.4}As$
(2 µm)   (5 nm)   (2 µm)

valence band

Fig. 2.8   Energy band diagram of a graded index (GRIN) single quantum well laser.

MOCVD on a n^+-type GaAs substrate. The growth starts with a buffer layer of 2 µm, linear graded from GaAs to $Al_{0.6}Ga_{0.4}As$ and n-type doped at 2×10^{18} cm^{-3}. The n-type cladding layer of $Al_{0.6}Ga_{0.4}As$ has a thickness of 2 µm, doped at 2×10^{18} cm^{-3}. The n-side GRIN region is undoped with a thickness of 0.2 µm linear graded from $Al_{0.6}Ga_{0.4}As$ to $Al_{0.18}Ga_{0.82}As$. The undoped GaAs QW is 50 Å thick. The p-side GRIN region is undoped with a thickness of 0.2 µm linear graded from $Al_{0.18}Ga_{0.82}As$ to $Al_{0.6}Ga_{0.4}As$. The p-type cladding layer of $Al_{0.6}Ga_{0.4}As$ has thickness of 2 µm, doped at 1×10^{18} cm^{-3}. The p-type GaAs cap layer is heavily doped at 1×10^{19} cm^{-3}, with a thickness of 0.5 µm. The GRIN layers in this structure are effective in improving both the carrier and the optical mode confinement. A threshold current density of only 200 A/cm^2 has been demonstrated from such a broad-area laser with a length of 400 µm and a stripe width of 180 µm [41].

Further reduction of laser threshold current density can be achieved by using a strained QW active layer, such as $In_xGa_{1-x}As$ sandwiched between the GaAs barriers with $0 < x < 0.25$, or between the InP barriers with $x > 0.25$. The biaxial strain caused by the slight lattice mismatch between the two material systems alters the valence band edge by removing the

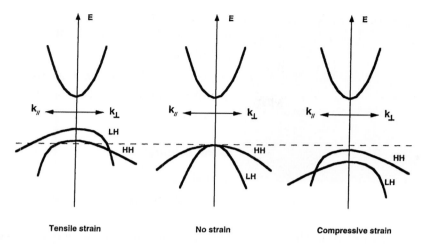

Tensile strain   No strain   Compressive strain

Fig. 2.9 Schematic energy band diagram in k space showing the removal of degeneracy between heavy hole (HH) valence band edge and light hole valence band edge for both compressive- and tensile-strained InGaAs gain medium.

degeneracy of the heavy hole and the light hole, resulting in reduced transparency carrier density and increased modal gain and thus reduced threshold current density [42]. The schematic energy band diagrams in k space for the strained gain medium are shown in Fig. 2.9. The strain in the active region also results in a larger differential gain, which helps improve the device operation bandwidth [40, 43]. Due to critical thickness constraints, the amount of strain in the active region and the active layer thickness, or the number of QWs with strained $In_x Ga_{1-x}As$, is limited. To overcome the critical thickness barrier, strain-compensated active layers are used to increase the net total active thickness and the differential gain [44–47], thereby improving the optical mode confinement and reducing the laser threshold current density. The strain-compensated active structure consists of compressively strained quantum wells and tensile-strained barriers, or vice versa, so that the compressive strain and the tensile strain are mutually compensating for each other. Active layers exceeding the critical thickness can therefore be demonstrated without forming any lattice misfit dislocations.

In long-haul telecommunication systems, long-wavelength semiconductor lasers are of interest because of the minimum fiber dispersion at 1.3 μm and the minimum fiber loss at 1.55 μm [48]. The dispersion-shifted fiber will have both the minimum dispersion and the minimum loss at

p - metal contact

p+ - InGaAsP (λ = 1.1 μm) contact layer

p - InP cladding

In$_{0.73}$Ga$_{0.27}$As$_{0.63}$P$_{0.37}$ active (0.2 μm)

n - InP cladding

n - InP buffer

n - InP substrate

n - metal contact

Fig. 2.10   Schematic diagram of a 1.3-μm InGaAsP/InP DH laser epitaxial structure.

1.55 μm [49–51]. The long-wavelength semiconductor lasers are based on In$_x$Ga$_{1-x}$As$_y$P$_{1-y}$ active layer lattice-matched to InP cladding layers [52, 53]. By varying the mole fractions x and y, any wavelength ranging between 1.1 and 1.6 μm can be selected. For example (Fig. 2.10), an InGaAsP/InP DH laser at 1.3 μm has an 0.2- to 0.3-μm-thick undoped In$_{0.73}$Ga$_{0.27}$As$_{0.63}$P$_{0.37}$ active layer, 3- or 4-μm n-type cladding layer doped by Sn at 2×10^{18} cm^{-3}, 2-μm p-type cladding layer doped by Zn at 1×10^{18} cm^{-3}, and 0.5-μm-thick p-type InGaAsP contact layer doped by Zn at 1×10^{19} cm^{-3} [54]. The contact layer bandgap corresponds to a wavelength of 1.1 μm. An etched-mesa buried-heterostructure (EMBH) laser [54] made with this epitaxial structure has a threshold current of approximately 15 mA at room temperature and a single-mode output power of approximately 10 mW per facet, as shown in Fig. 2.11. One problem with the long-wavelength semiconductor lasers is the threshold current sensitivity to temperature, at room temperature (small T_0), due to poor carrier confinement and large Auger nonradiative recombination [55]. Improved thermal characteristics [56] and higher modulation speed [39] have been demonstrated when a MQW active layer is used for the long-wavelength semiconductor lasers.

Red visible semiconductor lasers operating in the range of 635–700 nm can find applications in bar-code scanners, laser printers, and laser pointers. They can also be used for plastic fiber data links because of the minimum loss at 650 nm in the plastic fiber [57]. With the emergence of digital video disk (DVD) technology for data storage [58], the market demand for both 635- and 650-nm semiconductor lasers is expected to soon catch up with the demand for the 780-nm CD lasers. Several material systems, such as

Fig. 2.11   Laser output power vs current of a 1.3-μm InGaAsP/InP DH.

AlGaAs [59], InGaAsP [60, 61], and InAlGaP [62–64], have been demon-strated to work in this wavelength region, but InAlGaP is regarded as the most appropriate material because it has a large direct energy bandgap while completely lattice-matched to a GaAs substrate. For example (Fig. 2.12), an InAlGaP DH laser at 650 nm consists of an $In_{0.5}Ga_{0.5}P$ active layer sandwiched between the $In_{0.5}(Al_xGa_{1-x})_{0.5}P$ p-cladding layers grown on a GaAs substrate with $x = 0.7$. The active layer thickness is approximately 100 nm, and the cladding layer thickness is between 0.5 and 1 μm. Both the n-cladding and the p-cladding layers are doped at a concentration of 5×10^{17} cm^{-3}. Between the p-cladding layer and the p-GaAs contact layer is a p-type InGaP layer for improving the device I–V characteristics. This p-type InGaP intermediate layer is doped at 5×10^{17} cm^{-3}, and the p-type GaAs contact layer is doped as high as possible. The typical light output vs current characteristics is shown in Fig. 2.13.

High-temperature performance and high-power operation have been the concerns for InAlGaP visible semiconductor lasers due to carrier leak-age into the p-cladding layer. Methods utilizing components, such as strained active layer [65, 66], off-angle substrate [67, 68], MQW active

p - metal contact
p⁺- GaAs contact layer
p - $In_{0.5}Ga_{0.5}P$ transition layer
p - $In_{0.5}(Al_{0.7}Ga_{0.3})_{0.5}P$ cladding
$In_{0.5}Ga_{0.5}P$ active (100 nm)
n - $In_{0.5}(Al_{0.7}Ga_{0.3})_{0.5}P$ cladding
n - GaAs buffer

n- GaAs substrate

n - metal contact

Fig. 2.12   Schematic diagram of a visible InGaP laser epitaxial structure.

Fig. 2.13   Laser output power vs current of a 650-nm InGaP DH laser.

structure [69], and multiquantum barrier structure [70], have been developed to improve the laser performance.

The new DVD standard has increased the data storage capacity from 650 Mb to 4.7 Gb on a single-sided disk 12 cm in diameter. This storage capacity increase is attributed more to the tightening of system margin than to the shortening of laser wavelength from 780 to 635 or 650 nm. A 135-min high-definition motion picture, however, needs a storage capacity of 15 Gb. Because the current DVD standard has squeezed the system margin to the minimum, the future generation of DVD technology will rely to a great extent on laser wavelength shortening to expand the storage capacity in order to maintain the same disk size. Several groups have been investigating green/blue lasers using wide bandgap II–VI compound materials such as ZnCdSe/ZnSSe/ZnMgSSe grown on GaAs substrate [71–73], and good performance lasers have been demonstrated. The device, however, suffers serious reliability problem because of stacking fault-like defects that occur at and near the heterointerface between the GaAs substrate and the II–VI materials. Most II–VI semiconductor lasers degrade rapidly within minutes when running at CW. By reducing the grown-in defects, the device CW operation lifetime has been extended to 100 h [74] but is not yet long enough for any commercial applications.

Recent advancement in blue LED devices based on III–nitride materials [75] has prompted research in the blue/violet semiconductor lasers using InGaN/AlGaN MQW [76, 77] or InGaN/GaN DH structures [78]. The III–nitride epitaxial structures are grown on c-plane or a-plane sapphire substrates with a thick GaN buffer layer in between because there are no lattice-matched substrates available. Lasers on a spinel ($MgAl_2O_4$) substrate have also been demonstrated [79]. Crystal quality, p-contact resistivity, carrier and current confinement, and facet mirror reflectivity have been the four major problems in the III–nitride semiconductor laser development [80]. Continuous wave operation at room temperature has been achieved at a wavelength at approximately 400 nm by improving the p-contact resistance and thus reducing the device operating voltage [81]. The device has a threshold current of approximately 3 or 4 kA/cm^2 and a lifetime of approximately 20 h at 1.5-mW constant power when running at room temperature. The improvement in reliability relies on further reducing the contact resistance and reducing the grown-in crystal defects. The search for a lattice-matched substrate will help accelerate the device development cycles.

2.4.2. *VERTICAL CAVITY SURFACE-EMITTING LASERS*

The majority of the VCSELs being developed today are in the near-infrared wavelength range based on either GaAs/AlGaAs or strained InGaAs active materials. GaAs VCSELs at 850 nm are preferred as the light sources for short-distance optical communications because either Si or GaAs positive–intrinsic–negative (PIN) detectors can be used in the receiver end to reduce the total system cost. A typical GaAs VCSEL epitaxial layer structure is shown in Fig. 2.14, and the correspondent etched-mesa-type device

Fig. 2.14   TEM photo of a GaAs VCSEL epitaxial layer structure.

Fig. 2.15   Cross-section SEM photo of an etched-mesa GaAs VCSEL structure.

structure is shown in Fig. 2.15. It includes three major portions: bottom DBR, active region, and top DBR.

Generally, the epitaxial material is grown by either MOCVD or MBE techniques. The bottom DBR is n-type doped and grown on an n-type doped GaAs substrate. Typically, there is a GaAs buffer layer grown between the n-DBR and the substrate. Silicon (Si) and selenium (Se) are two commonly used n-type dopants. The n-DBR is composed of 30.5 pair [82, 83] of $Al_{0.16}Ga_{0.84}As$/AlAs, which starts and stops with the AlAs layer alternated by the $Al_{0.16}Ga_{0.84}As$ layer. Each DBR layer has an optical thickness equivalent to a one-fourth of the designed lasing wavelength. The intrinsic cavity region is composed of two $Al_{0.3}Ga_{0.4}As$ spacer layers and three or four GaAs quantum wells with each quantum well sandwiched between two $Al_{0.3}Ga_{0.7}As$ barriers. With the quantum well width of 100 Å and the quantum barrier width of 70 Å, the lasing wavelength is approximately 850 nm. The $Al_{0.6}Ga_{0.4}As$ spacer can be replaced by $Al_xGa_{1-x}As$ spacer, with x graded from 0.6 to 0.3 to form a graded-index separate confinement heterostructure. The total thickness of the spacers is such that the laser cavity length between the bottom and the top DBRs is exactly one wavelength or its multiple integer. The p-type doped DBR is grown on top of the active region. It consists of 22 pairs of $Al_{0.16}Ga_{0.84}As$/AlAs alternating layers that start with the AlAs layer and stop with the $Al_{0.16}Ga_{0.84}As$. Like the n-DBR, each layer has an optical thickness of one-fourth the wavelength. The most common p-type dopants utilized are carbon (C), zinc (Zn), and Beryllium (Be). Typically, C is used for the p doping in the top DBR when using MOCVD growth techniques with the top several layers doped by Zn for better metalization contact [84]. Finally, a GaAs cap is used to prevent the top AlGaAs layer of the p-DBR from oxidization. The cap is highly p doped with Zn as the dopant and is kept to 100 Å thick because the material will be highly absorptive to the optical mode if it is too thick.

The function of a DBR in a VCSEL is equivalent to a cleaved facet in an edge-emitting laser: to reflect part of the laser emission back into the laser cavity and to transmit part of the laser emission as the output. It is in essence similar to a dielectric mirror. A pair of low-refractive index and high-refractive index layers with the optical thickness of each layer to be one-fourth of a specific wavelength will enhance the reflectivity to this wavelength. Many pairs of such alternating layers stacked together will form a mirror with its reflectivity reaching above 99% centered at the designed wavelength with a certain bandwidth. The mirror can be designed to achieve higher reflectivity and wider reflection bandwidth with a larger refractive index difference. $Al_xGa_{1-x}As$ is an ideal semiconductor material that can be monolithically grown on a GaAs substrate to provide a similar function as a dielectric mirror because the refractive index of this material can vary continuously from 3.6 to 2.9, with x varying from 0 to 1. For example, if one-fourth wavelength-thick layer is made of GaAs and the other one-fourth wavelength-thick layer is made of AlAs, 20 pairs of such alternating layers will provide a reflectivity of 99% at the desired wavelength with a bandwidth of larger than 70 nm as long as the desired wavelength is longer than the GaAs absorption band edge, which is approximately 875 nm at room temperature.

Fig. 2.16 Reflectivity spectrum of 22 periods of $Al_{0.16}Ga_{0.84}As/AlAs$ one-fourth wavelength DBR mirror stacks centered at a wavelength of 850 nm.

Although reflectivity of a natural cleaved facet in an edge-emitting laser is only approximately 30%, it is enough to ensure the lasing action due to the large net gain provided by the long active length, which is usually several hundred micrometers. The reflectivity of a VCSEL DBR has to be more than 99% to reduce the cavity loss due to the extremely short gain length in the active region. For the example shown in Fig. 2.14, the optical gain is provided by the three GaAs quantum wells, with each quantum well thickness of only 100 Å. The net gain length of the VCSEL is thus only one-tenth of 1000 of that of an edge-emitting laser. Due to the reasonable refractive index difference between $Al_{0.16}Ga_{0.84}As$ and AlAs, the 22-pair DBR in this VCSEL has a reflectivity of 99.9% at 850 nm with a bandwidth of about 70 nm (Fig. 2.16). The large bandwidth provides some tolerance to any variation in the lasing wavelength due to growth variation in quantum well thickness and laser cavity length, the center wavelength mismatching between the top and the bottom DBRs, and the growth reactor/tooling variance. A typical GaAs VCSEL output power vs input current is shown in Fig. 2.17 for a mesa diameter of 10 μm and a laser emission aperture of 7 μm.

Although the structure shown in Fig. 2.14 is for GaAs VCSELs, many others have been working with strained InGaAs VCSELs operating at approximately 980 nm. The strain in the active region provides higher gain, which allows lower lasing threshold, and higher differential gain, which allows larger intrinsic modulation bandwidth. In addition, the InGaAs VCSEL has a wavelength transparent to the GaAs substrate, allowing the

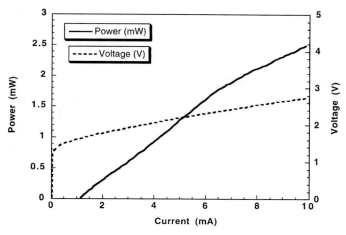

Fig. 2.17 A GaAs VCSEL output power vs input current for an etched-mesa structure with a mesa diameter of 10 μm and an emission aperture of 7 μm. The laser wavelength is approximately 850 nm.

light emission toward the substrate side and thus the epitaxial side down packaging scheme. In this way, heat generated in the p-DBR mirror and the active junction region can be dissipated more efficiently, resulting in lower junction temperature and higher output power. A disadvantage of this type of VCSELs is that the low-cost Si or GaAs PIN detectors cannot be used when the VCSELs are used as the light sources for data communications. A typical InGaAs VCSEL structure consists of two DBR mirrors with reflection bands centered at approximately 970 nm [17, 85]. Each mirror is composed of one-quarter wavelength-thick layers alternating between AlAs and GaAs. The top p-doped DBR mirror contains 15 periods, whereas the bottom n-doped DBR mirror contains 18.5 periods. The cavity between the mirrors is filled by spacer layers of $Al_{0.3}Ga_{0.7}As$ that are used to center three 8-nm $In_{0.2}Ga_{0.8}As$ quantum wells (almost the thickness limit for coherently strained materials) separated by 10-nm GaAs barriers to form a one-wavelength-long cavity. The $Al_{0.3}Ga_{0.7}As$ spacer on the p-DBR mirror side is p-type doped, and the $Al_{0.3}Ga_{0.7}As$ spacer on the n-DBR mirror side is n-type doped. Above the p-DBR mirror stack, a heavily p-doped (3×10^{19} cm^{-3}) GaAs phase-matching layer is deposited to provide a non-alloyed Ohmic contact to the hybrid Au mirror, which also acts as a p contact. The DBR mirrors are uniformly doped to 1×10^{18} cm^{-3} except for the digital grading region, which is uniformly doped to 5×10^{18} cm^{-3}.

The n dopant is Si and the p dopant is carbon. The whole epitaxial structure is grown by either MBE or MOCVD technique on a n-doped GaAs substrate. The device is designed to emit laser toward the substrate side.

Clearly, highly reflective semiconductor DBR mirrors are necessary to ensure low cavity loss, thus allowing the VCSEL to reach lasing threshold at a reasonable threshold carrier density level in the gain medium. Although the refractive index difference between the two constituents of the DBR structures is responsible for high optical reflectivity, the accompanied energy bandgap difference that scales approximately linearly with the index difference results in electrical potential barriers in the hetero-interfaces. These potential barriers impede the carrier flowing in the DBR structures and result in a large series resistance, especially in the p-type doping case. The large series resistance gives rise to thermal heating and thus deteriorates the laser performance.

The series resistance due to the heterojunctions in the DBR mirror can be minimized by grading and selectively doping the interfaces. In practice, the simplest approach to grade the interface is to introduce an extra intermediate-composition layer between the two alternating DBR constituents [86]. Instead of transiting directly from GaAs to AlAs, for example, a thin layer of $Al_{0.5}Ga_{0.5}As$ layer can be grown between the two to help smooth the interfaces. With the introduction of more $Al_xGa_{1-x}As$ layers of intermediate Al composition at each interface, greater performance over a single intermediate transition layer can be achieved. The advancement of MOCVD technology has allowed the continuous grading of an arbitrary composition profile. Very low-resistance DBRs have been achieved using this technique [87]. In the structure shown in Fig. 2.14, the hetero-interfaces are linearly graded from 15% Al composition to 100% over a distance of 12 nm. The total optical thickness of a pair of alternating layers is maintained to be one-half wavelength. The effect of this interface linear grading on the DBR mirror reflectivity is minimal. Likewise, heterointerface parabolic grading [88–90] and sinusoidal grading [91] have also been used in some cases to flatten the valence band and therefore reduce the p-DBR series resistance.

The graded interfaces can sometimes be heavily doped (5×10^{18} cm^{-3}), whereas the remainder of the mirror is lightly doped (1×10^{18} cm^{-3}) to reduce scattering and free-carrier absorption loss in the mirror, and there is also a reduction in series resistance [92]. A simplified delta doping scheme [93, 94] has worked successfully to reduce the series resistance in the p-DBR mirror. In this scheme, p doping is carried out at interfaces where the nodes of the optical intensity are located, at levels as high as the crystal can incorporate. This heavy doping at the heterointerfaces causes the valence band edge to shift upward and the thermionic emission current to increase. The excess resistance at the higher bandgap side (AlAs) of the heterointerfaces is also reduced together with the relaxation of the carrier depletion in this region. Furthermore, the delta doping introduces a thinner potential barrier that allows for an increased tunneling current. Because the carrier density is increased only locally, the excess free-carrier absorption in the DBR mirror is minimized.

The intracavity metal contact technique [95–97] is an alternative approach to achieve low series resistance. This technique allows the electrical contact to bypass the resistive p-DBR mirror stack layers. Furthermore, the mirror stack above the metal contact does not require any doping, thereby reducing the intracavity free-carrier absorption loss and the optical scattering loss. Good performance VCSELs using this technique have been demonstrated in the research environment.

The intracavity metal contact structure starts with a conventional VCSEL design similar to that shown in Fig. 2.14 [95]. p-Type and n-type bulk layers that are multiples of half wavelengths are inserted on either side of the active region to provide electrical paths for the current to reach the active region from the ring contacts that are deposited on top of the inserted layers. A current blocking region must be formed to force the current into the optical mode. This current blocking region can be formed by ion implantation or undercutting using wet etching between the p-type insertion layer and the active region. A resistive layer is further introduced between the conductive current distribution layer and the active region to overcome any residual current crowding effects near the contact periphery so that the injected current can be more uniformly distributed in the optical mode area. The current blocking layer in the p-type region can also be a thin n-type GaAs or AlGaAs layer that is inserted into the top p-doped cladding layer [96]. A second growth is needed to complete the epitaxial structure after a current flow path is opened by either wet or dry etching. The final VCSEL device will have a reverse-biased pn junction inside the cavity in the p-doped region for current blocking (Fig. 2.18).

One of the simple approaches to make intracavity metal contact is to start with a VCSEL with an only partially grown p-DBR mirror stack (Fig. 2.19). After the metal contact has been deposited onto the p-DBR, a dielectric mirror stack consisting of quarter-wavelength-thick alternating layers TiO_2/SiO_2 is used to complete the device [97]. In this instance,

Fig. 2.18   Schematic diagram of an intracavity contact VCSEL structure (after Ref. [96]).

Fig. 2.19   Schematic diagram of a VCSEL structure with partial monolithic semiconductor DBR and partial dielectric DBR on the *p*-doped contact side. (Reprinted with permission from Ref. [97]. Copyright 1995 American Institute of Physics.)

resistance of 50 Ω has been achieved with a laser emission aperture of 5 μm and a proton implantation aperture of 20 μm.

Within the microcavity structure of a VCSEL, only one Fabry–Perot cavity mode exists in the designed DBR reflective bandwidth. A laser can only be sustained at the wavelength of the cavity mode. Clearly, a temperature-insensitive VCSEL can be demonstrated by taking advantage of the microcavity mode characteristics [16]. Typically, the peak of the active gain profile shifts with the temperature at a rate of 3–5 Å/°C, and the cavity resonant mode shifts at a rate of 0.5–1 Å/°C [98]. If the resonant cavity mode is designed to initially sit at the longer wavelength side of the gain profile, the gain peak will gradually walk into the cavity mode with the rise of temperature (Fig. 2.20). Conversely, the gain peak usually decreases with the temperature. Together, the actual gain for the VCSEL cavity mode will vary little with temperature, and the VCSEL threshold current will stay almost constant within a certain temperature range. This temperature

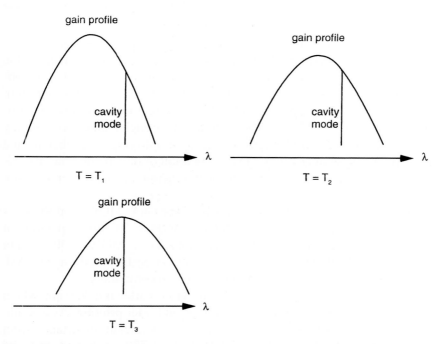

Fig. 2.20   VCSEL cavity mode vs gain profile for temperatures T_1, T_2, and T_3, with $T_3 > T_2 > T_1$.

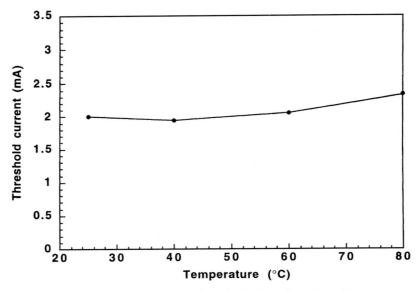

Fig. 2.21   Threshold current of a typical GaAs VCSEL varying with ambient temperature with minimum threshold current at 40°C.

insensitivity allows the VCSEL to be designed to operate optimally at the system temperature midpoint region. For example, with modest speed optical data links, the VCSELs can be designed to operate with minimum threshold current at approximately 40°C in a required working range of 0–70°C (Fig. 2.21) [17, 99]. The system can be implemented without any auto power control (APC) circuitry, thereby simplifying the packaging and reducing the system cost [14]. The application of this method has also allowed the demonstration of VCSELs operating at a record high temperature of 200°C [83].

Apart from VCSELs at 830–870 nm based on GaAs MQWs and VCSELs at 940–980 nm based on strained InGaAs MQWs, VCSELs operating at other wavelengths, such as 780 nm based on AlGaAs MQWs, 650–690 nm based on InAlGaP MQWs, and 1.3–1.5 nm VCSELs based on InGaAsP MQWs, have received attention in the research community.

The vast majority of the semiconductor laser market is at 780 nm, which is predominantly used for CD data storage and laser printing. As a result, the development of VCSELs at 780 nm is of strategic importance from a commercial standpoint. A typical VCSEL at 780 nm has an epitaxial layer structure similar to that of a VCSEL at 850 nm [100–102]. The larger

Fig. 2.22   Etched-mesa structure VCSEL output power vs input current at a wavelength of 780 nm.

bandgap requirement for 780 nm drives the MQW active region to the AlGaAs ternary system. The active region usually consists of three or four periods of $Al_{0.12}Ga_{0.88}As$ quantum wells sandwiched between the $Al_{0.3}Ga_{0.7}As$ barriers. The DBR mirror stack consists of 27 pairs of p-type doped $Al_{0.25}Ga_{0.75}As$/AlAs and 40 pairs of n-type doped $Al_{0.25}Ga_{0.75}$As/AlAs, with the bandwidth centered at 780 nm. The laser performance of a 780-nm VCSEL is similar to that of an 850-nm GaAs VCSEL (Fig. 2.22). The increased aluminum concentration in both the active region and the DBR mirror stack over that used in the 850-nm VCSEL raises a concern with the 780-nm VCSEL device reliability because of the poor edge-emitting semiconductor laser performance at 780 nm. There has been no reliability data published so far for the 780-nm VCSELs, and study is ongoing to address the issue.

Red visible VCSELs are of interest because of their potential applications in plastic fiber, bar-code scanner, pointer, and most recently the DVD format optical data storage. The epitaxial structure of a red visible VCSEL is grown on a GaAs substrate misoriented 6° off (100) plane toward the nearest |111 > A or on a (311) GaAs substrate [103–106]. It consists of three or four periods of $In_{0.56}Ga_{0.44}P$ QWs with InAlGaP or InAlP as barriers, InAlP as both p-type and n-type cladding layers, and two DBR

Fig. 2.23   A visible VCSEL structure. (Reprinted with permission from Ref. [105]. Copyright 1995 American Institute of Physics.)

mirrors (Fig. 2.23). The active QW layer is either tensile or compressive strained to enhance the optical gain. Typically, the QW thickness is 60–80 Å and the barrier thickness is 60–100 Å. The total optical cavity length including the active region and the cladding layers ranges from one wavelength or its multiple integer up to eight wavelengths. The DBR mirrors are composed of either InAlGaP/InAlP or $Al_{0.5}Ga_{0.5}As$/AlAs. The $Al_{0.5}Ga_{0.5}As$/AlAs DBR mirror performs better because of a relatively larger index difference between the two DBR constituents — thus a higher reflectivity and a wider bandwidth. In general, because the index difference between $Al_{0.5}Ga_{0.5}As$ and AlAs is much smaller than that used for the 850-nm VCSELs, more mirror pairs are needed to achieve the required DBR reflectivity. Typically, 55 pairs are needed for the n-DBR and 40 pairs are needed for the p-DBR to ensure a reasonable VCSEL performance. As a rule of thumb, the more pairs in the DBR mirror, the higher series resistance and thus more heat generated in the active region. This implies that the active junction temperature will be higher. Currently, sub-mA threshold red VCSELs have been demonstrated. More than 5-mW output power from a red VCSEL has also been reported. Unfortunately, the carrier confinement of the red visible VCSELs is poor because of the smaller bandgap offset between the quantum well and the barrier and between the active and the cladding. Therefore, the red visible VCSELs are extremely temperature sensitive, and more studies are needed to improve the red visible VCSEL high-temperature performances. VCSELs with wavelengths shorter than 650 nm pose more problems because of even worse carrier confinements.

Designing a VCSEL that can effectively confine the carriers in the active region is a challenging topic in today's research community.

Long-wavelength VCSELs at 1.3 and 1.55 μm have drawn attention because of their potential applications in telecommunications and medium-to long-distance data links, such as local area networks and wide area networks, where single-mode characteristics are required. The long-wavelength VCSELs are based on an InP substrate, with InGaAsP MQWs used as the active region. However, the lattice-matched monolithic InGaAsP/InP DBR mirrors do not have sufficient reflectivity for the long-wavelength VCSELs because of the small index difference between the two DBR mirror pair constituents, InGaAsP and InP. In addition, the Auger recombination-induced loss becomes evident due to smaller energy bandgap for the long-wavelength VCSELs. To overcome the difficulty, dielectric mirrors with 8.5 pairs of MgO/Si multilayers and Au/Ni/Au on the p side and 6 pairs of SiO$_2$/Si on the n side have been used instead of the semiconductor DBR. A continuous-wave 1.3-μm VCSEL has therefore been demonstrated at 14°C [107]. To further improve the device performance, wafer fusing techniques have been adopted to bond GaAs/AlAs DBR mirrors onto a structure with an InGaAsP MQW active layer sandwiched between the InP cladding layers that are epitaxially grown on the InP substrate [108, 109]. The InP substrate is removed to allow the GaAs/AlAs DBRs to be bonded onto one or both sides of the InGaAsP active region (Fig. 2.24).

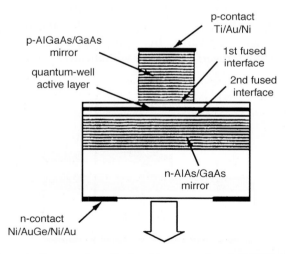

Fig. 2.24   Schematic diagram of a wafer-fused long-wavelength VCSEL (after Ref. [28]).

Because the DBR mirrors are either n-type or p-type doped, the completed fused wafer can be processed like a regular GaAs VCSEL wafer. In this way, a 1.5-μm VCSEL has been successfully fabricated that operates CW up to 64°C [28, 29]. Manufacturing yield and reliability are still currently unknown with the VCSEL wafer fusion technique. For commercial interest, the CW operation must be driven to at least the 100°C range for the junction, in addition to a number of other issues such as wall-plug efficiency, reliability, consistency, etc.

APC is one of the important features that is easily accomplished with edge-emitting lasers because of the backward emission that can be monitored from the cleaved facet. With VCSELs of wavelength shorter than 870 nm, the laser beam emits only toward the top epitaxy side. The backward emission is absorbed by the GaAs substrate, unless the substrate is removed. However, due to the unique vertical stacking feature of VCSELs, a detector can be integrated underneath or above the VCSEL structure during the epitaxial growth [102, 110–113] (Fig. 2.25). For example, a VCSEL can start with a p-type GaAs substrate, with a PIN detector structure grown first on top of the substrate. The PIN detector has a GaAs intrinsic layer of approximately 1 μm and p-doped AlGaAs cladding of approximately 2000 Å between the substrate and the intrinsic absorption layer. The detector structure stops at a n-type doped cladding layer of approximately 2000 Å. A regular GaAs VCSEL epitaxial structure follows the PIN detector, with layers of n-DBR, n cladding, active, p cladding, and p-DBR grown in order. The detector cathode in this structure shares a common contact with the VCSEL cathode, with two independent anodes for both the PIN detector and the VCSEL. In practical applications, the anode of the detector can be either reverse biased or without any bias if detector speed is

Fig. 2.25   Schematic diagram of a VCSEL with integrated detector (after Ref. [102]).

Fig. 2.26 VCSEL output power in relationship with current response of an integrated detector (after Ref. [102]).

not a major concern. The VCSEL backward emission transmitted through the n-DBR is normally in proportion to the VCSEL forward emission. It will be received by the integrated PIN detector and generate a current. The VCSEL output power and the integrated PIN detector response are shown in Fig. 2.26. There is a one-to-one relationship between the PIN detector current and the VCSEL output power up to certain point when the VCSEL output power saturates, but the detector current keeps rising due to the effect of spontaneous emission. Consequently, VCSEL operation with APC can be accomplished by monitoring the current variation generated in this detector when the VCSEL operates below the saturation [102, 110].

Super-low-threshold microcavity-type VCSELs have been proposed that utilize the spontaneous emission enhancement due to more spontaneous emission being coupled into the lasing mode [114, 115]. Although a thresholdless laser is theoretically possible when the spontaneous emission coupling effciency β is made approaching unity, the proposed structures are difficult to make in practice. One of the successful examples in research today is the use of oxidized lateral carrier confinement blocks by oxidizing an AlAs layer in the DBR or the cladding regions [23, 24, 116] (Fig. 2.27). This technology will be discussed in more detail in Chapter 15. Typically, sub-100-μA threshold can be achieved with this technique. A VCSEL with an extremely low threshold of 8.7 μA has been reported with an active area of 3 μm^2 [24]. It should be noted that there is still a debate on the exact mechanism that has generated this result. VCSELs with oxidized mirrors have been demonstrated with extremely simple epitaxy layers [117, 118]. In this structure, only four to six pairs of GaAs/AlAs DBR stacks are grown on one or both sides of an active region that is made of strained InGaAs MQWs at 970 nm (Fig. 2.28). The AlAs layers in the DBR mirrors are

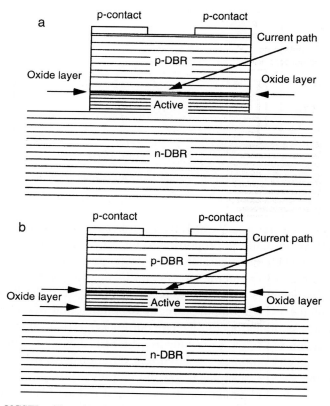

Fig. 2.27   VCSEL with native aluminum oxide for lateral current confinement. (a) Current confinement on *p* side, and (b) current confinement on both *p* side and *n* side.

oxidized during the fabrication procedure. The extremely large index difference between GaAs and the oxidized AlAs layer makes it possible that only four pairs of GaAs/AlAs stacks will provide sufficiently high reflectivity with very large bandwidth for proper device operation. The VCSEL electrical contacts in this case will have to be made laterally inside the cavity as opposed to at the top of the DBR mirror stacks because the electrical conduction through the DBR mirror is prohibited once the AlAs constituent of the mirror is oxidized.

High-speed data transmission requires that a VCSEL be modulated at multi-GHz. The cavity volume of a VCSEL is significantly smaller than that of an edge-emitting laser, resulting in a higher photon density in the VCSEL cavity. The resonance frequency of a semiconductor laser typically

Fig. 2.28   VCSEL with an AlAs oxide–GaAs DBR mirror (after Ref. [117]).

Fig. 2.29   Small signal modulation response of a 3-μm VCSEL at various bias current. The maximum 3-dB bandwidth is approximately 15 GHz (after Ref. [120]).

scales as the square root of the photon density, thus indicating that a VCSEL has a potential advantage in high-speed operation. However, the parasitic series resistance caused by the semiconductor DBR and the device heating limit the maximum achievable VCSEL modulation bandwidth. Currently, a modulation speed of larger than 16 GHz has been reported with an oxide-confined VCSEL at a current of 4.5 mA [119]. Modeling results indicate that a gain compression limited-oxide VCSEL with a diameter of 3 μm has an intrinsic 3-dB bandwidth of 45 GHz [120] and a measured 3-dB bandwidth of 15 GHz at 2.1 mA due to the parasitic resistance and the device heating (Fig. 2.29).

References

1. Gowar, J. 1984. *Optical communication systems*. Englewood Cliffs, N.J.: Prentice Hall.
2. Miller, S. E., and A. G. Chynoweth, eds. 1979. *Optical fiber telecommunications*. New York: Academic Press.
3. Lasky, R., U. Osterberg, and D. Stigliani, eds. 1995. *Optoelectronics for data communication*. New York: Academic Press.
4. Hall, R. N., G. E. Fenner, J. D. Kingsley, T. J. Soltys, and R. O. Carlson. 1962. Coherent light emission from GaAs junctions. *Phys. Rev. Lett.* 9:366.
5. Nathan, M. I., W. P. Dumke, G. Burns, F. H. Dill, Jr., and G. Lasher. 1962. Stimulated emission of radiation from GaAs pn junctions. *Appl. Phys. Lett.* 1:62.
6. Holonyak, N., Jr., and S. F. Bevacqua. 1962. Coherent (visible) light emission from $Ga(Al_{1-x}P_x)As$ junctions. *Appl. Phys. Lett.* 1:82.
7. Quist, T. M., R. H. Rediker, R. J. Keyes, W. E. Krag, B. Lax, A. L. McWhorter, and J. J. Zeiger. 1962. Semiconductor maser of GaAs. *Appl. Phys. Lett.* 1:91.
8. Kroemer, H. 1963. A proposed class of heterojunction injection lasers. *Proc. IEEE* 51:1782.
9. Hayashi, I., M. B. Panish, P. W. Foy, and S. Sumuski. 1970. Junction lasers which operate continuously at room temperature. *Appl. Phys. Lett.* 17:109.
10. Alferov, Zh. I., V. M. Andreev, D. Z. Garbuzov, Yu. V. Zhilyaev, E. P. Morozov, E. L. Portnoi, and V. G. Triofim. 1971. Investigation of the influence of the AlAs-GaAs heterostructure parameters on the laser threshold current and the realization of continuous emission at room temperature. *Sov. Phys. Semiconductor* 4:1573.
11. Anderson, S. G. 1996. Annual review of laser markets. *Laser Focus World* 32:50.
12. Soda, H., K. Iga, C. Kitahara, and Y. Suematsu. 1979. GaInAsP/InP surface emitting injection lasers. *Jpn. J. Appl. Phys.* 18:2329.
13. Iga, K., F. Koyama, and S. Kinoshita. 1988. Surface emitting semiconductor lasers. *IEEE J. Quantum Electron.* QE-24:1845.
14. Lebby, M., C. A. Gaw, W. B. Jiang, P. A. Kiely, C. L. Shieh, P. R. Claisse, J. Ramdani, D. H. Hartman, D. B. Schwartz, and J. Grula. 1996. Use of VCSEL arrays for parallel optical interconnects. *Proc. SPIE* 2683:81.
15. Orenstein, M., A. C. Von Lehmen, C. Chang-Hasnain, N. G. Stoffel, J. P. Harbison, and L. T. Florez. 1991. Matrix addressable vertical cavity surface emitting laser array. *Electron. Lett.* 27:437.
16. Shieh, C. L., D. E. Ackley, and H. C. Lee. 1993. Temperature insensitive vertical cavity surface emitting laser. U.S. Patent No. 5,274,655.

17. Young, D. B., J. W. Scott., F. H. Peters, B. J. Thibeault, S. W. Corzine, M. G. Peters, S. L. Lee, and L. A. Coldren. 1993. High-power temperature-insensitive gain-offset InGaAs/GaAs vertical-cavity surface-emitting lasers. *IEEE Photon. Tech. Lett.* 5:129.
18. Tai, K., G. Hasnain, J. D. Wynn, R. J. Fischer, Y. H. Wang, B. Weir, J. Gamelin, and A. Y. Cho. 1990. 90% coupling of top surface emitting GaAs/AlGaAs quantum well laser output into 8 μm diameter core silica fibre. *Electron. Lett.* 26:1628.
19. Iga, K., S. Ishikawa, S. Ohkouchi, and T. Nishimura. 1984. Room temperature pulsed oscillation of GaAlAs/GaAs surface emitting laser. *Appl. Phys. Lett.* 45:348.
20. Koyama, F., S. Kinoshita, and K. Iga. 1988. Room-temperature CW operation of GaAs vertical cavity surface emitting laser. *Trans. Inst. Electron. Commun. Eng. Jpn.* E71:1089.
21. Peters, F. H., M. G. Peters, D. B. Young, J. W. Scott, B. J. Thibeault, S. W. Corzine, and L. A. Coldren. 1993. High power vertical cavity surface emitting lasers. *Electron. Lett.* 29:200.
22. Grabherr, M., B. Weigl, G. Reiner, R. Michalzik, M. Miller, and K. J. Ebeling. 1996. High power top-surface emitting oxide confined vertical-cavity laser diodes. *Electron. Lett.* 32:1723.
23. Huffaker, D. L., J. Shin, and D. G. Deppe. 1994. Low threshold halfwave vertical-cavity lasers. *Electron. Lett.* 30:1946.
24. Yang, G. M., M. H. MacDougal, and P. D. Dapkus. 1995. Ultralow threshold current vertical-cavity surface-emitting lasers obtained with selective oxidation. *Electron. Lett.* 31:886.
25. Lear, K. L., K. D. Choquette, R. P. Schneider, S. P. Kilcoyne, and K. M. Geib. 1995. Selectively oxidised vertical cavity surface emitting lasers with 50% power conversion efficiency. *Electron. Lett.* 31:208.
26. Jäger, R., M. Grabherr, C. Jung, R. Michalzik, R. Reiner, B. Weigl, and K. J. Ebeling. 1997. 57% wallplug efficiency oxide-confined 850 nm wavelength GaAs VCSELs. To be published.
27. Iga, K. 1992. Surface emitting lasers. *Opt. Quantum Electron.* 24: S97.
28. Babic, D. I., K. Streubel, R. P. Mirin, N. M. Margalit, J. E. Bowers, E. L. Hu, D. E. Mars, L. Yang, and K. Carey. 1995. Room-temperature continuous-wave operation of 1.54-μm vertical-cavity lasers. *IEEE Photon. Tech. Lett.* 7:1225.
29. Margalit, N. M., D. I. Babic, K. Streubel, R. P. Mirin, R. L. Naone, J. E. Bowers, and E. L. Hu. 1996. Submilliamp long wavelength vertical cavity lasers. *Electron. Lett.* 32:1675.
30. Vakhshoori, D., J. D. Wynn, G. J. Zydik, and R. E. Leibenguth. 1993. 8 × 18 top emitting independently addressable surface emitting laser arrays

with uniform threshold current and low threshold voltage. *Appl. Phys. Lett.* 62:1718.

31. Uchiyama, S., and K. Iga. 1985. Two-dimensional array of GaInAsP/InP surface-emitting lasers. *Electron. Lett.* 21:162.

32. Deppe, D. G., J. P. van der Ziel, N. Chand, G. J. Zydzik, and S. N. G. Chu. 1990. Phase-coupled two-dimensional $Al_xGa_{1-x}As$-GaAs vertical-cavity surface-emitting laser array. *Appl. Phys. Lett.* 56:2089.

33. Orenstein, M., E. Kapon, N. G. Stoffel, J. P. Harbison, L. T. Florez, and J. Wullert. 1991. Two-dimensional phase-locked arrays of vertical-cavity semiconductor lasers by mirror reflectivity modulation. *Appl. Phys. Lett.* 58:804.

34. Iizuka, K., K. Matsumaru, T. Suzuki, H. Hirose, K. Suzuki, and H. Okamoto. 1995. Arsenic-free GaAs substrate preparation and direct growth of GaAs/AlGaAs multiple quantum well without buffer layers. *J. Cryst. Growth* 150:13.

35. Cheng, W. H., and J. H. Bechtel. 1993. High-speed fibre optic links using 780 nm compact disc lasers. *Electron. Lett.* 29:2055.

36. Soderstrom, R. L., S. J. Baumgartner, B. L. Beukema, T. R. Block, and D. L. Karst. 1993. CD lasers optical data links for workstations and midrange computers. *ECTC'93*, 505, June, Orlando.

37. Nakata, N. 1987. Laser diodes have low noise and low astigmatism. *JEE,* August, 49.

38. Wang, S. 1989. *Fundamentals of semiconductor theory and device physics,* 51. Englewood Cliffs, N.J.: Prentice Hall.

39. Morton, P. A., R. A. Logan, T. Tanbunek, P. F. Sciortino, A. M. Sergent, R. K. Montgomery, and B. T. Lee. 1993. 25 GHz bandwidth 1.55-μm GaInAsP p-doped strained multiquantum-well lasers. *Electron. Lett.* 29:136.

40. Ralston, J. D., E. C. Larkins, K. Eisele, S. Weisser, S. Buerkner, A. Schoenfelder, J. Daleiden, K. Czotscher, I. Esquivias, J. Fleissner, R. E. Sah, M. Maier, W. Benz, and J. Rosenzweig. 1996. Advanced epitaxial growth and device processing techniques for ultrahigh-speed (>40 GHz) directly modulated semiconductor lasers. *Proc. SPIE* 2683:30.

41. Hersee, S. D., B. de Cremoux, and J. P. Duchemin. 1984. Some characteristics of the GaAs/GaAlAs graded-index separate-confinement heterostructure quantum well laser structure. *Appl. Phys. Lett.* 44:476.

42. Coleman, J. J. 1995. Quantum-well heterostructure lasers. In *Semiconductor lasers: Past, present, and future,* G. P. Agrawal, Chapter 1. Woodbury, N.Y.: AIP Press.

43. Nagarajan, R., T. Fukushima, J. E. Bowers, R. S. Geels, and L. A. Colden. 1991. High-speed InGaAs/GaAs strained multiple quantum well lasers with low damping. *Appl. Phys. Lett.* 58:2326.

44. Miller, B. I., U. Koren, M. G. Young, and M. D. Chien. 1991. Strain-compensated strained-layer superlattices for 1.5 μm wavelength lasers. *Appl. Phys. Lett.* 58:1952.

45. Zhang, G., and A. Ovtchinnikov. 1993. Strain-compensated InGaAs/GaAsP/GaInAsP/GaInP quantum well lasers ($1 \sim 0.98 \,\mu$m) grown by gas-source molecular beam epitaxy. *Appl. Phys. Lett.* 62:1644.
46. Tsuchiya, T., M. Komori, R. Tsuneta, and H. Kakibayashi. 1994. Investigation of effect of strain-compensated structure and compensation limit in strained-layer multiple quantum wells. *J. Cryst. Growth* 145:371.
47. Bessho, Y., T. Uetani, R. Hiroyama, K. Komeda, M. Shono, A. Ibaraki, K. Yodoshi, and T. Niina. 1996. Self-pulsating 630 nm band strain-compensated MQW AlGaInP laser diodes. *Electron. Lett.* 32:667.
48. Nagel, S. R., J. B. MacChesney, and K. L. Walker. 1985. Modified chemical vapor deposition. In *Optical fiber communications,* ed. T. Y. Li, Vol. 1, Chap. 1. Orlando: Academic Press.
49. Cohen, L. G., C. Lin, and W. G. French. 1979. Tailoring zero chromatic dispersion into the 1.5–1.6 μm low-loss spectral region of single-mode fibres. *Electron. Lett.* 15:334.
50. Tsuchiya, H., and N. Imoto. 1979. Dispersion-free single-mode fibre in 1.5 μm wavelength region. *Electron. Lett.* 15:476.
51. Okamoto, K., T. Edahiro, A. Kawana, and T. Miya. 1979. Dispersion minimisation in single-mode fibres over a wide spectral range. *Electron. Lett.* 15:729.
52. Hsieh, J. J., J. A. Rossi, and J. P. Donnelly. 1976. Room-temperature cw operation of GaInAsP/InP double-heterostructure diode lasers emitting at 1.1 μm. *Appl. Phys. Lett.* 28:709.
53. Nelson, R. J., P. D. Wright, P. A. Barnes, R. L. Brown, T. Cella, and R. G. Sobers. 1980. High-output power InGaAsP ($1 = 1.3 \,\mu$m) strip-buried hetero-structure lasers. *Appl. Phys. Lett.* 36:358.
54. Hirao, M., S. Tsuji, K. Mizuishi, A. Doi, and M. Nakamura. 1980. Long wavelength InGaAsP/InP lasers for optical fiber communication systems. *J. Opt. Commun.* 1:10.
55. Dutta, N. K., and R. J. Nelson. 1982. The case for Auger recombination in $In_{1-x}Ga_xAs_yP_{1-y}$. *J. Appl. Phys.* 53:74.
56. Dutta, N. K., S. G. Napholtz, R. Yen, T. Wessel, T. M. Shen, and N. A. Olsson. 1985. Long wavelength InGaAsP ($1 \sim 1.3 \,\mu$m) modified multiquantum well laser. *Appl. Phys. Lett.* 46:1036.
57. Bates, R. J. S., and S. D. Walker. 1992. Evaluation of all-plastic optical fibre compute data link dispersion limits. *Electron. Lett.* 28:996.
58. Gwynne, P. 1996. Digital video disk technology offers increased storage features. *R&D Magazine* 38:40.
59. Yamamoto, S., H. Hayashi, T. Hayakawa, N. Miyauchi, S. Yano, and T. Hijikata. 1982. Room-temperature cw operation in the visible spectral range of 680–700 nm by AlGaAs double heterojunction lasers. *Appl. Phys. Lett.* 41:796.
60. Usui, A., T. Matsumoto, M. Inai, I. Mito, K. Kobayashi, and H. Watanabe. 1985. Room temperature cw operation of visible InGaAsP double

heterostructure laser at 671 nm grown by hydride VPE. *Jpn. J. Appl. Phys.* 24:L163.

61. Chong, T. H., and K. Kishino. 1990. Room temperature continuous wave operation of 671-nm wavelength GaInAsP/AlGaAs VSIS lasers. *IEEE Photon. Tech. Lett.* 2:91.

62. Kobayashi, K., S. Kawata, A. Gomyo, I. Hino, and T. Suzuki. 1985. Room-temperature cw operation of AlGaInP double-heterostructure visible lasers. *Electron. Lett.* 21:931.

63. Ikeda, M., Y. Mori, H. Sato, K. Kaneko, and N. Watanabe. 1985. Room-temperature continuous-wave operation of an AlGaInP double heterostructure laser grown by atmospheric pressure metalorganic chemical vapor deposition. *Appl. Phys. Lett.* 47:1027.

64. Ishikawa, M., Y. Ohba, H. Sugawara, M. Yamamoto, and T. Nakanisi. 1986. Room-temperature cw operation of InGaP/InGaAlP visible light laser diodes on GaAs substrates grown by metalorganic chemical vapor deposition. *Appl. Phys. Lett.* 48:207.

65. Hatakoshi, G., K. Nitta, Y. Nishikawa, K. Itaya, and M. Okajima. 1993. High-temperature operation of high-power InGaAlP visible laser. *Proc. SPIE* 1850:388.

66. Hashimoto, J., T. Katsuyama, J. Shinkai, I. Yoshida, and H. Hayashi. 1991. Effects of strained-layer structures on the threshold current density of AlGaInP/GaInP visible lasers. *Appl. Phys. Lett.* 58:879.

67. Honda, S., H. Hamada, M. Shono, R. Hiroyama, K. Yodoshi, and T. Yamaguchi. 1992. Transverse-mode stabilised 630 nm-band AlGaInP strained multiquantum-well laser diodes grown on misoriented ubstrates. *Electron. Lett.* 28:1365.

68. Tanaka, T., H. Yanagisawa, S. Yano, and S. Minagawa. 1993. High-temperature operation of 637 nm AlGaInP MQW laser diodes with quaternary QWS grown on misoriented substrates. *Electron. Lett.* 29:24.

69. Ueno, Y., H. Fujii, H. Sawano, K. Kobayashi, K. Hara, A. Gomyo, and K. Endo. 1993. 30-mW 690-nm high-power strained-quantum-well AlGaInP laser. *IEEE J. Quantum Electron.* QE-29:1851.

70. Arimoto, S., M. Yasuda, A. Shima, K. Kadoiwa, T. Kamizato, H. Watanabe, E. Omura, M. Aiga, K. Ikeda, and S. Mitsui. 1993. 150mW fundamental-transverse-mode operation of 670 nm window laser diode. *IEEE J. Quantum Electron.* QE-29:1874.

71. Nakayama, N., S. Itoh, K. Nakano, H. Okuyama, M. Ozawa, A. Ishibashi, M. Ikeda, and Y. Mori. 1993. Room temperature continuous operation of blue-green laser diodes. *Electron. Lett.* 29:1488.

72. Gaines, J. M., R. R. Drenten, K. W. Haberern, T. Marshall, P. Mensz, and J. Petruzzello. 1993. Blue-green injection lasers containing pseudomorphic

$Zn_{1-x}Mg_xS_ySe_{1-y}$ cladding lasers and operating up to 394 K. *Appl. Phys. Lett.* 62:2462.

73. Haase, M. A., P. F. Baude, M. S. Hagedorn, J. Qiu, J. DePuydt, H. Cheng, S. Guha, G. E. Hofler, and B. J. Wu. 1993. Low-threshold buried-ridge II–VI laser diodes. *Appl. Phys. Lett.* 63:2315.

74. Taniguchi, S., T. Hino, S. Itoh, K. Nakano, N. Nakayama, A. Ishibashi, and M. Ikeda. 1996. 100 h II–VI blue-green laser diode. *Electron. Lett.* 32:552.

75. Nakamura, S., T. Mukai, and M. Senoh. 1991. High-power GaN p–n junction blue-light-emitting diodes. *Jpn. J. Appl. Phys.* 30:L1998.

76. Nakamura, S., M. Senoh, S. Nagahama, N. Iwasa, T. Yamada, T. Matsushita, H. Kiyoku, and Y. Sugimoto. 1996. InGaN MQW structure laser diodes with cleaved mirror facets. *Jpn. J. Appl. Phys.* 35:L217.

77. Itaya, K., M. Onomura, J. Nishio, L. Sugiura, S. Saito, M. Suzuki, J. Rennie, S. Y. Nunoue, M. Yamamoto, H. Fujimoto, Y. Kokubun, Y. Ohba, G. Hatakoshi, and M. Ishikawa. 1996. Room temperature pulsed operation of nitride based multi-quantum-well laser diodes with cleaved facets on conventional c-face sapphire substrates. *Jpn. J. Appl. Phys.* (Part 2) 35:L1315.

78. Akasaki, I., S. Sota, H. Sakai, T. Tanaka, M. Koike, and H. Amano. 1996. Shortest wavelength semiconductor laser diode. *Electron. Lett.* 32:1105.

79. Nakamura, S., M. Senoh, S. Nagahama, N. Iwasa, T. Yamada, T. Matsushita, H. Kiyoku, and Y. Sugimoto. 1996. InGaN multi-quantum-well structure laser diodes grown on $MgAl_2O_4$ substrates. *Appl. Phys. Lett.* 68:2105.

80. Akasaki, I., and H. Amano. 1996. Progress and future prospects of group III nitride semiconductors. *LEOS'96,* Plen2, November, Boston.

81. Nakamura, S., M. Senoh, S. I. Nagahama, N. Iwasa, T. Yamada, T. Matsushita, Y. Sugimoto, and H. Kiyoku. 1996. First room-temperature continuous-wave operation of InGaN multi-quantum-well-structure laser diodes. *LEOS'96,* PD1.1, November, Boston.

82. Hasnain, G., K. Tai, J. D. Wynn, Y. H. Wang, R. J. Fischer, M. Hong, B. E. Weir, G. J. Zydzik, J. P. Mannaerts, J. Gamelin, and A. Y. Cho. 1990. Continuous wave top surface emitting quantum well lasers using hybrid metal/semiconductor reflectors. *Electron. Lett.* 26:1590.

83. Morgan, R. A., M. K. Hibbs-Brenner, T. M. Marta, R. A. Walterson, S. Bounnak, E. L. Kalweit, and J. A. Lehman. 1995. 200 degrees-C, 96-nm wavelength range, continuous-wave lasing from unbonded GaAs MOVPE-grown vertical cavity surface-emitting lasers. *IEEE Photon. Tech. Lett.* 7:441.

84. Zhou, P., J. L. Cheng, C. F. Schaus, S. Z. Sun, K. Zheng, E. Armour, C. Hains, W. Hsin, D. R. Myers, and G. A. Vawter. 1991. Low series resistance high-efficiency GaAs/AlGaAs vertical-cavity surface-emitting lasers with continuously graded mirrors grown by MOCVD. *IEEE Photon. Tech. Lett.* 3:591.

85. Tan, M. R. T., K. H. Hahn, Y. M. D. Houng, and S. Y. Wang, 1995. Surface emitting laser for multimode data link applications. *HP J.*, February: 67.
86. Tai, K., L. Yang, Y. H. Wang, J. D. Wynn, and A. Y. Cho. 1990. Drastic reduction of series resistance in doped semiconductor distributed Bragg reflectors for surface-emitting lasers. *Appl. Phys. Lett.* 56:2496.
87. Zhou, P., J. Cheng, C. F. Schaus, S. Z. Sun, K. Zheng, E. Armour, C. Hains, W. Hsin, D. R. Myers, and G. A. Vawter. 1991. Low series resistance high-efficiency GaAs AlGaAs vertical-cavity surface-emitting lasers with continuously graded mirrors grown by MOCVD. *IEEE Photon. Tech. Lett.* 3:591.
88. Schubert, E. F., L. W. Tu, G. J. Zydzik, R. F. Kopf, A. Benvenuti, and M. R. Pinto. 1992. Elimination of heterojunction band discontinuities by modulation doping. *Appl. Phys. Lett.* 60:466.
89. Peters, M. G., D. B. Young, F. H. Peters, J. W. Scott, B. J. Thibeault, and L. A. Coldren. 1994. 17.3-percent peak wall plug efficiency vertical-cavity surface-emitting lasers using lower barrier mirrors. *IEEE Photon. Tech. Lett.* 6:31.
90. Peters, M. G., B. J. Thibeault, D. B. Young, J. W. Scott, F. H. Peters, A. C. Gossard, and L. A. Coldren. 1993. Band-gap engineered digital alloy interfaces for lower resistance vertical-cavity surface-emitting lasers. *Appl. Phys. Lett.* 63:3411.
91. Lear, K. L., S. A. Chalmers, and K. P. Killeen. 1993. Low threshold voltage vertical cavity surface-emitting laser. *Electron. Lett.* 29:584.
92. Young, D. B., J. W. Scott, F. H. Peters, M. G. Peters, M. L. Majewski, B. J. Thibeault, S. W. Corzine, and L. A. Coldren. 1993. Enhanced performance of offset-gain high-barrier vertical-cavity surface-emitting lasers. *IEEE J. Quantum Electron.* QE-29:2013.
93. Schubert, E. F., A. Fischer, Y. Horikoshi, and K. Ploog. 1985. GaAs sawtooth superlattice laser emitting at wavelength $l > 0.9$ μm. *Appl. Phys. Lett.* 47:219.
94. Kojima, K., R. A. Morgan, T. Mullaly, G. D. Guth, M. W. Focht, R. E. Leibenguth, and M. T. Asom. 1993. Reduction of p-doped mirror electrical resistance of GaAs/AlGaAs vertical-cavity surface-emitting lasers by delta doping. *Electron. Lett.* 29:1771.
95. Scott, J. W., B. J. Thibeault, D. B. Young, L. A. Coldren, and F. H. Peters. 1994. High efficiency submilliamp vertical cavity lasers with intracavity contacts. *IEEE Photon. Tech. Lett.* 6:678.
96. Rochus, S., M. Hauser, T. Röhr, H. Kratzer, G. Böhm, W. Klein, G. Tränkle, and G. Weimann. 1995. Submilliamp vertical-cavity surface-emitting lasers with buried lateral-current confinement. *IEEE Photon. Tech. Lett.* 7:968.
97. Morgan, R. A., M. K. Hibbs-Brenner, J. A. Lehman, E. L. Kaiweit, R. A. Walterson, T. M. Marta, and T. Akinwande. 1995. Hybrid dielectric/AlGaAs mirror spatially filtered vertical cavity top-surface emitting laser. *Appl. Phys. Lett.* 66:1157.

98. Dudley, J. J., D. L. Crawford, and J. E. Bowers. 1992. Temperature dependence of the properties of DBR mirrors used in surface normal optoelectronic devices. *IEEE Photon. Tech. Lett.* 4:311.

99. Lebby, M., C. A. Gaw, W. B. Jiang, P. A. Kiely, P. R. Claisse, and J. Ramdani. 1996. Vertical-cavity surface-emitting lasers for communication applications. OSA annual meeting, WR1, October, Rochester, N.Y.

100. Lee, Y. H., B. Tell, K. F. Brown-Goebeler, R. E. Leibenguth, and V. D. Mattera. 1991. Deep-red CW top surface-emitting vertical-cavity AlGaAs superlattice lasers. *IEEE Photon. Tech. Lett.* 3:108.

101. Shin, H. E., Y. G. Ju, J. H. Shin, J. H. Ser, T. Kim, E. K. Lee, I. Kim, and Y. H. Lee. 1996. 780 nm oxidised vertical-cavity surface-emitting lasers with $Al_{0.1}Ga_{0.89}As$ quantum wells. *Electron. Lett.* 32:1287.

102. Kim, T., T. K. Kim, E. K. Lee, J. Y. Kim, and T. I. Kim. 1995. A single transverse mode operation of top surface emitting laser diode with an integrated photo-diode. *Proc. LEOS'95* 2:416.

103. Schneider, R. P., Jr., K. D. Choquette, J. A. Lott, K. L. Lear, J. J. Figiel, and K. J. Malloy. 1994. Efficient room-temperature continuous-wave AlGaInP/AlGaAs visible (670 nm) vertical-cavity surface-emitting laser diodes. *IEEE Photon. Tech. Lett.* 6:313.

104. Choquette, K. D., R. P. Schneider, M. H. Crawford, K. M. Geib, and J. J. Figiel. 1995. Continuous wave operation of 640–660 nm selectively oxidised AlGaInP vertical-cavity lasers. *Electron. Lett.* 31:1145.

105. Schneider, R. P., Jr., M. H. Crawford, K. D. Choquette, K. L. Lear, S. P. Kilcoyne, and J. J. Figiel. 1995. Improved AlGaInP-based red (670–690 nm) surface-emitting lasers with novel C-doped short-cavity epitaxial design. *Appl. Phys. Lett.* 67:329.

106. Crawford, M. H., R. P. Schneider, Jr., K. D. Choquette, and K. L. Lear. 1995. Temperature-dependent characteristics and single-mode performance of AlGaInP-based 670–690-nm vertical-cavity surface-emitting lasers. *IEEE Photon. Tech. Lett.* 7:724.

107. Baba, T., Y. Yogo, K. Suzuki, F. Koyama, and K. Iga. 1993. Near room temperature continuous wave lasing characteristics of GaInAsP/InP surface emitting laser. *Electron. Lett.* 29:913.

108. Dudley, J. J., M. Ishikawa, B. I. Miller, D. I. Babic, R. Mirin, W. B. Jiang, J. E. Bowers, and E. L. Hu. 1992. $144°C$ operation of 1.3 μm InGaAsP vertical cavity lasers on GaAs substrates. *Appl. Phys. Lett.* 61:3095.

109. Dudley, J. J., D. I. Babic, R. Mirin, L. Yang, B. I. Miller, R. J. Ram, T. Reynolds, E. L. Hu, and J. E. Bowers. 1994. Low threshold, wafer fused long wavelength vertical cavity lasers. *Appl. Phys. Lett.* 64:1463.

110. Shin, H. K., I. Kim, E. J. Kim, J. H. Kim, E. K. Lee, M. K. Lee, J. K. Mun, C. S. Park, and Y. S. Yi. 1996. Vertical-cavity surface-emitting lasers for optical data storage. *Jpn. J. Appl. Phys.* (Part 1), 35:506.

111. Hasnain, G., and K. Tai. 1992. Self-monitoring semiconductor laser device. U.S. Patent No. 5,136,603.
112. Hasnain, G., K. Tai, Y. H. Wang, J. D. Wynn, K. D. Choquette, B. E. Weir, N. K. Dutta, and A. Y. Cho. 1991. Monolithic integration of photodetector with vertical cavity surface emitting laser. *Electron. Lett.* 27:1630.
113. Hibbs-Brenner, M. K. 1995. Integrated laser power monitor. U.S. Patent No. 5,475,701.
114. Bjork, G., and Y. Yamamoto. 1991. Analysis of semiconductor microcavity lasers using rate equations. *IEEE J. Quantum Electron.* QE-27:2386.
115. Ram, R. J., E. Goobar, M. G. Peters, L. A. Coldren, and J. E. Bowers. 1996. Spontaneous emission factor in post microcavity lasers. *IEEE Photon. Tech. Lett.* 8:599.
116. Huffaker, D. L., D. G. Deppe, and K. Kumar. 1994. Native-oxide ring contact for low threshold vertical-cavity lasers. *Appl. Phys. Lett.* 65:97.
117. MacDougal, M. H., P. Daniel Dapkus, V. Pudikov, H. M. Zhao, and G. M. Yang. 1995. Ultralow threshold current vertical-cavity surface-emitting lasers with AlAs oxide-GaAs distributed Bragg reflectors. *IEEE Photon. Tech. Lett.* 7:229.
118. MacDougal, M. H., G. M. Yang, A. E. Bond, C. K. Lin, D. Tishinin, and P. D. Dapkus. 1996. Electrically-pumped vertical-cavity lasers with Al_xO_y-GaAs reflectors. *IEEE Photon. Tech. Lett.* 8:310.
119. Lear, K. L., A. Mar, K. D. Choquette, S. P. Kilcoyne, R. P. Schneider, Jr., and K. M. Geib. 1996. High frequency modulation of oxide-confined vertical cavity surface emitting lasers. *Electron. Lett.* 32:457.
120. Thibeault, B. J., K. Bertilsson, E. R. Hegblom, E. Strzelecka, P. D. Floyd, R. Naone, and L. A. Coldren. 1997. High-speed characteristics of low-optical loss oxide-apertured vertical-cavity lasers. *IEEE Photon. Tech. Lett.* 9:11.

Chapter 3 | Detectors for Fiber Optics

Carolyn J. Sher DeCusatis

*Formerly from Lighting Research Center, Rensselaer
Polytechnic Institute, Troy, New York 12180**

Ching-Long (John) Jiang

*Amp Incorporated, Lytel Division, Somerville,
New Jersey 08876*

3.1. Introduction

Detectors are an integral part of all fiber optic communication systems. They demodulate optical signals, that is, convert the optical variations into electrical variations, which are subsequently amplified and further processed. For such applications the detectors must satisfy stringent requirements such as high sensitivity at operating wavelengths, high response speed, and minimum noise. In addition, the detector should be compact in size, use low biasing voltages, and be reliable under operating conditions. There are many different types of optical detectors available, which are sensitive over the entire optical spectra, the infrared and the ultraviolet, including such devices as photomultiplier tubes, charge coupled devices (CCDs), and thermal and photoconductive detectors [1]. However, fiber optic communication systems most commonly use solid state semiconductor devices, such as the PIN photodiode. A general semiconductor detector has basically three processes: first, carrier (electron–hole pair) generation by incident light; second, carrier transport and/or multiplication by whatever current-gain mechanism may be present; and third, interaction of current with the external circuit to provide the output signal. In this chapter, we will begin with an overview of detector terminology. This is followed by a detailed description of the types of detectors used extensively in fiber optics.

*May be contacted through the editor.

HANDBOOK OF FIBER OPTIC
DATA COMMUNICATION
$35.00

The PIN photodiode is the most commonly used fiber optic detector, and almost exclusively used in transceivers. We will also discuss the avalanche photodiodes, photodiode arrays, Schottky-barrier photodiodes, metal-semiconductor-metal (MSM) detectors, and resonant-cavity photodetectors. These detectors are not used in transceivers, but are becoming more common in applications such as wavelength-divison multiplexing (WDM) and parallel optical interconnects (POI). Finally, we will briefly describe some noise sources in datacom receivers and their impact on the detection electronics.

3.2. Detector Terminology and Characteristics

In this section we will define the terminology used in a typical detector specification. Every detector specification should include a picture and/or physical description of the part, including dimensions and construction (i.e., plastic housing). We have tried to be inclusive in the list of terms, which means not all of these quantities will apply to every detector specification. Because specifications are not standardized, it is impossible to include all possible terms used; however, most detectors are described by certain standard figures of merit, which will be discussed in this section. It is important to consider the manufacturer's context for all values; a detector designed for a specific application may not be appropriate for a different application even though it seems as if it would be.

There are several figures of merit used to characterize the performance of different detectors. Responsivity, or Response, is the sensitivity of the detector to input flux. It is given by

$$R(\lambda) = I/\phi(\lambda), \tag{3.1}$$

where I is the detector output signal (in amps) and ϕ is the incident light signal on the detector (in watts). Thus, the units of responsivity are amps per watt. Even when the detector is not illuminated, some current will flow; this dark current may be subtracted from the detector output signal when determining detector performance. Dark current is the thermally generated current in a photodiode under a completely dark environment; it depends on the material, doping, and structure of the photodiode. It is the lowest level of thermal noise. Dark current in photodiodes limits the sensitivity (minimum detectable power). The reduction of dark current is important for the improvement of the minimum detectable power. It is usually simply measured and then subtracted from the flux, like background, in most specifications. However, the dark current is temperature dependent, so care must be taken to evaluate it over the expected operating conditions. It is not a good idea for the anticipated signal to be a small fraction of the dark current; root mean square (rms) noise in the dark current may mask the signal. Responsivity is defined at a specific wavelength; the term spectral responsivity is used to describe the variation at different wavelengths. Responsivity vs. wavelength is often included in a specification as a graph, as well as placed in a performance chart at a specified wavelength.

Quantum efficiency (*QE*) is the ratio of the number of electron–hole pairs collected at the terminals to the number of photons in the incident light. It depends on the material from which the detector is made, and is primarily determined by reflectivity, absorption coefficient, and carrier diffusion length. As the absorption coefficient is dependent on the incident light wavelength, the quantum efficiency has a spectral response. Quantum efficiency is the fundamental efficiency of the diode for converting photons into electron–hole pairs.

For example, the quantum efficiency of a PIN diode can be calculated by,

$$QE = (1 - R)T(1 - e^{-\alpha W}), \tag{3.2}$$

where R is the surface reflectivity, T is the transmission of any lossy electrode layers, W is the thickness of the absorbing layer, and α is the absorption coefficient.

Quantum efficiency affects detector performance through the responsivity (R), which can be calculated from quantum efficiency:

$$R(\lambda) = QE\lambda q/hc, \tag{3.3}$$

where q is the charge of an electron (1.6×10^{-19} coulomb), λ is the wavelength of the incident photon, h is Planck's constant (6.626×10^{-34} W), and c is the velocity of light (3×10^8 m/s). If wavelength is in microns and R is responsivity flux, then the units of quantum efficiency are amperes per watt. Responsivity is the ratio of the diode's output current to input optical power and is given in amperes per watt (A/W). A PIN photodiode typically has a responsivity of 0.6 to 0.8 A/W. A responsivity of 0.8 A/W means that incident light having 50 microwatts of power results in 40 microamps of current; in other words,

$$I = 50\,\mu\text{W} \times 0.8\,\text{A/W} = 40\,\mu\text{A}, \tag{3.4}$$

where I is the photodiode current. For an avalanche photodiode (APD), a typical responsivity is 80 A/W. The same 50 microwatts of optical power now produces 4 mA of current:

$$I = 50\,\mu W \times 80\,A/W = 4\,mA \tag{3.5}$$

The minimum power detectable by the photodiode determines the lowest level of incident optical power that the photodiode can detect. It is related to the dark current in the diode, because the dark current will set the lower limit. Other noise sources are factors, including those associated with the diode and those associated with the receiver. The noise floor of a photodiode, which tells us the minimum detectable power, is the ratio of noise current to responsivity:

$$\text{Noise floor} = \text{noise/responsivity} \tag{3.6}$$

For initial evaluation of a photodiode, we can use the dark current to estimate the noise floor. Consider a photodiode with $R = 0.8$ A/W and a dark current of 2 nA. The minimum detectable power is

$$\text{Noise floor} = (2\,nA)/(0.8\,nA/nW) = 2.5\,nW \tag{3.7}$$

More precise estimates must include other noise sources, such as thermal and shot noise. As discussed, the noise depends on current, load resistance, temperature, and bandwidth.

Response time is the time required for the photodiode to respond to an incoming optical signal and produce an external current. Similarly to a source, response time is usually specified as a rise time and a fall time, measured between the 10 and 90% points of amplitude (other specifications may measure rise and fall times at the 20–80% points, or when the signal rises or falls to $1/e$ of its initial value). The bandwidth of a photodiode can be limited by either its rise time and fall time or its RC time constant, whichever results in the slower speed or bandwidth. The bandwidth of a circuit limited by the RC time constant is

$$B = 1/2\pi RC \tag{3.8}$$

where R is the load resistance and C is the diode capacitance. Figure 3.1 shows the equivalent circuit model of a photodiode. It consists of a current source in parallel with a resistance and a capacitance. It appears as a low-pass filter, a resistor-capacitor network that passes low frequencies and attenuates high frequencies. The cutoff frequency, which is the frequency

Fig. 3.1 Small-signal equivalent circuit for a reversed biased photodiode.

that is attenuated 3 dB, marks the 3 dB bandwidth. Photodiodes for high-speed operation must have a very low capacitance. The capacitance in a photodiode is mainly the junction capacitance formed at the pn junction, as well as any capacitance contributed by the packaging.

Bias voltage refers to an external voltage applied to the detector, and will be more fully described in the following section. Photodiodes require bias voltages ranging from as low as 0 V for some PIN photodiodes to several hundred volts for APDs. Bias voltage significantly affects operation, because dark current, responsivity, and response time all increase with bias voltage. APDs are usually biased near their avalanche breakdown point to ensure fast response.

Active area and effective sensing area are just what they sound like: the size of the detecting surface of the detection element. (This is a figure of merit that is important to consider when modifying a single mode detector for use on multimode fiber.) The uniformity of response refers to the percentage change of the sensitivity across the active area. Operating temperature is the temperature range over which a detector is accurate and will not be damaged by being powered. However, read the manual: there may be changes in sensitivity and dark current, which must be taken into account. Storage temperature will have a considerably larger range; basically, it describes the temperature range under which the detector will not melt, freeze, or otherwise be damaged or lose its operating characteristics.

NEP stands for Noise Equivalent Power. It is the amount of flux that would create a signal of the same strength as the rms detector noise. In other words, it is a measure of the minimum detectable signal; for this reason, it is the most commonly used version of the more generic figure of merit, Noise Equivalent Detector Input. More formally, it may be defined as the optical power (of a given wavelength or spectral content) required to produce a detector current equal to the root mean square (rms) noise in a unit bandwidth of 1 Hz:

$$\text{NEP}(\lambda) = i_n(\lambda)/R(\lambda), \tag{3.9}$$

where i_n is the rms noise current and R is the Responsivity, defined above. It can be shown [2] that to a good approximation,

$$\text{NEP} = 2hc/QE\lambda, \tag{3.10}$$

where this expression gives the NEP of an ideal diode when $QE = 1$. If the dark current is large, this expression may be approximated by

$$\text{NEP} = hc(2qI)^{1/2}/QEq\lambda, \tag{3.11}$$

where I is the detector current. Sometimes it is easier to work with detectivity, which is the reciprocal of NEP. The higher the detectivity, the smaller the signal a detector can measure; this is a convenient way to characterize more sensitive detectors. Detectivity and NEP vary with the inverse of the square of active area of the detector, as well as with temperature, wavelength, modulation frequency, signal voltage, and bandwidth. For a photodiode detecting monochromatic light and dominated by dark current, detectivity is given by

$$D = QEq\lambda/hc(2qI)^{1/2}. \tag{3.12}$$

The quantity specific detectivity accounts for the fact that dark current is often proportional to detector area, A; it is defined by

$$D^* = DA^{1/2}. \tag{3.13}$$

Normalized detectivity is detectivity multiplied by the square root of the product of active area and bandwidth. This product is usually constant, and allows comparison of different detector types independent of size and bandwidth limits. This is because most detector noise is white noise (Gaussian power spectra), and white noise power is proportional to the bandwidth of the detector electronics; thus the noise signal is proportional to the square root of bandwidth. Also, note that electrical noise power is usually proportional to detector area and the voltage that provides a measure of that noise is proportional to the square root of power. Normalized detectivity is given by

$$D_n = D(AB)^{1/2} = (AB)^{1/2}/\text{NEP}, \tag{3.14}$$

where B is the bandwidth. The units are $(\text{cm Hz})^{1/2}$ W^{-1}. Normalized detectivity is a function of wavelength and spectral responsivity; it is often quoted as normalized spectral responsivity.

Bandwidth, B, is the range of frequencies over which a particular instrument is designed to function within specified limits. Bandwidth is often adjusted to limit noise; in some specifications it is chosen as 1 Hz, so NEP is quoted in watts/Hz. Wide bandwidth detectors required in optical datacom often operate into a low resistance and require a minimal signal current much larger than the dark current; the load resistance, amplifier, and other noise sources can make the use of NEP, D, D^*, and D_n inappropriate for characterizing these applications.

Linearity range is the range of incident radiant flux over which the signal output is a linear function of the input. The lower limit of linearity is NEP, and the upper limit is saturation. Saturation is when the detector begins to form less signal output for the same increase of input flux. When a detector begins to saturate, it has reached the end of its linear range. Dynamic range can be used to describe nonlinear detectors, such as the human eye. Although datacom systems do not typically use filters on the detector elements, neutral density filters can be used to increase the dynamic range of a detector system by creating islands of linearity, whose actual flux is determined by dividing output signals of the detector by the transmission of the filter. Without filtering, the dynamic range would be limited to the linear range of the detector, which would be less because the detector would saturate without the filter to limit the incident flux. The units of linear range are incident radiant flux or power (watts or irradiance). Measuring the response of a detector to flux is known as calibration. Some detectors can be self calibrated, others require manufacturer calibration. Calibration certificates are supplied by most manufacturers for fiber optic test instrumentation; they are dated, and have certain time limits. The gain, also known as the amplification, is the ratio of electron–hole pairs generated per incident photon. Sometimes detector electronics allow the user to adjust the gain. Wiring and pin output diagrams tell the user how to operate the equipment, by schematically showing how to connect the input and output leads.

3.3. PN Photodiode

Photodiode detectors used in data communications are solid state devices. To understand their function, we must first describe a bit of semiconductor physics. For the interested reader, other introductory references to solid state physics, semiconductors, and condensed matter are available [2]. In a solid state device, the electron potential can be described in terms of

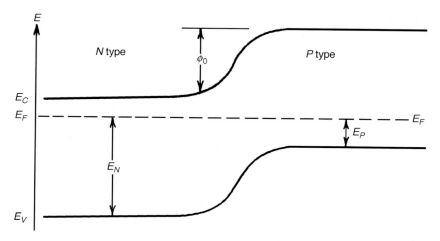

Fig. 3.2   Conduction and valance bands of a semiconductor.

conduction bands and valence bands, rather than individual potential wells (see Fig. 3.2). The highest energy level containing electrons is called the Fermi level. If a material is a conductor, the conduction and valence bands overlap and charge carriers (electrons or holes) flow freely; the material carries an electrical current. An insulator is a material for which there is a large enough gap between the conduction and valence bands to prohibit the flow of carriers; the Fermi level lies in the middle of the forbidden region between bands, called the band gap. A semiconductor is a material for which the band gap is small enough that carriers can be excited into the conduction band with some stimulus; the Fermi level lies at the edge of the valence band (if the majority of carriers are holes) or the edge of the conduction band (if the majority of carriers are electrons). The first case is called a *p*-type semiconductor, the second is called an *n*-type. These materials are useful for optical detection because incident light can excite electrons across the band gap and generate a photocurrent.

The simplest photodiode is the pn photodiode shown in Fig. 3.3. Although this type of detector is not widely used in fiber optics, it serves the purpose of illustrating the basic ideas of semiconductor photodetection. Other devices — the PIN and avalanche photodiodes — are designed to overcome the limitations of the pn diode. When the pn photodiode is reverse biased (negative battery terminal connected to *p*-type material), very little current flows. The applied electric field creates a depletion region on either side of the pn junction. Carriers — free electrons and holes — leave

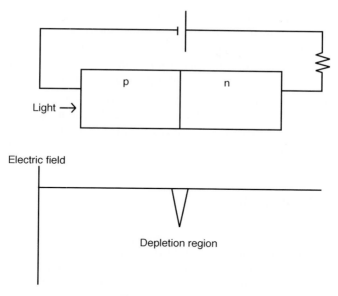

Fig. 3.3 PN diode.

the junction area. In other words, electrons migrate toward the negative terminal of the device and holes toward the positive terminal. Because the depletion region has no carriers, its resistance is very high, and most of the voltage drop occurs across the junction. As a result, electrical fields are high in this region and negligible elsewhere.

An incident photon absorbed by the diode gives a bound electron suffi-cient energy to move from the valence band to the conduction band, creating a free electron and a hole. If this creation of carriers occurs in the depletion region, the carriers quickly separate and drift rapidly toward their respective regions. This movement sets an electron flowing as current in the external circuit. When the carriers reach the edge of the depletion region, where electrical fields are small, their movement, however, relies on diffusion mechanism. When electron–hole creation occurs outside of the depletion region, the carriers move slowly toward their respective regions. Many car-riers recombine before reaching it. Their contribution to the total current is negligible. Those carriers remaining and reaching the depleted area are swiftly swept across the junction by the large electrical fields in the region to produce an external electrical current. This current, however, is delayed with respect to the absorption of the photon that created the carriers because of the initial slow movement of carriers toward their respective regions.

Current, then, will continue to flow after the light is removed. This slow response, due to slow diffusion of carriers, is called slow tail response.

Two characteristics of the pn diode make it unsuitable for most fiber optic applications. First, because the depletion area is a relatively small portion of the diode's total volume, many of the absorbed photons do not contribute to the external current. The created free electrons and holes recombine before they contribute significantly to the external current. Second, the slow tail response from slow diffusion makes the diode too slow for high-speed applications. This slow response limits operations to the kilohertz range.

3.4.   PIN Photodiode

The structure of the PIN diode is designed to overcome the deficiencies of its pn counterpart. The PIN diode is a photoconductive device formed from a sandwich of three layers of crystal, each layer with different band structures caused by adding impurities (doping) to the base material, usually indium gallium arsenide, silicon, or germanium. The layers are doped in this arrangement: p-type (or positive) on top, intrinsic, meaning undoped, in a thin middle layer, and n-type (or negative) on the bottom. For a silicon crystal a typical p-type impurity would be boron, and indium would be a p-type impurity for germanium [2–6]. Actually, the intrinsic layer may also be lightly doped, although not enough to make it either p-type or n-type. The change in potential at the interface has the effect of influencing the direction of current flow, creating a diode; obviously, the name PIN diode comes from the sandwich of p-type, intrinsic, and n-type layers.

The structure of a typical PIN photodiode is shown in Fig. 3.4. The p-type and n-type silicons form a potential at the intrinsic region; this potential gradient depletes the junction region of charge carriers, both electrons and holes, and results in the conduction band bending. The intrinsic region has no free carriers, and thus exhibits high resistance. The junction drives holes into the p-type material and electrons into the n-type material. The difference in potential of the two materials determines the energy an electron must have to flow through the junction. When photons fall on the active area of the device, they generate carriers near the junction, resulting in a voltage difference between the p-type and n-type regions. If the diode is connected to external circuitry, a current will flow that is proportional to the illumination. The PIN diode structure addresses the main problem with pn diodes, namely providing a large depletion region for the absorption of photons.

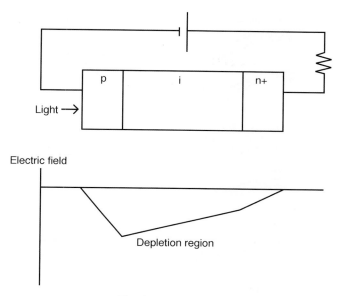

Fig. 3.4 PIN diode.

There is a tradeoff involved in the design of PIN diodes. Most of the photons are absorbed in the intrinsic region, so a thick intrinsic layer is desirable to improve photon–carrier conversion efficiency (to increase the probability of a photon being absorbed in the intrinsic region). On the other hand, a thin intrinsic region is desirable for high-speed devices, because it reduces the transit time of photogenerated carriers. These two conditions must be balanced in the design of PIN diodes.

Photodiodes can be operated either with or without a bias voltage. Unbiased operation is called the photovoltaic mode; certain types of noise, including 1/f noise, are lower and the NEP is better at low frequencies. Signal-to-noise ratio is superior to the biased mode of operation for frequencies below about 100 kHz [6]. Biasing (connecting a voltage potential to the two sides of the junction) will sweep carriers out of the junction region faster and change the energy requirement for carrier generation to a limited extent. Biased operation (photoconductive mode) can be either forward or reverse bias. Reverse bias of the junction (positive potential connected to the n-side and negative connected to the p-side) reduces junction capacitance and improves response time; for this reason it is the preferred operation mode for pulsed detectors. A PIN diode used for photodetection

Fig. 3.5   Forward and reverse bias of a diode [2].

may also be forward biased (the positive potential connected to the *p*-side and the negative to the *n*-side of the junction) to make the potential scaled for less current flow, or in other words to increase the sensitivity of the detector (Fig. 3.5).

An advantage of the PIN structure is that the operating wavelength and voltage, diode capacitance, and frequency response may all be predetermined during the manufacturing process. For a diode whose intrinsic layer thickness is w with an applied bias voltage of V, the self-capacitance of the diode, C, approaches that of a parallel plate capacitor,

$$C = \varepsilon_0 \varepsilon_1 A_0 / w \tag{3.15}$$

where A_0 is the junction area, ε_0 the free space permittivity (8.849×10^{-12} farads/m), and ε_1 the relative permittivity. Taking typical values of $\varepsilon_1 = 12$, $w = 50$ microns, and $A_0 = 10^{-7}$ m^2, $C = 0.2$ pF. Quantum efficiencies of 0.8 or higher can be achieved at wavelengths of 0.8–0.9 microns, with dark currents less than 1 nA at room temperature. Some typical responsivities for common materials are given in Table 3.1.

Table 3.1   **Common Semiconductor PIN Diode Properties**

Material	Wavelength range (μm)	Peak responsivity (A/W)
Silicon	0.3–1.1	0.5 A/W at 0.8 μm
Germanium	0.5–1.8	0.8 A/W at 0.7μm
InGaAs	1.0–1.7	1.1 A/W at 1.7 μm

The sensitivity of a PIN diode can vary widely by quality of manufacture. A typical PIN diode size ranges from 5 mm × 5 mm to 25 mm × 25 mm. Ideally, the detection surface will be uniformly sensitive (at the National Institute for Standards and Technology (NIST) there is a detector profiler which, by using extremely well-focused light sources, can determine the sensitivity of a detector's surface [3]). For most applications, it is required that the detector is uniformly illuminated or overfilled. The spectral responsivity of an uncorrected silicon photodiode is shown in Fig. 3.6. The typical QE curve is also shown for comparison. An ideal silicon detector would have zero responsivity and QE for photons whose energies are less than the band gap, or wavelengths much longer than about 1.1 micron. Just below the long wave limit, this ideal diode would have 100% QE and responsivity close to 1 A/W; responsivity vs. wavelength would be expected to follow the intrinsic spectral response of the material. In practice, this does not happen; these detectors are less sensitive in the blue region, which can sometimes be enhanced by clever doping, but not more than an order of magnitude. This lack of sensitivity is because there are fewer short wavelength photons per watt, so responsivity in terms of power drops off, and because more energetic blue photons may not be absorbed in the junction region. For color-sensitive applications such as photometry, filters are used so detectors will respond photometrically, or to the standardized CIE color coordinates; however, the lack of overall sensitivity in the blue region can potentially create noise problems when measuring a low-intensity blue signal. In the deep ultraviolet, photons are often absorbed by detector windows or surface coatings on the semiconductor before they reach the sensitive region. The departure from 100% QE in real devices is typically due to Fresnel reflections from the detector surface. The long wavelength cutoff is more gradual than expected for an ideal device because the absorption coefficient decreases at long wavelengths, so more photons pass though the photosensitive layers and do not contribute to the QE. As a result, QE tends to roll off gradually near the band gap limit. This response is

Fig. 3.6 PIN diode spectral response and quantum efficiency. (a) silicon; (b) InGaAs.

typical for silicon devices, which make excellent detectors in the wavelength range 0.8–0.9 microns. A common material for fiber optic applications is InGaAs, which is most sensitive in the near infrared (0.8 to 1.7 microns). Other PIN diode materials include HgCdZnTe for wavelengths of 2 to 12 microns. In the 1940s a popular photoconductive material for infrared solid state detectors — lead sulfide (PbS) — was introduced; this material is still commonly used in the region from 1 to 4 microns [6].

The speed of the detector depends on the junction capacitance. Two of the most widely used methods of reducing junction area use are mesa structures and planar structures with selective diffusion of the contact area. Another method is increasing the thickness of the depletion region, but this increases the transit time of photogenerated carriers. Strategies to improve transit time include edge-illuminated structure and absorptive resonance in the i-layer (see Section 3.5.5. Resonant-Cavity Enhanced Photodetectors (RECAP)) [7].

PIN diode detectors are not very sensitive to temperature (-25 C to $+80$ C) or shock and vibration, making them an ideal choice for a data communications transceiver. It is very important to keep the surface of any detector clean. This becomes an issue with PIN diode detectors because they are sufficiently rugged than they can be brought into applications that expose them to contamination. Both transceivers and optical connectors should be cleaned regularly during use to avoid dust and dirt buildup.

A sample specification for a photodiode is given in Table 3.2. Bias voltage (V) is the voltage applied to a silicon photodiode to change the potential that photoelectrons must scale to become part of the signal. Bias voltage is basically a set operating characteristic in a prepackaged detector, in this case -24 V. Shunt resistance is the resistance of a silicon photodiode when not biased. Junction capacitance is the capacitance of a silicon photodiode when not biased. Breakdown voltage is the voltage applied as a bias, large enough to create signal on its own. When this happens, the contribution of photoelectrons is minimal, so the detector cannot function. However, once the incorrect bias is removed the detector should return to normal.

To increase response speed and quantum efficiency, a variation on the PIN diode, known as the Schottky-barrier diode, can be used. This approach will be treated in a later section (3.5.3.). For wavelengths longer than about 0.8 microns, a heterojunction diode may be used for this same reason. Heterojunction diodes retain the PIN sandwich structure, but the surface layer is doped to have a wider bandgap and thus reduced absorption. In this case, absorption is strongest in the narrower bandgap region at the heterojunction, where the electric field is maximum; hence, good quantum efficiencies can be obtained. The most common material systems for heterojunction diodes are InGaAsP on an InP substrate, or GaAlAsSb on a GaSb substrate.

A typical circuit for biased operation of a photodiode is shown in Fig. 3.7, where C_a and R_a represent the input impedance of a post-amplifier. The photodiode impulse response will differ from an ideal square wave for several reasons, including transit time resulting from the drift of carriers

Table 3.2   **Sample Specifications for a PIN Diode**

Receiver Section						
Parameter	Symbol	Test Conditions	Min.	Typical	Max.	Units
Data rate (NRZ)	B	—	10	—	156	Mb/s
Sensitivity (avg)	P_{OH}	62.5 μm fiber .275 NA, BER ≤10^{-10}	-32.5	—	-14.0	dBm
Optical wavelength	λ_{IN}	—	1270	—	1380	nm
Duty cycle	—	—	25	50	75	%
Output risetime	t_{TLH}	20 - 80% 50Ω to V_{cc}-2V	.5	—	2.5	ns
Output falltime	t_{THL}	80 - 20% 50Ω to V_{cc}-2V	.5	—	2.5	ns
Output voltage	V_{OL}	—	V_{cc}-1.025	—	V_{cc}-.88	V
	V_{OH}		V_{cc}-1.81	—	V_{cc}-.1.62	V
Signal detect	V_A	$P_{IN} > P_A$	V_{cc}-1.025	—	V_{cc}-.88	V
	V_D	$P_{IN} < P_D$	V_{cc}-1.81	—	V_{cc}-.1.62	V
P_{IN} power levels:						
Deassert	P_B	—	-39.0 or P_B —	—	-32.5	dBm
Assert	P_A	—	-38.0	—	-30.0	dBm
Hysteresis	—	—	1.5	2.0	—	dB
Signal detect delay time:						
Deassert	—	—	—	—	50	μS
Assert	—	—	—	—	50	μS
Power supply voltage	V_{cc}-V_{EE}	—	4.75	5.0	5.25	V
Power supply current	I_{cc} or I_{EE}	—	—	—	150	mA
Operating temperature	T_A	—	0	—	70	°C
Absolute Maximum Ratings: Transceiver						
Parameter	**Symbol**	**Test Conditions**	**Min.**	**Typical**	**Max.**	**Units**
Storage temperature	—	—	-40	—	100	°C
Lead soldering limits	—	—	—	—	240/10	°C/s
Supply voltage	V_{cc}-V_{EE}	—	-.2	—	7.00	V

across the depletion layer, delay caused by the diffusion of carriers generated outside the depletion layer, and the RC time constant of the diode and
its load. If we return to the photodiode equivalent circuit of Fig. 3.1 and
insert it into this typical bias circuit, we can make the approximation that
R_s is much smaller than R_a to arrive at the equivalent small signal circuit
model of Fig. 3.8. In this model the resistance R is approximately equal to
R_a, and the capacitance C is the sum of the diode capacitance, amplifier
capacitance, and some distributed stray capacitance and j is the square root

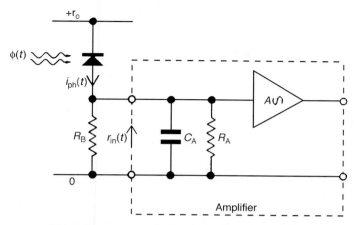

Fig. 3.7 PIN diode and amplifier equivalent circuit.

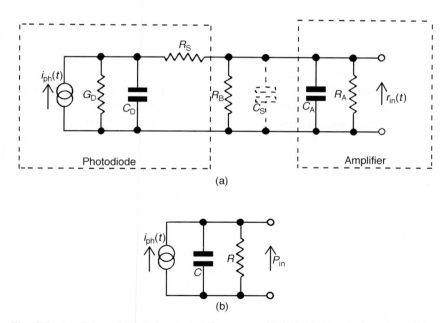

Fig. 3.8 Small-signal equivalent circuit for a normally biased photodiode and amplifier: (a) complete circuit; (b) reduced circuit obtained by neglecting R_s and lumping together the parallel components.

of negative one. In response to an optical pulse falling on the detector, the load voltage V_{in} will rise and fall exponentially with a time constant RC. In response to a photocurrent I_p, which varies sinusoidally at the angular frequency,

$$\omega = 2\pi f, \qquad (3.16)$$

the response of the load voltage will be given by

$$V_{in}(f)/I_p(f) = R/(1 + j(2\pi f)CR). \qquad (3.17)$$

To obtain good high-frequency response, C must be kept as low as possible; as discussed earlier, the photodiode contribution can normally be kept well below 1 pF. There is ongoing debate concerning the best approach to improve high-frequency response; we must either reduce R or provide high-frequency equalization. As a rule of thumb, no equalization is needed if

$$R < 1/2\pi C \Delta f \qquad (3.18)$$

where Δf is the frequency bandwidth of interest. We can also avoid the need for equalization by using a transimpedance feedback amplifier, which is often employed in commercial optical datacom receivers. These apparently simple receiver circuits can exhibit very complex behavior, and it is not always intuitive how to design the optimal detector circuit for a given application. A detailed analysis of receiver response, including the relative noise contributions and tradeoffs between different types of photodiodes, is beyond the scope of this chapter; the interested reader is referred to several good references on this subject [4–13].

3.5. Other Detectors

The PIN photodiode is definitely the workhorse detector of the fiber optics industry. For direct transceiver connections, it is dependable and cheap. But there are a number of detectors that offer advantages in more complex situations.

Parallel Optical Interconnects (POI) require detector arrays, which can be formed from either PIN diodes or MSM photoreceivers.

In wavelength-division multiplexing (WDM), the optical signal traveling over one fiber optic cable is wavelength separated, coded, and recombined. To properly receive and interpret this combined signal there is a need for more sensitive detectors, such as avalanche photodiodes, as well as position-sensitive detectors, such as photodiode arrays and MSM photore-ceiver arrays, and also wavelength-sensitive detectors, such as resonant-cavity enhanced photodetectors.

The rise of WDM hints at the importance of growing data rates. Schottky-barrier photodiodes have always been considered for situations that required faster response than PIN diodes (100 Ghz modulation has been reported), but less sensitivity. Metal-semiconductor-metal (MSM) detectors, which a Schottky diode-based planar detector is, are cheap and easy to fabri-cate, making them desirable for some low-cost applications where there are a number of parallel channels and dense integration. Resonant-cavity enhancement (RECAP) can increase the signal from Schottky-barrier and MSM detectors, as well as making high-speed, thin i-layer, PIN diode detectors feasible.

3.5.1. AVALANCHE PHOTODIODE

PIN diodes can be used to detect light because when photon flux irradiates the junction, the light creates electron–hole pairs with their energy deter-mined by the wavelength of the light. Current will flow if the energy is sufficient to scale the potential created by the PIN junction. This is known as the photovoltaic effect [2, 4–6]. It should be mentioned that the material from which the top layers of a PIN diode is constructed must be transpar-ent (and clean!) to allow the passage of light to the junction. If the bias voltage is increased significantly, the photogenerated carriers have enough energy to start an avalanche process, knocking more electrons free from the lattice to contribute to amplification of the signal. This is known as an avalanche photodiode (APD); it provides higher responsivity, especially in the near infrared, but also produces higher noise due to the electron avalanche process. For a PIN photodiode, each absorbed photon ideally creates one electron–hole pair, which, in turn, sets one electron flowing in the external circuit. In this sense, we can loosely compare it to an LED. There is basically a one-to-one relationship between photons and carriers and current. Extending this comparison allows us to say that an avalanche photodiode resembles a laser, where the relationship is not one-to-one. In a laser, a few primary carriers result in many emitted photons. In an avalanche photodiode (APD), a few incident photons result in many carriers and appreciable external current.

The structure of the APD, shown in Fig. 3.9, creates a very strong elec-trical field in a portion of the depletion region. Primary carriers — the free electrons and holes created by absorbed photons — within this field are

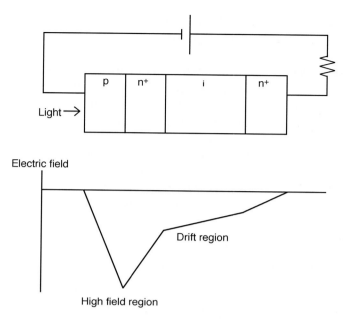

Fig. 3.9 Avalanche photodiode.

accelerated by the field, thereby gaining several electron volts of kinetic energy. A collision of these fast carriers with neutral atoms causes the accelerated carrier to use some of its energy to raise a bound electron from the valence band to the conduction band. A free electron–hole pair is thus created. Carriers created in this way, through collision with a primary carrier, are called secondary carriers. This process of creating secondary carriers is known as collision ionization. A primary carrier can create several new secondary carriers, and secondary carriers themselves can accelerate and create new carriers. The whole process is called photomultiplication, which is a form of gain process. The number of electrons set flowing in the external circuit by each absorbed photon depends on the APD's multiplication factor. Typical multiplication ranges in the tens and hundreds. A multiplication factor of 50 means that, on the average, 50 external electrons flow for each photon. The phrase "on the average" is important. The multiplication factor is an average, a statistical mean. Each primary carrier created by a photon may create more or less secondary carriers and therefore external current. For an APD with a multiplication factor of 50, for example, any given primary carrier may actually create 44 secondary carriers or 53 secondary carriers. This variation is one source of noise that limits the sensitivity of a receiver using an APD.

The multiplication factor M_{DC} of a APD varies with the bias voltage as described in equation (3.19)

$$M_{DC} = 1/(1 - V/V_B)^n \tag{3.19}$$

where V_B is the breakdown voltage, and n varies between 3 and 6, depending on the semiconductor.

The photocurrent is multiplied by the multiplication factor,

$$I = M\,QEq\phi(\lambda)\lambda/hc, \tag{3.20}$$

as is the responsivity,

$$R(\lambda) = M\,QE\lambda q/hc. \tag{3.21}$$

The shot noise in an APD is that of a PIN diode multiplied by M times an excess noise factor the square root of F, where:

$$F(M) = \beta M + (1 - \beta)(2 - 1/M). \tag{3.22}$$

In this case, β is the ration of the ionization coefficient of the holes divided by the ionization coefficient of the electrons. In III–V semiconductors $F = M$. It is also important to remember that the dark current of APDs is also multiplied, according to the same equations as the shot noise (see Section 3.6.2. Shot Noise).

Because the accelerating forces must be strong enough to impart energies to the carriers, high bias voltages (several hundred volts in many cases) are required to create the high-field region. At lower voltages, the APD operates like a PIN diode and exhibits no internal gain. The avalanche breakdown voltage of an APD is the voltage at which collision ionization begins. An APD biased above the breakdown point will produce current in the absence of optical power. The voltage itself is sufficient to create carriers and cause collision ionization. The APD is often biased just below the breakdown point, so any optical power will create a fast response and strong output. The tradeoffs are that dark current (the current resulting from generation of electron–hole pairs even in the absence of absorbed photons) increases with bias voltage, and a high-voltage power supply is needed. Additionally, as one might expect, the avalanche breakdown process is temperature sensitive, and most APDs will require temperature compensation in datacom applications.

Modern APD design uses separate absorption and multiplication layers (SAM). This technique allows for high electric fields in the avalanche multiplication region, but lower fields in the low bandgap absorbing ternary layer to prohibit tunneling. But the step in the field can cause hole trapping. This is solved by creating a transition region consisting of one or more intermediate steps, known as separate absorption grading and multiplication (SAGM) structure. In order to both make it easier to fabricate SAGM APDs, and at the same time avoid premature edge breakdown, a partial charge sheet layer is introduced between the multiplication and the grading region. This structure is known as separate absorption, grading, charge sheet, and multiplication (SAGCM) structure. It is fairly difficult for available SAGM and SAGCM APDs to respond reliably to 10 GB/s range signals. This is due to the speed limitations inherent in the avalanche process, the electron and hole ionization rates.

Superlattice avalanche photodiodes (SL-APDs) use the band gap discontinuity to enhance the ionization rate, and can overcome the limitation on conventional APD performance. This in turn leads to a multiple quantum well (MQW) structure. A gain bandwidth product of 150 Ghz, with a dark current of 6 nA, has been reported for a layer structure with a 0.23-μm-thick SL multiplication layer with 8-nm-thick InAlGaAs wells (Eg = 1 eV) and 13-nm-thick InAlAs barriers, a p-InP field buffer layer ($\sim$52 nm) and a 1-μm-thick p-InGaAs absorption layer [7].

3.5.2.   PHOTODIODE ARRAY

Photodiode arrays are beginning to be used as detectors in parallel optics for supercomputers. They are also the detector used for spectrometers (along with charge-coupled detector (CCD) arrays), which makes them useful to test fiber optics systems. Photodiode arrays are also being considered for wavelength division multiplexing (WDM).

In the previous sections (3.3. PN Photodiode and 3.4. PIN Photodiode) we have discussed the physics guiding photodiode operation. In a photodiode array, the individual diode elements respond to incident flux by producing photocurrents, which charge individual storage capacitors. Indium gallium arsenide (InGaAs) photodiode arrays are the materials used in spectroscopic applications. They can be embedded in erbium-doped fiber amplifiers. Cross-talk, or signal leakage between neighboring pixels, is normally a concern in array systems. In InGaAs photodiode arrays it is limited to nearest-neighbor interactions. However, the erbium-doped fiber amplifiers do exhibit cross-talk, so there is some discussion of using different amplifiers, such as trans-impedance amplifiers (TIA arrays). In the not-too-distant future, this type of detector will operate at 10 Gbit/s [15]. (See Fig. 3.10.)

Fig. 3.10 An optical demultiplexing receiver array, where data channels are focused onto an array of high-speed InGaAs photodiodes. The signals are then amplified using a trans-impedance amplifier (TIA) array [15].

While some wavelength-division multiplexers (WDM) filter out the wavelength dependent signal, another common way to separate the signal is by using a grating to place different wavelengths at different physical positions, as is done in spectrometers. In dense WDM as many as 100 optical channels can be used. Photodiode arrays are an obvious detector choice. In addition to detecting the signal, they offer performance feedback to the tunable lasers. At this time, photodiode arrays have not been used in WDM, but their use would be similar to spectrometer design.

3.5.3. SCHOTTKY-BARRIER PHOTODIODES

A variation from the PIN diode structure is shown in Fig. 3.11. This is known as a Schottky-barrier diode; the top layer of semiconductor material has been eliminated in favor of a reverse biased, metal-semiconductor-metal (MSM) contact. The metal layer must be thin enough to be transparent to incident light, about 10 nm; alternate structures using interdigital metal transducers are also possible. The advantage of this approach is improved quantum efficiency, because there is no recombination of carriers in the surface layer before they can diffuse to either the ohmic contacts or the depletion region.

Fig. 3.11   Schottky-barrier diode.

When you want to make a PIN detector respond faster, your only choice is to make the inductor layer thinner. This decreases the signal. Some architectures have been designed to try to get around this problem, such as resonant-cavity photodetectors (see Section 3.5.5. Resonant-Cavity Enhanced Photodetectors following), or placing a number of PIN junctions into an optical waveguide, known as traveling wave photodetectors. But Schottky-barrier photodiodes have been there as another traditional alternative.

Schottky-barrier photodiodes are not PIN based, unlike the prior detectors discussed. Instead, a thin metal layer replaces one half of the pn junction. However, it does result in the same voltage characteristics — when the an incident photon hits the metal layer, carriers are created in the depletion region, and their movement again sets current flowing. The similarity in the band structure can be seen in comparing the band structure of a forward-biased PIN diode in Fig. 3.5 to the band structure of a forward-biased n-type Schottky-barrier diode in Fig. 3.12. The voltage and field relations derived

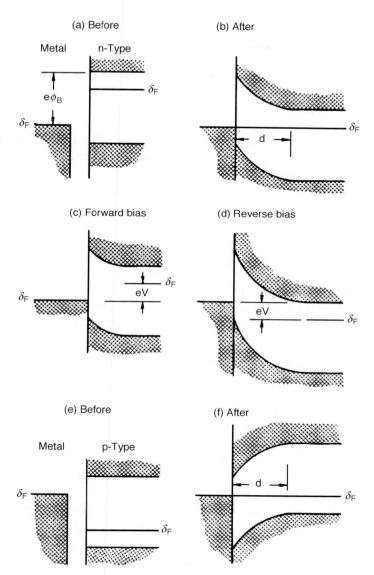

Fig. 3.12 The band diagram of a Schottky-barrier. This shows the metal-semiconductor ohmic contact with an n-type semiconductor (a) before and (b) after contact (c) forward biased (d) reverse biased. (e) and (f) are the same as (a) and (b), but the semiconductor is p-type [2].

for PIN junctions can be applied to Schottky-barrier diodes by treating the metal layer as if it were an extremely heavily doped semiconductor.

The metal layer is a good conductor, and so electrons leave the junction immediately. This results in faster operation — up to 100 Ghz modulation has been reported [14]. It also lowers the chance of recombination, increasing efficiency, and increases the types of semiconductors that can be used, as you are less limited by lattice-matching fabrication constraints. However, their sensitivity is lower than PIN diodes with the usual sized intrinsic layer. Resonant enhancement (see Section 3.5.5. Resonant-Cavity Enhanced Photodetectors) has been experimented with to increase sensitivity.

The main problem in designing Schottky-barrier photodiodes for fiber optic wavelengths is that suitable metals for use on InGaAs substrates are not available. Usually, Schottky-barrier diodes use a thin gold layer, surrounded by a current collecting low resistance thick ring of gold around the edge of the active region. But to avoid low Schottky-barrier height for InGaAs, a thin top layer of AlInAs is used instead. Intermediate bandgap material AlGaInAs is sandwiched between InGaAs and AlInAs layers to prevent band discontinuity [7].

Single Schottky-barrier photodiodes are not currently used in fiber optics applications, but are being considered as communication speeds increase.

3.5.4. METAL-SEMICONDUCTOR-METAL (MSM) DETECTORS

A metal-semiconductor-metal (MSM) photodetector is made by forming two Schottky diodes on top of a semiconductor layer. The top metal layer has two contact pads and a series of "interdigitated fingers" (Fig. 3.13), which form the active area of the device. One diode gets forward biased, and the other diode gets reverse biased. When illuminated, these detectors create a time-varying electrical signal. MSM photoreceiver arrays have been used as parallel optical interconnects. They are also being considered for use in smart pixels.

Fig. 3.13   A diagram of an MSM detector. [MSM Photodetectors by Pekka Kuivalainen, Nina Hakkarainen, Katri Honkanen, Helsinki University of Technology Electron Physics Laboratory.]

Like Schottky diodes, speed is an advantage of MSM detectors. The response speed of MSM detectors can be increased by reducing the finger spacing, and thus the carrier transit times. However, for very small finger spacing, the thickness of the absorption layer becomes comparable to the finger spacing and limits the high-speed performance.

The other advantages of MSM detectors are their ease in fabrication and integration into electrical systems. The MSM photodiodes do not require doping, and they are essentially identical to the gate metallization of field effect transistors (FET) [7]. They are being placed in systems using arrays and smart pixels.

MSM detectors have the same disadvantages as other Schottky-based systems — weak signal and noise. They have some advantages in systems with a number of parallel channels and dense integration of detectors, which could eventually be applied to WDM systems.

As with Schottky-barrier photodiodes, experiments have been done to increase their signal using resonant enhancement (see Section 3.5.5. Resonant-Cavity Enhanced Photodetectors).

For a MSM detector deposited on a photoconductive semiconductor with a distance L between the electrodes, the photocurrent and time-varying resistance can be calculated the same as for any photoconductive detector [14]. Let's assume the detector is irradiated by an input optical flux P, at a photon energy $h\nu$.

$$I_{\mathrm{ph}} = GP\,QEq\phi(\lambda)\lambda/hc, \tag{3.23}$$

where QE is the quantum efficiency, and G is the photoconductive gain, which is the ratio of the carrier lifetime to the carrier transit time. It can be seen that increasing the carrier lifetime decreases the speed, but increases the sensitivity, which is what you would expect.

The time varying resistance $R(t)$, is dependent on the photoinduced carrier density $N(t)$.

$$N(t) = QE\phi(\lambda)\lambda/hc. \tag{3.24}$$

The time-varying resistance is

$$R(t) = L/(eN(t)\mu wd), \tag{3.25}$$

where μ is the sum of the electron and whole mobilities, w is the length along the electrodes excited by the light, and d is the effective absorption depth into the semiconductor.

Table 3.3   **Sample Specifications for a MSM Detector**

Specification

Detector type MSM (metal-semiconductor-metal)

Active material InGaAs

Bandwith (−3 dB electrical) 20 GHz resp. 35 Ghz

Rise time (10%–90%) < 11 ps

Pulse width (FWHM) < 18 ps

Wavelength range 400 nm–1.6 μm

Responsivity 0.21 A/W @ 670 nm, 0.24 A/W @ 810 nm,

0.19 A/W @ 1.3 μm, 0.12 A/W @ 1.5 μm

Bias voltage 2 V–10 V

Bias input connector SMC male

RF signal output connector K-type female

Optical input connector FC/PC on 9 μm single mode fiber

Max. optical peak input power 200 mW @ 20 ps, 1.3 μm

Max. optical average input power 2 mW @ 1.3 μm

Options

Bias input connector MSM battery case with cable

RF signal output connector K-type male

Optical input connector on request (SMA, SC, ST...)

Optical input 50 μm multimode fiber, free space with collimation optics

---

On silicon and GaAs photodiodes, the transparent metal layer is made from indium tin oxide (ITO) or cadmium tin oxide (CTO). CTO can also be used for InGaAs photodetectors, but ITO is not appropriate due to the large absorption in the 1.0–1.6 μm wavelength region. Other variations using gold and SiN_x antireflection coatings have also been fabricated [7].

The specifications for a typical MSM detector are shown in Table 3.3.

3.5.5.   *RESONANT-CAVITY ENHANCED PHOTODETECTORS (RECAP)*

The principle of resonant-cavity enhanced photodetectors is that you put a fast photodetector into a Fabry-Perot (FP) cavity to enhance the signal magnitude (Fig. 3.14). In practice, the Fabry-Perot cavity's end mirrors are made of quarter-wave stacks of GaAs/AlAs, InGaAs/InAlAs, or any other

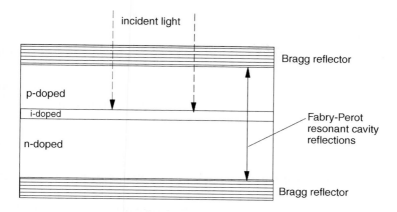

Resonant Cavity Photodiode

Fig. 3.14   The diagram of a resonant-cavity photodetector. Note how the PIN diode is located between the Bragg Reflector Stacks, creating a Fabry-Perot resonant cavity.

semiconductor or insulator materials that have the desired refractive index contrast. A typical example of a fast photodetector is a PIN photodetector with a very thin i-layer. By making the inductor layer thinner, you increase the speed of the detector, but decrease the signal. Avalanche photodiodes (Section 3.5.1.), Schottky-barrier photodiodes (Section 3.5.3.), and MSM detectors (Section 3.5.4.) can also have their signal magnitude increased using RECAP techniques.

By placing the detector in a FP cavity, you pass the light through it several times, which results in an increase of signal. However, due to the resonance properties of the FP cavity, this detector is highly wavelength selective. It is also highly sensitive to the position of the detector in the FP cavity, as it is a resonance effect.

The maximum quantum efficiency QE_{max} of a RECAP detector can be calculated by

$$QE_{\mathrm{max}} = \{(1 + R_2 e^{-\alpha d})(1 - R_1)/(1 - \sqrt{R_1 R_2} e^{-\alpha d})^2\} * (1 - e^{-\alpha d}),$$

$$(3.26)$$

where R_1 and R_2 are the reflectivity of the top and bottom mirrors, α is the absorption of the active area of the detector, and d is thickness of the active area of the detector [16]. If the reflection of the bottom mirror $R_2 = 0$, this reduces to the quantum efficiency of a photodiode, only with an ideal total transmission through the top electrode (Equation 3.2),

$$QE = (1 - R)T(1 - e^{-\alpha W}).$$ (3.2)

The resonant cavity enhancement, RE, can allow RCE detectors to have quantum efficiencies of nearly unity at their peak wavelength.

$$RE = \{(1 + R_2 e^{-\alpha d})/(1 - \sqrt{R_1 R_2} e^{-\alpha d})^2\}$$ (3.27)

The theoretical wavelength dependence of the quantum efficiency is shown in Fig. 3.15.

Because WDM applications are by their very nature wavelength dependent, the wavelength selectivity of RECAP may turn out be a useful design feature. RECAP is not yet being used in the field, but reported experimental quantum efficiencies have reached 82% for PIN diodes, and 50% improvements of photocurrent for Schottky-barrier photodiodes [16].

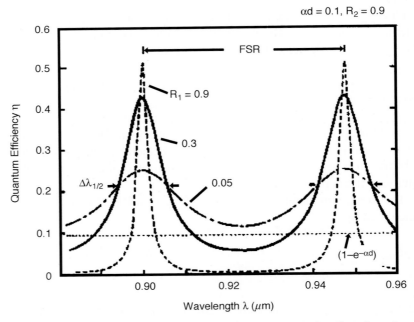

Fig. 3.15 The quantum efficiency of resonant cavity detectors as a function of wavelength [16].

3.6. Noise

Any optical detection or communication system is subject to various types of noise. There can be noise in the signal, noise created by the detector, and noise in the electronics. A complete discussion of noise sources has already filled several good reference books; because this is a chapter on detectors, we will briefly discuss noise created by detectors. For a more complete discussion, the reader is referred to treatments by Dereniak and Crowe [6], who have categorized the major noise sources. The purpose of the detector is to create an electrical current in response to incident photons. It must accept highly attenuated optical energy and produce an electrical current. This current is usually feeble because of the low levels of optical power involved, often only in the order of nanowatts. Subsequent stages of the receiver amplify and possibly reshape the signal from the detector. Noise is a serious problem that limits the detector's performance. Broadly speaking, noise is any electrical or optical energy apart from the signal itself. Although noise can and does occur in every part of a communication system, it is of greatest concern in the receiver input. The receiver works with very weak signals that have been attenuated during transmission, and the noise level, although very small compared to the signal levels in most circuits, is significant in relation to the weak detected signals. The same noise level in a transmitter is usually insignificant because signal levels are very strong in comparison. Indeed, the very limit of the diode's sensitivity is the noise. An optical signal that is too weak cannot be distinguished from the noise. To detect such a signal, we must either reduce the noise level or increase the power level of the signal. In the following sections, we will describe in detail several different noise sources; in practice, it is often assumed that the noise in a detection system has a constant frequency spectrum over the measurement range of interest; this is so-called white noise or Gaussian noise, and is often a combination of the effects we will describe here.

3.6.1. NOISE AND AMPLIFICATION

The amplification stages of the receiver amplify both the signal and the noise. Some AC techniques, such as lock-in amplification, can help separate the signal from the noise. To illustrate the principle, we will digress briefly to describe the example of the lock-in amp even though it is not commonly used in datacom systems. Lock-in amplification, or chopping, is a technique used to limit noise, although it can also be used to detect dc signals that are deliberately encoded with a known modulation. A mechanical light chopper (which looks somewhat like a fan) is placed between the signal and the detector. The amplifier is connected to the chopper, and so the amplifier "knows" from this reference signal when the signal to be detected is on and when it is off. As you can imagine, this makes for very precise background subtraction. However, that is not all the lock-in amplifier does.

The secret of the lock-in amplifier is that it narrows the bandwidth of the detector, and the narrower that bandwidth, the more precise the measurement. It does this by creating a weak periodic signal, and low pass filtering it (or narrowing the bandwidth). This helps eliminate 1/f noise (flicker noise, drifts, etc.). The signal you have after lock-in amplification is a differential signal, not a linear signal. It is good for determining changes in signal, not signal magnitude. Horowitz and Hill discuss lock-in amplification in "The Art of Electronics" [17, p. 631].

In order to illustrate the power of lock-in detection, we usually set up a small demonstration for our students. We use a lock-in to modulate a small LED of the kind used for panel indicators, with a modulation rate of a kilohertz or so. The current is very low, and you can hardly see the LED glowing in normal room light. Six feet away a phototransistor looks in the general direction of the LED, with its output fed to the lock-in. With the room lights out, there's a tiny signal from the phototransistor at the modulating frequency (mixed with plenty of noise), and the lock-in easily detects it, using a time constant of a few seconds. Then we turn the room lights on (fluorescent), at which point the signal from the phototransistor becomes just a huge messy 120 Hz waveform, jumping in amplitude by 50 db or more. The situation looks hopeless on the oscilloscope, but the lock-in just sits there, unperturbed, calmly detecting the same LED signal at the same level. You can check that it's really working by sticking your hand in between the LED and detector. It's darned impressive.

Similar tricks to narrow the bandwidth are used in signal averaging, boxcar integration, multichannel scaling, pulse height analysis, and phase-sensitive detection. For example, boxcar averaging takes its name from gating the signal detection time into a repetitive train of N pulse intervals, or "boxes," during which the signal is present. Because noise that would have been accumulated during times when the gating is off is eliminated, this process improves the signal-to-noise ratio by a factor of the square root of N for white, Johnson, or shot noise. (This is because the integrated signal contribution increases as N, while the noise contribution increases only as the square root of N [6].) Narrowing the bandwidth of a fiber optic receiver can also have beneficial effects in controlling RIN and modal noise (see Chapter 7). The use of differential signaling is also common in datacom receiver circuits, although they typically do not use lock-in amps but rather solid state electronics such as operational amplifiers (see Chapter 16).

3.6.2. SHOT NOISE

Shot noise occurs in all types of radiation detectors. It is due to the quantum nature of photoelectrons. Individual photoelectrons are created by absorbed photons at random intervals, so the resulting signal has some variation with time. The variation of detector current with time appears as noise; this can be due to either the desired signal photons or background flux (in the latter case, the detector is said to operate in a Background Limited in Performance, or BLIP, mode). To study the shot noise in a photodiode, we will consider the photodetection process. An optical signal and background radiation are absorbed by the photodiode, whereby electron–hole pairs are generated. These electrons and holes are then separated by the electric field and drift toward the opposite sides of the pn junction. In the process, a displacement current is induced in the external load resister. The photocurrent generated by the optical signal is I_p. The current generated by the background radiation is I_b. The current generated by the thermal generation of electron–hole pairs under completely dark environment is I_d. Because of the randomness of the generation of all these currents, they contribute shot noise given by a mean square current variation of

$$I_s^2 = 2q(I_p + I_b + I_d)B, \qquad (3.28)$$

where q is the charge of an electron ($1.6 \times 10{-19}$ coulomb) and B is the bandwidth. The equation shows that shot noise increases with current and with bandwidth. Shot noise is at its minimum when only dark current exists, and it increases with the current resulting from optical input.

3.6.3. THERMAL NOISE

Also known as Johnson or Nyquist noise, thermal noise is caused by randomness in carrier generation and recombination due to thermal excitation in a conductor; it results in fluctuations in the detector's internal resistance, or in any resistance in series with the detector. These resistances consist of R_j, the junction resistance, R_s the series resistance, R_l the load resistance, and R_i the input resistance of the following amplifier. All the resistances contribute additional thermal noise to the system. The series resistance R_s

is usually much smaller than the other resistances and can be neglected. The thermal noise is given by

$$I_t^2 = 4kTB((1/R_j) + (1/R_l) + (1/R_i)) \tag{3.29}$$

where k is Boltzmann's constant ($1.38 \times 10\text{--}23$ J/K), T is absolute temperature (kelvin scale), and B is the bandwidth.

3.6.4. *OTHER NOISE SOURCES*

Generation recombination and 1/f noise are particular to photoconductors. Absorbed photons can produce both positive and negative charge carriers, some of which may recombine before being collected. Generation recombination noise is due to the randomness in the creation and cancellation of individual charge carriers. It can be shown [4] that the magnitude of this noise is given by

$$I_{gr} = 2I(\tau B/N(1 + (2\pi f \tau)^2))^{1/2} = 2qG(\varepsilon EAB)^{1/2}, \tag{3.30}$$

where I is the average current due to all sources of carriers (not just photo-carriers), τ is the carrier lifetime, N is the total number of free carriers, f is the frequency at which the noise is measured, G is the photoconductive gain (number of electrons generated per photogenerated electron), E is the photon irradiance, and A is the detector active area.

Flicker or so-called 1/f noise is particular to biased conductors. Its cause is not well understood, but it is thought to be connected to the imperfect conductive contact at detector electrodes. It can be measured to follow a curve of $1/f^{\upsilon}$, where β is a constant that varies between 0.8 and 1.2; the rapid falloff with $1/f$ gives rise to the name. Lack of good ohmic contact increases this noise, but it is not known if any particular type of electrical contact will eliminate this noise. The empirical expression for the noise current is

$$I_f = \alpha(iB/f^{\upsilon})^{1/2} \tag{3.31}$$

where α is a proportionality constant, i is the current through the detector, and the exponents are empirically estimated to be $\alpha = 2$ and $\beta = 1$. Note this is only an empirical expression for a poorly understood phenomena; the noise current does not become infinite as f approaches zero (DC operation).

There may be other noise sources in the detector circuitry, as well; these can also be modeled as equivalent currents. The noise sources described here are uncorrelated and thus must be summed as rms values rather than a linear summation (put another way, they add in quadrature or use vector addition), so that the total noise is given by

$$I_{tot}^2 = I_f^2 + I_{gr}^2 + I_s^2 + I_t^2. \tag{3.32}$$

Usually, one of the components in the preceding expression will be the dominant noise source in a given application. When designing a data link, one must keep in mind likely sources of noise, their expected contribution, and how to best reduce them. Choose a detector so that the signal will be significantly larger than the detector's expected noise. In a laboratory environment, cooling some detectors will minimize the dark current; this is not practical in most applications. There are other rules of thumb that can be applied to specific detectors, as well. For example, although Johnson noise cannot be eliminated, it can be minimized. In photodiodes, the shot noise is approximately 3 x greater than Johnson noise if the dc voltage generated through a transimpedance amplifier is more than about 500 mV. This results in a higher degree of linearity in the measurements while minimizing thermal noise. The same rule of thumb applies to PbS- and PbSe-based detectors, though care should be taken not to exceed the maximum bias voltage of these devices or catastrophic breakdown will occur. For the measured values of NEP, detectivity, and specific detectivity to be meaningful, the detector should be operating in a high impedance mode so that the principle source of noise is the shot noise associated with the dark current and the signal current. Although it is possible to use electronic circuits to filter out some types of noise, it is better to have the signal much stronger than the noise by either having a strong signal level or a low noise level. Several types of noise are associated with the photodiode itself and with the receiver; for example, we have already mentioned multiplication noise in an APD, which arises because multiplication varies around a statistical mean.

3.6.5. SIGNAL-TO-NOISE RATIO

Signal-to-noise ratio (SNR) is a parameter of describing the quality of signals in a system. SNR is simply the ratio of the average signal power, S, to the average noise power, N, from all noise sources.

$$\text{SNR} = S/N. \tag{3.33}$$

SNR can also be written in decibels as

$$\text{SNR} = 10 \log 10(S/N). \tag{3.34}$$

If the signal power is 20 mW and the noise power is 20 nW, the SNR ratio is 1000, or 30 dB. A large SNR means that the signal is much larger than the noise. The signal power depends on the power of the incoming optical power. The specification for SNR is dependent on the application requirements.

For digital systems, bit error rate (BER) usually replaces SNR as a performance indicator of system quality. BER is the ratio of incorrectly transmitted bits to correctly transmitted bits. A ratio of 10^{-10} means that one wrong bit is received for every 10 billion bits transmitted. Similarly with SNR, the specification for BER is also dependent on the application requirements. SNR and BER are related. In an ideal system, a better SNR should also have a better BER. However, BER also depends on data-encoding formats and receiver designs. There are techniques to detect and correct bit errors. We cannot easily calculate the BER from the SNR, because the relationship depends on several factors, including circuit design and bit error correction techniques. For a more complete overview of BER and other sources of error in a fiber optic data link, see Chapter 7.

3.7. Conclusions

In this chapter, we have introduced the fundamentals of fiber optic detectors for data communications. The pn diode was used to illustrate the basic detection mechanisms before proceeding to a description of the more practical PIN diode. Then we discussed other less common detectors, such as the avalanche photodiode, photodiode array, Schottky-barrier photodiode, metal-semiconductor-metal (MSM) detectors, and resonant-cavity enhanced photodetectors (RECAP). These detectors are becoming more common in applications such as multiplexing, parallel optical interconnects, and smart pixels. We have described the terminology needed to read a basic photodiode specification, including such topics as phase lock loops for clock recover; we conclude with an example of how the detector and receiver design may dictate other properties of the datacom link.

A fiber optic receiver consists of many other components besides the photodiode, such as an integrated preamplifier and quantizer. The photodetector's signal-to-noise ratio and minimum detectable signal level are factors in the design of other receiver components. For example, most high-speed receivers compare the photodiode output signal with a running average of signal levels to set a threshold that determines whether a 1 or 0 was transmitted. However, this threshold can drift up or down if the incoming data depart from a 50% duty cycle (an equal number of 1's and 0's received). The quantizer may be designed to compensate for duty cycle distortion or pulse-width variation of the incoming signal. In addition, most data links specify an encoding technique for the data that affects the receiver's duty cycle. Manchester codes inherently produce a 50% duty cycle but require a baud rate twice that of the bit rate (a 100-Mbit/s data stream would require a 200-Mbaud transceiver). By contrast, nonreturn-to-zero (NRZ) encoding requires a lower bandwidth but can result in higher duty cycle variations. Because the long-term duty cycle depends on the encoding method, different applications will specify their own encoding rules. The FDDI standard, for example, uses NRZI and 4B/5B encoding to achieve a duty cycle between 40 and 60%. Fiber Channel Standard uses NRZ and 8B/10B encoding to ensure a duty cycle much closer to 50%, as does the ESCON/SBCON specification. These standards will be presented in more detail in other chapters.

ACKNOWLEDGMENTS

The contributions of Bill Reysen, AMP Incorporated, Lytel Division, to the preparation of this chapter are gratefully acknowledged.

References

1. DeCusatis, C. J. Sher. 1997. Fundamentals of Detectors. *Handbook of Applied Photometry*, ed. C. DeCusatis. Chap. 3 pp. 101–132. New York: AIP Press.
2. Burns, G. 1985. *Solid State Physics*. Chap. 10 pp. 323–334. Orlando: Academic Press.
3. NBS Special Publication SP250-17, "The NBS Photodetector Spectral Response Calibration Transfer Program."
4. Lee, T. P., and T. Li. 1979. Photodetectors. Chap. 18 in *Optical Fiber Communications,* eds. S. E. Miller and A. G. Chynoweth. New York: Academic Press.
5. Gowar, J. 1984. *Optical Communication Systems*. Englewood Cliffs, N.J.: Prentice Hall.
6. Dereniak, E., and D. Crowe. 1984. *Optical Radiation Detectors*. New York: John Wiley & Sons.

7. Bandyopadhyay, A., and M. Jamal Deen. 2001. Photodectors for Optical Fiber Communications. Chap. 4 in *Photodetectors and Fiber Optics,* ed. H. S. Nalwa. New York: Academic Press.

8. Graeme, J. 1994. Divide and conquer noise in photodiode amplifiers. *Electronic Design* (June 27): 10–26.

9. Graeme, J. 1994. Filtering cuts noise in photodiode amplifiers. *Electronic Design* (November 7): 9–22.

10. Graeme, J. 1987. FET op amps convert photodiode outputs to usable signals. *EDN* (October 29): 205.

11. Graeme, J. 1982. Phase compensation optimizes photodiode bandwidth. *EDN* (May 7): 177.

12. Bell, D. A. 1985. *Noise and the Solid State.* New York: John Wiley & Sons.

13. Burt, R., and R. Stitt. 1988. Circuit lowers photodiode amplifer noise. *EDN* (September 1): 203.

14. Garmire, Elsa. 2001. Sources, Modulators and Detectors for Fiber Optic Communication Systems. In *Handbook of Optics IV.* New York: McGraw-Hill.

15. Cohen, Marshall J. 2000. Photodiode arrays help meet demand for WDM. *Optoelectronics World, Supplement to Laser Focus World,* August.

16. Ünlü, M. Selim, and Samuel Strite. 1995. Resonant-Cavity Enhanced (RCE) Photonic Devices. At http://photon.bu.edu/selim/papers/apr-95/node1.html.

17. Horowitz, P., and W. Hill. 1980. *The Art of Electronics.* Chap. 14 pp. 628–631. Cambridge, England: Cambridge University Press.

Chapter 4 | Logic and Drive Circuitry

Ray D. Sundstrom

Motorola, Incorporated, Chandler, Arizona 85248

Eric Maass

Motorola, Incorporated, Tempe, Arizona 85284

4.1. System Overview

In data communications applications, the data and address information is generally supplied through parallel lines. The parallel information can be directly converted, line for line, into parallel optical signals; alternatively, the parallel information can first be converted into a serial bit stream and then transmitted through a serial optical link.

The parallel optical approach is illustrated through a block diagram in Fig. 4.1a, and the serial optical approach is illustrated in Fig. 4.1b. In the parallel optoelectrical interface, illustrated in Fig. 4.1a, each parallel electrical line has a laser driver and laser associated with it. The light is sent through parallel channels, such as fiber ribbon composed of several optical fibers. The light from each channel is guided to a photo detector, and the signal from each photo detector is amplified through a transimpedance and postamplifier, one for each channel. The array of postamplifiers, including conversion to appropriate digital signal levels, provides a parallel electrical output similar to the parallel electrical lines that were the original inputs.

In the serial optical–electrical interface, illustrated in Fig. 4.1b, the parallel electrical lines are multiplexed to provide a serial signal output. The parallel to serial conversion also requires clock generation. The serialized electrical signal is provided through a laser or light-emitting diode (LED) driver circuit to the laser or LED. The driver circuit may involve feedback

127

HANDBOOK OF FIBER OPTIC
DATA COMMUNICATION
$35.00

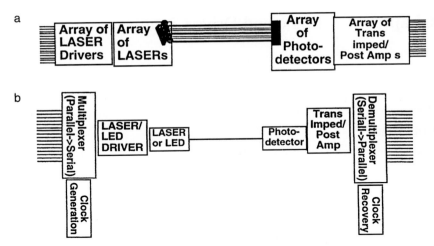

Fig. 4.1 (a) Block diagram for parallel optoelectrical interface, involving arrays of optoelectronic interfaces. (b) Block diagram for serial optoelectrical interface, involving parallel–serial–parallel conversions.

circuitry from the laser or LED to keep the drive current at an appropriate level for the laser or LED operation. The resulting serial optical signal is coupled to optical fiber or other light-transmitting medium.

On the receive side, the serial optical signal is coupled to a single photo detector. The signal from the photo detector is generally a very small current, which is converted into a voltage through the transimpedance amplifier, and amplified with a post- or limiting amplifier. The resulting serial electrical signal must then be converted to a parallel signal through a demultiplexer. This serial–parallel conversion requires clock-recovery circuitry. The clock-recovery circuitry involves special requirements, such as recovering clock signals from a variety of bit patterns; these special requirements and the design considerations involved are described later in this chapter and elaborated on further in Chapter 16.

In Fig. 4.2a and 4.2b, the customer requirements associated with the parallel and serial fiber optics systems are listed and mapped to the design and process issues involved in meeting those requirements. The basic customer requirements of distance, frequency or data rate, cost, clean data, low power, and reliability are similar for both approaches. The clean data requirement involves additional jitter requirements associated with the clock-generation and clock-recovery circuitry in the serial approach.

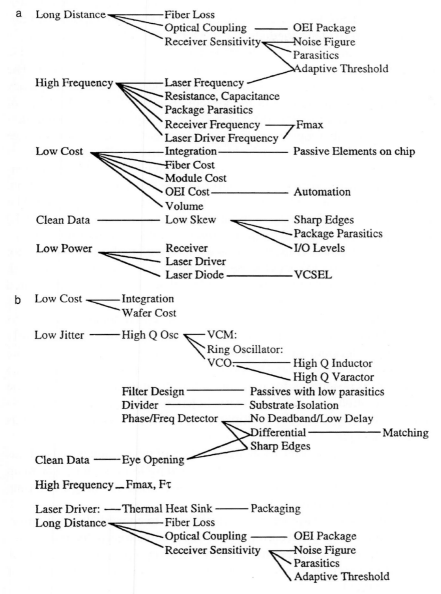

Fig. 4.2 (a) Customer requirements for parallel fiber optics interface related to design and process issues. (b) Customer requirements for serial fiber optics interface related to design and process issues.

The circuitry associated with these systems can be implemented in various semiconductor technologies, including GaAs, silicon bipolar, complimentary metal oxide semiconductor (CMOS), and BiCMOS. GaAs has a particular advantage for the transimpedance function on the receive side in that the photodiode can readily be integrated with this function. The high mobility associated with GaAs facilitates meeting the high-frequency requirements of any circuit in the system. The low transconductance of GaAs field effect transistors (FETs) is a disadvantage for the postamplifier function, generally requiring more amplifier stages. A major disadvantage of GaAs historically has been the high cost relative to silicon solutions.

Bipolar technology has advantages in high bandwidth and high transconductance, which is particularly helpful for the transimpedance function, and high voltage gain, which is helpful for the postamplifier function. The high-frequency performance of bipolar using emitter-coupled logic (ECL) design approaches is useful for the multiplexer and demultiplexer functions. The high-static power consumption associated with high-speed ECL is not an issue if data are expected to be continually transmitted and received. Bipolar integrated circuits tend to cost less than similar GaAs-integrated circuits but more than CMOS circuits.

Small geometry CMOS could be used for several if not all the functions and would be expected to provide the lowest cost solution. The relatively low transconductance of MOS transistors is a disadvantage for both the transimpedance amplifier and the postamplifier function. CMOS would also encounter disadvantages for high-speed, quadrature clock signals for the clock-generation and clock-recovery functions. Using differential techniques with CMOS for the multiplexer and demultiplexer functions has some advantages.

BiCMOS has an advantage at higher levels of integration because it has the flexibility associated with both bipolar and MOS transistors. The main disadvantage of BiCMOS is the relatively high cost and long cycle time due to the process complexity.

The purpose of an optoelectrical interface is to convert electrical information from voltages and currents to a stream of photons in the transmitter and back to voltages and currents in the receiver. For a digital data link, high and low voltage levels are converted to a higher or lower number of photons emitted.

A typical transmitter would include a driver circuit that accepts a standard logic input and converts a high (1) to some known high current and a low (0) to some known low current. This current switch is connected to an LED or Laser diode.

For LEDs, the low current is usually 0 mA, meaning that the diode is completely shut off and emits no light when it is low. The high current is set by the system designer and depends on the system requirements. The higher the current, the higher the light output of the LED. Of course, there is a maximum current limit for every device (beyond which the part's output or reliability will begin to degrade). A normal "on" current for LED's is in the range of 20–100 mA.

For lasers, the low current can also be 0, but it is usually some value large enough to keep the laser active. The advantage of keeping the laser active is that it reduces the turn-on delay of the device. Depending on device size and system requirements, the laser may switch from low = 1 mA to high = 5 mA or as great as low = 100 mA and high = 300 mA.

A typical receiver would include a p-type–intrinsic–n-type (PIN) diode or an avalanche photodiode (APD) to receive the photons and convert them into a proportional current. This current is fed into a transimpedance amplifier (TZA), which converts the current into a voltage.

A PIN diode is similar to a normal diode, in which a layer of p-type material forms a junction with n-type material, except for the addition of a layer of very lightly doped, or intrinsic material between the p-type and n-type layers. When the diode is reverse biased, a large depletion region is created. If a photon of sufficient energy is absorbed in the depletion region, an electron will be liberated to assist in current flow.

An APD performs a similar function as a PIN diode, except that the APD also performs the function of current amplification. There are more design issues to account for when designing with APDs, and this discussion will be limited to PIN diodes.

4.2. Electrical Interface

The interface between the electric components and the optical components may be any of the standard logic configurations or a novel interface that solves some particular system requirement such as power consumption, data rate, noise, signal skew, etc.

Transistor–transistor logic (TTL) and CMOS are very common interfaces that are easy to use and do not require termination if the technologies are not pushed to the limit of their bandwidth capabilities. For higher performance, TTL and CMOS can be transmitted over controlled impedances and terminated correctly to reduce problems with noise and reflections. The TTL interface can be used up to approximately 100 MHz, and low-voltage

CMOS can be used up to approximately 250 MHz. However, the rail-to-rail voltage swings of these two logic families can consume much power and generate much noise at such high frequencies. Some CMOS designs now include low-voltage differential interfaces to alleviate the noise and power problems.

Differential ECL and CML interfaces are better suited for very high-performance electrical interconnect. These two logic configurations were designed to drive controlled impedance transmission lines and have very good noise immunity due to the differential operation. These two interfaces also have smaller voltage swings, in the range from 0.25 to 1.0 V, that can consume less power at high frequencies. The lower voltage swing also reduces electromagnetic interference. These logic interfaces have been used at data rates as high as 10 Gb/s.

Different parts of a system may be best suited to different logic technologies. In a high-speed serial link, the interface to the parallel data source may be CMOS. The parallel source is often a CMOS microprocessor running less than 200 MHz. A CMOS chip operating at a similar frequency can process the data, encode it, frame it, and prepare it for transmission. For high-speed links, the multiplexer chip is usually bipolar or GaAs to achieve data rates of 1 Gb/s or above. With a parallel 16-bit input at 100 mb/s, the serial output would have to be 1.6 Gb/s. If there is error detection and correction, such as extra parity checking, or framing information, the serial rate must be higher to include this extra information. The small geometries and dense circuit layout of CMOS make it best suited for the lower frequency parallel processing. The high bandwidth of bipolar and GaAs make these processes best suited for the high-frequency serial processing. BiCMOS processes, which combine bipolar and CMOS transistors on the same substrate, make it possible to combine the best technologies on a single chip and provide the designer with greater flexibility — at the expense of greater processing complexity, longer cycle times, and higher wafer cost.

4.3. Optical Interface

The optical devices used in the transmitter can have a large effect on the system performance. All devices have parasitic properties that the circuit designer must take into account when designing an optical link. Light output of LEDs and lasers varies with temperature and age. Photodiode efficiency is affected by the reverse-bias voltage. Parasitics of the device packages and circuit board can play a major factor in system performance at high frequencies. Higher performance systems usually require that the driver and laser be packaged in a module to reduce lead inductance. The photodiode and TZA also need to be packaged in a module for the same reason.

On the receiver side, noise current at the input of the TZA is crucial in determining how small a signal the amplifier can reliably receive. It is a common practice to specify noise at the input of an amplifier. Input noise is determined by adding the contribution of each noise source in a circuit and calculating an equivalent value as if it were at the input. The noise must be about an order of magnitude less than the minimum input signal expected to be received; this ratio may vary, dependent on the level of errors that can be tolerated. The number of errors observed divided by the number of bits transmitted is commonly know as the bit error rate (BER). A BER of 10E-14 may be a reasonable goal.

There is always a compromise between bandwidth and input noise current when designing a TZA. The parasitic capacitance of the PIN diode is also one of the dominant factors in determining the bandwidth of the receiver. The diode capacitance and input impedence of the TZA often form the dominant pole in the receiver.

On the transmitter side, LEDs are very common and relatively inexpensive. They are usually turned on hard and shut off completely, such that the zero level coincides with no light emission. LEDs generally operate at lower frequencies than lasers and turn on relatively quickly for the frequencies used but shut off more slowly. Unless some method is used to actively turn off the device, shutting off the current results in a long decay in the optical output of the device as the parasitic capacitors discharge through the diode. If the temperature varies, the circuit must vary the current to maintain constant light output. As the device ages, perhaps accelerated by high temperature, the optical output is reduced. If a system has enough margin to allow for temperature and aging effects without compromising system performance, these effects can be left uncontrolled. However, if a higher performance system is required, active feedback circuitry that monitors the temperature or light output may have to be implemented to keep the optical signal at a constant amplitude.

Lasers are usually used for the highest performance systems. Edge-emitting lasers were developed first and have been used very successfully. Edge-emitting lasers cannot be tested in wafer form because the laser does not operate until mirror facets are created when the die is cleaved.

The vertical cavity surface-emitting laser (VCSEL) is a promising new development. The mirror facets are developed as layers during wafer processing, and the VCSEL emits light from the top surface of the die, so it can be tested on the wafer. The cleaving operation is no longer a critical process.

Both laser devices have similar design challenges. Lasers have a long turn-on delay if shut off completely. To minimize this turn-on delay, lasers are generally prebiased with a DC bias current and modulated with a switching current. This configuration greatly enhances optical performance but complicates the design of the transmitter and receiver: There are two separate currents that need precise control for optimal performance. The objective is to bias the DC current precisely at the threshold current (where the laser begins to emit light). This allows the maximum difference between a high level and a low level for the least amount of power. Maintaining proper biasing of the laser is not a trivial task.

4.3.1. DRIVER CIRCUITRY

The optical driver can vary from a simple switch to a complicated feedback system accounting for many variables. The simple driver may consist of a CMOS or TTL output tied to a resistor and a LED or laser diode, as shown in Fig. 4.3. If system performance is not pushed to the limits, this might be the simplest and therefore perhaps the best solution.

Lasers can be switched on and off, but their performance increases significantly if they are not shut off completely. The performance can be

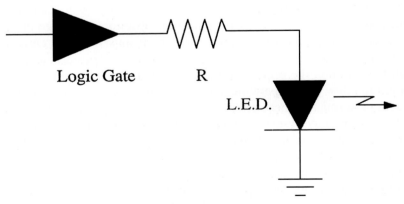

Fig. 4.3 Diagram of simple output.

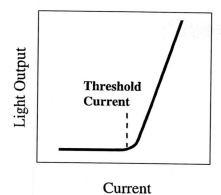

Fig. 4.4   Current vs light curve.

improved by providing both a DC bias current and a switching or modulation current. A differential gate can provide a controlled current source for modulation and another controlled current source continuously passed through the laser diode for the bias current.

Bias current is a constant current flowing through the laser. The bias current should be set at the threshold current — the point at which the laser just starts to emit light, as illustrated in Fig. 4.4. If the laser is biased below the threshold current, the turn-on delay will increase. If it is biased above the threshold current, the difference between the light emission corresponding to a 0 and a 1 will be decreased. It is important to maximize the difference between the light emission corresponding to a 0 and a 1 because this is one of the factors that determine how far the link can transmit — the larger the difference in the light signals for a 0 and a 1, the larger the signal will be at the receiver.

Figure 4.5 is a transistor-level diagram of an output driver that implements the bias and modulation currents for the laser. A bias control voltage is forced on the base of Q_1, which puts a voltage across R_1. The voltage across R_1 is the bias control voltage minus the base-emitter voltage of Q_1. The current in the emitter of Q_1 is the voltage across R_1 divided by the resistance of R_1. This current minus the base current is the current in the collector of Q_1, which will always flow through the laser diode. This is one method of obtaining the laser bias current. The bias control voltage can be either fixed or controlled by some other monitor circuit.

The modulation current is controlled in a similar fashion by the modulation current control voltage. The modulation current, however, is not always

Fig. 4.5 Simplified diagram of laser output driver.

directed through the laser diode. Q_3 and Q_4 make a differential gate that determines whether the current flows through the laser diode. If the voltage on the base of Q_3 is higher than the voltage on the base of Q_4, the current will flow through the laser diode, through Q_3, and into the current source created by Q_2 and R_2. Because the modulation current is flowing through the laser, the optical output is increased corresponding to an optical high level. Alternatively, if the voltage on the base of Q_4 is high and that on Q_3 is low, the current will flow from the power supply through Q_4 and into Q_2. Because the current is not flowing through the laser, the optical output is decreased, corresponding to an optical low state.

As data rates and transmission distances increase, the output of the laser diode needs to be more tightly controlled. Because the laser output power can vary with temperature and lifetime, some method of monitoring the light output and for feedback to the driver is generally incorporated in higher performance systems.

With edge-emitting lasers that emit light from both the front and the back facets, a photodetector or monitor diode is commonly mounted on the back side of the laser (Fig. 4.6). The signal from the monitor diode determines the bias and modulation current adjustments in a closed-loop feedback system.

**Photo Diode
for Monitor**

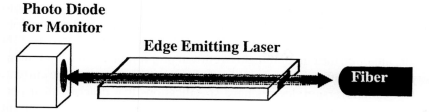

Fig. 4.6 Diagram of laser with monitor diode.

Fig. 4.7 Diagram of PIN diode.

4.3.2. RECEIVER CIRCUITRY

The optical receiver design can be virtually an art. The challenge involves trade-offs in improving some parameters at the expense of other parameters. The design is optimized by appropriate compromise so that the receiver has the best combination of high sensitivity, correct bandwidth, low noise, large dynamic range, and low power.

Until direct processing of optical information becomes a technology, optical information must be converted to electrical information for processing. The conversion commonly uses a PIN diode (Fig. 4.7). The intrinsic layer between the p-type and n-type layers allows the formation of a very large depletion region when a reverse DC bias is applied. Photons with sufficient energy are absorbed in the depletion region, liberating electrons. The electric field from the reverse bias on the diode sweeps the liberated electrons across the junction, providing a current proportional to the quantity of photons absorbed. The large depletion region of the PIN diode improves the probability of absorbing photons. Intrinsic region thicknesses of approximately 4–15 μm are common.

As alternatives to PIN photodiodes, avalanche photodiodes and photo transistors can also be used for the optical to electrical signal conversion. The mechanism of these devices is somewhat different from that of the PIN diode, although the output remains a current proportional to the light intensity. The main difference is that these two devices also amplify the current.

Optical conversion devices have specifications that describe their performance, limitations, and parasitics. Conversion efficiency is the current produced for a given amount of light. A conversion efficiency of 0.5 A/W means that a PIN diode would produce 0.5 A of current when receiving 1 W of light. Generally, light received is in the mW or μW range, so the output current is in the mA or μA range. Large photodiodes facilitate alignment to the fiber but have more parasitic capacitance. This parasitic capacitance can limit the frequency because it must be charged and discharged by the current produced by the photodiode. As with most high-frequency devices, capacitances should be minimized.

The output from the PIN diode, current proportional to the intensity of light shining on it, is coupled into a TZA. The purpose of the TZA is to convert the current to a voltage. Conceptually, the TZA is an operational amplifier (op amp) with a negative feedback resistor. In the ideal case, the transimpedance is the value of the feedback resistor (measured in ohms).

If the PIN diode in Fig. 4.8 has a conversion efficiency of 0.5 A/W, and the incoming light is 1 mW, it will produce 500 μA of current. The op amp with a 1 kohm transimpedance resistance will convert the 500 μA to produce a voltage of 500 mV. Similarly, if the transimpedance resistance were increased to 10 kohm, the output would be increased to 5 V.

The two most important specifications for a TZA are the bandwidth and input noise level, which involve a trade-off: Increasing bandwidth will increase noise. The key is to first choose the largest transimpedance possible to achieve the bandwidth required because the larger the transimpedance, the lower the input noise of the receiver. Of course, the larger transimpedance increases the input impedance, which reduces bandwidth. The amplifier design should use minimal current because diode shot noise increases with current. Of course, the bandwidth of the amplifier is lowered as the current is reduced.

The range from the smallest detectable signal to the largest allowable signal is called the dynamic range. The largest allowable signal is the maximum that can be received without saturating the circuitry. The smallest detectable signal is determined by the input noise of the TZA. For example, if the input noise of the receiver integrated over the bandwidth is 250 nA,

Fig. 4.8   Diagram of PIN diode and TZA.

the smallest detectable signal might be 10 times that current, or 2.5 μA. If the photo detector conversion efficiency is 0.5 A/W, the minimum difference between a low level and a high level would be 2.5 μA/(0.5 A/W) = 5 μW. If this same receiver had a dynamic range spec of 30 dB, the maximum input would be 5 μW(1000) = 5 mW. A 5-mW input corresponds to an input current of 2.5 mA. If the part operates at 5 V, the maximum transimpedence gain is 2 kohm (2.5 mA $\times$ 2 kohm = 5 V).

Often, a larger transimpedence is desired to provide more gain at low-input currents and to reduce noise. To achieve this, some method of gain compression can be used at high-input currents. Power supply noise rejection and substrate noise feedback are also extremely important for amplifiers with high gain and high bandwidth. If these are not properly controlled, the part will likely oscillate at low-input signal levels — perhaps being re-named the "transimpedance oscillator."

Figure 4.9 is a diagram of a simple transimpedance amplifier and PIN diode. When light input to the PIN diode increases, base drive is conducted away from Q_1. This starts to shut off Q_1, reducing current I_1 so the voltage across R_1 decreases, which causes the voltage on the base of Q_2 to rise.

Fig. 4.9   Diagram of TZA and PIN diode.

Transistor Q_2 is an emitter follower that increases current drive capability. When the base of Q_2 rises, the output voltage also rises. When the output voltage rises, the current through R_2 flows into the PIN diode. The voltage change on the output will be approximately equal to the current change in the PIN diode multiplied by the resistance of R_2. This is the voltage change required across R_2 to replace the current being conducted by the PIN diode. Note in this diagram that the PIN diode is biased to a negative voltage. This is necessary in this configuration because the base of Q_1 will be biased to approximately 0.8 V, and the PIN diode generally requires greater than 3 V of reverse bias to operate properly.

If the signal is transmitted over long distances, the optical signal is attenuated and the receiver may receive an extremely small signal. The attenuation is due to fiber loss and coupling losses at the interface between the optical components and fiber. The output voltage from the TZA must be further amplified and compared to a threshold voltage to determine if it is high or low. This is a difficult challenge, particularly if the signal has a substantial difference in the quantity of zeros and ones. If the data are encoded for an average 50% duty cycle, an analog feedback loop similar to that shown in Fig. 4.10 can be used to set the threshold.

If the input data are known to have a 50% duty cycle, output and $\overline{\text{output}}$ (OUT and OUTB) should have an average voltage in the center of the

Fig. 4.10 Analog feedback loop.

output swing. The resistors and capacitors average the output swings. If the threshold voltage is too high, OUT will tend to be low more than 50% of the time, whereas OUTB will tend to be high more than 50% of the time. The operational amplifier will sense that the noninverting input is lower than the inverting input, and the output will decrease. This will lower the threshold voltage for comparison to the TZA output by the postamplifier. This negative feedback loop will correct the threshold voltage until the outputs have a 50% duty cycle.

If the input data do not have a 50% duty cycle, the feedback circuit will still behave as if it is driving the outputs to a 50% duty cycle, and the threshold will be incorrectly set. Setting the comparison threshold is very challenging if the data are not encoded to maintain a 50% duty cycle.

A low-pass filter is often placed between the TZA and postamp. For a 1-GHz bandwidth, the TZA needs a nominal bandwidth of approximately 1.2 GHz to allow for a 20% process variation. A lot with processing such that the devices are unusually fast may have a bandwidth of 1.4 GHz, and another lot with slower devices may just meet the 1-GHz requirement. The "fast" lot will have higher noise when integrated over the 1.4-GHz bandwidth. A filter can be constructed with tight-tolerance discrete components to cut off the bandwidth at 1 GHz, eliminating the extra noise contribution from the excess bandwidth.

Once the threshold is correctly set, the differential data on the outputs of the postamp are the digital equivalent of the data at the input of the laser driver. If this is one channel of a parallel system (as in Fig. 4.1a), the data are ready to use. If it is the output of a serial link (Fig. 4.1b), a clock signal has to be recovered for clocking in the incoming data, and the data need to be demultiplexed (converted from serial to parallel) to recover the parallel data that were present at the transmitter. Because all the data transitions occur at regular intervals, a phase-locked loop (PLL) can be locked to the edges for clock recovery. Once the PLL is locked to the incoming data, a clock is available that is synchronous with the data so that it can be used to clock the data stream into a register. If those data are shifted serially in a shift register and read out at the regular intervals of the clock (every eight clock cycles for an eight-bit system), the data can be converted back to a parallel data stream.

The design of the optoelectrical interface is a challenge of compromises dictated by customer and system requirements. Equations, models, and simulators are absolute necessities. The layout of high-frequency optoelectronics is something of an art, and the designer gains skill through high-frequency circuit design/layout experience. Time spent in the lab troubleshooting circuits and observing the principles first hand gives the designer insight that cannot be learned from books or simulators.

In Chapter 16, some of the design issues and approaches are covered in further detail.

Chapter 5 | Optical Subassemblies

Herwig Stange

Infineon Technologies, Fiber Optics, Berlin, Germany

5.1. Function of the Optical Subassembly

Within the fiber optic link the optical subassembly converts the data signal from an electrical current to optical radiation in the fiber and vice versa. The optical subassembly comprises the electrooptical converter, optic, and connector components (Fig. 5.1).

Key components of the optical subassemblies are the electrooptic active elements. In the case of the transmitter this is the light-emitting diode (LED) or the laser, edge-emitting laser diode (EELD), or vertical cavity surface-emitting laser (VCSEL). The optical subassembly of the receiver comprises the photodetector (PD), mostly a PIN diode or an avalanche photodiode (APD). These elements are the actual electrooptical converters within the optical subassembly [1].

The optics of the optical subassembly couples the radiation from the electrooptic active element into the fiber. In many optical subassemblies the optics is a lens element. At the transmitter side this element collects the emitted radiation of the laser or LED and focuses it into the fiber. At the other side of the fiber optic link at the receiver the optics couples the divergent output beam of the fiber into the photodiode.

Finally the fiber has to be fixed to the optical subassembly. Optical subassemblies with a pigtail have a permanently fixed fiber mostly with a connector at the end. Other optical subassemblies have a receptacle. This is

HANDBOOK OF FIBER OPTIC
DATA COMMUNICATION
$35.00

E/O Converter	Optics	Connector	Fiber
LED	lens	SC	
VCSEL	mirror	ST	
EELD	waveguide	LC	
PIN Diode	taper	VF 45	
APD	fiber stub	MU	

Fig. 5.1   Structure of an optical subassembly.

a mechanical socket for a connector. The receptacle guides the connector with the fiber to the coupling position.

There are different names for the optical subassembly. Sometimes it is called a coupling unit, emphasizing the function of optical coupling. But mostly the name optical subassembly or the short form OSA is used.

5.2.   Basic Properties of the Transmitter OSA

The main function of the transmitter OSA is to convert current to radiation, which is launched into the optical fiber. The key parameters describing this property of the transmitter OSA are the efficiency of the electrooptic conversion and the optical coupling. This is the slope efficiency or the differential quantum efficiency of the laser combined with the losses of the optical coupling. The dimension of the efficiency is Watt per Ampere (W/A). That implies that the differential quantum efficiency of the laser or LED chip defines the upper limit for the total efficiency of the transmitter OSA. In particular, the optical coupling efficiency for LEDs can be very low due to the highly divergent radiation emitted by the LED. Also the optical coupling of edge-emitting lasers with SM fibers can result in strong losses. This type of laser with an emission wavelength of 1300 nm or 1550 nm is used for intermediate and long-reach datacom applications that demand more launched power than short-haul applications.

The electrooptic active element mainly defines the parameters for the specification of the OSA. The wavelength of the emitted radiation of the OSAs is given by the spectral output of the semiconductor device. Anti-reflection coatings at the optical surfaces can reduce the loss due to reflection and optimize the coupling efficiency of the OSA. A further reason to reduce reflections back to the laser is that they can disturb the laser emission and increase the output noise. Another possibility to prevent back reflections is to avoid optical surfaces exactly perpendicular to the laser beam. The relative intensity noise (RIN) in db/Hz measured to the average optical power characterizes the noise of the laser, which should be measured with a defined return loss. This parameter represents the reflections from connectors and the receiver. For example, the Fibre Channel Standard [2] demands a measurement at 12 dB return loss.

The optical rise and fall time of the transmitter OSA is mainly determined by the properties of the laser or LED chips. But again the integration of the chip within the OSA package can detrimentally effect these parameters. Parasitic capacitance and inductance due to long lines can affect the speed of the device electrically.

Threshold and differential quantum efficiency of lasers change with temperature and age. However, the fiber link demands a constant power. The problem is mostly solved by controlling the laser diode. A photodiode within the transmitter OSA detects a part of the laser radiation. This monitor diode current has to be proportional to the power launched into the fiber in order to control this value. Therefore, a constant ratio between the optical coupling of the fiber and the optical coupling of the monitor diode is necessary. The launched optical power should remain constant over the operating temperature range by controlling the monitor current. The tracking error describes the maximum deviation over temperature and is the ratio of minimum or maximum power (based on the maximum difference) and the power at the reference temperature in units of dB. Mostly it refers to the power at 25° C room temperature.

5.3. Basic Properties of the Receiver OSA

The responsivity of the receiver OSA summarizes the efficiency of the electrooptic conversion of the detector and the efficiency of the optical coupling between fiber and detector. It is the ratio of generated current to optical power with the dimension Ampere per Watt (A/W). Due to the spectral sensitivity of the detector it depends on the wavelength of the incident radiation. Many OSAs have an anti-reflection coating at the optical surfaces in order to improve the optical coupling and therefore the responsivity.

Not only a high transmission but a low reflectance of the receiver OSA is important. Reflected light could disturb the laser and can cause an increase of laser noise. A high return loss of the receiver prevents strong back reflection to the laser of the transmitter. The return loss is defined as the ratio of incident optical power to the reflected optical power in units of dB.

Often the receiver OSA includes a pre-amplifier. The advantage of integrating the pre-amplifier in the OSA package is that this design enables a short connection between detector and amplifier. This is important because the low currents of the detector are very sensitive to electrical cross-talk. A package like a TO can efficiently protect this connection. Additionally, parasitic capacitance and inductance due to lead-throughs of the package would affect the photodetector and reduce the bandwidth of the receiver.

5.4. Coupling Radiation from a Laser Diode into a Fiber

For most applications the first design goal is to launch the radiation from the laser or LED into the fiber as efficiently as possible with least cost. The optics of the OSA has to match the diameter and beam divergence of the emitter with the fiber core diameter and the numerical aperture of the fiber. An area mismatch occurs if the beam diameter is larger than the core of the fiber. Reducing the distance between the emitter and the fiber reduces the loss due to the area mismatch. Positioning the fiber directly in front of the emitter surface (butt coupling) reduces the effect of the divergence of the beam to a minimum. Even if the beam diameter is smaller than the fiber core, losses can occur due to numerical aperture mismatch. A part of the beam exceeds the acceptance angle of the fiber and is not guided by the fiber [3, 4].

Besides the optical coupling efficiency, other requirements may have an impact on the optical design. The OSA or at least the transmitter has to be eye safe. A laser product classification is preferred that imposes the fewest constraints on the user of the transmitter (see Chapter 6). One approach is to have divergent emission, which reduces the energy density of the accessible radiation reaching the eye from the open transmitter port [5].

Some transmitters are used alternatively for single-mode fibers and for multi-mode graded-index fibers. It is possible to increase the bandwidth of the multi-mode fiber link by reducing the differential mode delay (DMD) in the multi-mode fiber. The optical design of the OSA is one approach to minimize the differential mode delay; another is the design of the multi-mode fiber profile itself. The optical coupling of laser and multi-mode fiber has a strong impact on the distribution of modes. The optical design can optimize the generation of an appropriate mode distribution in the multi-mode fiber.

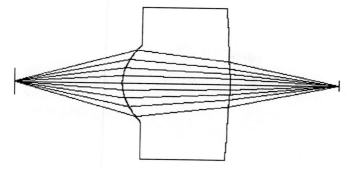

Fig. 5.2 Ray trace modeling of VCSEL coupling.

Coupling an edge-emitting laser with a beam divergence of 30° into a single-mode fiber with an accepting angle smaller than 6° demands an optic that transforms beam diameter and divergence. A magnifying lens shapes the beam into a slightly convergent beam that matches with the numerical aperture of the single-mode fiber.

One advantage of vertical cavity surface-emitting lasers is that fiber coupling is simplified. VCSELs are usually designed with a circular aperture, resulting in a circular beam. This, combined with the low divergence of the emitted radiation, implies an improved fiber launch efficiency compared to the edge-emitting components. Figure 5.2 shows the ray trace modeling of a VCSEL coupling with an aspherical plastic lens.

This lens can be integrated in a plastic flange. The plastic flange has to fulfill very different requirements. It has to be optical transparent and has to have surfaces in good optical quality at the lens. The tolerances of the inner diameter of the flange have to be very low. These properties should be stable at high temperatures. To date the precision achieved is only sufficient for multi-mode applications.

Figure 5.3 shows the TO can and plastic flange fixed together by epoxies. The VCSEL in the TO can is adjusted to the optimal position before the epoxy is cured. In order to reduce production time light-curing epoxies are used.

A crucial parameter of all receptacles is the inner diameter of the flange. The tolerance of the connector in the flange causes a variation of launched power. Figure 5.4 shows two different cases of coupling into a single-mode fiber. The variation of coupled power is very low if the receptacle is exactly aligned at the maximum of the coupled power. A tolerance of $+/-1$ μm of the connector position causes a power variation of only a few percent.

Fig. 5.3   VCSEL-OSA with plastic flange.

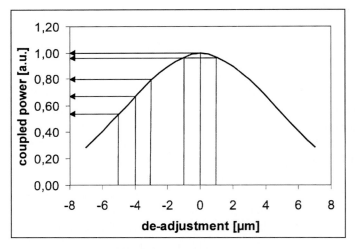

Fig. 5.4   Coupling curve.

The same tolerance of $+/-1$ μm leads to strong variation in launched power if the alignment is not perfect and the flange is de-adjusted. Figure 5.4 shows a de-adjustment of 4 μm. In this case the single plug repeatability is poor. It is defined by the variation of coupled power after several matings with the same connector.

The cause of de-adjustment is not only misalignment but also distortion during gluing or thermal expansion during operation. These effects can be minimized by designing a highly symmetrical component.

For applications with single-mode fibers, materials such as ceramics and stainless steel are mostly preferred. These materials can be manufactured

chip soldered
on submount

submount completed with
lens and monitor diode

Fig. 5.5 Laser submount module.

with high precision and have low coefficients of expansion. Steel parts can be welded almost free of distortion.

The tolerance of the connector position is given by the difference of inner flange diameter and outer connector diameter and the different eccentricities of connectors. These tolerances have to be considered if the cross plug repeatability shall be determined. The cross plug repeatability is a property of both the OSA and the connectors.

Edge-emitting lasers for 1300 nm or 1500 nm application demand a different assembly. Figure 5.5 shows a compact assembly on a Si submount including a lens and a monitor diode. The laser chip is placed at the center of a Si submount with two glass prisms on either side that reflect the front and rear emitted beams of the laser to the top. The laser beam emitted at the front facet of the laser chip is focused by a Si lens that is on top of the right glass prism (Fig. 5.6). On the left side, the rear beam is reflected to the monitor diode mounted on the left glass prism.

The hybrid integrated fiber optic module on a silicon submount chip allows a very compact design that brings the focusing lens almost directly in front of the laser chip. The total length of the beam path can be very short. The distance between laser chip and fiber or connector can be kept short, minimizing the size of the OSA.

Using standard semiconductor wafer technology allows efficient volume production of laser modules. Anodic bond processes and solder bonding techniques guarantee stable assemblies with high reliability. The assembly of the complete active fiber optic component on a wafer has further advantages. Burn-in and testing of the laser module can be done with the fully assembled module on the submount wafer. Figure 5.7 shows the wafer before the last production process, the assembly of the monitor diode [6].

Fig. 5.6   Beam path of the laser submount module.

Fig. 5.7   Bonded laser diodes on Si submount wafer with adjusted Si lens.

5.5.   Coupling Radiation from a Fiber into a Photodetector

On the other side of the optical link the light emitted out of the fiber has to be focused on the photodetector. The beam divergence is determined by the numerical aperture of the fiber. A 62.5 μm graded-index fiber with a numerical aperture of 0.275 results in a half angle of 16°. At this angle the power intensity has decreased to 5% of the maximum power intensity at the center of the far field. This diverging beam has to be focused in the OSA on the photodetector. Many applications allow a sensitive area that is bigger than the core diameter of the fiber. Therefore most optics achieve a sufficient coupling of light on the photodetector. The sensitive area of the photodetector is forced to become small only for high data rate applications in order to move the cutoff frequency of the device to high frequencies.

Contrary to the transmitter OSA, the optical coupling with a single-mode fiber is easier due to the lower numerical aperture of 0.1 and the smaller core diameter of about 9 μm. The half angle of the far field is below 6°.

Figure 5.8 shows the lens of a TO cap focusing the radiation out of the fiber onto the detector. These lenses can be produced in a simple melting process. The focal length varies strongly. Nevertheless, for many applications this component is sufficient. With the same assembly design as described for the laser the assembly of a very compact design is possible. The small distance between lens and detector allows a short length of the total beam path. The PIN diode is at the center of the submount between two posts that carry the Si lens (Fig. 5.9).

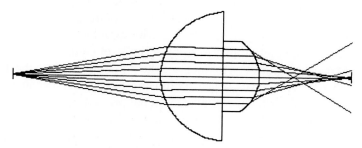

Fig. 5.8   Ray trace model of a TO lens focusing the light on the detector area.

Fig. 5.9   Receiver micromodule.

5.6. Packaging of Optical Subassemblies

The TO can is a standard housing with different functions. It protects the sensitive chips from mechanical damage. The hermiticity of the housing is a good protection against environmental impacts such as humidity or an aggressive atmosphere. In particular, ionic impurities in a humid atmosphere can cause severe damages to the chips. The header of the TO can dissipates the heat of the device arising from the laser and reduces the thermal heating (Fig. 5.10).

The beam is emitted through a glass window. It can be coated with an anti-reflection coating to optimize the optical coupling and to prevent disturbing reflections back to the laser.

The metal parts of the TO can allow highly stable welding processes. The tolerances for adjustment are in the range of 1 μm. Figure 5.11 shows different assemblies. The receptacle OSA combines a TO can with a flange that fits with a fiber connector. The second is a pigtail OSA with a connectorized fiber. Typically the first one is used for short-range datacom applications that demand pluggable connectors at the device. The pigtail solution is favored for long-range telecom applications that demand high optical coupling efficiencies with low return losses. This is achieved by the slightly tilted surface of the pigtail fiber in front of the laser element.

The TO package allows us to integrate the pre-amplifier into the TO can (see Fig. 5.12). This saves space within the transceiver because otherwise the pre-amplifier had to be placed on the printed circuit board of the transceiver. Additionally, the TO can builds an efficient shielding against electromagnetic emission and makes a short connection between photodetector and pre-amplifier possible (see Section 5.3.).

Fig. 5.10   TO can housing of a laser and a detector micromodule.

Fig. 5.11   Optical subassemblies with TO cans.

Fig. 5.12   PIN diode with Si lens and pre-amplifier on TO header.

Instead of packaging the electrooptic components in a TO can, a micro-module can be assembled on a lead frame (Fig. 5.13). A transfer molding process encapsulates the surface mount device. The SMD technology allows a large-scale fabrication at low cost.

Figure 5.14 shows a different packaging. The micromodules (center part) with the electrooptic components are assembled onto the PCB. Small metal caps prevent distortions from electromagnetic emission. The part shown in Fig. 5.14 was built for a VF45-connector. This connector guides the

Fig. 5.13   Optical subassemblies in SMD package.

Fig. 5.14   PCB assembly.

fibers of the connector via v-grooves to the coupling position. The optical adjustment of the v-grooves is fixed by light-curing epoxies. Adjustment tolerances of 2 μm can be achieved. Symmetrical design is important in order to prevent changes due to the shrinking epoxy. The advantage of this process is that it can be done with simple equipment that does not demand a high investment.

5.7.   Optical Subassemblies for Parallel Optical Links

One approach to reduce the size of fiber optic link modules is to bundle optical channels. Parallel optical links allow the miniaturization on the board and allow cost reduction (see Chapter 6). Figure 5.15 shows an example of how the number of channels can be increased on a single chip [7].

Fig. 5.15 Laser array with lenses.

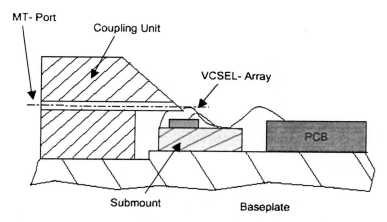

Fig. 5.16 Coupling design of a parallel optical link with VCSEL array.

The edge-emitting laser array is mounted on a submount. High-accuracy Si micro-bench technology achieves the precise spacing necessary to integrate a number of optical links at this level.

VCSEL arrays can be coupled with the fiber bundle without focusing optics and where the fiber axis is parallel to the PCB. The laser radiation emitted from the top surface is reflected into the axis of the fibers by the 45° angled surface (Fig. 5.16).

Fig. 5.17 Top view at VCSEL array (center) and high precision plastic part with fiber array (right).

The top view of the device (Fig. 5.17) shows the VCSEL array in the center and the plastic part with zigzag profile holding the fiber array. This part is made of a highly filled plastic material that is very temperature stable.

After assembly, a non-hermetic sealing is applied for protection against mechanical forces and environmental impacts.

5.8. Outlook

The goal of future developments is clear and can be easily extrapolated from the past. Fiber optic links will need optical subassemblies that enable transmission at increasingly higher data rates. The costs of the devices have to be reduced, the size will shrink, and the power consumption will decrease.

There is a strong demand to transmit more and more data with smaller and smaller devices. In particular, the cross-sectional dimension of the optical port has to be minimized, because the accessible front or rear panels of devices such as routers is a limiting factor. This leads directly to the requirement of small cross-sectional dimensions of optical subassemblies with electrical connections that are able to transfer the demanded high-frequency

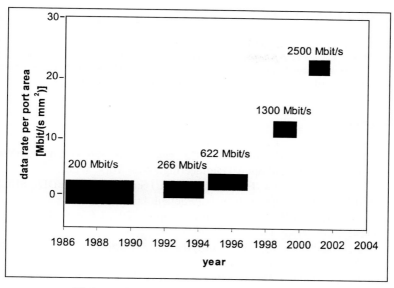

Fig. 5.18   Development of data rate and port area.

signal. The shrinkage of the transceiver port size for a duplex-connector is indicated in Fig. 5.18 by the size of the black rectangles. Within the last ten years the port area could be reduced by a factor of 5 whereas the data rate increased by a factor of 10, resulting in the strong increase of the data transfer rate per area.

There will be different solutions for different applications from short-reach to long-haul data communication. The merging of data and telecommunication will continue. Technologies from telecommunication such as wavelength division multiplexing (WDM) or bidirectional transmission on a single fiber will also be applied in datacom applications.

There are a variety of packaging technologies for optical subassemblies. For many applications OSAs with TO canned diodes and SMD-OSAs are beneficial as well as microbench technology with a fiber coupling between diode and fiber on submount level. In future the monolithic integration on the IC gives the possibility of further miniaturization. Today all these technologies are used. Depending on applications with different requirements and the volume of production, each technology has and will have an importance.

ACKNOWLEDGMENTS

Thanks to my colleagues at Infineon Technologies for supporting me in so many ways. In particular, thanks to Michael Langenwalter for coordinating our contributions. Special thanks to Thomas Murphy for providing me with many helpful suggestions for improvement. Thanks also to Klaus Schulz for fruitful discussions and to Rainer Ulrich for the many photographs he provided.

References

1. Agraval, Govind. 1997. *Fiber-Optic Communication Systems*, 2nd ed. (Wiley series in microwave and optical engineering.) New York: Wiley.
2. ANSI X3T9.x and T11.x, Fibre Channel (FC) Standards incl. FDDI, SBCON and HIPPI-6400, at http://web.ansi.org/default.htm.
3. Born, M., and E. Wolf. 1986. *Principles of Optics*, 7th ed. Cambridge, England: Cambridge University Press.
4. Pedrotti, F., and L. Pedrotti. 1993. *Introduction to Optics*, 2nd ed. Upper Saddle River, N.J.: Prentice Hall.
5. International Standard IEC 60825-1, 1993 incl. Amendment 2, January 2001, ISBN 2-8318-5589-6, Safety of laser products — Part 1: Equipment classification, requirements and user's guide.
6. Althaus, H., W. Gramann, and K. Panzer. 1998. Microsystems and Wafer Processes for Volume Production of Highly Reliable Fiber optic Components for Telecom- and Datacom-Application. *IEEE Transactions on Components, Packaging, and Manufacturing Technology — Part B*. 21: 2.
7. Karstensen, H., *et al.* 2000. Module Packaging for High-Speed Serial and Parallel Transmission. ECTC 2000, Las Vegas, Nev.

Chapter 6 | Fiber Optic Transceivers

Michael Langenwalter

Infineon Technologies AG, Fiber Optics Division,
Berlin, Germany

Richard Johnson

Infineon Technologies North America Corporation,
San Jose, California

6.1. Introduction

Fiber optic transceivers (TRX), each a combination of a transmitter (Tx) and a receiver (Rx) in a single common housing, have already replaced discrete solutions in most of the datacom applications with two-fiber, bidirectional transmission in the last 12 years. The transmitter unit, with a light-emitting diode (LED) or an edge- or vertical-emitting laser diode (LD) as its radiation source, and the receiver unit, with in most cases a positive-intrinsic-negative (PIN) photodiode, are connected together to the transmission medium, either multimode or single-mode fiber. Thus, contrary to typical telecom applications with single-fiber optic connectors at the ends of fiber pigtails, transceivers characteristically have duplex optical receptacles at one side of the housing, which fit to the corresponding duplex connectors. Depending on the application, more or less electronics with integrated circuits and passive components are implemented into the Tx and Rx units.

In this chapter we would like to discuss aspects of operation and application of these fiber optic transceivers in the physical layer (PHY), along with some innovative tendencies for integration of more data transport functions into these components or for a significant reduction of size and power consumption, respectively.

HANDBOOK OF FIBER OPTIC
DATA COMMUNICATION
$35.00

6.1.1. OVERVIEW OF TRANSMISSION SYSTEMS AND APPLICATION STANDARDS

The dominant advantage of fibers and fiber optic transceivers in datacom applications is the substantially higher distance–bandwidth product and reliability compared with other transmission media. Thus, a wide field of applications has developed in the past 10 to 15 years that covers several kinds of datacom networking technologies with transmission speeds from 10 Mbit/s up to 10 Gbit/s, using only one optical wavelength per single fiber. Transceivers are available for all established international and industrial or consortial standards, and can be found in complex networking systems as well as in single links, in supercomputers, mainframes, server clusters, bridges, routers, switches, and so on.

6.1.1.1.   Ethernet, Fast Ethernet, and Gigabit Ethernet

The 10 Mbit/s Ethernet as the networking standard with the largest installed base worldwide was upgraded to 100 Mbit/s (Fast Ethernet) by the mid nineties and to 1 Gbit/s (Gigabit Ethernet — GBE) by the end of the nineties. Thus, fiber is increasingly the transmission medium of choice when copper reaches or exceeds its fundamental limits for high-speed data transmission. Both 850 nm and 1300 nm wavelength bands are applicable with graded index 50/125 μm or 62.5/125 μm multimode or 9/125 μm standard single-mode fibers.

A group of fiber optic component manufacturers (Tyco — formerly AMP, Agilent — formerly Hewlett-Packard, Sumitomo Electric, and Infineon — formerly Siemens Semiconductors Division) has developed a *de-facto* standard outline and footprint for Fast Ethernet optical transceiver modules, also known as the multi-standard (MS) transceiver family with Fibre Channel standard (FCS) optical interface (i.e., duplex SC receptacle/connector) (Fig. 6.1) (see also Section 6.1.1.5.). These MS transceivers meanwhile also run in various different standard applications in order to empower all these network solutions efficiently.

A similar tendency can be seen for the next design generation of Extended Gigabit Ethernet transceivers, which have transmission rates of greater than 2.5 Gbit/s. Their small form factor (SFF) outline reflects increased worldwide demand for reduced board space requirements by ever smaller geometrical outlines with enhanced EMI performance (Fig. 6.2) (see also Sections 6.4. and 6.7.). These transceivers are available either with MU, LC, or VF-45 connector interfaces, which also have a significantly

Fig. 6.1 Multi-standard transceiver.

Fig. 6.2 SFF GBE transceiver.

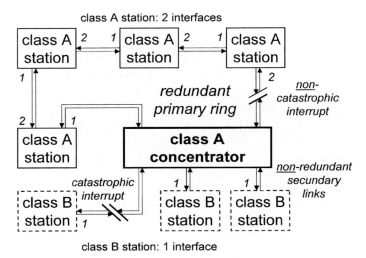

Fig. 6.3   FDDI network structure.

smaller outline than the FCS connector or just the FDDI media interface connector (see Section 6.1.1.2.) and the ESCON/SBCON duplex connector (see Section 6.1.1.3.).

Standardization of both the MS and SFF transceivers formerly driven by the activities of several component suppliers, meanwhile, is implemented into official international standardization procedure by IEC SC86C [6, 7].

6.1.1.2.   Fiber Distributed Data Interface (FDDI)

Operating at a nominal data rate of 100 Mbit/s, the characteristic of a FDDI network is the redundant/duplex ring architecture combined with its timed-token passing attachment method. FDDI is a substandard of Fibre Channel (FC) under the auspices of ANSI [5] (see also Section 6.1.1.5.).

With this architecture, in class A configuration, the FDDI network provides a self-healing structure in single fault conditions (Fig. 6.3). In case of the interruption of one of the two interface paths of one class A station, the system continues reliable operation, possibly by a reduced data rate depending on the number of attached stations. The simpler and cheaper alternative is the class B configuration without failure tolerance for the individual connected station, meaning that any physical link failure disconnects this station completely from the ring. Nevertheless, the performance of the remaining class A ring itself remains unaffected.

Fig. 6.4   FDDI media interface connector (MIC), multimode black, single-mode grey.

Transmission media are 62.5/125 μm graded index multimode or 9/125 μm standard single-mode fiber with transmission distances from 2 to 40 km at 1300 nm wavelength. A maximum of 500 stations can be connected. The original standardized fiber optic connector is the FDDI media interface connector (MIC) (Fig. 6.4). However, even for FDDI, the aforementioned multi-standard form factor is becoming increasingly popular due to its significantly smaller outline for modules as well as for connectors (see also Section 6.2.1.1.).

6.1.1.3.   Enterprise System Connection/Single Byte Command Code Sets Connection

Enterprise system connection (ESCON®) was announced and brought into the field by IBM in the late 1980s as an industrial standard, first for IBM's Enterprise System 390, the first system worldwide in which all mainframe system links switched from copper to fiber consequently. Therefore, ESCON played an important role as a pioneer, enabling many new technical developments and significantly pushing the penetration of fiber systems into datacom applications. So far, ESCON is still one of the most widely installed fiber optic network standards, with several millions of installed links worldwide. Consequently, ESCON was proposed as single byte command code sets connection architecture (SBCON) to ANSI for international standardization [5]. Meanwhile, SBCON has become an international ANSI standard as a substandard of Fibre Channel (FC) (see also Section 6.1.1.5.).

The ESCON/SBCON interface is defined as a point-to-point bidirectional serial optical link via two fibers. The specified data rate is 200 Mbit/s operating at 1300 nm wavelength with transmission distances up to 3 km via graded index multimode, and up to 10 km via standard single-mode fibers. The standardized optical interface is the ESCON duplex connector/receptacle, designed for highest reliability, robustness, and comfortable handling even by unskilled operators. Accordingly the fairly voluminous serial ESCON transceiver (Fig. 6.5) as well as its corresponding connector (Fig. 6.16) will be replaced by MS-sized modules and connectors (see also Sections 6.1.1.1. and 6.1.1.5.).

Fig. 6.5   Serial ESCON transceiver.

6.1.1.4.   Asynchronous Transfer Mode (ATM)

With asynchronous transfer mode (ATM), a novel switching and virtual network topology has been established. ATM will clearly tend to be the universal carrier for most of the future network services worldwide, connecting SANs, LANs, MANs, and WANs. ATM therefore has the potential to overcome the historic separation of the more or less separately grown worlds of telecom and datacom systems. The ATM network attachment method is asynchronous multiplexing. It supports broadband integrated services digital network (B-ISDN) and full services access network (FSAN) as well as synchronous digital hierarchy in synchronous optical networks (SDH/SONET) standard activities [10].

Variable data rates are specified, from 155 Mbits/s up to 2.5 Gbit/s. Common transmission media are both 62.5/125 μm multimode and 9/125 μm standard single-mode fibers. The transmission distance is related to the system architecture and may be almost unlimited. A wide variety of components are available commercially for transmission distances from a few hundred meters (LED-based) to more than 40 km (laser-diode-based).

6.1.1.5.   Fibre Channel Standard (FCS)

The fibre channel (FC) basically is a bidirectional point-to-point serial data transfer channel for high transmission capability. Even though "fiber" is used as a general term to cover all established (solid state) physical high-speed transport media, such as different optical fibers or copper coaxial cables, the fiber solutions have become increasingly dominant. FC inherently is able to run all IT and SCSI protocols, and thus supports a wide variety of applications in SCSIs, SANs, MANs, LANs, and links to WANs. FC is structured as a set of hierarchical functions called FC-0 to FC-4, where FC-0 to FC-1 are the basic physical transport layers, fairly well comparable to the commonly known 7-layer model for open systems interconnection (OSI), herein layers 1 and 2 (Fig. 6.6).

The FC-0 level covers a variety of media. The associated transceivers are capable of operating a wide range of speeds from nominal 100 Mbit/s to 4 Gbit/s at different wavelength bands from 780 nm to 1300 nm, thus providing a maximum of flexibility. Different fibers are specified — 50/125 μm or (preferably) 62.5/125 μm graded index multimode and 9/125 standard single-mode fibers — for various transmission distances up to 10 km [5].

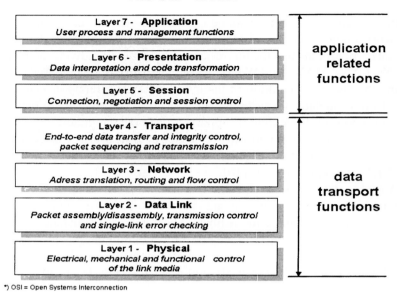

Fig. 6.6   The 7-layer OSI model.

The FCS/Duplex-SC connector is the standard connector for FC standard optical interface (Fig. 6.7). (FC standards also cover the high-performance parallel interface (HIPPI 6400) that defines parallel optical links. Parallel optical transmitters and receivers and their field of application are described in Section 6.8. in detail.)

6.1.2.   PHYSICAL LAYER INTERFACE

Figure 6.8 shows an example of a typical physical link structure with the original incoming 32-bit-wide transmit word. In the second stage, this transmit data byte is 8B/10B-coded and thus converted into a 40-bit-wide transmit word. Next, the parallel to serial conversion is performed at the serializer (sometimes also called the "multiplexer") and the serial data stream is fed into the physical media driver, e.g., a serial fiber optic transceiver. After passing through the transmission medium, the reverse of this process is carried out.

Fig. 6.7   FCS/duplex-SC connector.

Fig. 6.8   Physical link structure.

6.1.2.1.   Clock Oscillator and Regenerator

Clock generation/multiplication is mainly realized using a quartz oscillator for very stable clock sources. For example, the ANSI Fibre Channel requires a clock frequency accuracy of ± 100 ppm. Clock regeneration/ synchronization can be realized by either surface wave filter (SWF) or phase-locked loop (PLL), the latter being more common in today's applications.

6.1.2.2.   Serializer and Deserializer (or Multiplexer and Demultiplexer)

The serializer/multiplexer accepts parallel data from the encoder once per byte timeframe and shifts it into the serial interface output buffers using a PLL-multiplied bit clock. The deserializer/demultiplexer accepts serial bit-by-bit data from the serial receiver output, as clocked by the recovered receiver sync clock, and shifts it back into a parallel data stream.

6.1.2.3.   Encoding and Decoding

The dominant intention of coding for transmission of high-speed data is to maintain the DC balance by bounding the maximum run length of the code. Typical kinds of coding in data communication include 4B/5B coding (e.g., in FDDI) or 8B/10B coding (e.g., in ESCON/SBCON and Fibre Channel). This means high-speed receiver designs in fiber optic transceivers are normally AC-coupled so that each DC component inside the data stream reduces the signal-to-noise ratio in the pre-amplifier stages.

During clock recovery, the PLLs used in today's applications require a certain edge density to ensure that the receiving PLL remains synchronized to the incoming data. In addition, word alignment can be provided by special transmission characters, e.g., a K28.5 pattern. In general, two types of characters are defined: data characters and special characters.

6.2.   Technical Description of Fiber Optic Transceivers

When selecting a transceiver, the key parameters are optical output power, operating data rate, receiver saturation level, receiver sensitivity, transmit and receive pulse quality (defined by rise/fall times and over/undershoot), jitter, and extinction ratio. For transceivers, an optimal inter-operation with the associated circuitry is the key to success of the application. Thus, the high-speed characteristics and immunity against external influences have to be carefully evaluated during the design phase. Increasingly, users decide in favor of integrated transceiver solutions covering at a minimum the high-speed and synchronization functions and providing an easy-to-use lower-speed parallel interface at the user's board.

6.2.1. SERIAL TRANSCEIVERS

Serial fiber optic transceivers are the interface between the serial electrical signal and the transmission medium, the optical fiber. They are composed of the transmitter and the receiver functions, which operate in one common housing, but electrically independent of each other. The implementation of serial transceivers requires careful design of filter circuitry as well as sophisticated transmission line considerations for data input and output tracks to connect to the internal or additional external circuitry.

Figure 6.9 shows a simplified block diagram of a fiber optic transceiver containing a transmitter side with a laser driver circuit and a laser diode, together with its monitor diode, for tracking the laser output. Also shown is an alarm circuitry that activates if the laser crosses a preset upper or lower optical power threshold. A picture of the corresponding transceiver is shown in Fig. 6.10.

The receiver side shows a positive-intrinsic-negative (PIN) photodiode as opto-electric converter, a trans-impedance pre-amplifier that converts extremely low AC currents (in the range of nA) into differential voltage

Fig. 6.9   Simplified block diagram of a serial laser transceiver.

Fig. 6.10   Serial laser transceiver.

signals with some mV amplitude, and the buffer stage with integrated post-amplifier for ECL/PECL line driving.

In most of the standards mentioned in Section 6.1 — such as Ethernet, FDDI, ATM, and Fibre Channel (FC) including ESCON/SBCON — pure serial fiber optical transceiver solutions are described, all including so-called multi-standard transceivers. In order to minimize efforts in design, manufacturing, and testing of boards, a couple of parallel transceiver designs have also arisen in the interim.

6.2.2.   PARALLEL TRANSCEIVERS

Transceivers (TRX) with parallel electrical interfaces have become increasingly popular in fiber optic link applications. Parallel TRXs provide high potential for cost saving due to

- significant saving of board space due to higher functional integration
- much simplified board design by avoiding the need to manage high frequencies on the PC board
- low risk for development effort and design-in
- much reduced overall power consumption
- logistical benefits on all levels.

Fig. 6.11   Block diagram of parallel ESCON/SBCON transceiver.

6.2.2.1.   Parallel ESCON/SBCON Transceiver

Figure 6.11 shows the block diagram of the circuitry of the parallel ESCON/SBCON [5] transceiver, which contains, in addition to the previously described circuitry, functions of serializing/deserializing of 10 single-ended parallel data inputs/outputs, each of 20 Mbit/s, as well as clock transfer and synchronization (compare to Fig. 6.9). This solution reduces the board space required by roughly a factor of 4 compared to discrete solutions (e.g., with serial ESCON/SBCON transceivers). In Fig. 6.12 a picture of this parallel transceiver is shown. The parallel ESCON/SBCON was the first high-volume product in the market with this performance.

6.2.2.2.   Transponders

The electrooptical functions of transponders are basically very similar to parallel transceivers. But, contrary to the typical design of transceivers with fiber optic connector receptacles, transponders have connectorized

Fig. 6.12   The parallel ESCON/SBCON transceiver.

fiber pigtails. They are also plugged into the board, again contrary to the majority of transceivers, which are solder attached to the board by through hole or surface-mounting techniques.

Figure 6.13 shows as an example the OC-48 transponder, which supports ATM/SDH/SONET applications by converting 4 parallel LVDS lines of 622.08 Mbit/s input/output data stream (equal to the STM-4/OC-12 hierarchy) into 2.48832 Gbit/s serial optical output/input data stream (equal to the STM-16/OC-48 hierarchy) [10–12]. This transponder is also compliant with Proposal 99.102 of the optical inter-networking forum OIF.

Due to high functional integration with serializing/deserializing and additionally 8B/10B coding/decoding, the power consumption is the order of 3 W. Thus, heat dissipation is a serious challenge which can only be reliably managed by numerous cooling studs on top of the all-metal housing and by forced airflow inside the system device. A block diagram of the circuitry of the 4:1 OC-48 transponder is shown in Fig. 6.14.

Fig. 6.13   The 4:1 OC-48 transponder.

6.3.   The Optical Interface

An interface is the meeting of two objects. In the case of the optical interface, the fiber comes together with the optical transceiver elements: the transmitter diode and the receiver diode receptacle unit. The operating link demands that the mechanical and optical properties of these two objects match. These attributes are defined in various standards to maintain a reliable interoperability of products of different manufacturers [5–8].

6.3.1.   FIBER OPTIC CONNECTORS AND ACTIVE DEVICE RECEPTACLES

A brief closer look shows that a transceiver has not one but two optical interfaces: first, the connection of the outgoing fiber at the output of the transmitter, and second the connection of the incoming fiber at the input of the receiver. The optical subassembly of the transmitter launches the radiation into the fiber, which will be taken up by the optical subassembly of the receiver at the other side of the link.

For a flexible configuration or a reconfigurable network, it is necessary to have repeatedly connectible and disconnectible devices. This function is realized by a single, duplex, or multifiber connector at the end of the fiber optical cable and a port at the transceiver that accepts the fiber connector.

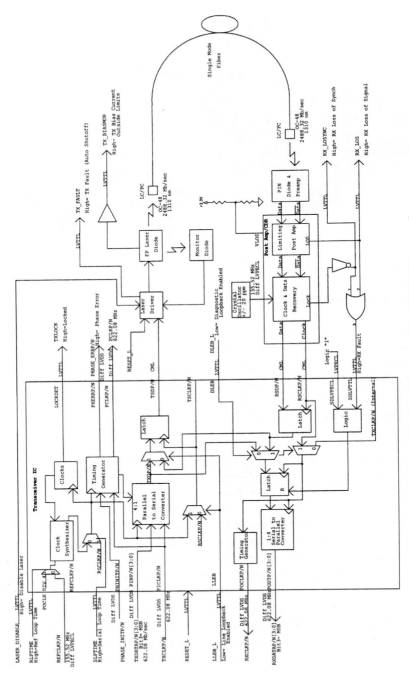

Fig. 6.14 Block diagram of the 4:1 OC-48 transponder.

174

There are various different design families on the market, such as the EC, ESCON/SBCON, FC/PC, FDDI MIC, LSA (screwed), LSG (push-pull version of LSA), LSH (E2000), SC, SMA, ST/BFOC, LC, MF, MU (mini-SC), MT, MT-RJ, and VF-45 international standardized connector families [8]. Most of the single-fiber connectors suitable for single-mode fiber application are available in either a version with physical contact (PC) or angled polished contact (APC). Nearly all push-pull or latch-type single-fiber connectors are also available in a duplex version. ESCON/SBCON and FDDI MIC were originally duplex connectors.

Most connector systems use a ferrule — a cylindrical tube containing the fiber end — that fits within a precise mechanical tolerance into the transceiver's optical receptacle. The VF-45 connector is the only one with a totally different concept based on V-grooves instead of ferrules.

For ferrule-based systems the most popular diameter of the cylindrical ferrule is 2.5 mm, while 1.25 mm has established itself for small form factor (SFF) designs such as MU- and LC-type connector families (see also Section 6.1.1.5.). The SMA connector is the only one with a 3.125-mm ferrule diameter. However, a ferrule diameter tolerance within a few microns must be maintained in all cases in order to achieve a satisfactory coupling of light and to ensure intermatability to receptacles of different suppliers.

Besides these single and duplex connectors, it is also possible to integrate up to 12 fibers into one connector to be used for parallel optical links (see Section 6.8.). The most popular multifiber connectors all are based on the MT design with its rectangular-shaped plastic ferrule. In general, any fiber connector has to be locked to the receptacle, for example, by screwing on a coupling ring (e.g., LSA) or snapping on a bayonet (ST) or a latch mechanism. Moreover, some connectors can be provided with a key to distinguish multimode and single-mode connectors.

For bidirectional links in datacom applications, cross-plugging must be avoided by using asymmetric-shaped connector/receptacle (e.g., ESCON/SBCON; Fig. 6.15 right, bottom view, and Fig. 6.16, top view) or corresponding key structures both inside the receptacle and at the connector (e.g., FDDI MIC; Fig. 6.15 center).

Pigtail solutions are preferred for telecom applications. These modules do not have an optical port, but instead have one or more permanently fixed fibers of short length, i.e., on the order of 1 meter. In this case the optical interface of the transceiver/transponder is situated at the connectorized end of the pigtail (see also Section 6.2.2.2. and Fig. 6.13).

Fig. 6.15   Different connectors.

Fig. 6.16   ESCON/SBCON connector, top view.

6.3.2. THE OPTICAL FIBER

When considering the optical aspects of the interface, it is useful to look at the parameters of the fiber, the transmission medium between transmitter and receiver. An optical fiber can be described by the properties of core diameter, refractive index profile and numerical aperture, bandwidth length product and attenuation. These interdependent parameters more or less directly influence the requirements for the output of the transmitter and the input of the receiver.

The core diameter of the fiber defines the maximum waist of the output beam at the transmitter, because only light inside this waist will be guided by the fiber. This is why the numerical aperture, which is determined by the index profile of the fiber, limits the maximum divergence of the input beam. Launched radiation exceeding this maximum divergence will be lost. An efficient optical coupling between transmitter and fiber can only be achieved if these two pairs of parameters — core diameter and beam waist — as well as numerical aperture and divergence match.

At the receiver side, the optical coupling is again mainly influenced by these two fiber parameters, namely, the waist and divergence of the beam leaving the fiber, which is determined by the core diameter and the numerical aperture of the fiber. The optical subassembly of the receiver must focus this beam onto the sensitive area of the photodiode (see also Chapter 5).

In some cases it is necessary to consider not only the coupling efficiencies at the optical interfaces of the transceiver but also the fraction of light that is reflected back to the light source. Lasers can be very sensitive to back-reflections that can disturb the laser emission and cause some noise on the optical signal. To prevent these effects the return loss at the optical interfaces has to be high. The back-reflections can be kept low by using angled polished physical contacts at the connectors or by inserting an optical isolator in front of the laser. Another method uses a design wherein the axis of the laser beam is slightly tilted to the axis of the optical fiber. In this case, the reflected light does not reenter the laser.

The fiber bandwidth is determined by the modal dispersion and the spectral dispersion. Modal dispersion occurs in multimode fibers and causes a pulse broadening due to the different propagation speeds of the different fiber modes. Therefore, the excitation of many fiber modes leads to a pulse broadening in comparison to the excitation of few modes with low order. This excitation of fiber modes depends strongly on the optical coupling at the transceiver.

Pulse broadening caused by chromatic dispersion limits the spectral bandwidth of the transmitter. For low bit rates and short distances, LEDs with a typical full width at half maximum (FWHM) of 40 to 60 nm are sufficient. However, higher bit rates and longer distances require the application of lasers with a typical bandwidth of a few nanometers down to the subnanometer region, which is achieved by distributed-feedback (DFB) lasers with only one spectral mode.

The transparency of the fiber material at the different wavelengths of radiation determines the spectral parameters of transmitter and receiver. In the past three "classical" optical windows at 850 (1st window), 1310 (2nd), and 1550 nm (3rd) of fused silica fibers were mostly used with only about ±30 nm wavelength deviation from the window's center (Fig. 6.17).

Today, these windows are broader due to the reduced OH concentration in fused silica fibers. With the availability of new types of emitters — dense wavelength division multiplexers/demultiplexers (DWDM) and different optical amplifiers (OA) — the new wide window from 1440 to 1625 nm is now applicable especially for high-speed long-haul telecom systems. The wavelength region between the 2nd window and the 5th window, named 2e-window, may also be used in the near future by the application of special "zero-OH" fibers. Moreover, other materials — for example,

(DDOA: Dysprosium-doped optical amplifier; EDFA: Erbium-doped fiber amplifier)

Fig. 6.17 Typical spectral attenuation of silica optical fibers (DDOA: Dysprosium-doped optical amplifier; EDFA: Erbium-doped fiber amplifier).

polymethylmethacrylate (PMMA) plastic optical fibers (POF) — open new windows in the visible region.

Basically, the wavelength of the transmitter and the spectral sensitivity of the receiver have to fit with these windows of low attenuation. The attenuation of the fiber, together with the sensitivity of the receiver, determines the minimum launched power in the fiber. This defines the minimum optical power of the transmitter at the optical interface if the coupling efficiency of the transmitter is provided.

6.3.3. EYE SAFETY

There is an upper limit for the optical power of laser transmitters because lasers can be dangerous for the human eye. This disadvantage must be considered as optical transceivers find more and more applications in the general public. There are benefits in the use of lasers in data communication: for example, high data rates and long operating distances. The challenge is how to use these benefits without generating an optical hazard.

In principle the answer is simple. The power of accessible laser radiation must not exceed a critical limit that is defined by the mechanisms of eye damage. These mechanisms depend heavily on the wavelength of the laser radiation. Many systems of optical data communication use one of the three mentioned "classical" optical windows of fused silica glass fibers in the infrared region. The human eye cannot see this infrared radiation. To be more precise, the eye's receptor cells, the rods and cones of the retina, are not sensitive to this region of the optical spectrum. This infrared radiation penetrates the cornea and the vitreous humor like visible light and it is focused by the lens of the eye. To estimate possible hazards, the energy density of this focused radiation on the retina must be considered. The mechanisms of eye damage are well known because of the experience and use of lasers in medical applications. The standards that define the maximum limit of laser light are based on this knowledge.

Two important standards must be observed:

1. The international standard IEC 60825-1 of the International Electrotechnical Commission [13] if a laser product or component is to be distributed worldwide.
2. The U.S. national regulation 21 CFR, Chapter 1, Subchapter J, of the Center for Devices and Radiological Health (CDRH), a subgroup of the U.S. Food and Drug Administration (FDA), if the laser product is for the U.S. market.

Fig. 6.18   Power/irradiance limits according to IEC and FDA.

Both standards categorize all laser products into four main classes, some of them with subclasses. This classification gives the user an idea of the hazards of the laser and lays down the precautions to which the user must adhere — for example, the duty to train the staff and/or to wear eye protection. Figure 6.18 shows the limits of irradiance or optical power, respectively, versus wavelength both for IEC and FDA. A remarkable difference is obvious between the class 1 limits given by IEC and those given by FDA.

The safest category defined by these standards is the laser class 1. Therefore, the laser class 1 products are the least restrictive with respect to the manufacturer and user requirements. The very wide distribution of CD players and laser printers demonstrates that the application of lasers is not an impossible hurdle if the product meets class 1.

Laser safety is also an important issue for optical fiber data links. The easiest way to obtain a safe laser class 1 data link is to use components that are inherently safe. In such a case the laser transceiver ensures the eye safety of the complete system. However, it is a difficult task to design a laser transceiver that meets the different class 1 categories of both standards. The eye safety of transceivers with optical connectors can be achieved in the following three ways:

1. Use mechanical solutions such as shutters or spring-loaded flaps, which prevent access to the laser radiation from the transceiver, if no connector is in the port. However, transceivers should remain small and this additional mechanical component would in most cases increase the size of the port, depending on the applied connector. Only two of the international standardized connectors, the LSH (E2000) and the VF-45 connector, include a shutter or a

flap, for both the connector and the receptacle/coupler. However, these two connector families are not widespread for datacom applications. Nowadays, the most popular connectors/receptacles, which are installed in millions, have not incorporated the function of mechanical shutters/flaps (see also Section 6.1.). However, the basic disadvantage of any single mechanical solution (i.e., a shutter/flap either inside the receptacle or at the connector's front end) is their intrinsic lack of redundancy. Thus, safe operation even in "reasonable foreseeable single fault conditions" cannot be guaranteed. Therefore, purely mechanical solutions haven't become popular worldwide.

2. Utilize optical solutions aimed at high coupling efficiency and a divergent beam to reduce the accessible optical power of the unplugged transceiver. This can be achieved by a fiber stub and a split sleeve, which enables a physical contact between stub and connector. However, pollution by dirt or mechanical damage of the physical contact could cause high losses or laser disturbing back-reflections.

3. Choose an electronic solution. The advantage is that electronic safety interlocks are easily integrated, usually in the laser driver circuit. This type of solution can also be much less expensive than any mechanical or optical solution. The laser power of the transceiver is checked electronically by a monitor diode. This is inherently required anyway for controlling the laser power to compensate for temperature and aging effects on the laser. In case of excessive emission, the laser is shut down to prevent a hazard. The transceiver has a redundant safety interlock, which ensures eye safety in case of a single-fault condition or any subsequent failures.

Point-to-point optical data links enable the application of another electronic safety interlock. The open fiber control (OFC) system detects link disruptions and forces the transceiver into a repetitive pulsing mode of operation with very low duty cycle. Only after a proper reconnection handshake has checked that the link is closed will the OFC system allow the transceiver to return to regular data traffic. The specification of the OFC system is part of the ANSI fibre channel standard [5]. The system can operate with higher optical power that would normally exceed the limits of class 1 lasers. The advantage is a higher power budget for the optical transmission.

If the transceivers are inherently safe, the design of the data communication systems is not restricted by any laser safety requirements. The manufacturer does not have to classify the product, is not obliged to submit reports about laser safety, and is not required to perform additional tests during production. This allows the safe and simple application of the transceivers.

6.4. Noise Testing of Transceivers

In everyday life, the dramatic increase in the use of electronic devices in general and mobile telecommunication components in particular has evoked a strong demand for electromagnetic compatibility (EMC) and electromagnetic immunity (EMI) of electronic devices. With growing efforts in the field of high-speed data transmission along with ever-increasing computer or processor integration and speed, this demand has also gained major influence on design, production, and testing of today's fiber optic components and transmission equipment.

In close cooperation with computer, telecom, and datacom equipment manufacturers, active fiber optic components of high EMC/EMI performance have been designed. As an important foundation for success, a common understanding of the really relevant phenomena had to be reached and reliable and repeatable test methods had to be evaluated. This gave the basis necessary to define, then to test and to meet standardized EMC/EMI requirements for system application of fiber optic transceivers, and finally to guarantee reliable operation of these components.

This section describes established measurement procedures for the EMC/EMI performance of active fiber optic components. The methods presented here deal with noise on supply voltage, with emission of electromagnetic noise into the surrounding environment, and with immunity against electromagnetic noise from outside.

6.4.1. DESCRIPTION OF THE DEVICE UNDER TEST

For all described measurements, the device under test (DUT) is a fiber optic transmission module, e.g., a transceiver or a combination of a transmitter and a receiver. The DUT is connected to a printed circuit board (PCB), which is populated with power supply filtering elements as described in the transmitter/receiver datasheet (Fig. 6.19).

The electrical data output lines of the receiver are connected directly to the electrical data input lines of the nearby transmitter. The top and bottom

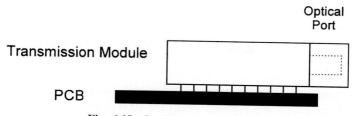

Fig. 6.19   Device under test (DUT).

Fig. 6.20   Test set-up for noise on Vcc.

layers of the PCB are connected to ground. Connections to the measurement setup are fiber optic jumper cables and power supply lines.

6.4.2.   NOISE ON VCC

Fiber optic transceivers usually operate in an environment with fast-switching circuitry. This results in high-frequency noise on the DC power supply Vcc. High-frequency noise on Vcc cannot always be sufficiently suppressed by capacitors or inductors because these components show resonant behavior of their own, with resonant frequencies lying close to the noise frequencies. Therefore, it is essential for fiber optic transceivers to be able to withstand high-frequency noise on Vcc. Another noise effect, commonly known as ripple, is less harmful to fiber optic modules because ripple frequencies are lower than frequencies of noise caused by fast switching.

To perform a test on a DUT's behavior concerning noise on Vcc, a noise signal has to be coupled into the Vcc line using a bias tee as shown in Fig. 6.20. When transceivers are tested, the receiver part should act directly as a driver for the transmitter part.

If the DUT contains internal decoupling elements, noise of certain frequencies may be sufficiently suppressed. In these cases, noise voltages at

Vcc device pins are close to zero. At frequencies at which noise suppression is insufficient, noise voltages can be measured at the pins.

At frequencies that are sufficiently suppressed, high noise generator amplitudes would be necessary to keep the noise voltage at the pins at a constant level. Testing a DUT under such rather artificial conditions is overly demanding. It is therefore recommended to merely keep the noise generator amplitude at a constant level for all frequencies.

For the measurement, regular data transmission is established and then optical power is attenuated to a level that causes a certain bit error rate (BER) value. When noise is applied on Vcc the BER increases. This increase in BER is a measure of the DUT's sensitivity to noise. The optical power is then increased until the DUT reaches the former BER. The device meets the requirements if the difference in optical power does not exceed a defined level. The measurement is repeated for different noise frequencies. It is always important to use identical test set-ups to ensure that measurements are consistent.

6.4.3.   ELECTROMAGNETIC COMPATIBILITY

6.4.3.1.   Emission

The emission of electromagnetic radiation may cause improper operation of components located near a radiating device. Therefore, maximum allowable electromagnetic noise levels (depending on noise frequency) are defined. For information technology equipment the emission limits are defined in standard EN 55022, derived from IEC CISPR 22, and in FCC CFR 47, part 15, classes A and B. A common measurement setup for noise emission is shown in Fig. 6.21. Figure 6.22 also shows a photo of the set-up inside an anechoic chamber.

The measurements are done in an anechoic chamber up to 40 GHz for any definded orientation of the DUT, or for continuous 360-degree rotation of the DUT on a wooden turntable. During testing the DUT is battery powered and the data pattern used is a Fibre Channel test pattern at nominal data rate. The DUT's emission is measured in a range of 250 MHz up to 18 GHz and, if necessary, up to 40 GHz. The DUT's emissions are tolerated if noise remains 6 dB below these limits in the appropriate range. An example of a measurement result is given in Fig. 6.23, wherein plot A shows the radiated field strength [dBμV/m] versus frequency for one full rotation of 360° and plot B shows the maximum radiation amplitude [dBμV] for a certain frequency versus angle of rotation.

Fig. 6.21   Set-up for testing noise emission.

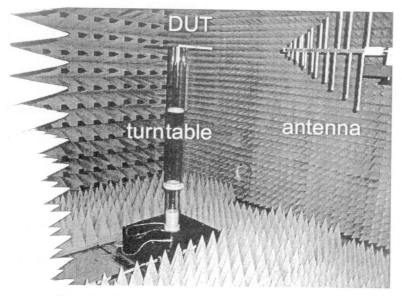

Fig. 6.22   Photo of test set-up inside an anechoic chamber.

6.4.3.2.   Immunity

When external electromagnetic fields are applied to the DUT, bit errors can occur. The most sensitive part of a DUT is the receiver section around the transimpedance amplifier. In this section, very low input currents (on the order of nA) are converted into low voltage signals. In typical transceiver

A   B

Fig. 6.23   Noise emission test results.

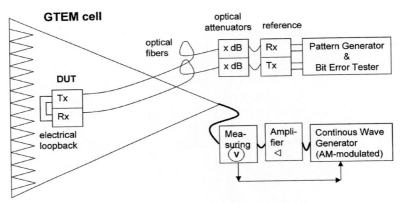

Fig. 6.24   Set-up for testing RF noise immunity.

designs this part of the receiver circuitry is located near the fiber optic connector receptacle.

Two measurement methods are established to determine the DUT's immunity: radio frequency (RF) immunity and electrostatic discharge (ESD) immunity.

6.4.3.2.1.   Noise Immunity Against Radio Frequency Electromagnetic Fields

This immunity test can be performed based on IEC 61000-4-3 standard. A usual measurement set-up is shown in Fig. 6.24.

At the start of the measurement, no noise is applied. Regular data transmission is established (2^7-1 pseudo-random bit sequence at nominal data rate), then the optical attenuator is set to a value at which a bit error rate of 10^{-6} is detected. At this operating point the DUT is more sensitive to external noise than during normal operation. The sine wave noise carrier source is switched on and regulated to generate a field strength of 3 or 10 V/m root mean square (RMS) at the DUT's location. Then amplitude modulation is applied to the carrier signal (modulation factor 80%, sine wave modulation 1 kHz). The noise carrier frequency starts at 80 MHz and is increased in steps of 1 or 10 MHz up to 2 GHz. The DUT is tested at its worst case position (evaluated experimentally) regarding its orientation and polarization of the noise field. Bit error rate (BER) values are measured for the different noise carrier frequencies.

Test results show that BER values depend on the noise frequency applied. Certain values of BER can be achieved by reducing optical power, so there is a corresponding optical power level for each BER value. Using this correspondence, test results can finally be presented as the minimum optical power necessary to achieve a certain constant BER value depending on the noise frequency. Figure 6.25 shows a possible test result. The DUT meets the requirements if the sensitivity values given in the DUT's specification are reached for all noise frequencies.

For a more detailed investigation, the difference between the minimum optical input power in the undisturbed case and the worst measurement result with noise applied can be required to be lower than a specified value. (This is shown as ΔP_{opt} in Fig. 6.27. The "baseline" of the graph can be

Fig. 6.25   Noise response of a DUT.

assumed to be similar to that of an undisturbed case.) This measurement can be used in the development process to indicate improvement or degradation of a certain design.

6.4.3.2.2.   *Immunity Against Electrostatic Discharge (ESD)*

ESD immunity tests deal with noise caused by fast transients from electrostatic discharges. A transient electromagnetic field based on the human body model is generated and influences the module behavior with respect to bit errors. Two tests are proposed: The first one is based on EN 61000-4-2 = IEC 61000-4-2 and the second one is suitable for a more detailed analysis. In both cases the source of the discharges is a so-called ESD gun (Fig. 6.26).

The DUT is operated with a 2^7-1 pseudo-random bit sequence at nominal data rate. The optical input power is set to the lower specification limit. The DUT's receptacle is fed through a grounded plate simulating a rack panel in a possible application environment. The following are the two steps involved in this test:

1. Without vertical coupling plate: Electrostatic discharges (both polarities) at any point on the rack panel or at the receptacle of the DUT are applied to induce bit errors during a defined time interval

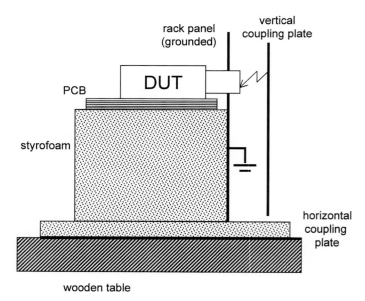

Fig. 6.26   ESD test set-up according IEC 61000-4-2.

of discharge. After the discharge the DUT continues error-free operation.

2. With vertical coupling plate: Electrostatic discharges are applied at any point on the vertical coupling plate and thereafter to the horizontal coupling plate to induce bit errors.

The discharge voltage and the number of bit errors in the time interval observed describe the ESD immunity of the DUT.

6.5.   Packaging of Transceivers (TRX)

6.5.1.   BASIC CONSIDERATIONS

As also discussed in Chapter 5, the most serious task of any packaging is to maintain functional integrity of a device over its total lifetime under all specified conditions. One of these tasks is to protect the electronic circuitry inside the transceiver from electromagnetic influences radiated or conducted into the module as well as to prevent electronics outside the module from being influenced by the module itself. This EMI shielding may be performed by means of a conductive metallic housing, or a conductive plating on a nonconductive housing, or even separate shielding structures inside the housing. Naturally, the electronic circuitry itself should also be carefully decoupled by properly selected capacitors and other filter components where needed (see also Section 6.4.2.).

The next task is to avoid possible damage from corroding chemical agents and/or humidity to sensitive devices and surfaces during the card assembly process or during normal operation. Otherwise, for example, corrosion effects in the presence of ionic impurities and humidity will occur. These effects are particularly pronounced with electronic circuitry where electro-migration in the presence of DC power occurs. Therefore, the appearance of intermittent short circuits or even other catastrophic failures due to complete destruction of joints or components is almost inevitable.

Another task of the TRX housing is to avoid influences of external mating or withdrawal forces applied by the fiber optic connector or the cable. This is a question of mechanical stability based on design, materials and techniques used for assembling the transceiver itself and the transceiver to a motherboard. Some typical features may be explained in more detail by the following examples.

Figure 6.27 shows the 2nd design generation of the ESCON/SBCON transceiver, with a housing containing a transmitter optical subassembly

Fig. 6.27 Exploded view of serial ESCON transceiver (connector shell not shown).

(OSA) and a receiver OSA. Also included is a chip on board (COB) electronic circuit with a driver IC for the infrared radiation emitting diode (IRED) on the transmitter side and a photodiode pre-amplifier IC and a separated buffer IC on the receiver side (compare to Section 6.1.1.3. and Fig. 6.5, 3rd generation of ESCON/SBCON). The plastic housing has ground pins that are directly molded in during injection molding. After molding the housing is plated with copper and nickel and a flash of gold, including the ground pins. Therfore, in combination with the metallized plastic cover, a completely closed metallic shielding and an extremely low ohmic-resistance connection to ground pins is achieved. Also, the Au-flashed nickel avoids oxidation or corrosion effects on any surface of the housing and the pins, which guarantees proper solderability.

Functional pins are press-fitted into the printed circuit board (PCB) and penetrate corresponding holes in the module housing's bottom. Also, the OSAs are press-fitted into noses on the front side of the housing. The PCB and cover are attached to the housing by conductive epoxy. To achieve washable seal, the housing's underside, as well as the nose area, are sealed by a suitable epoxy potting material.

As an aside, the 1st design generation of ESCON/SBCON transceiver had separate hermetically sealed metallic housings both for the transmitter and the receiver. The functional pins had fused glass feed-throughs. Hermeticity was achieved by seam-welding of base plate and lid to the housing's frame. The optical ports were hermetically sealed by press-fitting ball lenses into the so-called module noses, which were resistance welded to the said frame. An expensive solution, but the best method available during the development period in the mid-eighties.

6.5.2. TECHNIQUES FOR ASSEMBLING TRANSCEIVERS TO MOTHERBOARD

The transceiver shown in Fig. 6.27 is a typical example for a module under assembly by classic through-hole technique to a motherboard. After insertion of the module pins into the corresponding holes in the motherboard, the connector shell or the transceiver housing is screwed to the board. The transceiver is then connected to the board by a common wave soldering or fountain soldering technique followed by water-jet washing, rinsing, and hot-air blower drying. During this procedure the optical ports in the connector receptacle are protected from processing chemicals and water by a process plug.

The surface-mount technique (SMT) is especially used for transceivers with higher pin counts. Also, the possibility of having components on both sides of a motherboard is a basic advantage of this technique. Figure 6.28 shows a typical example for a transceiver with gull-wing leads

Fig. 6.28   Transceiver for SMT attachment (3rd generation of parallel ESCON/SBCON).

Fig. 6.29 Exploded view of an open module design (parallel ESCON/SBCON transceiver).

for SMT solder attachment. The corresponding exploded view is shown in Fig. 6.29.

Gull-wing leads in principle allow partial soldering techniques such as hot ram (thermode) or hot gas or even laser soldering. The important advantage of these processes is that only the leads are heated up to the liquid temperature of the solder and not the transceiver itself. Also, no washing procedure after soldering is needed if less agressive fluxes are used. Moreover, a visual inspection of the solder joint quality is easily possible.

The main reason J-leaded transceivers are not widely established is their basic disadvantage with respect to partial soldering techniques and inspection. This is also a principal concern for the ball grid array (BGA) technique for transceivers with high external lead counts due to very high functional integration.

However, in recent years a clear tendency in the market has been toward pluggable connection of transceivers to motherboards or backplanes. An obvious advantage is the possibility to change a failed transceiver quickly and easily without any special tools. The transponder mentioned in Section 6.2.2.2. is an example of such a solution. Other examples are described in Sections 6.5.4. and 6.5.5.

6.5.3. WASHABLE SEALED MODULE DESIGN

The previously mentioned soldering processes with washing and rinsing suggest a module that should be designed for hermeticity or a washable seal, which protects against intrusion of aqueous liquids during washing/rinsing. That also means that, on the one hand, no additional care must be taken to protect single electronic devices inside the module, which certainly will reduce running cost. On the other hand, the total design and nearly all assembling agents and processes have to be adapted to the demands of the soldering process. For example, the application of a standard soldering process with agressive fluxes may be strictly forbidden inside a sealed package. Additionally, during manufacturing, a suitable and cost-saving procedure for testing the seal must be implemented in a production line. The transceiver shown in Fig. 6.27 is an example of a washable sealed design.

6.5.4. OPEN MODULE DESIGN

The alternative to the sealed module is an open design. Such designs are well established worldwide and are becoming increasingly popular. This design is based on the simple consideration that any possibly damaging gaseous, liquid, or dusty agents going into an open housing may also be washed out or evaporate. This assumption may in general be correct, however, there are design complications and challanges to open packages.

In open packages, individual protection of single electronic devices such as ICs by separate housing or encapsulation by, e.g., globe top is necessary in the case of a chip-on-board (COB) technique. Additionally, special care must be taken regarding the circuit layout and the circuit assembly and protection in order to prevent electro-migration effects. Finally, the EMI shielding must be managed individually by separate structures such as metallic sheets over sensitive areas of the circuitry.

With open module design, the transceiver housing itself may now be a simple injection-molded plastic part and the final assembly of the TRX could be done by a snap-together technique. No additional sealing or corresponding test procedures are necessary. Figure 6.29 shows as an example of an open design, an exploded view of the 3rd generation of the parallel ESCON/SBCON transceiver. This view corresponds to the photo shown in Fig. 6.28.

6.5.5.   OPEN CARD MODULE DESIGN

As mentioned previously, open card module designs are taking their place in fiber optic datacom. The mechanical basis of a card module typically is a PCB combined with a peripheral plastic frame. This frame also contains the optical subassemblies (OSAs) (see also Chapter 5) and the connector receptacle on its front side. For the design of such open cards, the considerations for the previously mentioned open module designs generally apply.

Figure 6.30 shows as an example the Gigalink Card (GLC), which is a laser-based transceiver card compliant to ANSI FC-0 100-SM-LL-I standard [5] with 1.0635 GBd at 1300 nm for single-mode fiber application and 20-bit parallel electrical TTL in/out. Therefore, the GLC works as a transponder. The GLC is connected to a motherboard by an 80-pin connector on the rear area of the card's bottom side and secured by a pair of snapping legs. It is interesting to note that total insertion and withdrawal forces for that connector are on the order of nearly 100 N. Therefore, a special withdrawal lever is designed as an integral part of the frame in order to prevent damage of the module card and the motherbord during withdrawal.

Fig. 6.30   The Gigalink Card (left: bottom view; right: top view).

The GLC developed during the first half of the nineties for storage area network (SAN) applications in mainframe computer systems represents the first product worldwide with the described properties.

Another pluggable module card design family is the Gigabit Interface Card (GBIC), which is becoming increasingly popular. Characteristic for the GBIC design is that the insertion/withdrawal direction is identical for both the optical and the electrical connection.

6.5.6.   SMALL FORM FACTOR PLUGGABLE (SFP) DESIGN

The clear tendency toward pluggable transceivers mentioned in Section 6.5.2. combined with the other clear tendency toward increased port densities mentioned in Section 6.1.1.1. leads to the SFP design family. The SFP transceivers have at the bottom of the rear end an electrical connector specially designed to support hot pluggability. This means that during insertion in the motherboard, first the signal ground, then the power lines, and finally the data lines are connected. Disconnecting works in inverse order. Therefore, replacement of failed devices is possible without withdrawal of the motherboard or shutdown of the data-stream of other links in operation on this motherboard. Figure 6.31 shows a SFP transceiver with VF-45 optical connector interface.

6.6.   Series Production of Transceivers

6.6.1.   BASIC CONSIDERATIONS FOR PRODUCTION PROCESSES AND THEIR RELIABILITY

For every series production a basic precondition for cost savings and high-quality output is a continuously running production with qualified equipment as well as qualified and reliable processes. Correct equipment and processes mean that within a defined variation of process parameters, the so-called process window, a uniform high-quality result is achieved with a high yield. Naturally, a process is safer as its window is made wider.

Therefore, all equipment and processes must be qualified before the start of series production. This will typically be performed by careful and adequate testing of a statistically significant number of parts (i.e., 40 pieces at the absolute minimum) that had previously been produced on the related equipment by the corresponding process. The test results must be evaluated

Fig. 6.31   SFP transceiver with VF-45 optical connector interface.

in a statistical manner in which the number of tested parts defines the confidence level of the result.

6.6.1.1.   Qualification of Processes, the Process Capability

A brief example is provided as an explanation: A mechanical part A shall be attached to another part B by an epoxy adhesive. The key parameter for the process may be the mechanical strength characterized by, e.g., the pull force needed for destruction of the epoxy attachment between parts A and B. During tests the values for the forces are monitored and then their mean value, F_{mean}, and the standard deviation, σ, are calculated. These values are now put into a simple equation for calculating the so-called process capability:

$$C_{pk} = \frac{F_{mean} - F_{min}}{n\sigma} \geq S,$$

where S is the capability index, F_{min} is the minimum force specified for the attachment, and $1 \leq n \leq 6$ is the value for the statistical sharpness. For $n = 3$, for instance, S is 1.33 in order to reach a statistical safety of $\pm 3\sigma$ or 99.73%, or a failure probability of 2700 ppm. In the case of $n = 4$ and $C_{pk} \geq S = 1.66$, the statistical safety will be $\pm 4\sigma$ or 99.9937%, or the failure probability will be 63 ppm. Therefore, an improvement of a process is simply determined by increasing its C_{pk}.

The statistical sharpness indicated by n and the correlated value of S are defined indirectly by the requested product reliability or the permitted maximum number of failures in time (FIT) in the product specification, respectively.

6.6.1.2. Correlated Environmental Tests on Subcomponents

Here it makes sense to combine the evaluation of process capability with environmental tests that are also typically defined in a product specification. These tests normally assess long-term temperature and/or humidity stress, temperature cycling, shock stress, mechanical shock, vibration stresses, and also some additional stresses determined by the customer's special application. Upper and lower stress limits or values are mostly defined by commonly known MIL or Telcordia (formerly Bellcore) or IEC generic standards named in the product specification.

The major advantage in performing tests as early as possible during development phase and on subcomponent level is clear: Any weakness of a construction detail, a subcomponent, or a manufacturing process will be detected earlier, resulting in time and cost savings, and improvement or general change may be more easily implemented in the design itself and in the production flow.

Naturally, all tests will also be performed at the end of product development on the completed and final product. Normally, no major weakness will occur if all pretesting has been performed carefully during the design phase.

6.6.1.3. Equipment Qualification, the Machine Capability

A similar procedure to C_{pk} may be used for calculating the so-called machine capability, C_{mk}, which is defined as the statistical safety for processes performed by a tool or a machine. For example, the uniformity of a dispensed epoxy volume will be qualified by evaluation of C_{mk} of the automated dispenser. If C_{mk} exceeds the indicated value of capability index S, then the equipment would be safe and capable of its task. A combination with environmental stress for this evaluation does not make sense.

6.6.1.4. Frozen Process

After qualification and release of equipment and a corresponding process this production step generally should be frozen. Frozen means that in case of major changes of equipment — type or location of the equipment, process parameters, agents, or even parts — a partial or total requalification of the production step must be performed. The meaning of major change or even minor change should be defined in close communication with the customer. Also, the release of a changed production step for series production typically should be with customer agreement.

6.6.2. STATISTICAL PROCESS CONTROL AND RANDOM SAMPLING IN A RUNNING SERIES PRODUCTION

During running production some evident quality parameters should be fully monitored and some others checked by statistical random sampling in order to continuously assess the uniformity of processes and the quality of their output. Any deviation from the expected output quality or creeping degradation of process parameters can be detected in time for corrective action to be taken. For key process steps, a higher percentage of checks should be performed or a higher number of samples tested than would be for proven stable and uniform processes.

A commonly known tool for monitoring the quality level of any production process is the so-called statistical process control (SPC) card. Here, the actual level of a parameter is compared statistically to previously defined lower and upper limits for this most characteristic parameter that determines the quality of a component or a process. Additionally, the actual level may also be compared to the so-called warning limits that typically are defined as 1σ closer than the lower and upper limit. If during production an actual value goes close to a warning limit or even exceeds it, then an early corrective action may happen immediately. Therefore, components will almost never be produced out of specification. Figure 6.32 shows an example of a SPC card.

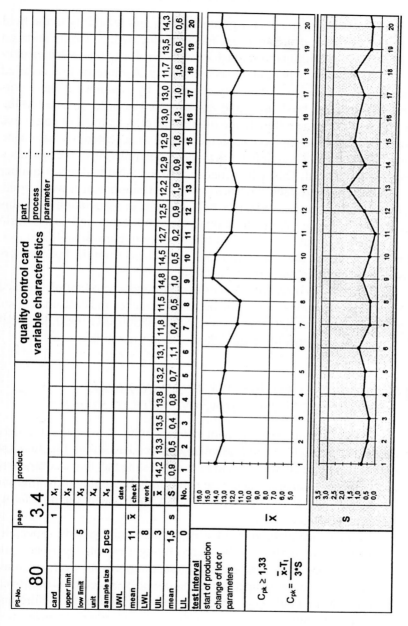

Fig. 6.32 Example of an SPC card.

199

6.6.3. DOCUMENTATION AND TRACEABILITY

Due to the high requirements for product safety and reliability in the field of telecom as well as datacom, every failed product will generally undergo a failure analysis. This analysis must answer questions about what exactly failed, how, possibly when, and especially why. A more or less detailed failure analysis report is commonly requested by customers. Then, corrective actions must be defined and implemented in order to rule out the occurrence of the same failure in the future. The effectiveness of corrective actions must be proven and documented as well as accepted by customers.

In answering why a part failed, a detailed documentation of the product's history is very helpful. This will be managed by an individual series number for each delivered product combined with a lot number or a week code. Additional information is typically held in store in association with this number, for example, about individual subcomponents and their origin and lot number, about date and shift of production, burn-in and final test, and finally about the individual values of typical electrooptical parameters measured during final outgoing inspection.

The extent of documentation and time of storage are commonly defined in delivery specifications contracted between producer and customer. Also, the generic quality managment system specification ISO 9000 describes procedures and contents for documentation and traceability.

6.6.4. ZERO FAILURE QUALITY, BURN-IN, FINAL OUTGOING INSPECTION, AND SHIP TO STOCK

Any technical system or component behaves with respect to its failure probability according to a well-known time-dependent failure function, the so-called bathtub life curve. This curve describes the fact that a component will most probably fail at a higher rate at the early beginning of its lifetime (BOL) and very late at end of lifetime (EOL). In between, the probability of failure is low and constant.

A well-established method for identifying early failing parts, especially in electronics, is to perform a burn-in. During burn-in, the fiber optic transceivers are operated at an elevated temperature level over a defined period of time. The time, temperature, and possibly some DC power overload will define the confidence level for effective screening. Before and after burn-in the transceivers will run their normal complete inspection of all relevant electrooptical parameters and the measured values will be compared. Individual transceivers will be rejected if there is a *delta* in any parameter exceeding a defined maximum value. Therefore, the failure rate for field-installed components can be reduced dramatically and the goal of a real zero-failure quality is approached.

In the case of a high level of confidence by customers, one additional step may be performed. If a producer organizes production according to the previously mentioned considerations and performs burn-in and documented tests of the products by intermediate and final outgoing inspection, then customers may install the delivered products in their systems without any additional incoming inspection.

In this manner, the producer will be certified to ship the products directly into custumer's stock with only a simple plausibility check. This procedure is an opportunity to reduce cost and time for both customer and supplier.

6.7. Transceivers Today and Tomorrow

6.7.1. TRANSCEIVERS TODAY

Fiber optic transceivers for applications in the field of datacom are mostly characterized by a couple of established international standards as described in detail in Section 6.1.1. These standards define the electrooptical performance of a transceiver/transponder as well as its pinout and its physical outline and package, including the corresponding fiber optic connector interfaces [6, 7, 8].

Fiber optic transceivers meeting these standards are operating worldwide in numerous applications in mainframes, server clusters, SANs, WANs, and LANs, and currently around 20 to 30 worldwide competing suppliers have been established. As an example, Fig. 6.33 demonstrates the increase in the number of involved partners in some important multi-sourcing agreements during years (see also Section 6.1.1.1.). This is also indicative of the increasing importance of industrial associations where both suppliers and applicators are represented. This speeds up the market penetration of novel components, systems, and applications. Nowadays, this seems not to generate conflicts with the commonly agreed normative power of international standardization organizations such as ISO, IEC, and ITU.

The demand for these transceivers has continuously increased during the past 10 years and the prices have shown dramatic decreases of the order of 25% per year. Consequently, the goal of all manufacturers is to offer a high level of performance, reliability, quality, and serviceability while

Fig. 6.33 Number of MSA partners (XGP: Multi-Gigabit Ethernet pluggable).

maintaining cost-effective production in the face of drastically increased volumes to meet the market pricing.

6.7.2.   SOME ASPECTS OF TOMORROW'S TRANSCEIVERS

The bit rates of fiber optic transceivers are continuously increasing in order to meet the worldwide demand for ever higher bandwidths. Therefore, a serious issue must be adressed due to EMI risks caused by electromagnetic radiation in or out of the receptacle opening in a front panel or backplane of a mainframe computer sub-unit, for example. To minimize EMI, the geometrical outline of the receptacle opening and consequent size of the connectors had to become as small as reasonably possible (i.e., in order to maintain proper handling of fiber optic connectors without special tools (see also Section 6.4. and Chapter 5)). In support of this size reduction trend, applicators' demands are following the same trends for much smaller transceivers in order to reduce space and volume on their PCBs and to increase the bit rate per unit area of PCB, corresponding with an increase of the port-density per unit area of rack panel.

6.7.2.1.   Geometrical Outline of Transceivers

In the past 10 years a significant reduction of module/transceiver size was possible due to significant progress in downsizing of optical sub-assemblies (see Chapter 5) and associated passive and active electronic

Fig. 6.34   Comparison of the outlines of different transceiver generations (left: ESCON/ SBCON; center: multi-standard; right: small form-factor).

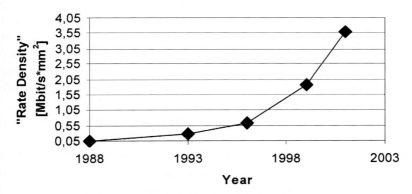

Fig. 6.35   "Rate-density" of transceivers.

components and circuitry. Figure 6.34 shows an in-scale comparison of the ESCON/SBCON outline (left) with the outlines of multi-standard (MS) (center) and small form-factor (SFF) transceivers (right). The function of the shown transceivers is described in detail in Section 6.1.1.

If one combines the increase of bit rates with the reduction of size, the success of the development efforts of the past 10 years is obvious. Figure 6.35 shows a graph for the bit rate per square millimeter, named "rate-density," versus the years of introduction of the products to the market. The dots represent, from left to right: ESCON/SBCON, MS 155 Mbit/s, MS 622 Mbit/s, SFF 1 Gbit/s, SFF 2.5 Gbit/s. The first dot differs from the last dot by a factor of 100. There is no obvious reason why this trend should change in the near future.

6.7.2.2.   Functional Integration

Another direction for the next generation of transceivers is the inclusion of additional electronic functions in a common module housing, such as

Serialization and deserialization of parallel digital bit-streams.

Encoding and decoding of serial bit-streams.

Clock synchronization/regeneration on receiver side.

Laser control and laser safety functions in laser-based transceivers/transponders as previously discussed in Section 6.2.2.

The main advantage for the user of such higher levels of integration is that no high-speed signals are on the system board, with the related cost savings. For the developers one of the challanges to solve is the issue of heat dissipation caused by such increased integration of a large number of high-speed digital electronic functions within the package. The only reasonable solution is to reduce the power consumption by application of low-power IC technologies with a supply voltage of 3.3 V or less. Such ICs have been already introduced, and will be continuously improved for reduced power consumption.

An additional complication is when some of the ICs in transceivers have to operate mixed signals, which means pure digital signals combined with analog signals, and additionally DC bias voltages and control functions. In the case of laser-driver ICs, the band gap of the laser's active radiating material defines the absolute minimum of the supply voltage, given by fundamental physical laws. The lower the wavelength emitted by the radiation source, the higher the band-gap energy and, consequently, the higher the required bias voltage. Therefore, the supply voltages of fiber optical transceivers may not completely follow the general tendency in digital electronics toward continuously decreasing supply voltages.

6.7.2.3.   Edge-emitting Lasers and VCSELs as Optical Sources

If one exceeds the bit rate of approximately 300 Mbit/s in a fiber optic intermediate-range multimode fiber link, the commonly used IRED on the transmitter side will be too slow and must be replaced by a laser diode (LD).

However, the very fast (up to 10 Gbit/s with direct modulation) conventional edge-emitting laser diode (EELD) is more complicated in application than an IRED because of the following:

1. An EELD needs a control circuit that monitors the output optical power and compensates for temperature and aging effects.
2. Accurate and reliable optical coupling of an EELD into a single-mode fiber is much more difficult and therefore more expensive compared to coupling an IRED into a multimode fiber. The position accuracy and stability needed for EELD in the range of 0.1 micrometers is approximately an order of magnitude higher than for coupling an IRED.
3. Laser safety due to potentially high optical output power has to be taken into account. The limitation of radiated optical power can be achieved by optical means or by electrical limitation of LD power output. Nevertheless, products for most datacom applications are unlikely to be successful in the market without certification as laser class 1 safe according to IEC 60825-1 or corresponding regulations such as those of the FDA (see also Section 6.3.3.).

Currently, a new type of laser is becoming dominant in some specific applications, the vertical cavity surface-emitting laser (VCSEL). This source was originally developed as 980 nm pumping of erbium-doped fiber amplifiers for long-haul telecom transmission lines. In Fig. 6.36 a schematic vertical cross-section of a VCSEL is shown.

850 nm Top-Emitting Vertical-Cavity Surface-Emitting Laser Diode

Fig. 6.36   Design of a VCSEL (schematic).

Table 6.1   **Comparison of Features**
**Edge-emitting laser diode (EELD) vs vertical cavity surface emitting
laser (VCSEL)**

Feature	EELD	VCSEL
Wavelength bands	650, 850, 1300 to 1660 nm	(650), 850, (1300, 1550) nm
Spectral bandwidth	Very narrow	Narrow
Size of active area	Typically 0.5–1× 2–10 μm	Variable, 5-50 μm diameter
Beam geometry	Strong elliptic	Circular
Beam divergence	High, up to 60° × 20°	Low, ca. 5°
Number of modes	Typically 1 or few	1 or even up to many 10s
Coupling to fiber	Difficult and sensitive	Easy
Coupling efficiency	Moderate	High
Threshold current	Approximately 10 mA	Some mA
Direct modulation bandwidth	High, up to 10 Gbit/s	High, up to 10 Gbit/s
Temperature drift of P_{opt}	Fairly high	Tendentially low
Environmental sensitivity	Extremely high	Moderate
Processing of chip	Very specific	Similar to LED
Final processing	Single bar	On wafer
Burn-in and functional test	Single on heatsink	On wafer

One of the key advantages of a VCSEL compared to an EELD is the IRED-like technology. This allows one to produce VCSELs with all processing steps, including burn-in and final testing, completely at a wafer level. Some additional advantages of VCSELs with respect to the EELD are listed in Table 6.1. A disadvantage of VCSELs is that not all of the wavelength-bands covered by EELDs are available with VCSEL technology. Currently only VCSELs for the 850-nm band are available for volume production with proven reliability and lifetime. VCSELs for the 1300-nm and the 1550-nm bands are still under basic research and design development. The experts estimate that possibly in the next years 1300-nm VCSELs will also be available in small volumes with acceptable yield.

6.7.2.4.   Laser Diodes for Multimode Fibers, Mode Underfill

Worldwide there are many miles of graded index (GI) multimode fibers installed in buildings and campuses. However, the speed and transmission field length of fiber optic links with GI multimode fibers combined with IREDs is limited due to power budget and bandwidth-length limits.

In order to safeguard this investment and use this current cabling even for higher speed transmission over distances of more than 100 m, the concept emerges of using laser diodes as sources for GI multimode fibers. There are groups studying the idea of extending the limits of GI multimode fibers by means of the so-called mode underfill launch condition. That would mean that coupling of optical power from a LD with limited focal diameter and numerical aperture into a GI multimode fiber would establish only a few low-order propagation modes near the center of the fiber core. The result would be a significant increase of the fiber's bandwidth-length limit.

Experimental investigations have confimed the theoretical assumptions. Therefore, transmission of up to 1 Gbit/s with 1300-nm wavelength over more than 500 m of standard graded index multimode fibers would work well. This direction is still receiving intensive discussion in the related standardization groups.

However, this technique will only establish itself if the price and performance for laser-based products is drastically improved. One key component would be an inexpensive laser optical subassembly (see also Chapter 5) with a laser diode that operates uncooled over the temperature range of category C, controlled environment ($-10°C$ to $+60°C$) according to IEC 61300-2-22, a typical office or building environment.

6.8.   Parallel Optical Links

6.8.1.   *HIGH DENSITY POINT-TO-POINT COMMUNICATIONS*

Fiber optic transceivers have become very well established for applications requiring high bandwidth transmission of data. Such applications include high-end routers, backbone switching for telecommunications, and data flow for disk clusters. Consider, for example, a high-end router. The switching function is typically distributed over a number of equipment racks, necessitating point-to-point communications from one rack to the next. Fiber optic transceivers are attractive because they exhibit a substantially higher distance-bandwidth product and reliability compared with other transmission media.

The point-to-point communications are often configured as "patch panels," in which the fiber optic transceivers are mounted onto a front panel, with the fiber sockets accessed through holes in the front panel. A duplex fiber cord is routed from one transceiver in one rack to the next desired transceiver in an adjacent rack. The number of fiber cords that a given panel can support, and hence the total aggregate bandwidth available from standard fiber optic transceivers, is practically limited by the number of fiber sockets that can be installed on the panel. Consider, for example, a Small Form Factor transceiver with a width of approximately 14 mm, operating at a data rate of 2.5 Gbit/s. Such a transceiver offers a bandwidth per front panel width of almost 200 Mbit/s per millimeter; let us call this the bandwidth density.

As bandwidth density requirements increase, the density limit imposed by single-channel transceivers becomes increasingly burdensome. This density constraint can be significantly relaxed by using a combination of multiple-channel fiber optic modules and multi-fiber ribbon cable. Consider, for example, Fig. 6.37, which shows the aggregate data throughput for a multiple-fiber ribbon cable, as a function of the per-fiber data rate. The simplest case is a single fiber, for which the aggregate throughput is identical to the per-channel data rate. The most common size of multi-fiber cable is a 12-fiber ribbon cable. The middle line shows its aggregate throughput. Even higher throughput performance (the top line in Fig. 6.37) is possible with a 96-fiber cable, arrayed 12 fibers wide stacked 6 arrays high.

6.8.2.   COMMON PARALLEL OPTIC MODULE CONFIGURATIONS

Just as multi-fiber cables improve the bandwidth density of the front panel, twelve-channel fiber optic modules dramatically improve the area utilization of the printed circuit board (bandwidth per unit board area). For simplicity, these parallel optics modules are typically configured as separate transmitter and receiver units, with a 12-fiber ribbon cable connecting the two (see Fig. 6.38). A transmitter module consists of a linear array of 12 lasers plus associated drive electronics; a receiver module consists of a linear array of 12 PIN diodes plus associated transimpedance amplifiers. The operation of each channel is independent of that of the next adjacent channel. Current state of the art for per-channel bandwidth is 2.5 Gbit/s,

Fig. 6.37   Aggregate data rate advantage of parallel optics.

which for a 12-fiber cable enables an aggregate data rate of 30 Gbit/s. The module width is only slightly wider than a Small Form Factor transceiver, on the order of 18 mm instead of 14 mm. For an accurate comparison we should consider two parallel optics modules, one transmitter cluster and one receiver cluster. Thus we can calculate a factor of 5 increase in the bandwidth density, while the electrical power consumption can be as small as a fifth that of a dozen single-channel transceivers.

VCSELs are by far the most common choice for laser in the transmitter modules because of low cost and ease of launching laser light into the optical fiber. Currently the state of the art is 850-nm emission of multimode light; thus the fiber cable should also be multimode. However, transmitters have also been fabricated from edge-emitting Fabry-Perot lasers at 1310 nm, launching into single-mode fiber. VCSEL arrays operating at 1310 nm are emerging from a number of laboratories. These will be attractive candidates for parallel optical links.

Interest is growing in an alternative configuration for a parallel optics module in which 4 transmitters and 4 receivers are combined into a single

Fig. 6.38   Photograph of a parallel optic module.

package. A 12-fiber ribbon cable is typically used, with the center 4 fibers "dark." This gives less aggregate bandwith density for a given module width compared with a 12-channel device. However, many applications can only use 8 of the 12 channels. For such applications, a 4-channel transceiver is an attractive alternative.

6.8.3.   LINK REACH

One of the most critical questions about a parallel optical link is "What is a reasonable link reach?" This means "What fiber cable length can be supported while still obtaining acceptable link performance in a low-cost installation?" This is a complex issue with at least three distinct questions.

The first question concerns technical feasibility: What link reaches can be demonstrated in a laboratory? The current state-of-the-art is for a per-channel bandwidth of 2.5 Gbit/s at an operating wavelength of 850 nm.

Such an optical signal propagating through multimode becomes degraded through one of three mechanisms: optical absorption (which is significantly higher at 850 nm than at 1310 nm), chromatic dispersion (which is much more severe at 850 nm that at 1310 nm), and modal dispersion. The optical attenuation and chromatic dispersion performance are largely defined by the glass materials system, and hence not likely to exhibit major improvements. However, the third mechanism, modal dispersion, is highly sensitive to the fiber manufacturing process. A number of fiber manufacturers are optimizing their processes to produce very low modal dispersion fiber for operation at 850 nm. Lucent's LazrSpeed and Corning's NGMM fiber are examples of this effort. A number of companies have demonstrated excellent link performance over a reach of at least a kilometer at 2.5 Gbit/s using such fiber.

The second question concerns prudent system design: What is a reasonable link budget that gives high assurance of successful operation under essentially all circumstances? Just because a particular link reach can be (routinely) demonstrated in the laboratory doesn't make this a prudent choice for system design. Typically a desired system design is expressed in terms of a link budget. The lowest expected laser power (over the life of the laser, for all allowed performance limits of temperature and voltage), minus the worst-case receiver sensitivity, defines a possible operating range. From this range must be subtracted expected losses such as worst-case fiber connector losses and fiber attenuation. Additional "penalties" are deducted to account for degradation mechanism such as laser residual intensity noise (RIN), etc.

Parallel optics modules are at a disadvantage compared with single-channel transceivers for achieving link budgets. Specifications on expected transmitter optical power need to be wider for parallel optics module to account for expected channel-to-channel power variations. Furthermore, laser safety limits are more restrictive for a parallel optics module because of the multiple channels. As performance goals become ever more aggressive, finding an optimum balance between laser safety constraints and a prudent link budget becomes an ever more difficult challenge for parallel optics.

The third question concerns the cost of the link, as fiber cable costs (especially for high-performance multimode ribbon fiber) can be a substantial fraction of the total link cost. Consider Fig. 6.39, which shows the cost trend for a typical parallel optic link as a function of link reach. For zero link reach the cost is essentially dominated by the cost of the parallel optics module pair (transmitter and receiver). As the link reach requirements

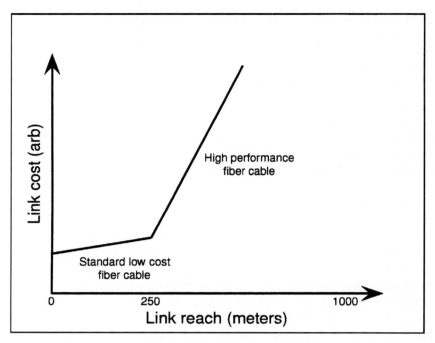

Fig. 6.39   Parallel optics link cost trend.

grow, additional cost is incurred because of the multifiber cable. But this cost growth is modest because standard low-cost fiber can be used for the short reach applications. After some break point (arbitrarily shown in this figure at 250 meters), the performance demands exceed the capabilities of standard fiber. For these applications, high-performance fiber must be deployed. However, this fiber has much higher cost per link length, so that the total link cost is dominated by the fiber cable cost.

6.8.4.   A LOOK TO THE FUTURE

The cost and performance analysis changes dramatically if the link were to be configured with single-mode fiber operating at 1310-nm wavelength. Mode dispersion is eliminated because of the single-mode fiber. Optical attenuation is significantly reduced, as is chromatic dispersion. Laser safety is much less restrictive at the longer wavelength, so higher optical powers can be considered. Thus significantly longer link reach can be realized at 1310 nm. But the most dramatic change is the cable cost. Single-mode multifiber cables should have the lowest cost of all multifiber possibilities.

However, such a link will require two key changes. One is the use of a long wavelength laser. For reasons of cost and ease of light launch, the laser array is preferably a VCSEL. The second major change is a dramatic tightening of optomechanical tolerances. Single-mode fiber has a core diameter that is almost an order of magnitude smaller than standard multimode fiber. This will make the manufacture of such a parallel optics module much more challenging.

Given the significant advantages of a single-mode 1310-nm parallel optics module, we are confident that such a design will become reality sometime in the next 2 or 3 years.

ACKNOWLEDGMENTS

Many thanks to all colleagues for their help in giving hints for corrections and updates, in particular:

- Thomas Murphy for careful check of grammer and wording in this chapter
- Herwig Stange for the update of currently valid laser safety limits
- Mario Festag for checking and updating section 6.4.3
- Ursula Armbrust and Renate Lindner for their help in preparation of figures, graphs, and photos

References

1. Agrawal, Govind P. *Fiber-Optic Communication Systems*, 2nd Ed. 1997. New York: Wiley.
2. Saleh, B.E.A, and M.C. Teich. *Fundamentals of Photonics*. 1991. New York: Wiley.
3. Proceedings of 26th ECOC. Sept. 3–7, 2000, Munich, Germany: VDE-Verlag.
4. IEC CA/1727/QP March 2000. SB4 FWG: Survey of future telecommunications scenario.
5. ANSI X3T9.x and T11.x, Fibre Channel (FC) Standards incl. FDDI, SBCON and HIPPI-6400, URLs: http://web.ansi.org/default.htm and http://www.fibrechannel.com.
6. IEC SC86C Drafts, released or midterm to be released IEC Standards, Group 62148-xx, Discrete/integrated optoelectronic semiconductor devices for fiber optic communication - Interface Standards, URL: http://www.iec.ch.
7. IEC SC86C Drafts, released or midterm to be released IEC Standards, Group 62149-xx, Discrete/integrated optoelectronic semiconductor devices for fiber optic communication including hybrid devices - Package interface standards.
8. IEC SC86B Drafts, released or midterm to be released IEC Standards, Group 61754-xx, Fibre Optic Connector Interfaces.

9. IEEE Projects 802.x, LAN/MAN Standards and Drafts URL: http://standards. ieee.org.

10. Telcordia Technologies (formerly BELLCORE) GR-253-CORE, Issue 3, Sept. 2000, Synchronous Optical Network (SONET) Transport Systems: Common Generic Criteria.

11. ITU-T G.957, June 1999, Optical interfaces for equipment and systems relating to the synchronous digital hierarchy (SDH).

12. ITU-T G.958, November 1994, Digital line systems based on the synchronous digital hierarchy (SDH) for use on optical fiber cables.

13. International Standard IEC 60825-1, 1993 incl. Amendment 2, January 2001, ISBN 2-8318-5589-6, Safety of laser products - Part 1: Equipment classification, requirements and user's guide.

Part 2 | The Links

Chapter 7 | Fiber Optic Link Requirements

Casimer DeCusatis

IBM Corporation, Poughkeepsie, New York 12601

7.1. Introduction

In previous chapters, an overview of the basic technologies used in opto-electronic communications has been given. With these building blocks, we can begin to assemble practical fiber optic data communication links. In this chapter, I will describe the performance requirements of a fiber link and consider some of the design trade-offs involved in link planning and implementation. For the purposes of this chapter, I will consider the performance of point-to-point data links as illustrated in Fig. 6.4; this is the most common building block of data communication networks, which is used to build up more complex logical topologies such as switched stars, rings, and so forth. Network management is a much more complex topic beyond the scope of this book; in any case, the performance of a network system depends on the interaction of many elements, including both hardware and software, and is thus highly application dependent. Most data links are full duplex, permitting two-way communication between attached devices; a separate transmitter/receiver pair is typically used for each direction. Datacom systems do not make efficient use of the fiber bandwidth; one user per link is typical. Data link design is not driven by optimization of the fiber bandwidth but rather by the cost of optoelectronic transceivers and adapters. By contrast, the telecommunications world makes extensive use of multiplexing to support multiple users on a single fiber. This minimizes

HANDBOOK OF FIBER OPTIC
DATA COMMUNICATION
$35.00

cabling costs and is most clearly illustrated by transoceanic fiber links that carry hundreds or thousands of voice signals on a single optical fiber. In other words, the number of electronic transceivers per kilometer of fiber is much higher in datacom applications than in telecom systems. Multiplexing techniques are only just beginning to be introduced in the datacom market and currently are only practical for large users (see the IBM 9729 Wavelength Division Multiplexer [1] for an example).

There are other important differences between the telecom and datacom applications, even though at first glance they appear to use the same technologies. First, datacom links must maintain a much lower bit error rate (BER), defined as the number of errors per second. This is because the consequences of an error in a data link may be very serious; bad data in the wrong place could potentially crash the computer operating system or cause small but significant errors such as moving the decimal point in a bank account balance. Although transmission rates vary, typical low-end fiber optic links operate around 10–15 Mb/s. By contrast, telecom systems transmit voice signals with a bandwidth of only about 4 kHz and the ultimate receiver is the human ear; excess line noise can often be ignored, such as the background static present on a long-distance call overseas. For this reason, datacom systems must support higher fidelity transmission and a BER of 10e-12 or better (approximately one error every hour for a 200 Mb/s link), as compared with about 10e-9 for telecom systems (approximately one error every 5 s). Some datacom links will support a BER as low as 10e-15; even testing such a link becomes challenging when an error occurs only once every 57 days! I will discuss optical link budgets and BER further in the following sections.

Another unique requirement of datacom systems is distance. Before the introduction of fiber optics in the data center, copper coax or parallel wires were limited to a maximum transmission distance of perhaps a few hundred meters (this is one reason why computer data centers tended to keep equipment clustered together in a single large room). Fiber optic links are capable of error-free transmission over tens of km at data rates significantly higher than those of copper systems (Table 7.1).

This increase in the data rate-distance product has opened up the data center and is partly responsible for recent distributed computing strategies; it has also enabled new applications, such as real-time data backup at a remote location for emergency disaster recovery. Finally, telecom systems are serviced only by trained personnel and are completely controlled by the local telephone company. In contrast, datacom systems are often serviced by the users themselves (who would never think of crawling into a

Table 7.1 **Comparison of Datacom and Telecom Requirements**

	Datacom	*Telecom*
BER	10e-12 to 10e-15	10e-9e
Distance	20–50 km	Varies with repeaters
No. transceivers/km	Large	Small
Signal bandwidth	00 Mb–1 Gb	3–5 Kb
Field service	Untrained users	Trained staff
No. fiber replugs	250–500	<100 over lifetime

Table 7.2 **Comparison of Serial Optical Fiber vs Parallel Copper Datacom**

	Fiber	*Copper*
Distance, unrepeated	20–50 km	0–100 m
Data rate	00–1000 Mb/s	<100 Mb/s
Weight	20 g/m	1.5 kg/m
Size (outer jacketed diameter)	<5 mm	10 mm–5 cm
Flexibility (bend radius)	10–20 mm	>2 cm (limited by cable thickness)
Noise immunity	Excellent	Good to poor

manhole and reconnecting wires if their telephone service failed to work). Thus, datacom equipment must be rugged enough to withstand a large number of connections and disconnections, high pulling forces on fiber cables, and other user-inflicted stress. The equipment should also be easy to use and maintain; for example, connectors are keyed to prevent misplugging, and receptacles can be designed to minimize the need for frequent cleaning. Industry standards are an important part of the multivendor datacom environment; however, they do not always guarantee that equipment will work together as intended or promised. Whereas most telecommunications carriers purchase their equipment from only a handful of suppliers, a large data center may be composed of a wide variety of equipment from different vendors. There are standardized procedures for testing and specifying the physical interface of fiber optic equipment; a partial list of some useful fiber optic test procedures (FOTP) standards is given in Table 7.2. However, this does not guarantee interoperability of equipment

from different sources; it is not possible to test every combination of equipment from different sources. A system composed of parts from multiple
sources will often fail at its weakest point; thus, different reliability requirements in fiber optic components should be considered (remember that even
if a product is intended for only a 40,000-hour lifetime, it must still ensure
at least 100,000-hours mean time between failures). This problem is somewhat more acute in data communication systems than in telecom systems
because data communication networks tend to have more active devices
per kilometer of fiber than telecommunications links. The requirements of
the international market further complicate interoperability. In most parts
of the world, it is possible to make an international phone call without
any special effort because the local phone companies have agreed to work
together passing calls along their respective networks. There are strict standards that must be met by datacom equipment in order to operate over the
public network; it is impossible to even export equipment to some countries,
regardless of the intended application, if it fails to meet certain standardized tests. This process is generally called type approval or homologation,
and the requirements vary greatly among different countries; Table 7.1
illustrates some requirements for shipping datacom products outside the
United States.

7.1.1.   TESTING STANDARDS

The standards for specification testing of fiber optic cable and components are well defined by the FOTP series of standards. There are more
than 170 such standard documents describing how to measure the specifications of fiber optic link components. I have provided a partial list of
some useful FOTP standards (Table 7.3); contact ANSI for a complete listing. In addition to testing of the link components, fiber optic equipment
must demonstrate compliance with national and international standards
intended to guarantee performance in typical applications (Table 7.4). In
particular, the standards required in order to export fiber optic datacom
products from the United States are especially detailed and can be difficult to meet. Table 7.5 lists some of the technical standards required
in order to obtain homologation and type approval to import or export
datacom equipment from the United States to various international countries. Not all nations require approval to all these standards, and the country by country requirements change frequently; consult the local telecom

Table 7.3   Selected FOTP Standards for Fiber Optic Testing

FOTP Standard Number	Purpose
27	Methods for measuring outside diameter of optical waveguide fibers
30	Frequency domain measurements of multimode optical fiber information transmission capacity
45	Microscopic method for measuring fiber geometry of optical waveguide fibers
47	Output far-field radiation pattern measurement
51	Pulse distortion measurement of multimode glass optical fiber information transmission capacity
54	Mode scrambler requirements for overfilled launching conditions to multimode fibers
80	Cutoff wavelength of uncabled single-mode fiber by transmitted power
107	Return loss for fiber optic components
127	Spectral characteristics for multimode lasers
168	Chromatic dispersion measurement of multimode graded-index and single-mode optical fibers by spectral group delay measurement in the time domain
170	Cable cutoff wavelength of single-mode fiber by transmitted power
176	Measurement method of optical fiber geometry by automated gray-scale analysis

Note. The table provides a partial list of industry standard FOTP published by the EIA. Documents can be ordered directly from EIA. Equivalent test procedures as specified in CCITT G.651 or G.652; all FOTP XXX are equivalent to EIA/TIA-455-XXX.

authority (PTT) for the latest information by country. However, the European Commission has recently produced a set of directives intended to harmonize legal product standards in Europe and create a single market by ensuring free sale of compliant products anywhere in Europe. The manufacturer must demonstrate compliance with these standards for all devices containing datacom equipment or adapters; failure to comply may result in denial of import–export permission, legal action, fines, or other actions by

Table 7.4   **Recent Homologation Requirements
for Leased Dark Fiber by Country**

Country	Dark Fiber
Argentina	x
Austria	c
Belgium	c
Brazil	—
Canada	x
China, PR	x
Denmark	—
Finland	—
France	—
Germany	c
Ireland	—
Israel	c
Italy	x
Japan	x
Korea	x
Mexico	x
Norway	—
Poland	c
Portugal	—
Spain	x
Switzerland	c
United Kingdom	x

Note. All information currrent as of January 1996; subject to change by local
telecom authorities without notice. c, available on case-by-case basis; —, no
homologation; x, service not available.

individual countries. There is also a European standard governing the shipment of electronic products to the European Community (EC) after January
1996 — the Communaute Europeenne (CE) mark is applied to products
that conform to the EC directives 89/336/EEC [radio frequency interference (RFI) and electrical safety for electronic equipment] or 91/263/EEC

Table 7.5   **General Technical Specifications Required for Datacom Product Homologation**

Safety compliance: IEC 950/EN-60950 (2nd ed.); LVD (low voltage directive) 73/23/EEC; csa-c22.2, no. 950-m89

ISDN only: ETS 300-046/047

Optical fiber only (optical safety): EN-60825/IEC 825

EMC compliance: EN-55022/CISPR-22 for information technology equipment
Related standards: EN-60555; EN-61010-1 (IEC 1010-1); EN-50082-1; FCC part 15

Analog only: VDE 0878

(RFI and electrical safety for telecommunications equipment). Products that fail to earn the CE mark by demonstrating compliance at a certified laboratory will not be allowed to enter the EC, and products that claim compliance and are found not to meet these standards are subject to both criminal and civil penalties. Note that this is a telecom CE mark, an additional requirement to the more common CE mark applied to products that only meet radiated noise requirements; it also requires products to have immunity to any signals likely in their normal environment, which has not been a legal requirement in the United States. These standards are strict, and datacom products not designed for EMC compliance are unlikely to meet them. Not only must the product itself comply but also it must comply when configured in a typical system, including cables and external interfaces. For example, this can impact fiber optic transceivers that may radiate switching noise outside the protective covers of a host box or fiber optic connectors that contain metal that can reradiate electrical noise when plugged into the transceiver. Manufacturers must sign a declaration of conformity specifying the product, the directives it meets, and the standards applied. If a product meets standards in full, self-certification may be acceptable; otherwise, a technical construction file explaining the compliance rationale must be approved by a designated competent body. Products must be supported by a design file containing test certificates/reports, design information, users guide, schematics, and related information; this documentation must be retained for at least 10 years after the manufacture of the last unit to which it applies. Manufacturers outside Europe must appoint an authorized representative within Europe to take responsibility for all CE mark issues. There may also be additional export standards for datacom and telecom products to meet, such as ISO 9000 series certification on development or manufacturing, the related QS-9000 standards, or the European Environmental Conformance Standard ISO 14000, which requires that all products entering the EC will not pollute the environment or endanger local wildlife habitats. Many other nations are also considering adopting similar standards, so these product requirements should not be considered specific to the European market. Datacom products may also be affected by other standards in the United States, such as Federal Communications Commission (FCC) compliance standards and Environmental Impact Assessments (EIA) on datacom products (because lasers and photodetectors commonly contain gallium arsenide compounds, which qualify as toxic substances for disposal). Because these standards change often, consult the standards bodies or approval labs for the latest information.

7.1.2.   BIT ERROR RATE

All fiber optic channels consist of an optical source or transmitter, the cable plant, and an optical receiver. These components have been described in previous chapters; I will concentrate on their interaction when designing a fiber link. The transmitter is capable of launching a limited amount of optical power into the fiber, and there is a limit to the weakness of a signal that can be detected by the receiver in the presence of noise. Thus, a fundamental consideration is the optical link power budget, or the difference between the transmitted and received optical power levels. Some power will be lost due to connections, splices, and bulk attenuation in the fiber. There may also be optical power penalties due to dispersion, model noise, or other effects in the fiber and electronics. The optical power levels define the signal-to-noise ratio at the receiver, Q; this is related to the bit error rate by the well-known Gaussian integral

$$\text{BER} = \frac{1}{\sqrt{2\pi}} \int_{Q}^{\infty} e^{-\frac{Q^2}{2}} dQ \cong \frac{1}{Q\sqrt{2\pi}} e^{-\frac{Q^2}{2}}. \qquad (7.1)$$

From Eq. (7.1), we see that a plot of BER vs received optical power yields a straight line on semilog scale, as illustrated in Fig. 7.1. Nominally, the slope is approximately 1.8 dB/decade; deviations from a straight line may indicate the presence of nonlinear or non-Gaussian noise sources. Some effects, such as fiber attenuation, are linear noise sources; they can be overcome by increasing the received optical power, as seen from Fig. 7.1, subject

Fig. 7.1 Bit error rate as a function of received optical power. Curve A shows typical performance, whereas curve B shows a BER floor [5].

to constraints on maximum optical power (laser safety) and the limits of receiver sensitivity. There are other types of noise sources, such as mode partition noise or relative intensity noise (RIN), that are independent of signal strength. When such noise is present, no amount of increase in transmitted signal strength will affect the BER; a noise floor is produced, as shown by curve B in Fig. 7.1. This type of noise can be a serious limitation on link performance.

As can be seen from Fig. 7.1 that receiver sensitivity is specified at a given BER, which is often too low to measure directly in a reasonable amount of time (for example, a 200 Mb/s link operating at a BER of 10e-15 will take only one error every 57 days on average, and several hundred errors are recommended for a reasonable BER measurement!). For practical reasons, the BER is typically measured at much higher error rates, where the data can be collected more quickly (such as 10e-4 to 10e-8) and then extrapolated to find the sensitivity at low BER. This assumes the absence of nonlinear noise floors, as cautioned previously. The relationship between optical input power, in watts, and the BER, is the complimentary Gaussian error function.

$$\text{BER} = 1/2 \text{ error function } (P_\text{out} - P_\text{signal}/\text{RMS noise}), \qquad (7.2)$$

where the error function is an open integral that cannot be solved directly. Several approximations have been developed for this integral, which can be developed into transformation functions that yield a linear least-squares fit to the data. The resulting fitted equation can then be used to extrapolate the receiver sensitivity. Examples of two transformation functions that have been developed are the following:

$$T(\text{BER}) = 10^{(0.526966\log(-1.837794\log(3.7678\,\text{BER})))} \qquad (7.3)$$

$$T(\text{BER}) = -C_2 + \sqrt{C_2^2 + 4C_1(C_3 + \ln(2\text{BER}))/2\,C_1}, \qquad (7.4)$$

where $C_1 = 0.4926119$, $C_2 = 0.2498322$, and $C_3 = 0.7912445$. The collected raw data are used to calculate the least-squares curve coefficients, which are used to assemble a curve fit according to

$$m = \sum XY - (\sum X \sum Y/N)/\sum X^2 - ((\sum X)^2/N) \qquad (7.5)$$

$$c = (\sum Y - m \sum X)/N, \qquad (7.6)$$

where N is the number of data points, X is optical power in watts (not dBm), $Y = T(\text{BER})$ is the transformed BER data, and

$$T(\text{BER}) = m(\text{optical power}) + c. \qquad (7.7)$$

Rearranging this equation, the receiver sensitivity cannot be extrapolated to the desired BER:

$$\text{Optical sensitivity} = 1/m(T(\text{Spec. BER}) - c). \qquad (7.8)$$

The same curve-fitting equations can also be used to characterize the eye window performance of optical receivers. Clock position/phase vs BER data are collected for each edge of the eye window; these data sets are then curve fitted with the previous equations to determine the clock position at the desired BER. The difference in the two resulting clock positions on either side of the window gives the clear eye opening. Further information on curve fitting of low BER can be obtained in the literature [2–4].

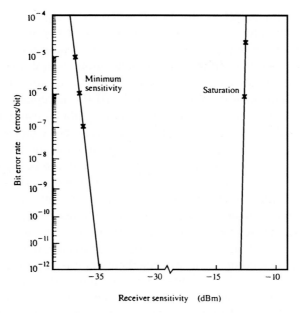

Fig. 7.2   Bit error rate as a function of received optical power illustrating range of operation from minimum sensitivity to saturation. [5].

After consulting Fig. 7.1, we might be tempted to conclude that in the absence of nonlinear noise effects, higher optical power is always more desirable at the receiver. If we plot BER vs receiver sensitivity for increasing optical power, we obtain a curve similar to that in Fig. 7.2, which shows that for very high power levels, the receiver will go into saturation. This can happen if a laser transmitter designed for long-distance applications is inadvertently connected over a short link; the situation also arises when using optical loopbacks or wrap plugs for transceiver diagnostics. We see from the characteristic "bathtub"-shaped curve that there is a window of operation with both upper and lower limits on the received power, using a powerful laser does not provide a brute-force solution to link distance problems.

In describing Figs. 7.1 and 7.2, I have also made some assumptions about the receiver circuit. Most data links are asynchronous and do not transmit a clock pulse along with the data; instead, a clock is extracted from the incoming data and used to retime the received data stream. I have made the assumption that the BER is measured with the clock at the center of the received data bit; ideally, this is when we compare the signal with a preset threshold to determine if a logical "1" or "0" was sent. When the clock is recovered from a receiver circuit such as a phase-lock loop, there is always some uncertainty about the clock position; even if it is centered on the data bit, the relative clock position may drift over time. The region of the bit interval in the time domain where the BER is acceptable is called the eye width; if the clock timing is swept over the data bit using a delay generator, the BER will degrade near the edges of the eye window. Eye width measurements are an important parameter in link design, and transceiver vendors will often show a sample eye pattern on their product data sheets. I will describe timing jitter as part of the discussion on optical link budget modeling. In order to design a proper optical data link, the contribution of different types of noise sources should be assessed when developing a link budget. There are two basic approaches to link budget modeling. One method is to design the link to operate at the desired BER when all the individual link components assume their worst-case performance. This conservative approach is desirable when very high performance is required or when it is difficult or inconvenient to replace failing components near the end of their useful lifetimes. The resulting design has a high safety margin; in some cases, it may be overdesigned for the required level of performance. Because it is very unlikely that all the elements of the link will assume their worst-case performance at the same time, an alternative is to model the link budget statistically. For this method, distributions of transmitter power output, receiver sensitivity, and other parameters are either obtained from the vendors or estimated by measuring a small sample of parts. They are then combined statistically using an approach such as the Monte Carlo method, in which many possible link combinations are simulated to generate an overall distribution of the available link optical power. The industry standard is a 3-σ design, in which the combined variations of all link components are not allowed to extend more than 3 standard deviations from the average performance target in either direction. The statistical approach results in greater design flexibility and generally increased distance compared with a worst-case model at the same BER. The statistical approach is only as valid as the input data, however, and these can be difficult to obtain; many vendors consider such data proprietary because it gives an indication of their overall manufacturing process performance. A good design compromise is to combine the two methods, using worst-case for some parameters and statistical for others. I will describe the link budget planning process in greater detail in the following sections.

7.2.   Link Budget Planning

I now discuss important factors in the modeling and design of fiber optic data links, with emphasis on the link budget requirements, following the treatment of ref. [5]. It is convenient to break down the link budget into two areas: installation loss and available power. Installation or DC loss refers to optical losses associated with the fiber cable plant such as connector loss, splice loss, and bandwidth considerations. Available optical power is the difference between the transmitter output and receiver input powers, minus additional losses due to optical noise sources on the link (also known as AC losses). With this approach, one may choose to treat the installation loss budget as statistical and the available power budget as worst case.

7.2.1.   INSTALLATION LOSS BUDGET

First, I consider the installation loss budget. Some of the important factors to be considered include the following:

Fiber cable loss

Fiber attenuation as a function of wavelength

Connector loss

Splice loss

Link measurement uncertainty

A typical point-to-point fiber optic link connecting two devices is shown in Fig. 7.3, consisting of both trunk and jumper cables. Trunk cable is used for long installations (hundreds of meters to several kilometers, usually between buildings). A single trunk cable may contain many individual fibers. Jumper cables are used for short connections between devices that are close together or between a device and a distribution panel; they typically contain only one or two fibers. Jumper cables are usually limited to less than 100 m in length and are therefore allowed to have a slightly higher attenuation loss than trunk cable. Typical trunk cable will have a loss of approximately 0.5 dB/km near 1300 or 1550 nm, whereas jumper cable in this range may average 0.8 dB/km or more. Fibers are available with loss as low as 0.2 dB/km for some applications.

Transmission loss is perhaps the most important property of an optical fiber or cable; it affects the link budget and maximum unrepeated distance. The number and separation between optical repeaters and regenerators

Fig. 7.3 Example of a point-to-point data link, ESCON environment [5].

are largely determined by this loss. The mechanisms responsible for this loss (material absorption and linear and nonlinear scattering) have been discussed in Chapter 1. It was also noted that fiber loss varies with operating wavelength, providing transmission windows for optical communication. In most optical transmitters, the central wavelength is subject to variation due to the manufacturing process as well as environment factors such as temperature. Even for laser sources with a spectral width of 6 nm or less, the center wavelength may vary by 80 nm or more. This causes additional loss that must be accounted for in the link budget. Although wavelength-dependent attenuation data are available from most fiber manufacturers, it is often not practical to perform measurements of the fiber at each wavelength of interest, especially during a repair activity. An accurate model for fiber loss as a function of wavelength has been developed by Walker [6]; this model accounts for the effects of linear scattering, macrobending, material absorption due to ultraviolet and infrared band edges, hydroxide (OH) absorption, and absorption from common impurities such as phosphorous. Using this model, it is possible to calculate the fiber loss as a function of wavelength for different impurity levels; an example plot is shown in Fig. 7.4. Using this method, the fiber properties can be specified along with the acceptable wavelength limits of the source to limit the fiber loss over the entire operating wavelength range; design trade-offs are possible between center wavelength and fiber composition to achieve

Fig. 7.4   Spectral loss characteristics of single-mode fiber for different impurity levels [5].

the desired result. Typical loss due to wavelength-dependent attenuation for laser sources on single-mode fiber can be held below 0.1 dB/km.

In Chapter 1, optical losses due to microbends and macrobends in optical fiber were discussed; clearly, these are to be avoided during the fiber installation. Another important environmental factor is exposure of the fiber to ionizing radiation damage, which is important in some environments such as research labs, nuclear power plants, medical facilities, military applications, and space satellites. Although optical fiber is by definition immune to electromagnetic interference, there is a large body of literature concerning the effects of ionizing radiation on fiber links [7–11]. There are many factors that can affect the radiation susceptibility of optical fiber, including the type of fiber, type of radiation (gamma radiation is usually assumed to be representative), total dose, dose rate (important only for higher exposure levels), prior irradiation history of the fiber, temperature, wavelength, and data rate. Optical fiber with a pure silica core is least susceptible to radiation damage; however, almost all commercial fiber is intentionally doped to control the refractive index of the core and cladding as well as dispersion properties. Trace impurities are also introduced that become important only under irradiation; among the most important are Ge dopants in the core of graded-index fibers, in addition to F, Cl, P, B, OH content, and the alkali metals [7]. In general, radiation sensitivity is worst at lower temperatures and is also made worse by hydrogen diffusion from materials in the fiber cladding.

Because of the many factors involved, there does not exist a comprehensive theory to model radiation damage in optical fibers. The basic physics of the interaction has been described [8, 9]; there are two dominant mechanisms — radiation-induced darkening and scintillation. First, high-energy radiation can interact with dopants, impurities, or defects in the glass structure to produce color centers that absorb strongly at the operating wavelength. Carriers can also be freed by radiolytic or photochemical processes; some of these become trapped at defect sites, thus modifying the band structure of the fiber and causing strong absorption at infrared wavelengths. This radiation-induced darkening increases the fiber attenuation; in some cases, it is partially reversible when the radiation is removed, although high levels or prolonged exposure will permanently damage the fiber. A second effect is caused if the radiation interacts with impurities to produce stray light, or scintillation. This light is generally broadband but will tend to degrade the BER at the receiver; scintillation is a weaker effect than radiation-induced darkening. These effects will degrade the BER of a link; they can be prevented by shielding the fiber or partially overcome by a third mechanism, photobleaching [10, 11]. The presence of intense light at the proper wavelength can partially reverse the effects of darkening in a fiber. It is also possible to treat silica core fibers by briefly exposing them to controlled levels of radiation at controlled temperatures; this increases the fiber loss, but makes the fiber less susceptible to future irradiation. These so-called radiation-hardened fibers are often used in environments where radiation is anticipated to play an important role.

Models for the fiber's response to high radiation levels have been proposed [11]. There have also been studies of the long-term reliability and lifetime effects of background radiation levels, such as those encountered in buried or undersea fibers [12]. Recently, several models have been advanced [13, 14] for the performance of fiber under moderate radiation levels; the effect on BER is a power law model of the form

$$\text{BER} = \text{BER}_0 + A \, (\text{dose})^b, \tag{7.9}$$

where BER_0 is the link BER prior to irradiation, the dose is given in rads, and the constants A and b are empirically fitted. I will not discuss in detail modeling work in this area, which is available in several references [13, 14]. In most applications, the link can be sufficiently shielded from exposure to mitigate these effects, or specially fabricated radiation-hardened fiber can be used to reduce the effects. The loss due to normal background radiation

MIC (FDDI) Connector

IBM Duplex Single-Mode Connector

FC Connector

ST Connector

Biconic Nonphysical-Contact Connector

Fiber Channel Standard SC-Duplex Connector

Fig. 7.5   Common fiber optic connectors: MIC (FDDI), ST, FC, ESCON, SC duplex, and biconic (nonphysical contact) [5].

exposure over a typical link lifetime can be held below approximately 0.5 dB.

There are also installation losses associated with fiber optic connectors and splices; both of these are inherently statistical in nature. There are many different kinds of standardized optical connectors, some of which are shown in Fig. 7.5. Generally, higher variability has been observed for screw-type connectors compared with push–pull or bayonette-type designs. By "making and breaking" the connection many times, a connector loss distribution can be measured; for large number of connections, the distribution is often approximately Gaussian and can be characterized by a mean and σ value as shown. Typically, published manufacturer's connection loss values assume similar fiber parameters on both sides of the connection. This may not always be the case; for example, due to different tolerances on jumper and trunk cables, there may be a mode field diameter mismatch between a jumper-to-trunk connection at a distribution panel. Usually, the variation due to mechanical tolerances on the connectors will dominate any excess connector loss due to mode field diameter mismatch; for some applications, this represents an addition statistical penalty that must be accounted for in the installation loss budget. Usually, the net effect is limited to a few tenths of a dB. There are many different models that have been published for estimating connection loss due to fiber misalignment [15–20]; most of these treat loss due to misalignment of fiber cores, offset of fibers on either side of the connector, and angular misalignment of fibers. The loss due to these effects is then combined into an overall estimate of the connector performance. Although some of these models have experimental verification, there is still no general model available to treat all types of connectors that then allows the modeling information to be used in improving the connector design. The development of optical connectors remains something of an art; even so-called "standardized" connectors exhibit widely different properties depending on the manufacturer. The standard merely ensures that connectors will be mechanically compatible and interoperate; it does not guarantee system performance or other desirable properties such as ease of plugging and durability.

Optical splices are required for longer links, because fiber is usually available in spools of 1–5 km, or to repair broken fibers. There are two basic types, mechanical splices (which involve placing the two fiber ends in a receptacle that holds them close together, usually with epoxy) and fusion splices (in which the fibers are aligned and then heated sufficiently to fuse the two ends together). Fusion splices have become much more common because of their low loss and the advent of new portable tools for efficiently performing fusion splices in the field. Typical splice loss values are given in Table 7.6.

If the optical fiber is improperly cabled or installed, additional loss can be experienced due to bending of the fibers. This falls into two categories, micro-bending (due to nanometer-scale variations in the fiber) and macro-bending (due to much larger, visible bends in the fiber). Because both types of bending loss may contribute to the attenuation of a single-mode or multimode fiber; it is obviously desirable to minimize bending in the application whenever possible. Most qualified optical fiber cables specify a maximum bend radius to limit macro-bending effects, typically around 10–15 mm, although this varies with the cable type and manufacturer's recommendations.

An additional effect of lossy connectors and splices is modal noise. Because high-capacity optical links tend to use highly coherent laser transmitters, different modes propagating in the fiber may interfere with one another. Random coupling between fiber modes causes fluctuations in the optical power coupled through splices and connectors; this phenomena is known as modal noise [18]. As one might expect, modal noise is worst when using laser sources in conjunction with multimode fiber; recent industry standards [19] have allowed the use of shortwave lasers on 50-μm fiber

Table 7.6   **Typical ESCON Cable Plant Optical Losses [5]**

Components	Description	Size (μm)	Mean Loss	Variance (dB^2)
Connector[a]	Physical contact	62.5–62.5	0.40 dB	0.02
		50.0–50.0	0.40 dB	0.02
		9.0–9.0[b]	0.35 dB	0.06
		62.5–50.0	2.10 dB	0.12
		50.0–62.5	0.00 dB	0.01
Connector[a]	Nonphysical contact (multimode only)	62.5–62.5	0.70 dB	0.04
		50.0–50.0	0.70 dB	0.04
		62.5–50.0	2.40 dB	0.12
		50.0–62.5	0.30 dB	0.01
Splice	Mechanical	62.5–62.5	0.15 dB	0.01
		50.0–50.0	0.15 dB	0.01
		9.0–9.0[b]	0.15 dB	0.01
Splice	Fusion	62.5–62.5	0.40 dB	0.01
		50.0–50.0	0.40 dB	0.01
		9.0–9.0[b]	0.40 dB	0.01
Cable	IBM multimode jumper	62.5	1.75 dB/km	NA
	IBM multimode jumper	50.0	3.00 dB/km at 850 nm	NA
	IBM single-mode jumper	9.0	0.8 dB/km	NA
	Trunk	62.5	1.00 dB/km	NA
	Trunk	50.0	0.90 dB/km	NA
	Trunk	9.0	0.50 dB/km	NA

[a]The connector loss value is typical when attaching identical connectors. The loss can vary significantly if attaching different connector types.
[b]Single-mode connectors and splices must meet a minimum return loss specification of 28 dB.

because of potential cost savings on this type of link hardware. Shortwave lasers are currently readily available because of their wide use in compact disk systems for optical storage, and the relaxed tolerances of multimode fiber makes connectorization problems simpler and less expensive to solve. The main disadvantage to such systems is modal noise, along with the limitation of using only 50-μm fiber (the bandwidth of 62.5-μm fiber seriously limits link distance). Modal noise is usually considered to be nonexistent in single-mode systems. However, modal noise in single-mode fibers can arise when higher order modes are generated at imperfect connections or splices. If the lossy mode is not completely attenuated before it reaches the next connection, interference with the dominant mode may occur. This is especially important in local area networks (LANS) in which large numbers of connectors and/or splices are used over relatively short distances. The effect of modal noise have been modeled previously [18], assuming that the only significant interaction occurs between the LP01 and LP11 modes for a sufficiently coherent laser. For N sections of fiber, each of length L in a single-mode link, the worst-case σ for a modal noise can be given by

$$\sigma_m = \sqrt{2}N\eta(1 - \eta)e^{-aL}, \tag{7.10}$$

where a is the attenuation coefficient of the LP11 mode, and η is the splice transmission efficiency, given by

$$\eta = 10^{-(\eta_0/10)}, \tag{7.11}$$

where η_0 is the mean splice loss (typically, splice transmission efficiency will exceed 90%). The corresponding optical power penalty due to modal noise is given by

$$P = -5\log\left(1 - Q^2\sigma_m^2\right), \tag{7.12}$$

where Q corresponds to the desired BER.

One final factor that deserves attention as part of the installation loss budget is the uncertainty in performing link loss measurements. There are many different standardized techniques to measuring link loss [21]. One approach involves using a known source and detector to measure the loss of a sample cable and then replacing the cable with the link under consideration and measuring the link loss. By taking the difference between these measurements, random factors such as the connection loss can be reduced. However, these effects cannot be totally eliminated, and the replacement loss technique will itself produce a distribution of measured loss values centered about the mean or average link loss. If this variability is small, as for the large tolerances associated with multimode fiber, then the statistics of the replacement loss technique can be measured and incorporated into the installation loss budget. However, for some applications (particularly single-mode fiber) this approach is not accurate enough. Instead of using typical jumper cables, special reference jumper cables must be used that are manufactured to very tight tolerances. Although reference jumpers are often expensive, they are required to provide the level of accuracy needed in most single-mode measurements. The requirements of reference jumpers must be determined on an individual basis for different applications; even a good reference jumper results in some link loss variability, which must be accounted for in the installation loss budget. Note that due to the nature of the measurement process, it is possible to measure link attenuation that is below the true value; enough link budget margin must be provided to allow for such an error. The design of reference jumper cables and dependence of the assembly loss on the type of cable have been the subject of much research [22], and ongoing work continues in this area.

7.2.2. OPTICAL POWER PENALTIES

Next, we will consider the assembly loss budget, which is the difference between the transmitter output and receiver input powers, allowing for optical power penalties due to noise sources in the link. The most important of these effects, and the most important fiber characteristic after transmission loss, is dispersion, which refers to the broadening of optical pulses as they propagate along the fiber. As pulses broaden, they tend to interfere with adjacent pulses; this limits the data rate and bandwidth of the optical link. In multimode fibers, there are two dominant kinds of dispersion, modal and chromatic. Modal dispersion refers to the fact that light can propagate along many different paths, or modes, in a multimode fiber; because not all these paths are the same length, different modes will travel at different velocities and cause pulse broadening. The fiber manufacturer will measure and specify modal bandwidth of the fiber in units of MHz/km. This has been characterized according to

$$BW_{modal} = BW_1/L^{\gamma}, \qquad (7.13)$$

where BW_{modal} is the modal bandwidth for a length L of fiber, BW_1 is the manufacturer-specified modal bandwidth of a 1-km section of fiber, and γ is a constant known as the modal bandwidth concatenation length scaling factor. The term γ usually assumes a value between 0.5 and 1, depending on details of the fiber manufacturing and design as well as the operating wavelength; it is conservative to take $\gamma = 1.0$. Modal bandwidth can be increased by mode mixing, which promotes the interchange of energy between modes to average out the effects of modal dispersion. Fiber splices tend to increase the modal bandwidth, although it is conservative to discard this effect when designing a link. There have been many attempts to fabricate fibers with enhanced modal bandwidth, most of which have not yet produced commercially viable products [23, 24].

The other major contribution is chromatic dispersion, BW_{chrom}, which occurs because different wavelengths of light propagate at different velocities in the fiber. For multimode fiber, this is given by an empirical model of the form

$$BW_{chrom} = \frac{L^{\gamma c}}{\sqrt{\lambda_w}(a_0 + a_1 |\lambda_c - \lambda_{eff}|)}, \tag{7.14}$$

where L is the fiber length in km, λ_c is the center wavelength of the source in nm; λ_w is the source full width at half maximum spectral width in nm; γ_c is the chromatic bandwidth length scaling coefficient, a constant; λ_{eff} is the effective wavelength, which combines the effects of the fiber zero dispersion wavelength and spectral loss signature; and the constants a_1 and a_0 are determined by a regression fit of measured data. From Ref. [25], the chromatic bandwidth for 62.5/125 μm fiber is empirically given by

$$BW_{chrom} = \frac{10^4 L^{-0.69}}{\sqrt{\lambda_w}(1.1 + 0.0189 |\lambda_c - 1370|)}. \tag{7.15}$$

For this expression, the center wavelength was 1335 nm and λ_{eff} was chosen midway between λ_c and the water absorption peak at 1390 nm; although λ_{eff} was estimated in this case, the expression still provides a good fit to the data. For 50/125 μm fiber, the expression becomes

$$BW_{chrom} = \frac{10^4 L^{-0.65}}{\sqrt{\lambda_w}(1.01 + 0.0177 |\lambda_c - 1330|)}. \tag{7.16}$$

For this case, λ_c was 1313 nm and the chromatic bandwidth peaked at $\lambda_{eff} = 1330$ nm. Recall that this is only one possible model for fiber bandwidth; refer to Refs [26] and [27] for further details. The total bandwidth capacity of multimode fiber BW_t is obtained by combining the modal and chromatic dispersion contributions, according to

$$\frac{1}{BW_t^2} = \frac{1}{BW_{chrom}^2} + \frac{1}{BW_{modal}^2}. \tag{7.17}$$

Once the total bandwidth is known, the dispersion penalty can be calculated for a given data rate. This is also an empirical process; the type of receiver used can also compensate to some extent for the fiber's bandwidth.

Fig. 7.6 Dispersion penalty vs distance for different fiber bandwidths: (a) 50-μm fiber; (b) 62.5-μm fiber.

One possible expression for the dispersion penalty in dB is

$$P_d = 1.22 \left[\frac{\text{Bit Rate (Mb/s)}}{\text{BW}_t \text{ (MHz)}} \right]^2 . \tag{7.18}$$

For typical telecommunication-grade fiber, the dispersion penalty for a 20-km link is approximately 0.5 dB. Figure 7.6 shows the dispersion penalty as a function of the fiber bandwidth and length.

By definition, single-mode fiber does not suffer modal dispersion. However, chromatic dispersion is an important effect, even given the relatively narrow spectral width of most laser diodes. The dispersion of single-mode fiber corresponds to the first derivative of group velocity τ_g with respect to

Fig. 7.7   Single-mode fiber dispersion as a function of wavelength [5].

wavelength, and is given by

$$D = \frac{d\tau_g}{d\lambda} = \frac{S_0}{4}\left(\lambda_c - \frac{\lambda_0^4}{\lambda_c^3}\right), \tag{7.19}$$

where D is the dispersion in ps/km-nm and λ_c is the laser center wavelength. The fiber is characterized by its zero dispersion wavelength, λ_0, and zero dispersion slope, S_0. Usually, both center wavelength and zero dispersion wavelength are specified over a range of values; it is necessary to consider both upper and lower bounds in order to determine the worst-case dispersion penalty. This can be seen from Fig. 7.7, in which I plot D vs wavelength for some typical values of λ_0 and λ_c; the largest absolute value of D occurs at the extremes of this region. Once the dispersion is determined, the intersymbol interference penalty as a function of link length, L, can be determined to a good approximation from a model proposed by Agrawal *et al.* [28]:

$$P_d = 5\log(1 + 2\pi(BD\Delta\lambda)^2 L^2), \tag{7.20}$$

where B is the bit rate and $\Delta\lambda$ is the RMS spectral width of the source. More detailed models of dispersion penalty have been proposed [29] taking into account the design details of the receiver and equalizer circuits. By maintaining a close match between the operating and zero dispersion wavelengths, this penalty can be kept to a tolerable 0.5–1.0 dB in most cases.

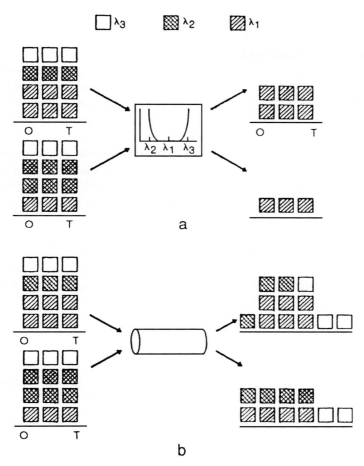

Fig. 7.8 Model for mode partition noise; an optical source emits a combination of wavelengths, illustrated by different color blocks: (a) wavelength-dependent loss; (b) chromatic dispersion [5].

Group velocity dispersion contributes to another optical penalty that remains the subject of continuing research — mode partition noise and mode hopping. This penalty is related to the properties of a Fabry–Perot-type laser diode cavity; although the total optical power output from the laser may remain constant, the optical power distribution among the laser's longitudinal modes will fluctuate. This is illustrated by the model depicted in Fig. 7.8: When a laser diode is directly modulated with injection current, the total output power stays constant from pulse to pulse; however, the power distribution among several longitudinal modes will vary between pulses. We must be careful to distinguish the behavior of the instantaneous laser spectrum, which varies with time, from that of the time-averaged spectrum, which is normally observed experimentally. The light propagates through a fiber with wavelength-dependent dispersion or attenuation, which deforms the pulse shape. Each mode is delayed by a different amount due to group velocity dispersion in the fiber; this leads to additional signal degradation at the receiver in addition to the intersymbol interference caused by chromatic dispersion alone. This is known as mode partition noise; it is capable of generating bit error rate floors, such that additional optical power into the receiver will not improve the link BER. This is because mode partition noise is a function of the laser spectral fluctuations and wavelength-dependent dispersion of the fiber, so the signal-to-noise ratio due to this effect is independent of the signal power. For this reason, mode partition noise has been closely studied by many authors; I will not attempt a complete treatment here, but rather summarize some of the recent models of this effect. The power penalty due to mode partition noise was first calculated by Ogawa [30] as

$$P_{mp} = 5 \log \left(1 - Q^2 \sigma_{mp}^2 \right), \tag{7.21}$$

where

$$\sigma_{mp}^2 = \frac{1}{2} k^2 (\pi B)^4 \left[A_1^4 \Delta \lambda^4 + 42 A_1^2 A_2^2 \Delta \lambda^6 + 48 A_2^4 \Delta \lambda^8 \right] \tag{7.22a}$$

$$A_1 = DL \tag{7.22b}$$

$$A_2 = \frac{A_1}{2(\lambda_c - \lambda_0)}. \tag{7.22c}$$

The mode partition coefficient, k, is a number between 0 and 1 that describes how much of the optical power is randomly shared between modes; it summarizes the statistical nature of mode partition noise. According to Ogawa [30], k depends on the number of interacting modes and root mean square (rms) spectral width of the source, the exact dependence being complex. However, subsequent work has shown that Ogawa's model tends to underestimate the power penalty due to mode partition noise because it does not consider the variation of longitudinal mode power between successive baud periods and because it assumes a linear model of chromatic dispersion rather than the nonlinear model given in the previous equation. A more detailed model has been proposed by Campbell [31], which is general enough to include effects of the laser diode spectrum, pulse shaping, transmitter extinction ratio, and statistics of the data stream. Whereas Ogawa's model assumed an equiprobable distribution of zeros and ones in the data stream, Campbell showed that mode partition noise is data dependent as well. Recent work based on this model [25] has rederived the signal variance:

$$\sigma_{\text{mp}}^2 = E_{\text{av}}\left(\sigma_0^2 + \sigma_{+1}^2 + \sigma_{-1}^2\right), \tag{7.23}$$

where the mode partition noise contributed by adjacent baud periods is defined by

$$\sigma_{+1}^2 + \sigma_{-1}^2 = \frac{1}{2}k^2(\pi B)^4\left[1.25A_1^4\Delta\lambda^4 + 40.95A_1^2A_2^2\Delta\lambda^6 + 50.25A_2^4\Delta\lambda^8\right], \tag{7.24}$$

and the time-average extinction ratio $E_{\text{av}} = 10 \log (P_1/P_0)$, where P_1 and P_0 represent the optical power by a "1" and "0," respectively. If the operating wavelength is far away from the zero dispersion wavelength, the noise variance simplifies to

$$\sigma_{\text{mp}}^2 = 2.25\frac{k^2}{2}E_{\text{av}}\left(1 - e^{-\beta L^2}\right)^2, \tag{7.25}$$

which is valid provided that,

$$\beta = (\pi BD\,\Delta\lambda)^2 \ll 1. \tag{7.26}$$

Mode partition effects deserve careful consideration because they can often limit the performance of a link or generate BER floors. However, many diode lasers have been observed to exhibit mode hopping or mode splitting in which the spectrum appears to split optical power between two or three modes for brief periods of time. The exact mechanism is not fully understood, but stable Gaussian spectra are generally only observed for CW operation and temperature-stabilized lasers [32]. During these mode hops the previous theory does not apply because the spectrum is non-Gaussian, and the model will overpredict the power penalty; hence, it is not possible to model mode hops as mode partitioning with $k = 1$. There is no currently published model describing a treatment of mode hopping noise, although several recent papers [33, 34] suggest approximate calculations based on the statistical properties of the laser cavity. In a practical link, some amount of mode hopping is probably unavoidable as a contributor to burst noise;

empirical testing of link hardware remains the only way to reduce this effect.

In the previous discussion, I mentioned the effect of receiver extinction ratio on mode partition noise. There is a much more direct effect of this parameter as well. The receiver BER is a function of the modulated AC signal power; if the laser transmitter has a small extinction ratio, the DC component of total optical power is significant. Gain or loss can be introduced in the link budget if the extinction ratio at which the receiver sensitivity is measured differs from the worst-case transmitter extinction ratio. If the extinction ratio, E_1, at the transmitter is defined as the ratio of optical power when a one is transmitted vs when a zero is transmitted,

$$E_t = \frac{\text{Power }(1)}{\text{Power }(0)}. \tag{7.27}$$

Then we can define a modulation index at the transmitter M_1 according to

$$M_t = \frac{E_t - 1}{E_t + 1}. \tag{7.28}$$

Similarly, we can measure the linear extinction ratio at the optical receiver input and define a modulation index M_r. The extinction ratio penalty is given by

$$P_{er} = -10 \log\left(\frac{M_t}{M_r}\right), \tag{7.29}$$

where the subscripts t and r refer to specifications for the transmitter and receiver, respectively. By careful choice of the extinction ratios, an effective gain of up to 1 dB or more can be realized in the link budget.

Furthermore, extinction ratio is used to calculate the optical modulation amplitude (OMA), which is sometimes specified in place of receiver sensitivity (for example, in the ANSI Fibre Channel Standard recent revisions). OMA is defined as the difference in optical power between logic levels of one and zero; in terms of average optical power (in microwatts) and extinction ratio, it is given by

$$\text{OMA} = 2P \text{ av }((E - 1)/(E + 1)) \tag{7.30}$$

where extinction ratio in this case is the absolute (unitless linear) ratio of average optical power (in microwatts) between a logic level one and zero, measured under fully modulated conditions in the presence of worst-case reflections. In the Fibre Channel Standard, for example, the OMA specified at 1.0625 Gbit/s for short-wavelength (850 nm) laser sources is between 31 and 2000 microwatts (peak-to-peak), which is equivalent to an average power of -17 dBm and an extinction ratio of 9 dB. Similarly, the OMA specified at 2.125 Gbit/s for short-wavelength (850 nm) laser sources is between 49 and 2000 microwatts (peak-to-peak), which is equivalent to an average power of -15 dBm and an extinction ratio of 9 dB.

Another important property of the optical link is the amount of reflected light from the fiber end faces that returns up the link back into the transmitter. Whenever there is a connection or splice in the link, some fraction of the light is reflected back; each connection is thus a potential noise generator because the reflected fields can interfere with one another to create noise in the detected optical signal. The phenomenon is analogous to the noise caused by multiple atmospheric reflections of radio waves and is known as multipath interference noise. To limit this noise, connectors and splices are specified with a minimum return loss. If there are a total of N reflection points in a link and the geometric mean of the connector reflections is α, then based on the model of Ref. [35] the power penalty due to multipath interference (adjusted for bit error rate and bandwidth) is closely approximated by

$$P_{mpi} = 10 \log(1 - 0.7N\alpha). \qquad (7.31)$$

Multipath noise can usually be reduced well below 0.5 dB with available connectors, whose return loss is often better than 25 dB. A far more serious effect occurs when stray light is reflected back into a Fabry–Perot-type laser diode, giving rise to intensity fluctuations in the laser output. This is a complicated phenomena, strongly dependent on the type of laser; it is called either reflection-induced intensity noise or RIN. This effect is very important because it can also generate BER floors. The power penalty due to RIN is the subject of ongoing research; because the reflected light is measured at a specified signal level, RIN is data dependent, although it is independent of link length. Because many laser diodes are packaged in windowed containers, it is difficult to correlate the RIN measurements on an unpackaged laser with those of a commercial product. There have been several detailed attempts to characterize RIN [36–38]; typically, the RIN noise is assumed Gaussian in amplitude and uniform in frequency over the receiver bandwidth of interest. The RIN value is specified for a given laser by measuring changes in the optical power when a controlled amount of light is fed back into the laser; it is signal dependent and is also influenced by temperature, bias voltage, laser structure, and other factors that typically influence laser output power [38]. If we assume that the effect of RIN is to produce an equivalent noise current at the receiver, then the additional receiver noise, σ_r, may be modeled as

$$\sigma_r = \gamma^2 S^{2g} B, \tag{7.32}$$

where S is the signal level during a bit period, B is the bit rate, and g is a noise exponent that defines the amount of signal-dependent noise. If $g = 0$, noise power is independent of the signal, whereas for $g = 1$ noise power is proportional to the square of the signal strength. The coefficient γ is given by

$$\gamma^2 = S_i^{2(1-g)} 10^{(RIN_i/10)}, \tag{7.33}$$

where RIN_i is the measured RIN value at the average signal level S_i, measured for a given application including worst-case backreflection conditions and operating temperaures. Clearly, the RIN model contains many parameters that are application dependent; if these can be determined, the Gaussian BER probability due to the additional RIN noise current is given by

$$P_{error} = \frac{1}{2}\left[P_e^1\left(\frac{S_1 - S_0}{2\sigma_1} \right) + P_e^0\left(\frac{S_2 - S_0}{2\sigma_0} \right) \right], \tag{7.34}$$

where σ_1 and σ_0 represent the total noise current during transmission of a digital 1 and 0, respectively, and P_e^1 and P_e^0 are the probabilities of error during transmission of a 1 or 0, respectively. The power penalty due to RIN may then be calculated by determining the additional signal power required to achieve the same BER with RIN noise present as without the RIN contribution. One approximation from Ref. [38] for the RIN power penalty is given by

$$P_{rin} = -5\log\left[1 - Q^2(BW)(1 + M_r)^{2g}(10^{RIN/10})\left(\frac{1}{M_r} \right)^2 \right], \tag{7.35}$$

where the RIN value is specified in dB/Hz, BW is the receiver bandwidth, M_r is the receiver modulation index, and the exponent g is a constant varying between 0 and 1 that relates the magnitude of RIN noise to the optical power level. Because g may vary depending on factors such as the bias point and extinction ratio, this expression provides only an approximation of the true penalty.

Another important area in link design deals with the effects of timing jitter on the optical signal. In a typical optical link, a clock is extracted from the incoming data signal that is used to retime and reshape the received digital pulse; the received pulse is then compared with a threshold to determine if a digital 1 or 0 was transmitted. So far, I have discussed BER testing with the implicit assumption that the measurement was made in the center of the received data bit; to achieve this, a clock transition at the center of the bit is required. When the clock is generated from a receiver timing recovery circuit, it will have some variation in time and the exact location of the clock edge will be uncertain. Even if the clock is positioned at the center of the bit, its position may drift over time. There will be a region of the bit interval, or eye, in the time domain at which the BER is acceptable; this region is defined as the eyewidth [39]. Eyewidth measurements are an important parameter for evaluation of fiber optic links; they are intimately related to the BER, as well as to the acceptable clock drift, pulse width distortion, and optical power. At low optical power levels, the receiver signal-to-noise ratio is reduced; increased noise causes amplitude variations in the received signal. These amplitude variations are translated into time domain variations in the receiver decision circuitry, which narrows the eyewidth. At the other extreme, an optical receiver may become saturated at high optical power, reducing the eyewidth and making the system more sensitive to timing jitter. This behavior results in the typical "bathtub" curve shown in Fig. 7.2; for this measurement, the clock is delayed from one end of the bit cell to the other, with the BER calculated at each position. Near the ends of the cell, a large number of errors occur; toward the center of the cell, the BER decreases to its true value. The eye opening may be defined as the portion of the eye for which the BER remains constant; pulse width distortion occurs near the edges of the eye, which denotes the limits of the valid clock timing. Uncertainty in the data pulse arrival times causes errors to occur by closing the eye window and causing the eye pattern to be sampled away from the center. This is one of the fundamental problems of optical and digital signal processing, and a large body of work has been done in this area [40–42]. For purposes of this chapter, I will provide only a brief outline of the problem; this remains an active area of research, and the reader is encouraged to follow the recent literature on this subject. Some international standards bodies have adopted conventions for the treatment of jitter in data links and often specify the maximum allowed jitter for a given application. For example, the ANSI Fibre Channel Standard contains recommendations on the maximum allowable jitter per bit time. In general, multiple jitter sources will be present in a link; these will tend to be uncorrected. However, jitter on digital signals, especially resulting from a cascade of repeaters, may be coherent. It is often difficult to determine how to assess jitter penalties against the individual components of a link.

The International Telecommunications Standards Body (CCITT) has adopted a standard definition of jitter [43] as short-term variations of the significant instants (rising or falling edges) of a digital signal from their ideal position in time. Longer-term variations are described as wander; in terms of frequency, the distinction between jitter and wander is somewhat unclear. The predominant sources of jitter include the following:

Phase noise in receiver clock recovery circuits, particularly crystal-controlled oscillator circuits; this may be aggravated by filters or other components that do not have a linear phase response. Noise in digital logic resulting from restricted rise and fall times may also contribute to jitter.

Imperfect timing recovery in digital regenerative repeaters, which is usually dependent on the data pattern.

Different data patterns may contribute to jitter when the clock recovery circuit of a repeater attempts to recover the receive clock from inbound data. Data pattern sensitivity can produce as much as .5 dB penalty in receiver sensitivity. Higher data rates are more susceptible (>1 Gb/s); data patterns with long run lengths of 1s or 0s, or with abrupt phase transitions between consecutive blocks of 1s and 0s, tend to produce worst-case jitter.

At low optical power levels, the receiver signal-to-noise ratio, Q, is reduced; increased noise causes amplitude variations in the signal, which may be translated into time domain variations by the receiver circuitry.

Low-frequency jitter, also called wander, results from instabilities in clock sources and modulation of transmitters.

Very low-frequency jitter caused by variations in the propagation delay of fibers, connectors, etc., typically results from small temperature variations (this can make it especially difficult to perform long-term jitter measurements).

Note that timing jitter may also indirectly contribute to excursions in the high and low data levels at the receiver; sometimes this is discussed in terms of chatter, which means any transient response greater than about 0.5 V in going between the signal detect negative level to its assert level, and of any duration that can be sensed by the channel logic. Many signal detect circuits are designed with a degree of hysteresis included to minimize chatter on the output signal lines, although in principle hysteresis alone does not guarantee chatter-free operations.

In general, jitter from each of these sources will be uncorrelated; jitter related to modulation components of the digital signal may be coherent, and cumulative jitter from a series of repeaters or regenerators may also contain some well-correlated components.

There are several parameters of interest in characterizing jitter performance. Jitter may be classified as either random or deterministic, depending on whether it is associated with pattern-dependent effects; these are distinct from the duty cycle distortion that often accompanies imperfect signal timing. Each component of the optical link (data source, serializer, transmitter, encoder, fiber, receiver, retiming/clock recovery/deserialization, and decision circuit) will contribute some fraction of the total system jitter. If we consider the link to be a "black box" (but not necessarily a linear system) then we can measure the level of output jitter in the absence of input jitter; this is known as the "intrinsic jitter" of the link. The relative importance of jitter from different sources may be evaluated by measuring the spectral density of the jitter. Another approach is the maximum tolerable input jitter for the link. Finally, because jitter is essentially a stochastic process, we may attempt to characterize the jitter transfer function of the link or estimate the probability density function of the jitter. When multiple traces occur at the edges of the eye, this can indicate the presence of data-dependent jitter or duty cycle distortion; a histogram of the edge location will show several distinct peaks. This type of jitter can indicate a design flaw in the transmitter or receiver. By contrast, random jitter typically has a more Gaussian profile and is present to some degree in all data links. All these approaches have their advantages and drawbacks; first, some simple attempts to model jitter in a real link will be considered.

One of the first attempts to model the optical power penalty due to jitter [44] considered the general case of a receiver whose input was a raised cos signal of the form

$$S_t = \frac{1}{2}(1 + \cos(\pi B t)), \tag{7.36}$$

where B is the bit rate. A decision circuit samples this signal at some interval $t = n/B$. In the presence of random timing jitter, the sampling point fluctuates and the jitter-induced noise depends on the probability density function (PDF) of the random timing fluctuations. Determination of the actual PDF is quite difficult; if we can approximate $Bt \ll 1$, then it has been derived [45] that for a uniform PDF, the jitter-induced noise, σ_j, is given by

$$\sigma_j^2 = \frac{4}{5}(\pi Bt/4)^4. \tag{7.37}$$

For the less conservative case of a Guassian PDF in the same limit,

$$\sigma_j^2 = 2(\pi Bt/4)^4. \tag{7.38}$$

The optical power penalty in dB due to jitter noise is then given by

$$P_j = -5\log\left(1 - 4Q^2\sigma_j^2\right), \tag{7.39}$$

where Q is the Gaussian error function. Based on this expression, the penalty for a Guassian system is much larger than that for a uniform PDF; when $Bt = 0.35$, a BER floor appears at le-9. However, it was subsequently shown [46] that the approximation on Bt is very restrictive, and the actual PDF is far from Gaussian; indeed, it is given by

$$P_{\text{error}} = \int_{-1/B}^{1/B} (\text{PDF}_j)Q\left(\frac{A\cos(\pi Bt)}{2\sigma_j}\right)dt, \tag{7.40}$$

where the probability density function of the jitter is included under the integral. Numerical integration of this PDF shows that the approximate results given previously tend to underestimate the effects of Gaussian jitter and overestimate the effects of uniform jitter by more than 2 dB! This points out one of the basic problems in attempting to model a power penalty due to jitter effects — namely, the limitations of the assumptions that must be made to reduce the expression to closed form vs the real-world jitter behavior. Using similar approximations, jitter power penalties have been derived for both positive–intrinsic–negative and avalanche photodiode receivers [47], although the limiting assumptions make it difficult to generalize these models. An alternative modeling approach has been to derive a worst-case distribution for the PDF, which will provide an upper bound on the performance of the optical link and can be more easily evaluated for analytical purposes [48].

The problem of jitter accumulation in a chain of repeaters becomes increasingly complex; however, we can state some general rules of thumb. It has been shown [49] that jitter can be generally divided into two components — one due to repetitive patterns and one due to random data. In receivers with phase-lock loop timing recovery circuits, repetitive data patterns will tend to cause jitter accumulation, especially for long run lengths. This effect is commonly modeled as a second-order receiver transfer function [50]; the jitter accumulates according to the relationship

$$\text{Jitter} \propto N + \left(\frac{N}{\zeta}L\right)^2, \qquad (7.41)$$

where N is the number of identical repeaters and ζ is the loop damping factor, specific to a given receiver circuit. For large ζ, jitter accumulates almost linearly with the number of repeaters, whereas for small ζ the accumulation is much more rapid. Jitter will also accumulate when the link is transferring random data; jitter due to random data is of two types — systematic and random. The classic model for systematic jitter accumulation in cascaded repeaters was published by Byrne *et al.* [51]. The Byrne model assumes cascaded identical timing recovery circuits; the general expression for jitter power spectrum is given by

$$J_s^N = J_s \left| \prod_{k=2}^{N} H_k(f) + \prod_{k=1}^{N} H_k(f) + \cdots + H_k(f) \right|^2, \qquad (7.42)$$

where J_s is the systematic jitter generated by each repeater and $H(f)$ is the jitter transfer function of each timing recovery circuit. The corresponding model for random jitter accumulation is

$$J_r^N = J_r \left[\prod_{K=1}^{N} |H_k(f)|^2 + \prod_{K=2}^{N} |H_k(f)|^2 + \cdots + |H_k(f)|^2 \right], \qquad (7.43)$$

where J_r is the random jitter generated by each repeater. The systematic and random jitter can be combined as rms quantities;

$$J_t^2 = \left(J_s^N\right)^2 + \left(J_r^N\right)^2, \qquad (7.44)$$

so that total jitter due to random jitter may be obtained. This model has been generalized to networks consisting of different components [52] and to nonidentical repeaters [53]. Despite these considerations, for well-designed practical networks the basic results of the Byrne model remain valid for N nominally identical repeaters transmitting random data:

Systematic jitter accumulates in proportion to $N^{1/2}$.

Random jitter accumulates in proportion to $N^{1/4}$.

Details of jitter accumulation modeling are provided in Refs. [43–53].

Except for those penalties that produce BER floors such as mode partitioning and RIN, most penalties can be reduced by increasing the transmitted or received optical power. This brute-force approach is subject to limitations such as maintaining class 1 laser safety; even if it were possible to increase optical power significantly (as I will discuss later using optical amplifiers), a new class of nonlinear optical penalties becomes important at high power levels. I will briefly discuss some of these here; the reader is referred to Ref. [54] for a more complete treatment of nonlinear phenomena. Class 1 laser systems typically do not experience these nonlinear effects, although they may be important for optical fiber amplifiers or systems using open fiber control (OFC), for which there may be significantly higher power levels present in the fiber.

At higher optical power levels, nonlinear scattering may limit the behavior of a fiber optic link. The dominant effects are stimulated Raman and Brillouin scattering. When incident optical power exceeds a threshold value, significant amounts of light may be scattered from small imperfections in the fiber core or by mechanical (acoustic) vibrations in the transmission media. These vibrations can be caused by the high-intensity electromagnetic fields of light concentrated in the core of a single-mode fiber. Because the scattering process also involves the generation of photons, the scattered light can be frequency shifted [55]. Put another way, we can think of the high-intensity light as generating a regular pattern of very slight differences in the fiber refractive index; this creates a moving diffraction grating in the fiber core, and the scattered light from this grating is Doppler shifted in frequency by about 11 GHz. This effect is known as stimulated Brillouin scattering (SBS); under these conditions, the output light intensity becomes nonlinear as well. Stimulated Brillouin scattering will not occur below the optical power threshold defined by

$$P_c = 21A/G_nL_c \text{ watts,} \tag{7.45}$$

where L_c is the effective interaction length, A is the cross-sectional area of the guided mode, and G_n is the Brillouin gain coefficient [55]. Brillouin scattering has been observed in single-mode fibers at wavelengths greater than cutoff with optical power as low as 5 mW; it can be a serious problem in long distance communication systems when the span between amplifiers is low and the bit rate is less than about 2 Gbit/s, in WDM systems up to about 10 Gbit/s when the spectral width of the signal is very narrow, or in remote pumping of some types of optical amplifiers. In general, SBS is worse for narrow laser linewidths (and is generally not a problem for channel bandwidth greater than 100 MHz), wavelength (SBS is worse near 1550 nm than near 1300 nm), and signal power per unit area in the fiber core. All of these factors are summarized in the above expression. SBS can be a concern in long distance communication systems, when the span between amplifiers is large and the bit rate is below about 2.5 Gbit/s, in WDM systems where the spectral width of the signal is very narrow, and in remote pumping of optical amplifiers using narrow linewidth sources. In cases where SBS could be a problem, the source linewidth can be intentionally broadened by using an external modulator or additional RF modulation on the laser injection current. However, this is a tradeoff against long distance transmission, as broadening the linewidth also increases the effects of chromatic dispersion.

When the scattered light experiences frequency shifts outside the acoustic phonon range, due instead to modulation by impurities or molecular vibrations in the fiber core, the effect is known as stimulated Raman scattering (SRS). The mechanism is similar to SBS, and scattered light can occur in both forward and backward directions along the fiber. The threshold below which Raman scattering will not occur is given by

$$P_t = 16A/G_r L_c \text{ watts,} \tag{7.46}$$

where G_r is the Raman gain coefficient [54]. Note that SRS is also influenced by fiber dispersion, and standard fiber reduces the effect of SRS by half (3 dB) compared with dispersion shifted fiber. As a rule of thumb, the optical power threshold for Raman scattering is about three times larger than for Brillouin scattering. Another good rule of thumb is that SRS can be kept to acceptable levels if the product of total power and total optical bandwidth is less than 500 GHz-W. This is quite a lot; for example, consider a 10-channel DWDM system with standard wavelength spacing of 1.6 nm (200 GHz). The bandwidth becomes $200 \times 10 = 2000$ GHz, so the total power in all 10 channels would be limited to 250 mW in this case (in most DWDM systems, each channel will be well below 10 mW for other reasons such as laser safety considerations). In single-mode fiber, typical thresholds for Brillouin scattering are about 10 mW and for Raman scattering about 35 mW; these effects rarely occur in multimode fiber, where the thresholds are about 150 mW and 450 mW, respectivly. In general, the effect of SRS becomes greater as the signals are moved further apart in wavelength (within some limits); this introduces a tradeoff with FWM, which is reduced as the signal spacing increases. Optical amplifiers can be constructed using the principle of SRS; if a pump signal with relatively high power (half a watt or more) and a frequency 13.2 THz higher than the signal frequency is coupled into a sufficiently long length of fiber (about 1 km), then amplification of the signal will occur. Unfortunately, more efficient amplifiers require that the signal and pump wavelengths be spaced by almost exactly the Raman shift of 13.2 THz, otherwise the amplification effect is greatly reduced. It is not possible to build high-power lasers at arbitrary signal wavelengths; one possible solution is to build a pump laser at a convenient wavelength, then wavelength shift the signal by the desired amount. However, another good alternative to SRS amplifiers are the widely used erbium-doped fiber amplifiers (EDFA). These allow the amplification of optical signals along their direction of travel in a fiber, without the need to convert back and forth from the electrical domain. While there are other types of optical amplifiers based on other rare earth elements such as praseodymium (Pd) or neodymium (Nd), and even some optical amplifiers based on semiconductor devices, the erbium-doped amplifiers are the most widely used because of their maturity and good performance at wavelengths of interest near 1550 nm. Note that there is still a good deal of research going on in this area, most recently including thulium-doped fiber amplifiers (TDFAs) based on flurozirconiate (ZBLAN) fiber for use at wavelengths between 1450 and 1510 nm (the S-band).

The final nonlinear effect we will consider is frequency chirping of the optical signal. Chirping refers to a change in frequency with time and takes its name from the sound of an acoustic signal whose frequency increases or decreases linearly with time. There are three ways in which chirping can affect a fiber optic link. First, the laser transmitter can be chirped as a result of physical processes within the laser [56]; the effect has its origin in carrier-induced refractive index changes, making it an inevitable consequence of high-power direct modulation of semiconductor lasers. For lasers with low levels of relaxation oscillation damping, a model has been proposed for the chirped power penalty [56]:

$$P = 10 \log \left(1 + \frac{\pi B^2 \lambda^2 L D a}{4c} \right), \tag{7.47}$$

where c is the speed of light, B is the fiber bandwidth, λ is the wavelength of light, L is the length of the fiber, D is the dispersion, and a is the linewidth enhancement factor (a typical value is -4.5); this model is only a first approximation because it neglects the dependence of chirp on extinction ratio and nonlinear effects such as spectral hole burning [57]. Second, a sufficiently intense light pulse will be chirped by the nonlinear process of self-phase modulation in an optical fiber [58]. The effect arises from the interaction of the light and the intensity-dependent portion of the fiber's refractive index; it is thus dependent on the material and structure of the fiber, polarization of the light, and the shape of the incident optical pulse. Based on a model from Ref. [58], the maximum optical power level before the spectral width increases by 2 nm is given by

$$P = \frac{n^2 A}{377 n_2 \kappa L_e} \text{ Watts,} \qquad (7.48a)$$

where

$$L_e = (1/a_0)(1 - \exp(-a_0 L)), \qquad (7.48b)$$

where n is the fiber's refractive index, κ is the propagation constant (a typical value is 7×10^4 at a wavelength of 1.3 μm) a_0 is the fiber attenuation coefficient, A is the fiber core cross-sectional area, n_2 is the nonlinear coefficient of the fiber's refractive index (a typical value is 6.1×10^{-19}), and L_e is the effective interaction length for the nonlinear interaction, which is related to the actual length of the fiber, L, by Eq. (7.48b). This expression should be multipled by $\frac{5}{6}$ if the fiber is not polarization preserving. Typically, this effect is not significant for optical power levels less than 950 mW in a single-mode fiber at 1.3-μm wavelength. Finally, there is a power penalty arising from the propagation of a chirped optical pulse in a dispersive fiber because the new frequency components propagate at different group velocities. This may be treated as simply a much worse case of the conventional dispersion penalty [59], provided that one of the first two effects exists to chirp the optical signal.

7.2.3. *LINK BUDGET EXAMPLE*

To illustrate some of the previous principles, the following example discusses the design of a data communication link using single-mode optical fiber operating at 1300 nm. The application is remote backup of data from a hard disk to a tape drive located in a building 10 km away; because of systemwide fault tolerance, the link has a target BER of 10e-12. The link adheres to an industry standard data rate of 531 Mb/s so that the equipment will interoperate. The transmitter is guaranteed to have a worst-case output of −9.0 dBm, accounting for temperature variation and end of life; the worst-case receiver sensitivity is −26.0 dBm. Optical fiber is chosen with parameters shown in Table 7.6, whereas Table 7.7 gives a complete specification for the transmitter and receiver. Note that the link center wavelength

Table 7.7   **Transmitter, Receiver, and Optical Fiber Specifications
Used in the Illustrated Example [5]**

Transmitter	
Power output	−3 to −9 dBm
Extinction ratio	6 dB
RIN	−112 dB/Hz
Center wavelength	1270–1340 nm
Spectral width	6 nm
Receiver	
Sensitivity	−27 dB
Saturation	−3 dBm
Fiber	
Zero dispersion wavelength	1310 nom. 1295–1322 (range) nm
Zero dispersion slope	0.095 ps/nm-km
Cutoff wavelength	280 nm
Attenuation	0.5 at 1310 nm dB/km
Max. attenuation delta (1270–1340)	0.06 dB/km

is sufficiently close to the zero dispersion wavelength of the fiber that the dispersion penalty is kept to 0.2 dB worst case. The transmitter is class 1, so we can neglect nonlinear effects in the link; connector return loss is held low enough so that multipath effects are not a concern either. Using the manufacturer's specification for RIN and assuming $g = 0.5$, we can estimate the RIN power penalty of 0.7 dB, which should be tolerable. By clever choice of the extinction ratios, we realize an approximately 0.5-dB gain in the link to recover part of the RIN penalty. The wavelength-dependent attenuation of the fiber was selected so that the penalty would be tolerable at the required 10-km distance. For this distance, we have chosen a single-mode link, so modal noise is not a concern provided we use sufficiently low loss splices and connectors to minimize modal noise. Although partition noise should be verified experimentally, we estimate that the power penalty is less than 1.5 dB (note that $\beta < 1$, so the approximate model is valid). This gives us a total available power of

$$P_{\text{avail}} = Tx(\text{output}) - Rx(\text{sens}) - \text{Penalties} = 14 \text{ dB}. \quad (7.49)$$

Next, we compute the installation loss. The link contains one distribution panel near the source end and uses previously installed connectors as shown in Table 7.6. We will consider a link with four connectors (mean loss = 0.35 dB; variance = 0.06 dB), two fusion splices (mean loss = 0.15 dB; variance = 0.01 dB), and a 20-km link with 240 m of jumper cable (loss = 0.8 dB/km for the jumpers and 0.5 dB/km for the trunk). If we perform a statistical calculation on the installed budget for connectors, splices, and cable transmission loss, the mean installed loss is 11.75 dB with a variance of 0.26 dB. If we use the industry-standard 3σ design, then the calculated installation loss is 13.5 dB. This allows us to operate the link with 0.5 dB to spare, the difference between the available power and installation loss budgets. A conservative design should always allow some link safety margin, as we have done, to allow for the many assumptions used in deriving the various link performance models we have used. Of course, there are other possible ways to compute the link budget; if we had obtained the statistics for all the link parameters, we could perform a Monte Carlo simulation for the entire link that might yield a different link budget margin. It should be apparent that link design requires not only the latest valid performance models but also a good amount of engineering judgment in how best to apply these tools. Also, real-world problems are often not as clear-cut as the example I have just provided; a link designer must be able to manage the trade-offs involved in frequently working with incomplete information (Table 7.8). Still, this example should illustrate some of the typical steps required to perform link planning and installation.

A summary of the important optical link power penalties is given in Table 7.9; more detailed descriptions of link budget modeling, including equations for all the link noise parameters and examples of a complete link budget calculation, are available in the literature [5, 25, 26].

Table 7.8   **Link Budget Loss Sources**

Sources	Typical Loss
Fiber attenuation	0.5 dB/km
Connector loss	0.25–0.50 dB
Splice loss	0.15–0.40 dB
Transmitter/receiver end-of-life degradation	0.5–1.0 dB

Table 7.9   **Typical Link Power Penalties**

Penalty	Typical at 20 Km, 200 Mb/s (dB)
Chromatic dispersion	1.2
Mode partition	1.5
RIN	0.8
Multipath	0.25

Table 7.10   **Typical Link Distances and Data Rates**

ESCON, multimode	200 Mb/s	2 km, 50-μm fiber, 8 dB 3 km, 62.5-μm fiber, 8 dB
ESCON, single mode		20 km, single-mode fiber, 14 dB
ATM, multimode	155 Mb/s	3 km, 62.5-μm multimode fiber, 7 dB 20 km, single-mode fiber, 15 dB
FDDI (multimode only)	125 Mb/s	2 km, 62.5-μm multimode fiber, 11 dB
FCS, multimode	266 Mb/s	2 km, 50-μm fiber 1 km, 62.5-μm fiber
	1 Gb/s	0.5 km, 50-μm fiber
FCS, single mode	266 Mb/s	10 km
	1 Gb/s	10 km

7.3.   Link Planning Considerations

There are many factors to consider when designing fiber optic data links. The maximum distance for a fiber optic data link is determined by the link budget calculations described in the previous section; it is also affected by established industry standards for the physical layer of common datacom protocols. Generally, there is a trade-off between link distance and data rate; some of the typical link distances are summarized in Table 7.10. In general, multimode fiber is limited to a distance of a few kilometers, whereas single-mode fiber enables distances of 10 km or more. In some cases, the difference in bandwidth between 50- and 62.5-μm fiber will result in a distance limitation; this depends on the application requirements and may not appear for some datacom standards. The SONET physical layer is assumed for all asynchronous transfer mode (ATM) protocols listed in

Table 7.10. A complete list of the datacom speeds allowed for each standard is provided in Chapters 11–14.

7.3.1. *LASER SAFETY REQUIREMENTS*

Laser safety is an important criteria for many data communication products because they are accessible to customers who may have little or no laser training. This limits the amount of optical power that can safely be launched into the link and may impact distance and BER performance. This is a very important area; infrared sources of less than 1 mW can cause serious damage to the unprotected eye. Typically, it is required that all datacom fiber optic products be class 1, or inherently safe; there is no access to unsafe light levels during normal operation or maintenance of the product or during accidental viewing of the optical source. This includes viewing of the optical fiber end face while the transmitter is in operation. In the United States, the Department of Health and Human Services of the Occupational Safety and Health Administration and the Food and Drug Administration (FDA) define the laser safety standards; a class 1 product is limited to −2.0 dBm output power. International standards defined by the IEC are somewhat different and limit class 1 exposure to −6.0 dBm. Following recent revisions to the laser safety standards, the FDA requirements currently permit lower optical power levels and high power densities than the IEC standards; however, the IEC requirements include the provision that a product must operate as a class 1 device even under a single point of failure. This requires redundancy in the hardware design; for example, although many ATM-type transceivers use a 1 × 9 pinout, some designs are standardizing around a 2 × 9 pinout to implement clock recovery and redundant laser safety. There is no consensus on how to deal with emerging transceiver technology such as vertical cavity surface-emitting laser (VCSEL) arrays; these lasers produce a beam of circular cross section rather than the elliptical beam profile typical of long-wavelength lasers. As a result, most of the optical power is launched into the lower order modes of an optical fiber and tends to remain there even after fairly long distances. Although this improves modal noise performance, it also makes laser safety compliance more difficult. It has not been determined whether VCSELs should be treated as extended sources for purposes of laser safety classification. A summary of the different laser safety classifications and power levels is given in Fig. 7.9 ; for details and current revisions of all standards, contact the

Fig. 7.9 Optical laser safety requirements.

Wavelength λ(nm) \\ Emission Duration t(s)	$<10^{-9}$	10^{-9} to 10^{-7}	10^{-7} to 1.8×10^{-5}	1.8×10^{-5} to 5×10^{-5}	5×10^{-5} to 10	10 to 10^{3}	10^{3} to 10^{4}	10^{4} to 3×10^{4}
200 to 302.5	2.4×10^{4} W	2.4×10^{-3} J (spanning)						
302.5 to 315	2.4×10^{4} W	$7.9\times10^{-7}\,C_1$ J $(t<T_1)$	$7.9\times10^{-7}\,C_1$ J	$7.9\times10^{-7}\,C_1$ J $(t>T_1)$	$7.9\times10^{-7}\,C_1$ J	$7.9\times10^{-7}\,C_1$ J	7.9×10^{-6} W	7.9×10^{-6} W
315 to 400	200 W	2×10^{-7} J	$7.9\times10^{-7}\,C_1$ J	$7.9\times10^{-7}\,C_1$ J	7.9×10^{-3} J	7.9×10^{-3} J	7.9×10^{-2} W	3.9×10^{-2} W
400 to 550	200 W	2×10^{-7} J	$7\times10^{-4}\,t^{0.25}$ J	$7\times10^{-4}\,t^{0.25}$ J	$7\times10^{-4}\,t^{0.75}$ J	3.9×10^{-3} J	3.9×10^{-3} J	3.9×10^{-2} W
and*	10^{11} W·m^{-2}·sr^{-1}	$10^{5}\,t^{0.33}$ J·m^{-2}·sr^{-1}	$10^{5}\,t^{0.33}$ J·m^{-2}·sr^{-1}	$3.9\times10^{4}\,t^{0.75}$ J·m^{-2}·sr^{-1}		2.1×10^{3} J·m^{-2}·sr^{-1}	2.1×10^{3} J·m^{-2}·sr^{-1}	21 W·m^{-2}·sr^{-1}
550 to 700	$200\,C_3$ W	2×10^{-7} J	$7\times10^{-4}\,t^{0.75}\,C_3$ J $(t<T_2)$	$7\times10^{-4}\,t^{0.75}\,C_3$ J	$3.9\times10^{-3}\,C_3$ J $(t>T_2)$	$7\times10^{-4}\,t^{0.75}\,C_3$ J	$3.9\times10^{-3}\,C_3$ J $(t>T_2)$	$3.9\times10^{-7}\,C_3$ W
and*	10^{11} W·m^{-2}·sr^{-1}	$10^{5}\,t^{0.33}$ J·m^{-2}·sr^{-1}	$10^{5}\,t^{0.33}$ J·m^{-2}·sr^{-1}	$3.9\times10^{4}\,t^{0.75}$ J·m^{-2}·sr^{-1} $(t<T_2)$	$2.1\times10^{5}\,C_3$ J·m^{-2}·sr^{-1} $(t>T_2)$		$2.1\times10^{5}\,C_3$ J·m^{-2}·sr^{-1} $(t>T_2)$	$21\,C_3$ W·m^{-2}·sr^{-1}
700 to 1050	$200\,C_4$ W	$2\times10^{-7}\,C_4$ J	$10^{5}\,t^{0.33}\,C_4$ J·m^{-2}·sr^{-1}	$7\times10^{-4}\,t^{0.75}\,C_4$ J	$7\times10^{-4}\,t^{0.75}\,C_4$ J	$7\times10^{-4}\,t^{0.75}\,C_4$ J	$1.2\times10^{-4}\,C_4$ W	$1.2\times10^{-4}\,C_4$ W
and*	$10^{11}\,C_4$ W·m^{-2}·sr^{-1}			$3.9\times10^{4}\,t^{0.75}\,C_4$ J·m^{-2}·sr^{-1}		$6.4\times10^{3}\,C_4$ W·m^{-2}·sr^{-1}	$6.4\times10^{3}\,C_4$ W·m^{-2}·sr^{-1}	$6.4\times10^{3}\,C_4$ W·m^{-2}·sr^{-1}
1050 to 1400	2×10^{3} W	2×10^{-6} J	$5\times10^{5}\,t^{0.33}$ J·m^{-2}·sr^{-1}	$3.5\times10^{-3}\times t^{0.75}$ J	$3.5\times10^{-3}\times t^{0.75}$ J	$1.9\times10^{5}\,t^{0.75}$ J·m^{-2}·sr^{-1}	6×10^{-4} W	6×10^{-4} W
and*	5×10^{11} W·m^{-2}·sr^{-1}	$5\times10^{5}\,t^{0.33}$ J·m^{-2}·sr^{-1}					3.2×10^{4} W·m^{-2}·sr^{-1}	3.2×10^{4} W·m^{-2}·sr^{-1}
1400 to 10^{5}	8×10^{4} W	8×10^{-5} J	8×10^{-5} J	$4.4\times10^{-3}\,t^{0.25}$ J	$4.4\times10^{-3}\,t^{0.25}$ J		8×10^{-4} W	8×10^{-4} W
10^{5} to 10^{6}	10^{7} W	10^{-2} J	10^{-2} J	$0.56\,t^{0.25}$ J	$0.56\,t^{0.25}$ J		0.1 W	0.1 W

* See Item *d)* of Sub-clause 9.3 for Class 1 dual limits requirements.

260

Wavelength λ (nm)	Emission duration t (s)	Class 2 AEL
400 to 700	$t < 0.25$	Same as Class 1 AEL
	$t < 0.25$	10^{-3} W

Fig. 7.9 (*Continued*).

Wavelength λ (nm) / Emission Duration t (s)	$<10^{-9}$	10^{-9} to 10^{-7}	10^{-7} to 1.8×10^{-5}	1.8×10^{-5} to 5×10^{-5}	5×10^{-5} to 0.25	0.25 to 10	10 to 10^{3}	10^{3} to 3×10^{4}
200 to 302.5	1.2×10^{-4} J and 30 J.m^{-2}							
302.5 to 315	1.2×10^{5} W and 3×10^{10} W.m^{-2}	$4\times C_1\times10^{-6}$ J and C_1 J.m^{-2}	$(t>T_1)$ / $(t<T_1)$			$4\times C_2\times10^{-6}$ J and C_2 J.m^{-2}	$4\times C_2\times10^{-6}$ J and C_2 J.m^{-2}	
315 to 400	1000 W and 5×10^{9} W.m^{-2}	10^{-6} J and 5×10^{-3} J.m^{-2}	$4\times C_1\times10^{-6}$ J and C_1 J.m^{-2}				4×10^{-2} J and 10^{4} J.m^{-2}	4×10^{-5} W and 10 W.m^{-2}
400 to 700	1000 W and 5×10^{9} W.m^{-2}	10^{-6} J and 5×10^{-3} J.m^{-2}	10^{-5} J and 5×10^{-3} J.m^{-2}	$3.5\times10^{-3}\times t^{0.75}$ J and $18\,t^{0.75}$ J.m^{-2}			5×10^{-3} W and 25 W.m^{-2} (Aversion responses protect for emission >0.25 s)	
700 to 1050	1000 W $\times C_4$ and $5\times C_4\times10^{8}$ W.m^{-2}	$10^{-6}\times C_4$ J and $5\times C_4\times10^{-3}$ J.m^{-2}			$3.5\times10^{-3}\times C_4\times t^{0.75}$ J and $18\times C_4\times t^{0.75}$ J.m^{-2}			$6\times10^{-4}\times C_4$ W and $3.2\times C_4$ W.m^{-2}
1050 to 1400	10^{4} W and 5×10^{7} W.m^{-2}	10^{-5} J and 5×10^{-2} J.m^{-2}				$1.8\times10^{-2}\times t^{0.75}$ J and $90\times t^{0.75}$ J.m^{-2}		3×10^{-3} W and 16 W.m^{-2}
1400 to 10^{3}	4×10^{5} W and 10^{11} W.m^{-2}	4×10^{-4} J and 100 J.m^{-2}			$2.2\times10^{-2}\times t^{0.25}$ J and $5600\times t^{0.25}$ J.m^{-2}		4×10^{-3} W and 1000 W.m^{-2}	
10^{3} to 10^{6}	5×10^{7} W and 10^{11} W.m^{-2}	5×10^{-2} J and 100 J.m^{-2}			$2.8\times t^{0.25}$ J and $5600\times t^{0.25}$ J.m^{-2}		0.5 W and 1000 W.m^{-2}	

Fig. 7.9 (Continued).

Wavelength λ(nm) / Emission duration t(s)	$<10^{-9}$	10^{-9} to 0.25	0.25 to 3×10^4
200 to 302.5	3.8×10^5 W	3.8×10^{-4} J	1.5×10^{-3} W
302.5 to 315	$1.25 \times 10^4\ C_2$ W	$1.25 \times 10^{-5}\ C_2$ J	$5 \times 10^{-5}\ C_2$ W
315 to 400	1.25×10^8 W	0.125 J	0.5 W
400 to 700	3.14×10^{11} W.m^{-2}	$3.14 \times 10^5\ t^{0.33}$ J.m^{-2} and $<10^5$ J.m^{-2}	0.5 W
700 to 1050	$3.14 \times 10^{11}\ C_4$ W.m^{-2}	$3.14 \times 10^5\ C_4\ t^{0.33}$ J.m^{-2} and $<10^5$ J.m^{-2}	0.5 W
1050 to 1400	1.57×10^{12} W.m^{-2}	$1.57 \times 10^6\ t^{0.33}$ J.m^{-2} and $<10^5$ J.m^{-2}	0.5 W
1400 to 10^6	10^{14} W.m^{-2}	10^5 J.m^{-2}	0.5 W

Fig. 7.9 (*Continued*).

references provided in this figure. Recent revisions in the standards apply to all optical sources, including light-emitting diodes (LEDs), both surface-emitting diodes and the edge-emitting diodes required for higher speed applications. Currently, there is an increasing amount of laser safety standards activity. The base standard, ANSI Z136.1 ("Standard for the safe use of lasers") was first released in 1988 and most recently reissued in 1993 with some revisions to the safe emission levels that made it easier to qualify as a class 1 laser product (Table 7.11). The related standard ANSI Z136.2 ("Standard for the safe use of optical fiber communication systems utilizing laser diode and LED sources") is being updated to reflect the changes to Z136.1 in 1993; this is expected to be available in late 1996. A related document, ANSI Z136.3 ("Standard for the safe use of lasers in health care facilities"), was reissued in April 1996 with some new applications and modest changes to control procedures. In addition, three new standards are being discussed for release sometime in 1997 or 1998: ANSI Z136.4 will cover the measurement of laser power and energy, Z136.5 will cover safety aspects of lasers used in educational institutions, and Z136.6 will cover outdoor use of lasers for applications such as surveying, laser displays, and military systems.

The laser safety standards presented in Table 7.9 undergo regular revisions; as this chapter goes to press, the IEC has pending changes to

Table 7.11   **ANSI Z136.1 (1993) Laser Safety Classifications**

Class 1 — Inherently safe; no viewing hazard during normal use or maintenance; 0.4 μW or less; no controls or label requirements

Class 2 — Human aversion response sufficient to protect eye; low power visible lasers with power less than 1 mW continuous; in pulsed operation, power levels exceed the class 1 acceptable exposure limit for the entire exposure duration but do not exceed the class 1 limit for a 0.25 second exposure; requires caution label

Class 2a — Low power visible lasers that do not exceed class 1 acceptable exposure limit for 1000 s or less (not intended for viewing the beam); requires caution label

Class 3a — Aversion response sufficient to protect eye unless laser is viewed through collecting optics; 1–5 mW; requires labels and enclosure/interlock; warning sign at room entrance

Class 3b — Intrabeam (direct) viewing is a hazard; specular reflections may be a hazard; 5–500 mW continuous; <10 J per square centimeter pulsed operation (less than 0.25 s); same label and safety requirements as class 3a plus power actuated warning light when laser is in operation

Class 4 — Intrabeam (direct) viewing is a hazard; specular and diffuse reflections may be a hazard; skin protection and fire potential may be concerns; >500 mW continuous; >10 J per square centimeter pulsed operation; same label and safety requirements as class 3b plus locked door, door actuated power kill, door actuated filter, door actuated shutter, or equivalent.

Note. Laser training is required in order to work with anything other than class 1 laser products.

their laser safety standards scheduled for approval in 2001. While products classified under the current system will likely be able to continue without reclassification for a number of years, the revisions also include several new laser safety classes. These changes are made in order to keep pace with new technology and new applications, such as ultrashort optical pulses. Under the new regulations, class 1, 2, 3b, and 4 lasers remain unchanged, although some products currently using these classifications may be eligible for reclassification under the new standards. A new class 1M laser is considered safe under reasonably foreseeable conditions if optical instruments are not used for viewing. There is no upper limit on the total output power of class 1M laser products; they are likely to include lasers or LEDs with highly divergent beams or lasers with wide aperture collimated beams. A new class 2M laser operates in the visible range only, and is considered safe if no optical instruments are used for viewing and the blink or aversion response operates. Some current class 3b laser products (capable of causing injury) are expected to qualify for the new class 1M or 2M; in particular, this may affect parallel optical interconnects. Also, shortwave lasers (400 nm–600 nm) will introduce angle of acceptance measurements and will have their maximum permissible exposure (MPE) revised for exposure times exceeding 10 seconds. Finally, a new class 3R laser will be defined as having accessible emissions that exceed the MPE for exposures of 0.25 seconds if they are visible and for 100 seconds if they are invisible. Total output power for a class 3R laser must not exceed the accessible emission limits for class 2 (visible) or class 1 (invisible) by more than a factor of 5.

Class 1 laser certification is granted by testing a product at an independent laboratory. In the United States, products are granted an accession number by the FDA to certify their compliance with class 1 limits. It is a popularly held misconception that the FDA issues "certification" for products such as fiber optic transceivers that contain laser devices. This is not the case. Manufacturers of such equipment provide a report and product description to the FDA and the manufacturer certifies through supporting test results that the product complies with class 1 regulations; the FDA then asigns an accession number in order to track this part and permit import–export as a class 1 laser device. International certification is usually performed by a recognized testing laboratory, such as TUV or VDE in Germany. These labs will assign approval numbers to the products in question based on their independent testing of samples provided by the manufacturer and assessment of any supporting manufacturer data. There may be a fee associated with this evaluation, and there is an annual fee that must be paid in order to keep an IEC registration number active; otherwise, it is withdrawn and the product can no longer claim to be IEC class 1 compliant. International safety standards define the requirements for labels that must be affixed to class 1 optical products and the terminology that must be used in product safety literature. In general, class 1 products comply with ANSI 136.X and FDA/CDRH 21 CFR subchapter J in the United States and with IEC/CEI 825 in the rest of the world. There may be additional local or state labor safety regulations as well; for example, the New York State Department of Labor enforces New York Code Rule 50, which requires all manufacturers to track their primary laser components by a state-issued serial number. Check with local and state authorities for applicable regulations.

In order to receive class 1 certification, a product must be single-fault tolerant; that is, the optical power cannot exceed recommended levels if a single point of failure occurs in the link. There are three common ways in which this can be guaranteed by the manufacturer. First, the optical transceiver can be designed to never emit more than the maximum allowed optical power level; this approach is the simplest and is implemented on most ATM/SONET transceivers. This is only possible due to recent relaxation in the laser safety standards; products manufactured prior to 1994 required alternate solutions. Some products use higher power laser transmitters but never couple more than the maximum power level into the optical fiber; this accounts for coupling loss in the transceiver and does not present a hazard for viewing the fiber end face. The transmitter must have some additional form of protection, such as mechanical shutters that snap into place covering the laser when not in use. This design was used on some single-mode ESCON transceivers prior to 1995. Finally, the ANSI Fibre Channel Standard defines a laser safety method called OFC. This implements an interlock at the transceiver that detects whether light is being received; both ends of a duplex link must detect light and exchange a handshake sequence before the laser transmitter is activated. If the link is opened for any reason, such as a broken fiber or pulled connector, the lasers turn off on both sides of the link. The link then enters a waiting mode, in which a low-power optical pulse is transmitted every 10 s. When the link is closed once again, this pulse is detected by both ends of the link, which exchange handshakes and activate the laser transmitters at full power again. Using this approach, the link can employ much higher powered lasers without violating class 1 certification limits; this can enable longer distances and improved BER performance. Many FCS products, including the 266 Mb/s and 1 Gb/s optical link cards from IBM, implement this form of laser safety.

7.3.2.   OPTICAL CABLE AND CONNECTOR TYPES

Many different types of fiber optic cables are available, as discussed in previous chapters. Selection of the proper fiber type is an important link design requirement. There are three classes of fiber cable, defined by their fire code rating for use within building environments; riser, plenum, and low halogen [there is actually a fourth, data processing (DP) center rated, as defined by the National Electrical Code, which I will discuss shortly]. Riser-rated cable (UL 1666, OFNR rating) has been used in data centers for many years but provides only nominal fire protection. Many insurance companies require the use of more advanced cables in new construction.

The most common in the United States is plenum-rated cable, which is certified not to burn until extremely high temperatures are reached (UL 910, OFNP rating). Plenum-rated cable is required for installation in air ducts by the 1993 National Fire Code 770-53; however, there is an exception in Section 3000-22 of the code for raised floor and data processing environments. Other local codes may require plenum cable even in a data processing environment; check with your local fire marshall or insurance carrier to determine the proper cable type for a given installation. There are two basic types of plenum cable, manufactured with either a Teflon-based jacket or PVC-based jacket. Although they are functionally equivalent, the Teflon-based jackets tend to be stiffer and less flexible, which may cause problems during installation. In Europe and some other nations, the emerging standard is low-halogen fiber, which burns at a lower temperature than plenum cable but does not give off toxic fumes. Currently, plenum-rated cable is more popular; there is no cable that meets both the plenum and low-halogen requirements, and demand for low-halogen cable is currently less than 5% of the world market. In fact, there is an exception in the National Electrical Code that allows some cables to be installed under the raised floor in a computer center to be DP rather than plenum rated ["Information Technology Equipment" section, Article 645-5 (d), exception 3]. Because the NEC standards change and are subject to interpretation by local agencies, always consult the code itself before installing any type of fiber optic cable in a building.

There are also hybrid cables available that combine both fiber and copper in a single cable jacket. These include Type 5 cables, which feature a pair of 100/140 μm fibers bound with a pair of copper conductors in the same cable body. Variations with other types of fiber, including 50/125 and 62.5/125 multimode, are available; there may be a large number of fibers or copper conductors in a single cable jacket. These hybrid cables must also meet appropriate safety codes of riser, plenum, or low halogen.

Fiber jackets are typically rated as either riser, plenum, low smoke, low/zero halogen, or a dual rated combination of the above; the IEC 1034 specification provides the dual rated jacket specifications, while the zero halogen jackets are typically free of chlorine, fluroine, and bromides. New cable jacket types are also emerging from the national fire and electrical codes, such as the "limited combustibility" designation for some types of plenum cables.

Many different types of commonly used optical connectors have been discussed in Chapter 1. There are many different fiber optic connectors that are used less frequently, including the DIN connector (47256), D-4, Diamond, E2000, HMS-10/HP, and SMA (905 and 906). I will discuss only the more widely used connector types detailed in Chapter 1; planning considerations for optical cable will be addressed in the next chapter. Most are mechanically keyed to prevent multimode fiber from plugging into single-mode receptacles; this is intended for optical safety and is not a guarantee of plugging the proper connector type into the proper receptacle (and there is no keying to prevent plugging single-mode fiber into a multimode receptacle). Some popular connector standards, such as the ATM Forum SC duplex, do not specify multimode or single-mode connector keying. For now, I will discuss an important requirement in planning datacom systems — connector color coding and keying. Optical fiber cables are usually color coded with orange jackets for multimode (either 50 or 62.5 μm) and yellow for single mode; these colors were chosen for their visibility in case of fire. Most cables are also labeled with details of the fiber type, diameter, etc. In those cases in which a duplex cable breaks out into a pair of simplex cables, it is conventional for the simplex connectors to be colored black if light is emerging from the connector and white if light is intended to enter the connector; to further assist in identification (and as a guide for the color-blind) the respective connector ends are often labeled "A" and "B" in addition to their color coding. With the wider adoption of short wavelength laser transmitters for Gigabit Ethernet, FICON, and Fibre Channel, the need has arisen to distinguish between 50 and 62.5 micron multimode fiber, since both use an orange cable jacket. Recent vintage cables have begun to adopt a convention of black connectors on 50 micron cable and beige connectors on 62.5 micron cable; there is also interest in adopting these colors for parallel optical connectors as part of the Infiniband standard. Single-mode cables are now color coded with blue connectors following recent ATM standards. This can be confusing, because some early IBM systems with SC duplex interfaces used blue for 50 micron multimode, black for 62.5 micron multimode, and gray or tan for single mode. Other standards for connector color are still emerging and subject to change; in some cases, there is not yet a clear consensus for the accepted coloring of optical connectors. In some cases, there may be technical reasons that interfere with obtaining a good color match; some plastic materials with a high glass fill content cannot be easily colored because this alters their thermal properties. Some colored plastics shrink during injection molding or require very tight process tolerances on temperature, pressure, etc. in order to produce a molded part with the required tight tolerances. Fortunately, this is usually an aesthetic issue; when the

Table 7.12   **Optical Connector Color Coding**

Application	Connector Type	Color	Remarks
ESCON	ESCON duplex	Black	200 Mb/s, 50/62.5 μm, MM, 2 or 3 km
ESCON	ESCON duplex	Gray	200 Mb/s, SM, 20 km
ESCON	SC duplex	Gray	200 Mb/s, SM, 20 km (ANSI SBCON)
IBM ISC	SC duplex	Black	531 Mb/s, 50 μm, MM, 1 km
IBM ISC	SC duplex	Gray or Blue	1 Gb/s, SM, 3 km
ATM	SC duplex	Beige	155 Mb/s or higher, 62.5 μm, MM, 3 km
ATM	SC duplex	Blue	155 Mb/s or higher, SM, 20 km
LC FDDI	SC duplex	Black	125 Mb/s, 62.5 μm, MM, 2 km

Abbreviations used: MM, multimode; SM, single mode; ISC, Intersystem Channel (trade name of IBM Corporation); ATM, asynchronous transfer mode; LC FDDI, low-cost FDDI (color code is an industry convention, not a standard, for LC FDDI); media interface connector.

connectors are viewed from 10 ft or more in indirect lighting, an exact color match is not often required. Table 7.12 lists the current published conventions, as defined by EIA/TIA 658 and the emerging ANSI SBCON standard. Other emerging standards have not yet defined connector specifications as of this writing; these include the ANSI X3T11 workgroup on gigabit interface converters, also known as fiber channel optical converters, and the IEEE 802.3 1.250 Gb/s Ethernet alliance. There is also no general agreement on matching the color of optical transceivers with the corresponding connector color, so it is not always possible to identify cable receptacles or patch panel inserts in this way. The fiber distributed data interface (FDDI) standard media interface connector (MIC) utilizes three field-installable color-coded keys to identify different types of connections; these are described Table 7.13. The FDDI connectors may also be identified with four colored labels corresponding to the type of keys used. The keys also function as mechanical stops to prevent misplugs of the connectors and avoid installation errors; without a key installed, the MIC connector can be inserted into any type of FDDI receptacle, but with correct keying only certain configurations are allowed. Examples of cabling problems that are

Table 7.13   **FDDI Network Keying**

FDDI Connection Type	Key Type
Workstation to wall	S/S
Distribution panel to concentrator M port	M/M
Distribution panel to concentrator A port	A/A
Distribution panel to concentrator B port	B/B
Concentrator A to concentrator M port	A/M
Concentrator B to concentrator M port	B/M
Concentrator 1A to concentrator 2B port	A/B
Concentrator 1B to concentrator 2A port	B/A
Workstation to concentrator M port	S/M

Note. There are four types of keys/labels in the FDDI MIC standard: port A (A), red key/label; port B (B), blue key/label; master (M), green key/label; slave (S), no key/white label.

difficult to detect without proper keying in an FDDI environment include the following.

> Connection of a single attachment station directly to the trunk ring through either an A- or B-type receptacle (this results in a break in the trunk ring)
>
> Connection of the M-type connector directly into a trunk ring by attaching it to either an A- or B-type receptacle (this results in a break in the trunk ring)
>
> Reversal of a dual-attach station within the trunk ring so that the intended positions of the A- and B-type connectors are reversed (this causes the station media access controls to be inserted in the opposite position of the intended trunk ring)

7.3.3.   COST/PERFORMANCE TRADE-OFFS

It is difficult to discuss the cost/performance trade-offs in fiber optic data communication systems, especially in a rapidly changing marketplace. In this section I will attempt to give an overview of this area along with some examples of how much a data link should cost; the reader should consult current product data sheets in this area, and draw his or her own conclusions based on his or her system requirements.

The cost of even basic fiber optic components can vary widely, making direct comparisons with copper links difficult. For example, in some applications fiber optic cable is practically free compared with the cost of attached datacom equipment. Some vendors will supply fiber at no additional charge with the purchase of switches or routers or will provide fiber optic jumper cables that attach to existing patch panels. However, the cost of installing and terminating fiber in new building construction can be prohibitively high; in some parts of the United States, contractors can charge 20–30% more for the installation of optical fiber in buildings compared with "standard" unshielded twisted pair wiring (UTP-3). Even "premium" grade copper wiring (UTP-5) can be 10% more expensive than UTP-3, or more in some markets; it is hoped that this will change with the increased proliferation of high-grade copper and fiber in new buildings. This apparently high cost is not slowing the installation of unterminated optical fiber, however, as part of the building infrastructure, particularly overseas. Japan, for example, has announced a government initiative to wire all new government buildings with optical fiber by the year 2000. Currently available options include hybrid cables with two pairs of optical fiber and two copper coax cables bound in a single jacket. If the cost of purchasing optical fiber is highly variable, so is the cost of renting or leasing it. For some time, telecommunications carriers resisted the idea of providing bandwidth-on-demand services; these are now available in the United States but can be expensive at $150–300 per kilometer per month. The cost of leased or "dark fiber" is even higher overseas and may not be available in some international markets. This has resulted in a demand for bandwidth from alternative carriers; in the United States, for example, many public utility companies are taking advantage of their ownership of the right-of-way into consumer's homes to provide communication services along with electrical power. As the U.S. government considers new deregulation legislation, other alternative carriers such as cable television are expected to begin offering bandwidth on a leased basis.

We should pause to consider the possible competition for optical fiber from other datacom carriers, such as copper wiring and wireless datacom. The Infrared Datacom Association (IrDA) finalized a standard for wireless infrared communications in 1995; more than 100 members, including Microsoft, IBM, Sun, DEC, Motorola, and others have endorsed this new standard and announced plans to build compliant equipment [60–61]. Products that comply with the IrDA standards (either the IrLAP or Ir LMP specifications) may be self-certified to carry the IrDA compliance trademark. This is an emerging market and price/availability vary widely; most applications do not compete directly with fiber because they are limited to data rates of a few hundred kb/s and a few meters distance. The primary applications for infrared datacom today are areas where it is too expensive or impractical to retrofit optical fiber, such as large warehouses using computerized data management. In the future this could be an important area. At least one company currently offers 155 Mb/s ATM service on an unguided line-of-sight laser beam connection; a possible application is skyscrapers in large cities where it is prohibitively expensive to rent even a short segment of dark fiber to go across the street. For additional information, contact the IrDA or review the standards and design guidelines.

There has also been some recent research into plastic optical fiber as a low-cost alternative for optical data links; because of the much higher volume of glass fiber being manufactured today, the price of plastic optical fiber remains higher than glass. This may change as demand for low-cost plastic fiber links improves, but today such fibers are limited by their high loss (120–150 dB/km) to fairly short distances (approximately 50 m) rather than tens of kilometers provided by glass fiber. Despite this, there is increasing interest in plastic fiber because these links can be operated using visible light sources that are potentially very low cost; a 680-nm (red) LED transceiver for plastic fiber links can be offered for as low as $5–10. Plastic fiber links have already found some applications in laser printers and the microprocessor control systems for automotive applications. The current areas of interest include the small office/home office environment, where there is a need for a low-cost, high-bandwidth interconnect solution. The electromagnetic noise environment in a typical home or small office is much noisier than a computer machine room because many home appliances (such as microwave ovens) produce a broad radiated noise spectrum that can interfere with sensitive data communication links. Existing home wiring, such as stereo audio cables, may not be sufficient in this environment, especially because the computer equipment itself must not interfere with existing protected services, such as television or ham radio broadcasts. Plastic fiber is also much easier to install than glass because it requires only minimal polishing when terminating a new connection; this could be done by a homeowner with minimal training. With this motivation, the ATM Forum has recently begun investigating new standards for residential broadband networks, sometimes called home area networks, as an alternative to UTP CAT 5 copper cables. Current standards under discussion include a 51.8 Mb/s link over plastic fiber, 50 m long, with 17 dBm link budget; this is considered adequate for even the largest home offices. Several new types

Table 7.14   **Copper vs Optical Data Link Solutions; Distance and Cost Considerations**

	Distance				
	Kilometers +	**10–100 m**	**1–10 m**	**0.1–1 m**	**<10 cm**
Cost	>$1000/Gb/s	$100/Gb/s	$10/Gb/s	$1/Gb/s	$0.1/Gb/s
Copper	Coax/ repeated	Coax/twisted pair	Coax TP	Backplane	Chip-on-board
Optics	Single mode	Multimode	Multimode	Waveguide	Free space

Note. From Ref. [1]. Costs are order of magnitude market cost required to compete at a given distance.

of plastic fiber connectors have been proposed, including variations on the F07 connector, SMA 905 and 906, EIAJ Digital Audio, and JIS Industrial; plastic fiber versions of the ST and FC are also available. Some vendors have even proposed that plastic fiber with red VCSEL lasers could operate up to 1 Gb/s over short distances, and this may be the solution for fiber to the desktop, PC/television convergence, and other areas where there is currently a speed bottleneck and cost is the primary driver to enter the market. This area is expected to see increased activity in the coming years.

A direct cost comparison between optical fiber and copper media is difficult, even accounting for the improved performance of optical fiber with respect to distance, noise immunity, and other factors (see Table 7.2). Many different cost estimates have been proposed; a comparison of the application areas where either fiber or copper are currently the dominant media is shown in Table 7.14, along with a typical cost comparison between fiber and copper links in different environments [62]. This table is intended as a guide only; application-specific costs may vary. In general, fiber is cost/performance competitive with copper for links longer than a few tens of meters, shown by the vertical line in this table (approximately $20 for a 50-m simplex data link, including optical transceivers on each end). Transceiver cost has also been falling rapidly, driven by 21% compound annual growth in this industry; a typical FDDI-type transceiver that cost more than $300 in 1988 was available in 1996 for approximately $60. Even at this growth rate, it is difficult to speculate as to whether datacom will ever match the huge volumes available in the telecommunications market; some observers have predicted that the U.S. telecom infrastructure will need to support an aggrate data rate of 1 Tb/s by the Year 2007.

For longer distances, the performance advantages and room for future bandwidth growth of optical fiber make it the preferred alternative. If we try to compare the end-to-end cost of a data link using fiber vs copper, the results again vary widely. For example, consider an application involving attachment of 10 workstations to a large server using FDDI links. If the server is a bipolar-based mainframe, it is possible to obtain converter boxes that attach FDDI links to either parallel copper or ESCON channel ports; one adapter is required per port, and the average cost per FDDI client can be $25,000 or more. However, if the server is a CMOS-based transaction server, adapters are available that allow direct attachment of an FDDI LAN port to the mainframe; the direct attach feature may cost approximately $15,000, but thereafter each additional client can be added for the cost of inserting a standard FDDI adapter card in the workstation — approximately $1,000 per client. The reduction in total cost of computing is significant enough for a large network that it may justify upgrading the bipolar mainframe to a CMOS version! As can be seen from this example, network attachment costs are clearly very application dependent; we have not even considered associated software costs and maintenance fees because they vary too much to permit a clear discussion within the scope of this chapter.

Future emerging technologies are likely to have an important affect on data link costs. One example is the development of parallel optical transmitters and receivers that incorporate between 10 and 32 channels in a compact package. To emphasize the improved performance of parallel optical links, these units are often priced in terms of dollars/Gb/km. Unfortunately, until the development of long-wavelength vertical cavity surface-emitting lasers or similar transceiver technology, these links will likely be limited in distance and bandwidth; using short-wavelength laser arrays, distances of only a few hundred meters are possible. Multiplexing technology has been used effectively by telecommunication carriers to reduce costs by sharing a single data channel among multiple users; a similar effort is under way in the datacom community. One example is the IBM 9672 Optical Wavelength Division Multiplexer; it is perhaps the first device of its kind developed for the datacom industry. This device provides a 20:1 multiplexing of datacom channels onto a single fiber at wavelengths in the 1550-nm range. Because optical fiber is protocol insensitive, many different types of data can be accomodated. This product has only recently been announced, and in small volume the cost remains high (approximately $250,000 per device or $500,000 for a link). Still, this is cost competitive with leased fiber; at a rate of $150/mile/month, assuming a 10-mile link with 10 independent

Table 7.15   **Proposed Versions of Digital Subscriber Line Technology**

Technology	Standard	Data Rate	Distance (ft)
DSL/ISDN	ANSI T1.219 (1991)	160 kbps	18,000
384DSL	None	384 kbps	18,000
HDSL	ANSI T1	1.544 Mbps	12,000
		2.048 Mbps	
SDSL	None	Same as HDSL	10,000
DMT ADSL	ANSI T1.413 (1995)	6.144 Mbps/640 kbps	12,000
RADSL	None	9 Mbps/1.5 Mbps	Varies with data rate
VDSL	Proposed ANSI	12.96, 25.92, or 51.84 Mbps/2–20 Mbps	
			1,000–4,500

Note. Speed indicated is full duplex unless otherwise indicated; variable rates are indicated as max. speed down-stream/upstream.

bidirectional channels, the 9729 link pays for itself in less than a year and actually results in a savings of more than $340,000 per year.

There are also emerging standards and technologies that could drastically change the face of datacom over the next decade. Wavelength multiplexing may face competition from various digital subscriber line (DSL) technologies that have recently become available; these promise to increase the data rate for attached modems on PCs, which in turn drives bandwidth requirements of the supporting network infrastructure. Because DSL approaches are fairly new, there are a number of proposed standards as listed in Table 7.15. Other emerging standards include gigabit ethernet; this standard is not yet finalized and there are many proposed options at this time, including a proposal for operating 62.5-μm fiber with long-wavelength lasers at gigabit speeds over hundreds of meters. Fiber optics is also adding function and driving down cost for existing standards by providing optical channel extensions for high-performance parallel interace (HIPPI) and serial storage architecture (SSA) links. HIPPI was developed by the ANSI X3T9.3 committee as an 800 MB/s, 25-m parallel copper link for high-performance computing; in 1991 an ad hoc ANSI subcommittee proposed serial HIPPI, a channel extension using long-wave lasers and single-mode fiber. The HIPPI data and control characters are serialized into a 1.1 Gb/s data stream, which is transmitted over the fiber; this extends the link to 10–13 km. Fiber optic channel extenders also exist for other peripheral attachment standards, including the SSA; in the future, SSA links may merge into an ANSI-approved 1 Gb/s version of fiber channel, tentatively called fiber channel enhanced loop. These and other developments are expected to influence the price and availability of fiber optic components in coming years.

Finally, it should be noted that although a performance comparison favors fiber over copper, the cost of fiber data links is volume driven. Fiber cost is more important for telecommunication carriers because the ratio of fiber distance to number of transceivers is large. Conversely, for datacom systems there are many more transceivers per kilometer of fiber, so the transceiver cost dominates the link cost. Datacom transceivers are available in two basic types, an integrated transceiver solution that is packaged by the manufacturer and a "bag of parts" that is available to equipment designers to construct their own data links. The bag of parts solution is cost effective and offers design flexibility by mixing and matching different transmitters, receivers, serializers, deserializers, and other electronics in the application; however, it does not address component interactions such as high frequency noise. For data rates greater than approximately 1 Gb/s, the integrated solution is more widely accepted because it addresses these system problems. Over time, the cost point between these two solutions will probably be driven higher by advanced technology, such that effective bag of parts solutions will be available by the Year 2000 in the 1 or 2 Gb/s range. With more than 700 miles of optical fiber installed in the United States each year and some analysts projecting annual volumes of 300,000–400,000 for ATM transceivers (with a per channel cost of under $300 for ATM adapter cards), fiber optics remains a growth area. The cost of fiber optics is likely to continue decreasing, especially because installed fiber provides the bandwidth requirements to meet the needs of emerging applications well into the next century.

Appendix A: Contact Information for Optoelectronics and Fiber Optics Information

PROFESSIONAL ORGANIZATIONS

American Institute of Physics (AIP) and American Physical Society (APS)
Headquarters, American Center for Physics
1 Physics Ellipse
College Park, MD 20740
http://www.aip.org or http://www.aps.org

Fiber 66

"The first virtual group of fiber optic teachers on the Internet." A worldwide team of professional instructors and educators working through private newsgroups and mailing lists to collaborate on the design of courses in fiber optics, which are then made available on-line as appropriate. To join, send e-mail to fiber66@optoroute.com or contact Marc Duchesne in France at http://www.optoroute.com/fiber66.

Fiber University (Fiber U)
Administered by Fotec Corporation
151 Mystic Ave
Medford, MA 02155
http://www.std.com/fotec

Provides training and professional certification on fiber optic installation and testing.

Institute of Electrical and Electronics Engineers (IEEE)
445 Hoes Lane, P.O. Box 1331
Piscataway, NJ 08855
http://www.ieee.org

Institute for Optical Data Communication (IODC)
Direct inquires to Dr. Casimer DeCusatis
IBM Corporation
522 South Road, MS P343
Poughkeepsie, NY 12601
casimer@vnet.ibm.com

Or see the Photonics Technical Home link from the Optical Society of America
(http://www.osa.org/photonics).

National Society of Professional Engineers (NSPE)
1420 King Street
Alexandria, VA 22314
Phone: (203) 684-4811
http://www.nspe.org.

The NSPE administers the licensed Professional Engineer certification programs across all engineering disciplines; contact the main office listed above or the Web site for information or a list of the state license boards.

New York Academy of Sciences
2 East 63rd Street

New York, NY 10021
http://www.nyas.org
National Institute of Standards & Technology (NIST)
(particularly Boulder, Colorado, laboratory for fiber optic research)
http://www.nist.gov

Optical Society of America (OSA)
2010 Massachusetts Ave NW
Washington, DC
http://www.osa.org

Society of Photooptic Instrumentation Engineers (SPIE)
P.O. Box 10
Bellingham, WA 98227
http://www.spie.org

LASER SAFETY CERTIFICATION INFORMATION

International Electrotechnical Commission (IEC): IEC International
Laser Safety & Compliance Laboratories
VDE (Verband Deutscher Electrotechniker)
Association of German Electrical Engineers
Merianstrabe 28 D-6050 Offenbach
Phone: +49 069 8306-0

TUV Rheinland of North America, Inc. (U.S. contact for TUV
Laboratory)
12 Commerce Road
Newton, CT 06470
Phone: (203) 426-0888

TUV Rheinland
Prufstelle fur Geratesicherheit
Am Grauen Stein
5000 Koln 91
Germany

U.S. Food and Drug Administration (FDA)
Center for Devices and Radiological Health
2098 Gaither Road
Rockville, MD 20850
Issues accession numbers to verify compliance of laser and optical
transmitter products with regulations for the administration and

enforcement of the Radiation Control for Health and Safety Act of 1968 (Title 21, CFR, Subchapter J); the U.S. laser safety certification is obtained from the standard FDA 21 CFR Part 1040, performance standards for light-emitting products Sections 1040.10 (laser products) and 1040.11 (specific purpose laser products)

Certified Laser Safety Officer Training
Laser Institute of America (LIA)
12424 Research Parkway
Suite 125
Orlando, FL 32826
Phone: (407) 380-1553
(In addition to providing laser safety training, copies of the ANSI Z136.X laser safety standards are available from the LIA).

STANDARDS BODIES

American National Standards Institute (ANSI)
1430 Broadway
New York, NY 10018
or
11 West 42nd Street
New York, NY 10036
http://www.ansi.org
The ANSI Tl Committee can be accessed on-line at http://www.tl.org.
Administers the laser safety standard document: ANSI Standard Z136.X (first issue 1988) for the safe use of optical fiber communication systems utilizing laser diode and LED sources; to purchase and ANSI/TIA/EIA standards, contact the publisher, Global Engineering Documents (see listing below for address).

ATM Forum
303 Vintage Park Drive
Foster City, CA 94404
(Phone: 1-415-578-6860)
http://www.atmforum.com
For discussion of cell-relay based networks, consult the Usenet newsgroup {comp.dcom.cell-relay}; there is also a mailing list, which is a virror of the newsgroup and is identified as {cell-relay}.

The Desktop ATM 25 Alliance can be accessed on-line at http://www.
atm25.com.

Bellcore Standards

Bell Communications Laboratories
331 Newman Springs Road
or
Bellcore Customer Service
8 Corporate Place, Room 3A183
Piscataway, NY 08854
http://www.bellcore.com
Phone: 1-800-521-CORE
   Bellcore has recently introduced a program to certify that
   third-party equipment conforms to recognized networking
   standards; they issue the Bellcore Certification Mark for Certified
   Excellence to products that meet these standards. They also provide
   a catalog of technical information, publication number SR-264.
Building Premises Wiring Standards Body (CICSI)
10500 University Center Drive
Suite 100
Tampa, FL 33612
   Provides certification to work with premises wiring, LAN
   equipment, and other fiber optic applications; administers a
   professional certification program for registered communication
   distribution designers (RCDD) and LAN specialists; recently began
   a program to certify fiber optic technicians. Also sponsors regular
   conference and training sessions and publishes the
   Telecommunications Methods Manual and *LAN Design Manual*
Committee on Optical Science & Engineering (COSE) of the National
Research Council (NRC)
http://www.nsa.edu
e-mail cose@nas.edu

Compliance Standards

In addition to contacting the standards bodies or the certification laboratories, copies of the compliance standards for EMC, ESD, Telecom, Product
Safety, and others including EMC resource manuals may be ordered from

the following:

Compliance Engineering
One Tech Drive
Andover, MA 01801
Phone: (508) 681-6673
  Other information is available on-line at http://world.std.com/
  ~techbook (hotlink to EMC regulatory compliance) from the
  newsgroup sci.engr.electrical.compliance
Electronic Industries Association (EIA)
Engineering Dept.
2500 Wilson Blvd.
Arlington, VA 22201
or see: EIA Headquarters
2001 Pennsylvania Ave. NW
Washington, DC 20006
also see: Electronic Industries Foundation (EIF)
919 18th St. NW Suite 900
Washington, DC 20006
  Publishes FOTP and JEDEC standards, which can be ordered
  through Global Engineering.
Fiber Channel Association (FCA)
12407 MoPac Expressway, North 100-357
P.O. Box 9700
Austin, TX 78766
http://www.amdahl.com/ext/CARP/FCA/FCA.html
Phone: 1-800-272-4618
Frame Relay Forum
http://frame-relay.indiana.edu/index.html
Gigabit Ethernet Alliance
http://www.gigabit-ethernet.org
Global Engineering Documents
3130 South Harbor Blvd. #330
Santa Ana, CA 92704
or
15 Inverness Way East
Englewood, CO 80112
Phone: 1-800-854-7179 (internationally call 303-397-7956)
  Order ANSI, JEDEC, EIA/TIA, and other standards.

HIPPI Network Forum
P.O. Box 10173
Albuquerque, NM 67184
http://www.esscom.com/hnf
IBM Corporation
http://www.ibm.com

International Electrotechnical Commission, Internationale
Elektrotechnische Kommission (IECCE/CB, IECQ [also known as
Commission Electrotechnique Internationale (CEI)
Bureau Central de la Commissions Electrotechnique Internationale
3 rue de Varembe
Geneva, Switzerland
http://www.hike.te.chiba-u.ac.jp/ikeda/iec/home.html
  Issues CEI/IEC 825-1 and 825-2, "Safety of Laser Products, Part 1:
  Equipment Classification, Requirements & Users Guide," and
  "Part 2: Safety of Optical Fiber Communication Systems,"
  frequently collaborates/coreferences documents with CENELEC-
  Europaisches Kommittee fur Electrotechniscne Normung-CCA,
  HAR, CECC.

Infrared Data Association (IrDA)
Walnut Creek, CA
Phone: (510) 943-6546
http://www.irda.org
e-mail irda@netcom.com

Institute for Interconnecting & Packaging Electronic Circuits (IPC)
2215 Sanders Road
Northbrook, IL 60062
http://www.ipc.org
  (Sponsors circuit board designer certification).

International Standards Organization (ISO)
Central Secretariat
1 Rue de Varembe
Case Postale 56
CH-1211 Geneve 20
Switzerland
http://www.iso.ch

International Telecommunications Union (ITU, formerly CCITT)
http://info.itu.ch

Internet standards are available for discussion on a number of on-line forums:

    ftp://ds.internic.net/ds/dspg1intdoc.html
    ftp://ftp.gmd.de/documents
    ftp://ftp.isi.edu

    gopher://gopher.nic.ddn.mil

Japanese-based telecom/datacom standards (JEDEC)

Contact JEDEC Excutive Secretary at EIA headquarters

Military specification standards

For information or to order copies of the military standards (Mil. Spec.), contact:

Superintendent of Documents

U.S. Government Printing Office

Washington, DC 20036

or

Commanding Officer

U.S. Naval Supply Depot

Att: code DMD

5801 Tabor Ave.

Philadelphia, PA 18201

    the Mil. Spec. standards are controlled by the U.S. Department of Defense, Washington, DC.

Network Computer Alliance & NC Reference Prolife:
http://www.nc.ihost.com

National Institute for Science and Technology (NIST)
Headquarters
Gaithersburg, MD 20899
http://www.nist.gov

or

NIST Boulder, Colorado (center for fiber optic research and development)
325 Broadway
Boulder, CO 80303
http://www.boulder.nist.gov

    NIST Boulder is also the headquarters for the Institute for Telecommunication Sciences (ITS) and the National Telecommunication and Information Administration (NTIA):
http://its.bldrdoc.gov/its.html.

Optoelectron Industry Development Association (OIDA)
2010 Massachusetts Ave. NW Suite 200
Washington, DC 20036
Phone: (202) 785-9926
http//www.oida.org

Siecor Corporation
http://www.siecor.com

Switched Megabit Data Services (SMDS) interest group
http://www.cerf.net/smds

DEFINITION OF TECHNICAL TERMS

The Information Technology Vocabulary, developed by subcommittee
1, joint technical committee 1, of the International Organization for
Standardization and the International Electrotechnical Commission
(ISO/IEC JTCl/SC1)
The ANSI/EIA Standard 440A: Fiber Optic Terminology (copyright
1989), available from the EIA, 2001 Pennsylvania Ave. NW,
Washington, DC 20006
*The American National Standard Dictionary for Information Systems,
ANSI X3.172* (1990), available from ANSI

REFERENCES ON TECHNOLOGY SUPPLIERS

Electronic Engineers Master Catalog, Vols. A–D, published annually
by
    Hearst Business Communications Inc.
    645 Stewart Ave.
    Garden City, NY 11530
Laser Focus World Buyer's Guide, published annually by
    Laser Focus World Magazine
    Penwell Publishing Co.
    Ten Tara Blvd. 5th Floor
    Nashua, NH 03062
Photonics Corporate Guide and Buyer's Guide, published annually by
    Photonics Spectra Magazine
    Laurin Publishing Co, Inc.
    Berkshire Common, PO Box 4949
    Pittsfield, MA 01202

Appendix B: Some Accredited Homologation Test Labs

Note. The following list is provided for reference purposes only and does not constitute an endorsement of the services provided by these laboratories. There are many different international standards for optical data communication, ISDN services, cabling products, voice service equipment, etc. Consult the standards bodies or testing labs for the latest information as to what approvals are required to ship a given product outside the United States. Another resource is the home page for Homologation & Type Approval (H&TA) department of IBM Corporation in LaGaude, France, http://w3.lagaude.ibm.com/homologation/wwwhta.htm.

United States
  ASL Inc.
  Santa Clara, CA
  (408) 244-2921

  AT&T Bell Laboratories
  International Regulatory Division
  11900 North Pecos Street
  Denver, CO 80234
  (303) 538-3906

  Certelecom
  820 Proctor Ave
  Ogdensberg, NY 13669
  (800) 348-9433

  Global Product Compliance Lab
  Room 11C-165, P.O. Box 3030
  Holmdel, NJ 07733
  (908) 834-1801

  Underwriters Laboratories Inc.
  333 Pfingsten Road
  Northbrook, IL 60062
  (847) 272-8000

Canada
  Bell Northern Research
  Standards & Accredation
  P.O. Box 3511
  Station C

Ottawa, Ontario, Canada K1Y 4H7
(613) 763-5380

Certelecom
3325 River Road, RR#5
KIV 1H2
Canada
(800) 563-6336

Germany
BZT Labs
Postfach 100443, 66004 Saarbrucken
Germany
49681 598 0

United Kingdom
Assessment Services Limited
Segenworth Road, Tichfield
Fareham
Hampshire PO15 5RH
United Kingdom
44(0) 1329 443322

British Standards Institute (BSI)
P.O. Box 375, Breklands
Linford Woods West
Milton Keynes MK14 6LL
United Kingdom
01908 312636

KTL Corporation
Newlands Science Park
Inglemire Lane
Kingston upon Hull HU6 7TQ
United Kingdom
44 (0)1482 801801

Telecommunications Services
Maylands Avenue
Hernel Hempstead, HERTS HP2 4SQ
United Kingdom
44 1 44 223 0442

Japan
The Japanese telecom specs are available on the Internet at
http://www.mpt.go.jp/telecom

AKZO Kashima Limited
Telecommunications
Yokahama Test Centre
2-30 Hatsunecho
Naka-ku
Yokahama, 231
Japan
81 45 251 8300

Japan Quality Assurance Organization
Telecommunications
21-25, Kinuta 1-Chrome
Setagaya-Ku
Tokyo 157
Japan
81 3 3416 0193

Australia

Telecom specs are available on the Internet at http://www.dca.au
Advantech Electronics Pty. Ltd.
3/14 Salisbury Road
Hornsby NSW 2077
Australia
61 2 477 7757

AUSTEST Laboratories
4/87 Reserve Road
Artamon NSW 2064
Australia
61 2 437 4500

New Zealand
Wakefield Laboratories
P.O. Box 4512
Auckland, New Zealand
64 9 379 5768

References

1. IBM Corporation 9729 Fiber Optic Multiplexer. 1996 (June). *Photon Spectra* (special issue on Photonics Circle of Excellence Awards), 89.
2. Wolke, A. 1994. Improved curve fit algorithm for receiver sensitivity and eye window extrapolation. Amp Lytel Engineering Report ER:94-3634-037R.

3. Wolke, A. 1994. Curve fitting for sensitivity and eye window extrapolation. Amp Lytel Engineering Report ER:94-3634-016R.

4. Chan, D. 1973. The extrapolation of error rate performance of baseband digital communication systems." Bell Northern Research Technical Report TR 1E90-673.

5. DeCusatis, C. 1995. Data processing systems and optoelectronics. In *Optoelectronics for data communication,* eds. R. Lasky, U. Österberg, and D. Stigliani, Chap. 6. New York: Academic Press.

6. Walker, S. S. 1996. Rapid modeling and estimation of total spectral loss in optical fibers. *IEEE J. Lightwave Tech.* 4:1125–1131.

7. Allard, F. C. 1990. *Fiber optics handbook for engineers and scientists.* New York: McGraw-Hill.

8. Frieble, E. J., *et al.* 1988. Interlaboratory comparison of radiation-induced atteunation in optical fibers, part I. *IEEE J. Lightwave Tech.* 6:165–171. Frieble, E. J., *et al.* 1990. Interlaboratory comparison of radiation-induced attenuation in optical fibers, part II. *IEEE J. Lightwave Tech.* 8:967–988.

9. Frieble, E. J., *et al.* 1984. Effect of low dose rate irradiation on doped silica core optical fibers. *Appl. Opt.* 23:4202–4208.

10. Frieble, E. J., and M. E. Gingrich. 1981. Photobleaching effects in optical fiber waveguides. *Appl. Opt.* 20:3448–3452.

11. Tsai, T. E., and H. B. Lin. 1990. Photoluminscence induced by 6.4 cV photons in high purity silica. *Proc. Mater. Res. Soc. Symp.* 172:125–131.

12. Haber, J. B., *et al.* 1988. Assessment of radiation-induced loss for AT&T fiber optic transmission systems in the terrestrial environment. *IEEE J. Lightwave Tech.* 6:150–154.

13. Amos, N. A., A. D. Bross, and M. C. Lundin. 1990. Optical attenuation length measurements of scintillating fibers. *Nucl. Inst. Methods* A297:396–403.

14. Kyoto, M., *et al.* 1992. Gamma-ray radiation hardened properties of pure silica core single-mode fiber and its data link system in radioactive environments. *IEEE J. Lightwave Tech.* 10:289–294.

15. Young, M. 1973. Geometrical theory of multimode optical fiber to fiber connections. *Opt. Commun.* 7:253–255.

16. Bisbee, D. L. 1971. Measurements of loss due to offsets and end separation of optical fibers. *Bell Sys. Tech. J.* 49:3159–3168.

17. Chu, T. C., and A. R. McCormick. 1978. Measurements of loss due to offset, end separation, and angular misalignment in graded index optical fibers excited by an incoherent source. *Bell Sys. Tech. J.* 57:595–602.

18. Gloge, D. 1975. Propagation effects in optical fibers. *IEEE Trans. Microwave Theor. Tech.* MTT-23:106–120.

19. Fiber channel—Physical layer and signaling specification, ANSI X3T9/91-062, X3T9.3/92-001 rev. 4.2, 1993.

20. Shanker, P. M. 1988. Effect of modal noise on single-mode fiber optic networks. *Opt. Commun.* 64:347–350.
21. Fiber Optic Test Procedure FOTP-171, Attenuation measurement by substitution method (also EIA-TIA-455-171), available from Electronics Industry Association, New York.
22. Kosiorska, H., *et al.* 1991. Statistical method for predicting fiber loss in the field using factory measurements. *Proc. OFC'91,* 138.
23. DeCusatis, C., and M. Benedict. 1992 (May). Method for fabrication of high bandwidth optical fiber. *IBM Tech. Disc. Bull.*
24. Marcuse, D., and H. M. Presby. 1975. Mode coupling in an optical fiber with core distortion. *Bell Sys. Tech. J.* 1:3.
25. Refi, J. J. 1986. LED bandwidth of multimode fiber as a function of source bandwidth and LED spectral characteristics. *J. Lightwave Tech.* LT-14:265–272.
26. Miller, S. E., and I. P. Kaminow, eds. 1988. *Optical fiber telecommunications II.* New York: Academic Press. Also see Kulikowski, A. 1991. *Fundamentals of geometric design & tolerancing.* New York: Delmar Press.
27. Green, P. 1993. *Fiber optic networks.* New York: Prentice Hall.
28. Agrawal, G. P., *et al.* 1988. Dispersion penalty for 1.3 micron lightwave systems with multimode semiconductor lasers. *IEEE J. Lightwave Tech.* 6:620–625.
29. Midwinter, J. E. 1977. A study of intersymbol interference and transmission media instability for an optical fiber system. *Opt. Quantum Electron.* 9:299–304.
30. Ogawa, K. 1982. Analysis of mode partition noise in laser transmission systems. *IEEE J. Quantum Electron.* QE-18:849–855. Also see Ogawa, K. 1985. Semiconductor laser noise: Mode partition noise. In *Semiconductors and semimetals,* eds. R. K. Willardson and A. C. Beer, Vol. 22C, Chap. 8. Lightwave Comm. Tech. Ser., Tsang, W. T. ed. New York: Academic Press.
31. Campbell, J. C. 1988. Calculation of the dispersion penalty of the route design of single-mode systems. *J. Lightwave Tech.* 6:564–573. Also see Jiang, W., R. Feng, and P. Ye. 1990. Improved analysis of dispersion penalty in optical fiber systems. *Opt. Quantum Electron.* 22:23–31.
32. Ohtsu, M., *et al.* 1988. Mode stability analysis of nearly single-mode semiconductor laser. *IEEE J. Quantum. Electron.* 24:716–723.
33. Ohtsu, M., and Y. Teramachi. 1989. Analysis of mode partition and mode hopping in semiconductor lasers. *IEEE. Quantum Electron.* 25:31–38.
34. Olshansky, R. 1976. Mode coupling effects in graded index optical fibers. *Appl. Opt.* 14:935–945.
35. Duff, D., *et al.* 1989. Measurements and simulations of multipath interference for 1.7 Gbit/s lightwave systems utilizing single and multifrequency lasers. *Proc. OFC 1989,* 128.

36. Radcliff, J. 1989. Fiber optic link performance in the presence of internal noise sources. IBM Tech. Report, Glendale Labs, Endicott, N.Y.

37. Xiao, L. L., C. B. Su, and R. B. Lauer. 1992. Increase in laser RIN due to asymmetric nonlinear gain, fiber dispersion, and modulation. *IEEE Photon. Tech. Lett.* 4:774–777.

38. Hakki, B. W., F. Bosch, and S. Lumish. 1989. Dispersion and noise of 1.3 micron lasers in microwave digital systems. *IEEE J. Lightwave Tech.* 7:804–812.

39. AT&T Technical Report. 1985. Methods of optical data link module testing, 14.

40. Manalino, A. A. 1987. Time-domain eyewidth measurement of an optical data link operating at 200 Mbit/s. *IEEE Trans. Inst. Meas.* IM-36:551–553.

41. Trischitta, P., and P. Sannuti. 1988. The accumulation of pattern dependent jitter for a chain of fiber optic regenerators. *IEEE Trans. Comm.* 36:761–765.

42. Walker, S. D., R. B. P. Carpenter, and P. Cochrane. 1983. Jitter accumulation with SAW timing extraction at 325 Mbaud. *Electron. Lett.* 19:193.

43. CCITT 1984. Recommendations G.824, G.823, O.171, and G.703 on timing jitter in digital systems.

44. Gardner, F. M. 1979. *Phaselock techniques,* 2nd ed. New York: Wiley.

45. Agrawal, G. P., and T. M. Shen. 1986. Power penalty due to decision-time jitter in optical communication systems. *Electron. Lett.* 22:450–451.

46. Schumacher, K., and J. J. O'Reily. 1986. Power penalty due to jitter on optical communication systems. *Electron. Lett.* 23:718–719.

47. Shen, T. M. 1986. Power penalty due to decision-time jitter in receivers using avalanche photodiodes. *Electron. Lett.* 22:1043–1045.

48. Schumacher, K., and J. J. O'Reily. 1989. Distribution-free bound on the performance of optical communication systems in the presence of jitter. *IEEE Proc.* 136:129–136.

49. Bates, R. J. S. 1983. A model for jitter accumulation in digital networks. *IEEE Globecon,* 145–149, San Diego, Calif.

50. Trischitta, P. R., and E. L. Varma. 1989. *Jitter in digital transmission systems.* Boston: Artech House.

51. Byrne, C. J., B. J. Karafin, and D. B. Robinson, Jr. 1963. Systematic jitter in a chain of digital regenerators. *Bell Sys. Tech. J.* 43:2679–2714.

52. Bates, R. J. S., and L. A. Sauer. 1985. Jitter accumulation in token-passing ring LANs. *IBM J. Res. Dev.* 29:580–587.

53. Chamzas, C. 1985. Accumulation of jitter: A stochastic model. *AT&T Tech. J.* 64.

54. Stolen, R., 1979. Nonlinear properties of optical fiber. In *Optical fiber communications,* eds. S. E. Miller and A. G. Chynoweth, Chap. 5. New York: Academic Press.

55. Cotter, D. 1982. Observation of stimulated Brillouin scattering in low loss silica fibers at 1.3 microns. *Electron. Lett.* 18:105–106.

56. da Silva, H. J. A., and J. J. O'Reily. 1989. System preformance implications of laser chirp for long haul high bit rate direct detection optical fiber systems, *Proc. IEEE Globecom*, 19.5.1–19.5.5.

57. Kurashima, T., T. Horiguchi, and M. Tateda. 1990. Thermal effects of Brillouin gain spectra in single-mode fibers. *IEEE Photon. Tech. Lett.* 2:718–720.

58. Stolen, R. H., and C. Lin. 1978. Self-phase modulation in silica optical fiber. *Phys. Rev. A* 17:1448.

59. Fisher, R. A., and W. K. Bischel. 1975. Numerical studies of the interplay between self-phase modulation and dispersion for intense plane-wave laser pulses. *J. Appl. Phys.* 46:4921.

60. IrDA Serial Interface Test Suite, Link Access Protocol (Ir LAP), Secondary Station Implementation Test Specification Ver. 3.0; and IrDA Serial Interface Test Suite, Link Management Protocol (Ir LMP), Secondary Station Implementation Test Specification Ver. 3.0; both available from Genoa Technology, Moorpark, Calif. (July 1995) or by anonymous FTP at ftp.gentech.com in the /pub directory.

61. *IrDA data link design guide.* 1995. Palo Alto, Calif.: Hewlett–Packard.

62. Hibbs-Brenner, M. K., Y. Liu, A. Cox, D. Greenlaw, L. Galarneau, J. Bristow, J. A. Lehman, T. Werner, and R. A. Morgan. 1996. Components and packaging for low-cost optoelectronic modules, Paper WB3, 132. Proceedings of the OSA Annual Meeting, Rochester, N.Y.

Chapter 8 | Planning and Building the Optical Link

R. T. Hudson
D. R. King
T. R. Rhyne
T. A. Torchia

Siecor Corporation, Hickory, North Carolina 28601

8.1. Introduction

With the rapidly advancing information age, the importance of building a reliable and long-lasting communications infrastructure is becoming critical. Organizations that best develop and understand information technology will have a distinct advantage over their competitors.

Significant planning and foresight is required to build a stable communications platform. Predicting both today's and the future requirements for a building or campus infrastructure is very difficult due to the tremendous rate of progress in information technology. Planning is essential in order to implement a practical solution meeting today's needs while providing a migration path to tomorrow's requirements.

Employing a structured cabling system utilizing optical fiber can meet these needs. Optical fiber is the only medium with the bandwidth to meet current and future needs. Fiber is a very attractive medium due to its ability to support the widest range of applications at the fastest speeds and for the longest distance. Other advantages that fiber offers include electromagnetic interference and radio frequency interference immunity. In addition, it is extremely secure because it is almost impossible to tap. Fiber optic cables are also smaller and lighter in weight than their copper counterparts.

This chapter contains a brief introduction to structured fiber optic cabling systems used in voice, video, and data communication applications.

292

HANDBOOK OF FIBER OPTIC
DATA COMMUNICATION
$35.00

It is divided into three sections: private network (premises) applications, standards for fiber optic building wiring, and installation/handling issues.

8.2. Private Networks

8.2.1. INTRABUILDING

This section deals with cabling systems within a common building. It includes backbone and horizontal applications as well as the specific types of cables used in each.

8.2.1.1. Intrabuilding Backbone

The intrabuilding backbone connects the main cross-connect in the building to each of the other cross-connects. An optical fiber intrabuilding backbone has emerged as the medium of choice due to its ability to support multiple high-speed networks in a smaller cable without cross-talk concerns. Also, more users are utilizing fiber to support voice and private branch exchange (PBX) applications by placing small remote PBX shelves within cross-connects supported by the intrabuilding backbone.

8.2.1.2. Intrabuilding Topologies

The intrabuilding backbone design between the building main cross-connect (MC) or intermediate cross-connect (IC) and the horizontal cross-connect (HC) is usually straightforward; however, various options exist. A single hierarchical star design between the building cross-connect and the HC is strongly recommended (Fig. 8.1). The only possible exception occurs in an extremely large building, such as a high-rise, where a two-level hierarchical star may be considered (Fig. 8.2).

The same options and decision processes apply here. Sometimes, building pathways link HC to HC in addition to IC to HC, especially in buildings with multiple HCs per floor. Only in specific applications should a user design an HC-to-HC link. For example, this pathway might be of value in providing a redundant path between HC and IC, although direct connection between HCs should always be avoided (Fig. 8.3).

8.2.1.3. Horizontal Cabling

The benefits of fiber optics (high bandwidth, low attenuation, and system operating margin) allow greater flexibility in cabling design. Because of the distance limitations of copper and small operating margin, passing tests

Fig. 8.1   Intrabuilding backbone (one-level hierarchy).

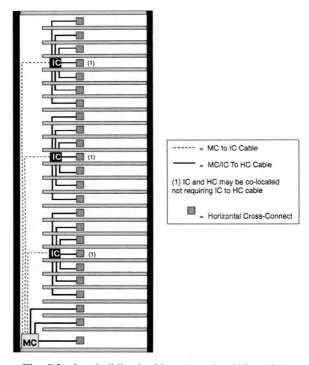

Fig. 8.2   Intrabuilding backbone (two-level hierarchy).

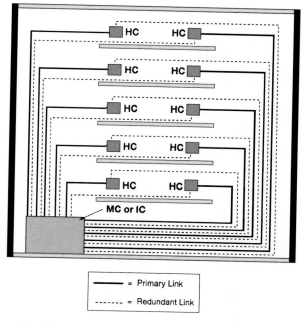

Fig. 8.3 Intrabuilding backbone (redundant routing).

over unshielded twisted pair (UTP) requires compliance with stringent design rules. In addition to the traditional horizontal cabling network, optical fibers support designs specifically addressing the use of open-systems furniture and/or centralized management.

The traditional network consists of an individual outlet for each user within 300 ft of the telecommunications closet. This network typically has data electronics equipment (hub, concentrator, or switch) located in each telecommunications closet within the building. Because electronics are located in each closet, this network is normally implemented with a small fiber count backbone cable (12 or 24 fiber count). This network uses a single cable per user in a physical star from the HC to the telecommunications outlet (Fig. 8.4).

For open-systems furniture, a multiuser outlet is recommended. The multiuser outlet calls for a high-fiber count cable to be placed from the closet to an area in the open office where there is a fairly permanent structure, such as a wiring column or cabinet, in a grid-type wiring scheme. At this structure a multiuser outlet, instead of individual outlets, is used (Fig. 8.5).

Fig. 8.4 Traditional horizontal cabling.

Fig. 8.5 Multiuser outlet cabling.

297

Fiber optic patch cords are then installed through the furniture raceways from the multiuser outlet to the office area. This allows the user to rearrange the furniture without disrupting or relocating the horizontal cabling. This type of network is supported in Annex G of TIA/EIA-568-A.

Centralized optical fiber cabling is intended as a cost-effective alternative to the optical HC when deploying 62.5-μm optical fiber cable in the horizontal in support of centralized electronics. Category 5 UTP systems are limited to 100 m total length and require electronics in each closet for high-speed data systems. Fiber optic systems do not require the use of electronics in closets on each floor, and therefore support a centralized cabling network. This greatly simplifies the management of fiber optic networks, provides for more efficient use of ports on electronic hubs, and allows for easy establishment of work group networks. Centralized cabling provides direct connections from the work areas to the centralized cross-connect by allowing the use of pull-through cables, a splice, or an interconnect in the telecommunications closet instead of a HC. Although each of these options has its benefits, splicing a low-fiber count horizontal cable to a higher fiber count intrabuilding cable in the telecommunications closet is often the best choice. This type of network is currently under study and being considered by the committee responsible for TIA/EIA-568-A (Fig. 8.6).

8.2.1.4. Applications of Intrabuilding Cables

Premises cables are generally deployed in one of three intrabuilding areas: backbone, horizontal, or interconnect. Higher fiber count tight buffered cables can be used as intrabuilding backbones that connect a MC or data center to IC and telecommunications closets. Likewise, lower fiber count cables can link an IC to a HC to multiple workstations.

Interconnect cables patch optical signals from the optical end equipment to the hardware that contains the main distribution lines. These cables can also provide optical service to workstations or transfer an optical signal from one patch panel to a second patch panel.

8.2.1.5. Specific Cables for Intrabuilding–Tight-Buffered Cable Construction

Tight-buffered cables have several beneficial characteristics that make them well suited for premises applications. Tight-buffered cables are generally small, lightweight, and flexible in comparison to outside plant cables. As a result, they are relatively easy to install and easy to terminate with optical

Fig. 8.6 Centralized optical fiber cabling.

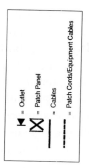

▼ = Outlet

✕ = Patch Panel

— = Cables

--- = Patch Cords/Equipment Cables

299

fiber connectors. Tight-buffered cables are capable of passing the most stringent flame and smoke generation tests and are normally listed as type OFNP (*o*ptical *f*iber *n*onconductive *p*lenum listing) or type OFNR (*o*ptical *f*iber *n*onconductive *r*iser listing). Tight-buffered cables were the first generation of premises cables. The name tight buffered is derived from the layer of thermoplastic or elastomeric material that is tightly applied over the fiber coating. This method contrasts sharply with the loose tube design cable, in which 250-μm fibers are loosely contained in an oversized buffer tube.

A 250- or 500-μm-coated optical fiber is overjacketed with a thermoplastic or elastomeric material, such as polyvinyl chloride (PVC), polyamide (nylon), or polyester, to an outer nominal diameter of approximately 900 μm. All tight-buffered cables will contain these upjacketed fibers. The 900-μm coating on the fiber makes it easier to terminate with an optical fiber connector. The additional material also makes the fiber easier to handle due its larger size.

The 900-μm tight-buffered fiber is the fundamental building block of the tight-buffered cable. Several different tight-buffered cable designs can be manufactured and the design depends on the specific application and desires of the user.

In one cable design, a number of 900-μm optical fibers are stranded around a central member, a serving of strength member (such as aramid yarn) is applied to the core, and a jacket is extruded around the core to form the cable (see Fig. 8.7). As an alternative, the fibers in the cable can be allocated into smaller groups, called subunits. Each subunit contains the same number of 900-μm tight-buffered fibers, a strength member, and a subunit jacket. Each of these individually jacketed groups of fibers (subunits)

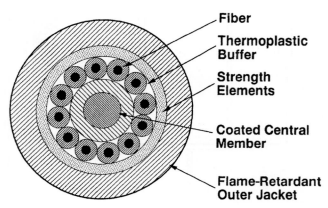

Fig. 8.7 Twelve-fiber tight-buffered cable.

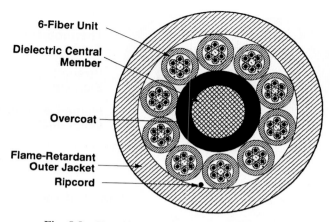

Fig. 8.8 Sixty-fiber unitized tight-buffered cable.

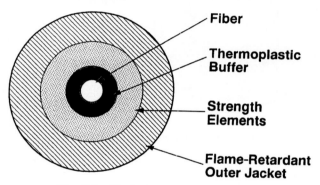

Fig. 8.9 Single-fiber Interconnect cable.

are stranded around another central member and the composite cable is enclosed in a common outer jacket (Fig. 8.8).

Lower fiber count cables are necessary to connect the optical end equipment to the optical fiber distribution system. They can also be used to bring fiber to a workstation. Tight-buffered interconnect cables with $900\text{-}\mu m$ fibers are again well suited for this application. One or two $900\text{-}\mu m$ tight-buffered fibers are independently jacketed with a serving of strength member to construct a tight-buffered interconnect cable commonly called a jumper (Figs. 8.9 and 8.10). For applications requiring higher fiber counts, a number of these individually jacketed fibers can be stranded around a central member to construct a fan-out cable (Fig. 8.11).

Materials are chosen for premises cables based on flame resistance, mechanical performance, chemical content, chemical resistance, cost, and

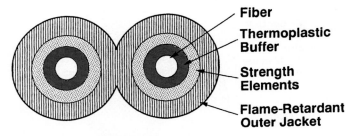

Fig. 8.10   Two-fiber zipcord cable.

Fig. 8.11   Tight-buffered fan-out cable.

so on. The selection of material is based on the application for which the cable is designed.

The cable jacket can be constructed from PVC, fluoropolymers, polyurethane, flame-retardant polyethylene (FRPE), or other polymers. The specific jacket material used for a cable will depend on the application for which the cable is designed. Standard, indoor cables for riser or general-purpose applications may use PVC for its rugged performance and cost-effectiveness. Indoor cables designed for plenum applications may use fluoropolymers or filled PVCs due to the stringent flame resistance and low smoke requirements. Filled PVCs contain chemical flame and smoke inhibitors and provide a more flexible cable that is generally more cost-effective than fluoropolymer-jacketed cables. Polyurethane is used on cables that will be subjected to extensive and unusually harsh handling but do not require superior flame resistance. FRPE provides a flame-resistant cable that is devoid of halogens. Halogen-free cables and cable jackets do not produce corrosive, halogen acid gases when burned.

8.2.2. INTERBUILDING

8.2.2.1. Interbuilding Backbone

In campus environments, such as universities and industrial parks, optical fiber is used extensively in the interbuilding backbone that provides communications between a number of buildings. This backbone can be used for voice, data, and video applications.

8.2.2.2. Interbuilding Topologies

The interbuilding backbone cabling is the segment of the network that typically presents the designer and user with the most options. It is also the most constrained by physical considerations such as duct availability, right-of-way, and physical barriers. For this reason, this section will present a number of options for consideration.

In a smaller network (both in number of buildings and in geographical area), the best design involves linking all the buildings requiring optical fibers to the MC. The cross-connect in each building then becomes the IC, linking the HC in each building to the MC. The location of the MC should be in close proximity to (if not colocated with) the predominant equipment, for example, the data center or PBX. Ideally, the MC is centrally located among the buildings being served, has adequate space for the cross-connect hardware and equipment, and has suitable pathways linking it with the other buildings. This network design would be compliant with the TIA/EIA-568-A standard. Some of the advantages of a single hierarchical star for the interbuilding backbone include:

- Provides a single point of control for system administration
- Allows testing and reconfiguration of the system's topology and applications from the MC
- Allows easy maintenance for security against unauthorized access
- Allows for the easy addition of future interbuilding backbones

Larger interbuilding networks (both in number of buildings and in geographical area) often found at universities, industrial parks, and military

Fig. 8.12   Interbuilding backbone (two-level hierarchy).

bases may require a two-level hierarchical star. This design provides an interbuilding backbone that does not link all the buildings to the MC. Instead,
it uses selected ICs to serve a number of buildings. The ICs are then linked
to the MC. This option may be considered when the available pathways do
not allow for all cables to be routed to an MC or when it is desirable to
segment the network because of geographical or functional communication
requirement groupings. In large networks, this often translates to a more
effective use of electronics, such as multiplexers, routers, and switches,
to better utilize the bandwidth capabilities of the fiber or to segment the
network (Fig. 8.12).

It is recommended that no more than five ICs be used unless unusual
circumstances exist. If the number of interbuilding ICs is kept to a minimum, the user can experience the benefits of segmenting the network
without significantly sacrificing control, flexibility, or manageability.

It is strongly recommended that when such a hierarchical star for the
interbuilding backbone is used, it be implemented by a physical star in all
segments. This ensures that flexibility, versatility, and manageability are
maintained. However, the following are two main situations in which the

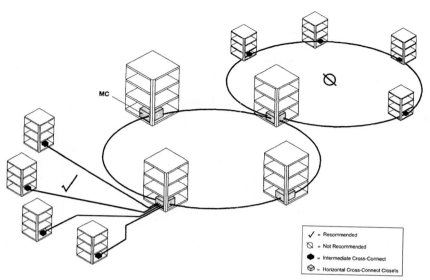

Fig. 8.13   Interbuilding main backbone ring.

user may consider a physical ring for linking the interbuilding ICs and MC:

- Existing conduit supports it
- Primary (almost sole) purpose of the network is fiber distributed data interface (FDDI) or token ring

It is seldom recommended that the outlying buildings use a second physical ring (Fig. 8.13).

The ideal design for a conduit system that provides a physical ring routing would dedicate X number of fibers of the cable to a ring and Y number of fibers to a star by expressing (not terminating) fibers through the ICs directly back to the MC. This design requires the end user to have a more exact knowledge of present and future communication requirements (Fig. 8.14).

8.2.2.3.   Applications of Interbuilding Cables

Interbuilding cables must be capable of withstanding a variety of environmental and mechanical extremes. The cable must offer excellent attenuation performance over a wide range of temperatures. The cable must also be sufficiently strong to endure the rigors of installation and provide protection against ultraviolet (UV) radiation, gnawing rodents, and other

Fig. 8.14   Main backbone ring and redundant backbone star combined.

mechanical disturbances. Furthermore, the cable should have a high packing density to maximize the use of available installation space.

8.2.2.4.   Specific Cables for Interbuilding–Loose-Tube Cable Construction

Loose-tube cables are designed primarily for outside plant environments
and interbuilding applications. Fibers are placed in gel-filled buffer tubes
to isolate them from external forces. The loose-tube design provides stable

and highly reliable optical transmission characteristics over a wide temperature range. Additionally, the loose tube design ensures long cable life by isolating the fibers from mechanical stresses.

After optical fibers have been placed in loose buffer tubes, the buffer tubes are then stranded around an antibuckling central member. The central member may be either steel or glass-reinforced plastic (GRP). GRP central members are used when the customer wants an all-dielectric cable (no metallic components).

For those designs that do not rely solely on the central member for its tensile strength, high tensile strength yarns, such as fiberglass and aramid yarns, are typically helically wrapped around the cable core before a sheath is placed on the cable. Other designs include embedding the strength elements in the cable jacket. These elements could be various sizes of steel wire or GRP rods. Embedding the strength elements in the sheath is typically employed in a central tube cable.

A water-blocking material protects the interstitial areas of the cable. This compound or tape blocks the ingress and migration of water in the cable. This allows the cable to be installed in up to 10 ft of standing water without special precautions.

The entire cable core is jacketed with an outer sheath. The jacket is the first line of defense for the fibers against any mechanical and chemical influences on the cable. Several jacket material options could be selected. The best material depends on the application. The most commonly used cable jacket material for outdoor cables is polyethylene. The polyethylene sheat contains carbon black to provide excellent UV resistance (Figs. 8.15 and 8.16).

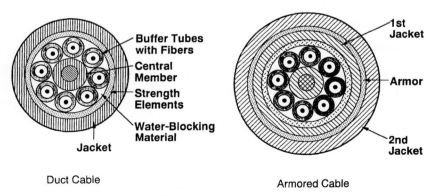

Duct Cable

Armored Cable

Fig. 8.15   Stranded loose-tube cable.

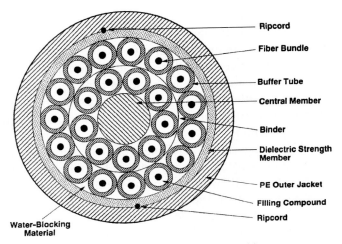

Fig. 8.16   288-fiber loose-tube cable.

8.2.2.5. Loose-Tube Cable Options

Several cable options are available for applications that require additional protection or special installation conditions.

a.   Armoring

Steel armoring provides additional mechanical strength and resistance to rodents (see Fig. 8.15).

b.   Self-Supporting

If no messenger is available, a self-supporting loose-tube cable can be installed for aerial applications (Fig. 8.17).

c.   Flame-Retardant

Some loose-tube cables are designed for use in both outside plant and premises applications. Variations of the loose-tube cables that are halogen free, flame-retardant, or that possess water-blocking characteristics are also available.

8.2.3. *DATA CENTERS*

A specific application of data communications that utilizes both interbuilding and intrabuilding cabling systems is the data center. Communication systems are a vital part of today's corporate structure. Most companies

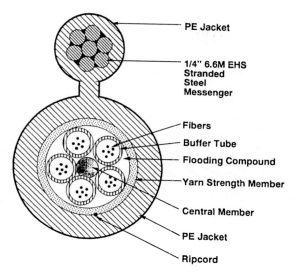

Fig. 8.17   Aerial cable.

require data communications supported by an efficiently structured and managed cabling system to support future growth.

A data center installation can have multimode or single-mode fibers going to different places within the site, including the following:

- Within a computer raised floor
- Local area network wiring closets
- Between floors in a building (intrabuilding)
- Between buildings in a campus (interbuilding)
- Between campuses

Data center equipment can be connected with either jumper cables, which directly connect two pieces of equipment, or through trunk cables and distribution panels. Jumper cables appear to be the least expensive approach until a reconfiguration occurs. Each jumper must be rerouted under a raised floor in this case. Usually, the most cost-effective connectivity solution is to use trunk cables and distribution panels. Individual equipment is connected to the distribution panel via short jumpers. In the event of a reconfiguration, one simply has to move the equipment and reconnect the jumpers while leaving the trunk cabling system in place. An example of a data center configuration is shown in Fig. 8.18.

Fig. 8.18

8.3. Standards

Building and fire codes must be considered not only with intrabuilding telecommunications cabling, but also, more importantly, when interbuilding telecommunications cabling enters the building. One of the most significant documents, although advisory in nature, is the "National Electrical Code" (NEC).

8.3.1. NATIONAL ELECTRICAL CODE

The "National Electrical Code" is a document issued every 3 years by the National Fire Protection Association. The document specifies the requirements for premises wire and cable installations to minimize the risk and spread of fire in commercial buildings and individual dwellings. Section 770 of the NEC defines the requirements for optical fiber cables. Although the NEC is advisory in nature, its content is generally adopted by local building authorities.

   The NEC categorizes indoor spaces into three general areas — plenums, risers, and general building. A plenum is an indoor space that handles air as a primary function. Examples of plenum areas are fan rooms and air ducts. A riser is an area that passes from one floor of a building to another floor. Elevator shafts and conduits that pass from one floor to another floor are examples of risers. Indoor areas that are not classified as plenums or risers are also governed by the NEC but are not specifically named.

These different installation locations — plenums, risers, and general buildings — require different degrees of minimum flame resistance. Because plenum areas provide a renewable source of oxygen and distribute environmental air, the flame-resistance requirements for cables installed in plenum areas are the most stringent. Likewise, riser cables must demonstrate the ability to retard the vertical spread of fire from floor to floor. Other indoor cables (not plenum or riser) must meet the least demanding fire-resistant standards.

Optical fiber cables that can be installed in plenums and contain no conductive elements are listed as type OFNP. These cables are generally referred to as plenum cables. Likewise, optical fiber cables that can be installed in riser applications and contain no conductive elements are listed as type OFNR. These cables are generally referred to as riser cables.

Optical fiber cables that can be installed indoors in nonplenum and nonriser applications and contain no conductive elements are listed as type OFN. These cables may be referred to as general-purpose cables and meet the least demanding of the fire-resistance standards.

Premises cables must be subjected to and pass standardized flame tests to acquire these listings. The tests are progressively more demanding as the application becomes more demanding. Therefore, plenum cables must pass the most stringent requirements, whereas general-purpose cables must meet less demanding criteria.

To obtain a type OFN listing, the cable must pass a standardized flame test, such as the UL 1581 flame test. The UL 1581 test specifies the maximum burn length the cable can experience to pass the test. Type OFN cables can be used in general-purpose indoor applications.

Similar to the type OFN listing is the type OFN-LS listing. To obtain this listing, the cable must pass the UL 1685 test. This test is similar to the UL 1581 test, but the UL 1685 test includes a smoke-generation test. The cable must produce limited smoke as defined in the standardized test to achieve the listing. A cable with a type OFN-LS listing is restricted to general-purpose indoor applications only.

To obtain a type OFNR listing, the cable must pass a standardized flame test, such as the UL 1666 flame test. The UL 1666 test contains specifications for allowable flame propagation and heat generation. The test is more stringent than the UL 1581 test. Type OFNR cables can be used in riser and general-purpose applications.

To obtain a type OFNP listing, the cable must pass a standardized flame test, such as the NFPA 262-1990 (UL 910) flame test. The NFPA 262 test contains a specification for allowable flame propagation and a specification for the average and peak smoke that can be produced during the test. The plenum test is the most severe of all flame tests. Type OFNP cables can be used in plenum, riser, and general-purpose applications.

Section 770 of the NEC contains a variety of other requirements on the placement and use of optical fiber cables in premises environments. The section also contains several exceptions that permit the use of unlisted cables under specific conditions. The NEC and the local building code should always be consulted during the design of the premises cable plant.

8.3.2. TIA/EIA-568-A, "COMMERCIAL BUILDING TELECOMMUNICATIONS CABLING STANDARD"

8.3.2.1. Brief Description of Cabling Topologies and Applications

One of the most challenging series of decisions a telecommunications manager makes is the proper design of an optical fiber cabling plant. Optical fiber cable, which has an extremely high bandwidth, is a powerful telecommunications media that supports voice, data, video, and telemetry/sensor applications. However, the effectiveness of the media is greatly diminished if proper connectivity, which allows for flexibility, manageability, and versatility of the cable plant, is not designed into the system.

The traditional practice of installing cables or cabling systems dedicated to each new application is all but obsolete. Instead, designers and users alike are learning that proper planning of a structured cabling system can save time and money. For example, a cable that is installed for point-to-point links should meet specifications, or later upgrades, as it becomes part of a much larger network. As a result, duplications of cable, connecting hardware, and installation labor can be avoided by anticipating future applications and providing additional fibers for unforeseen applications as well as interfacing with local service providers.

A structured cabling design philosophy that uses a physical hierarchical star topology is recommended. The hierarchical star offers the user a communications transport system that can be installed to efficiently support all logical topologies (star, bus, point-to-point, and ring). The guidelines presented here follow the general recommendations set forth by the TIA/EIA-568-A "Commercial Building Telecommunications Cabling Standard." This section will discuss the physical implementation of these logical topologies in the interbuilding backbones and intrabuilding backbones within the premises environments.

Logical   **Physical**

◄── = Data Flow   ─── = Cable Routing

Fig. 8.19   Networking topologies.

8.3.2.2.   Network Topologies

Network topologies are shown in Fig. 8.19.

8.3.2.3.   Logical Topology

To be universal for all possible applications, a structured cabling plant must support all logical topologies. These topologies define the electronic connection of the system's nodes. Fiber applications can support the following logical topologies:

- Point-to-point
- Star
- Ring
- Bus

a.   Point-to-Point

Point-to-point logical topologies are still common in today's customer premises installations (Fig. 8.20). Two nodes requiring communication are directly linked by the fibers — normally a fiber pair (one to transmit and one to receive). Common point-to-point applications include

- Fiber channel
- Terminal multiplexing

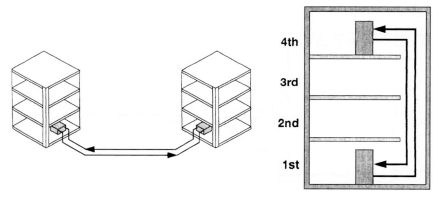

Fig. 8.20   Logical point-to-point topology.

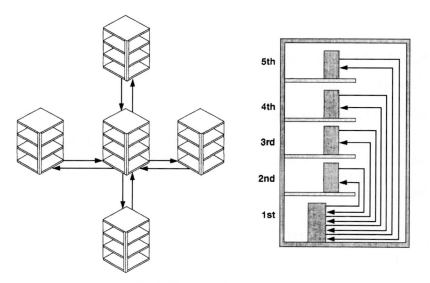

Fig. 8.21   Logical star topology.

- Satellite up/down links
- ESCON
- Telemetry/sensor

b.   Star

An extension of the point-to-point topology is the logical star topology (Fig. 8.21). This is a collection of point-to-point topology links, all of which

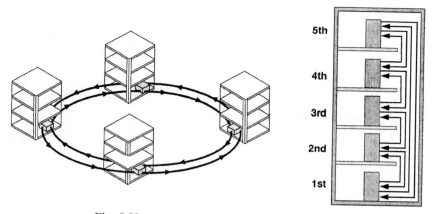

Fig. 8.22 Logical counter-rotating ring topology.

have a common node that is in control of the communications system. Common star applications include

- A switch such as a PBX, asynchronous transfer mode, or data switch
- A security video system with a central monitoring station
- An interactive video conference system serving more than two locations

c. Ring

In this topology, each node is connected to its adjacent nodes in a ring (Fig. 8.22). The logical ring topology, especially prevalent in the data communications area, is supported by two primary standards:

- Provides Token Ring (IEEE 802.5)
- FDDI (ANSI X3T9)

d. Logical Bus

The logical bus topology is also utilized by data communications and is supported by the IEEE 802.3 standards. All nodes share a common line rather than in one direction, as on a ring. When one node transmits, all the other nodes receive the transmission at approximately the same time.

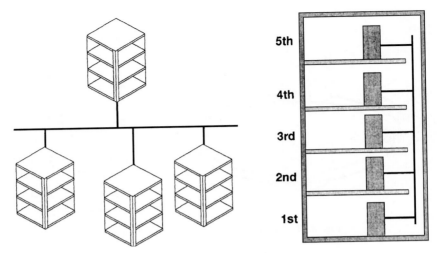

Fig. 8.23   Logical bus topology.

The most popular systems requiring a bus topology include Ethernet and Manufacturing Automation Protocol.

e.   Token Bus

A token bus topology is shown in Fig. 8.23.

8.3.2.4.   Physical Topologies and Logical Topology Implementation

All the logical topologies noted earlier are easily implemented with a physical star cabling scheme, as recommended by TIA/EIA-568-A "Commercial Building Telecommunications Cabling Standard." Implementing point-to-point and star topologies on a physical star is straightforward.

The use of data networks that utilize bus or ring topologies are very prominent in the market, i.e., Ethernet, Token Ring, and FDDI. With the benefits of physical star cabling, electronic vendors and standards have developed electronic solutions designed to interface with a star network. These applications are typically implemented with an "intelligent hub" or concentrator. This device, in simple terms, establishes the bus or ring in the back plane of the device and the connections are made from a central location(s). Therefore, from a physical connection, these networks appear to be a star topology and are quite naturally best supported by a physical star cabling system.

8.3.2.5. Physical Stars vs. Physical Rings

There are some situations in which a physical ring topology would seem appropriate. As seen in the following comparison of an optical fiber physical star topology to a physical ring topology, a physical star topology is recommended for supporting the varied requirements of a structured cabling system.

Physical star
  Advantages
      Flexible — supports all applications and topologies
      Acceptable connector loss with topology compatibility
      Centralized fiber cross-connects make administration and
      rearrangement easy
      Existing outdoor ducts are often configured for the physical star,
      resulting in easy implementation
      Easily facilitates the insertion of new buildings or stations into the
      network
      Supported by TIA/EIA-568-A, "Commercial Building
      Telecommunications Cabling Standard"
  Disadvantages
      Cut cable will result in node failure unless redundant routing is
      used
      More fiber length is used in a logical ring topology
Physical ring
  Disadvantages
      Less flexible
      Unacceptable connector loss if star or bus logical topologies are
      required
      Connectors are located throughout the system, making
      administration and rearrangement difficult
      Physical implementation can be difficult, often requiring new
      construction
      Insertion of new buildings or stations into a physical ring can
      cause disruptions to current applications or inefficient cable use

Even with ring topology applications, the majority of nodes, if not all, must be present and active
Node survivability in the event of cable cut, if cable truly utilizes redundant paths
Less fiber lengths are used between nodes

8.3.2.6. Physical Star Implementation

The more important recommendations of the TIA/EIA-568-A standard as they relate to optical fiber are summarized as follows: The standard is based on a hierarchical star for the backbone and a single star for horizontal distribution (Fig. 8.24).

The rules for backbone cabling include a 2000-m (3000 m for single mode) maximum distance between the MC and the HC for 62.5/125 μm and a maximum of one IC between any HC and MC. The MC is allowed to provide connectivity to any number of HCs.

The standard does not distinguish between interbuilding (outside) or intrabuilding (inside) backbones because these are determined by the facility size and campus layout. However, in most applications, one can envision the MC to IC being an interbuilding backbone link. The exceptions to this would be major high-rise buildings in which the backbone may be entirely inside the building or a campus containing small buildings with only one HC per building, thereby eliminating the requirement for an IC (Fig. 8.25).

Fig. 8.24 "Commercial Building Telecommunications Cabling Standard" (TIA/EIA-568-A). *Backbone distances based on optical fiber cable. When IC-to-HC distance is less than 500 m, then the MC to IC can be increased accordingly but the total MC to HC shall not exceed 2000 m for 62.5/125 μm or 3000 m for single mode. **Single-mode fiber can support distances up to 60 km (37 miles); however, this is outside the scope of TIA/EIA-568A.

Fig. 8.25   Backbone cabling.

The horizontal cabling is specified to be a single star linking horizontal cross-connect closet to the work-area telecommunications outlet with a distance limitation of 90 m (Fig. 8.26). This distance is not based on fiber capabilities but rather it is based on copper distance limitations to support data requirements. Currently, TIA is studying two alternate horizontal cabling topologies when using fiber in the horizontal: the multiuser outlet and centralized optical fiber cabling.

8.3.3.   TIA/EIA-598-A, "OPTICAL FIBER CABLE COLOR CODING"

A common denominator for all cable designs is identification/color coding of fibers and fiber units. TIA/EIA-598-A defines identification schemes for fibers, buffered fibers, fiber units, and groups of fiber units within outside-plant and premise optical fiber cables. This standard has been adopted by the Rural Utility Service within 7 CFR 1755.900 and Insulated Cable Engineers Association, Incorporated S-87-640-1992, "Standard for Fiber Optic Outside Plant Communications Cable."

Fig. 8.26   Horizontal cabling.

This standard allows for fiber units to be identified by means of a printed legend. This method can be used for identification of fiber ribbons and fiber subunits. The legend will contain a corresponding printed numerical position number and/or color for use in identification (Table 8.1).

8.4.   Handling and Installing Fiber Optics

8.4.1.   *CABLE HANDLING CONSIDERATIONS*

8.4.1.1.   Minimum Bend Radius

Optical fiber cable installation is as simple as, and in many cases much easier than, installing coaxial or UTP cable in the horizontal. The most important factor in optical fiber cable installations is maintaining the cable's minimum bend radius. Bending the cable tighter than the minimum bend radius may result in increased attenuation and broken fibers. If the elements of the cable are not damaged, when the bend is relaxed, the attenuation

Table 8.1 **Color Code**

Position Number	Base Color and Tracer	Abbreviation
1	Blue	BL
2	Orange	OR
3	Green	GR
4	Brown	BR
5	Slate	SL
6	White	WH
7	Red	RD
8	Black	BK
9	Yellow	YL
10	Violet	VI
11	Rose	RS
12	Aqua	AQ
13–24	Colors 1–12 repeated with black tracer	BL/BK, OR/ BK etc

should return to normal. Cable manufacturers specify minimum bend radii for cables under tension and for long-term installation (Table 8.2).

8.4.1.2. Maximum Tensile Rating

The cable's maximum tensile rating must not be exceeded during installation. This value is specified by the cable manufacturer. Tension on the cable should be monitored when a mechanical pulling device is used. Hand pulls do not require monitoring. Circuitous pulls can be accomplished through the use of backfeeding or centerpull techniques. For indoor installations, pull boxes can be used to allow cable access for backfeeding at every third 90° bend (Table 8.3).

8.4.1.3. Maximum Vertical Rise

All optical fiber cables have a maximum vertical rise that is a function of the cable's weight and tensile strength. This represents the maximum vertical distance the cable can be installed without intermediate support

Table 8.2   Minimum Bend Radii

Application	Fiber Count	Loaded cm	Loaded in.	Unloaded cm	Unloaded in.
Interbuilding backbone	2–84	22.5	8.9	15.0	5.9
	86–216	25.0	9.9	20.0	7.9
Intrabuilding backbone	2–12	10.5	4.1	7.0	2.8
	14–24	15.9	6.3	10.6	4.2
	26–48	26.7	10.5	17.8	7.0
	48–72	30.4	12.0	20.3	8.0
	74–216	29.4	11.6	19.6	7.7
Horizontal cabling	2	6.6	2.6	4.4	1.7
	4	7.2	2.8	4.8	1.9

Note. Specifications are based on representative cables for applications. Consult manufacturer for specifications of specific cables. For open-systems furniture applications where cables are placed, Siecor's 2- and 4-fiber horizontal cables have a 1-in. bend radius. Please contact your Siecor representative for further information.

Table 8.3   Maximum Tensile Load

Application	Fiber Count	Short Term N	Short Term lbs	Long Term N	Long Term lbs
Interbuilding backbone	2–84	2700	608	600	135
	86–216	2700	608	600	135
Intrabuilding backbone	2–12	1800	404	600	135
	14–24	2700	608	1000	225
	26–48	5000	1124	2500	562
	48–72	5500	1236	3000	674
	74–216	2700	600	600	135
Horizontal cabling	2	750	169	200	45
	4	1100	247	440	99

Note. Specifications are based on representative cables for applications. Consult manufacturer for specific information.

Fiber Optic Cable

Grip Pin

Pulling Grip

Fig. 8.27   Vertical cable.

points (Fig. 8.27). Some guidelines for vertical installations include the following:

- All vertical cable must be secured at the top of the run. A split mesh grip is recommended to secure the cable.
- The attachment point should be carefully chosen to comply with the cable's minimum bend radius while holding the cable securely.
- Long vertical cables should be secured when the maximum rise has been reached.

8.4.1.4.   Cable Protection

If future cable pulls in the same duct or conduit are a possibility, the use of innerduct to sectionalize the available duct space is recommended. Without this sectionalization, additional cable pulls can entangle an operating cable and could cause an interruption in service. Care should be exercised to ensure the innerduct is installed as straight as possible, without twists that could increase the cable pulling tension.

Fig. 8.28   Spare innerducts.

When the cable is installed in raceways, cable trays, or secured to other cables, consideration should be given to movement of the existing cables. Although optical fiber cable can be moved while in service without affecting fiber performance, it may warrant protection with conduit in places exposed to physical damage (Fig. 8.28).

8.4.1.5.   Duct Utilization

When pulling long lengths of cable through duct or conduit, less than a 50% fill ratio by cross-sectional area is recommended. For example, one cable equates to a 0.71-in. outside-diameter cable in a 1-in. inside-diameter duct. Multiple cables can be pulled at once as long as the tensile load is applied equally to all cables. Fill ratios may dictate higher fiber counts in anticipation of future needs. One sheath can be more densely packed with fiber than multiple cable sheaths. In short, for customer premises applications, the cost of extra fibers is usually small when these extra fibers are not terminated until needed. For a difficult cable pull, extra fibers installed now but not terminated may be the most cost-effective provision for the future (Fig. 8.29).

8.4.1.6.   Preconnectorized Assemblies

Of special consideration is the use of preconnectorized cables. Although the use of factory-terminated cross-connect and interconnect jumper assemblies is acceptable, the use of preconnectorized backbone and distribution cable presents special installation techniques. These connectors must be protected when installing the connectorized end of these cables. Protective pulling grips are available to protect connectors, but the grips' outside

Innerduct/Conduit

Cable

Fig. 8.29 Duct utilization: $\frac{d^2}{D^2} < 50\%$, where d is the cable diameter and D is the innerduct diameter.

diameter may prevent installation in small innerducts or conduits. The size of the preconnectorized assembly and pulling grip should be considered before ordering factory connectorized cables. There may also be additional installation requirements imposed on the grip by the manufacturer, in terms of minimum bend radius and tension, that would be the limiting parameters in an installation.

8.4.1.7. Slack

A small amount of slack cable (20–30 ft) can be useful in the event that cable repair or relocation is needed. If a cable is cut, the slack can be shifted to the damaged point, necessitating only one splice point in the permanent repair rather than two splices if an additional length of cable is added. This results in reduced labor and hardware costs and link loss budget savings.

Additional cable slack (approximately 30 ft) stored at planned future cable drop points will result in savings in labor and materials when the drop is finally needed. Relocation of terminals or cable plant can also take place without splicing if sufficient cable slack is available.

8.4.2. CABLE SPLICING METHODS

In the commercial building and campus environment, the designer/installer can often avoid the requirement of fiber-to-fiber splicing by installing a continuous length of cable. This is normally the most economical and convenient solution. However, because of the cable plant layout, length, raceway congestion, or requirements to transition between nonlisted and UL-listed cable types at the building entrance point, splices cannot always

Fig. 8.30   Fusion splicing.

be avoided. This requires splicing cables together, of which there are two basic techniques. Field splicing methods for optical fibers can be grouped into two major categories: fusion and mechanical.

Fusion splicing consists of aligning the cores of two clean (stripped of coating), cleaved fibers and fusing the ends together with an electric arc. The fiber ends are positioned and aligned using various methods. Alignment can be performed using fixed V-groove, profile alignment, or light injection. It can be manual or automatic and is normally accomplished with the aid of a viewing scope, video camera, or a specialized type of optical power meter for local injection and detection of lights. High-voltage electrodes, contained in the splicer, arc across the fiber ends as the fibers are moved together, thus fusing the fibers together (Fig. 8.30).

A mechanical splice, by comparison, is an optical junction, where two or more optical fibers are aligned and held in place by a self-contained assembly approximately 2 in. in length. Single-fiber mechanical splices rely on the alignment of the outer diameter of the fibers, making the accuracy of core/cladding concentricity critical to achieving low splice losses.

This method aligns the two fiber ends to a common centerline, thereby aligning the cores. The cleaned (stripped of coating) fiber ends are cleaved, inserted into an alignment tube, and butted together. The tube has factory-installed index-matching gel to reduce reflections and loss at the splice point. Usually, the fibers are held together by compression or friction,

Fig. 8.31 Mechanical splicing.

although some older methods rely on epoxy to permanently secure the fibers (Fig. 8.31).

8.4.3. HARDWARE

The selection of proper hardware for a commercial building or campus environment is essential to provide a flexible cabling system that will conform to the TIA/EIA-598-A "Commercial Building Telecommunications

Cabling Standard." The hardware selection process can be narrowed down to a few products based on the following four fundamental categories surrounding the connecting hardware application:

- Indoor or outdoor environment
- Field connectorize or splicing pigtails
- Wall- or rack-mounted
- Fiber count

8.4.4. INDOOR HARDWARE

Indoor hardware selection is complex because of the variety of applications and topologies encountered. Selecting the proper indoor hardware will ensure a flexible fiber optic cable plant. Indoor hardware can be divided into the application area's MC. The MC is the hub of the cable plant. Consequently, it is the main administrative point and is characterized by a high concentration of fibers as well as multiple indoor and outdoor cables. Adds, moves, and charges are also frequent. Hardware must have a high termination density and capacity. The capacity should be modular, however, to provide solutions that can grow as additional capacity is required. Different termination methods may be employed. Typically, indoor cable is terminated by field connectorization, outdoor cable is terminated by pigtail splicing, and interconnection cables (jumpers) are terminated by factory termination. The hardware must be readily compatible with the different methods, but must be flexible to provide only necessary components for the method employed. The MC generally requires rack-mounted hardware and equipment. Hardware should be wall-mounted if rack space is limited or not available. As the main administrative point, this area will contain a high number of jumpers for both cross-connection of backbone fibers and interconnection to the electronics. Hardware must provide ample room for jumper storage as well as neat, secure routing throughout the rack.

8.4.4.1. Intermediate Cross-Connect

This is the second hierarchical level of cross-connects in the backbone wiring. The IC is generally where the interbuilding backbone and intrabuilding backbone meet. In general, the IC has lower fiber counts than the MC. The required number of jumpers for cross-connection of backbone fibers and interconnection to the electronics is reduced. Racks with fiber routing provisions may be unnecessary.

8.4.4.2.   Telecommunications Closet

This is the location where the backbone wiring and horizontal wiring meet. Typically, this is the location in the network where the fiber backbone transitions to the horizontal distribution to the work areas. The backbone wiring is typically low-fiber count indoor cable. Thus, low-capacity termination housings are usually required. The standard termination method for indoor cables is field connectorization. Thus, only path panels are necessary — the purchase of a splice housing is rarely, if ever, required. The selection of wall- or rack-mountable hardware depends almost entirely on the architectural spaces provided. Jumpers will be required only to interconnect at the electronics. Jumper routing and storage capacities are not as critical.

8.4.4.3.   Work Area Telecommunications Outlet

This is the endpoint of the horizontal wiring. Again, the cables may be fiber only or a combination of fiber and copper. Although the work area usually requires only one outlet, the outlet must usually accommodate two or three different types of media. Mounting is usually in fixed-wall offices or in open-systems furniture; however, floor outlets are sometimes used. Hardware must be compatible with the most common mounting applications and allow a variety of cable entry locations. The work area is a high traffic zone and therefore hazardous to the cable, connectors, and jumpers. Units must be low profile to minimize the risk of damage. The hardware must provide adequate routing and storage space so that the minimum bend radius of the fiber is maintained.

Outdoor hardware consists of a line of splice closures, wall-mountable distribution centers, and pedestal-mountable cross-connects. These units provide environmental protection for splices, connectors, and jumpers in the outside plant environment, often required in industrial and other special applications.

In some indoor circumstances, space is limited for mounting hardware. Specially designed furcation (or fan-out) kits provide protection and pull-out strength for bare fibers, and they are direct connectorized. These are most useful when the fiber counts are low and all of the fibers will be patched into other hardware or electronics in the same area.

8.4.5.   CONNECTORS

8.4.5.1.   Field and Factory Connectorization

With the advent of easy-to-install field connectors and new methods to fan-out loose-tube cables, field connectorization has become the common method for terminating fiber optic cables in the premises market. There are numerous types of connectors on the market, each with slightly different installation procedures.

With regard to all connectors, the fiber must be epoxied into the connector. The epoxy keeps fiber movement over temperature at a minimum, allows polishing without fracturing the fiber, and seals the fiber from the effects of the environment. In addition, it allows the fiber to be aggressively cleaned on the end face.

In addition, the connector end face must be polished. A physical contact (PC) finish is recommended and is specified by TIA/EIA-598-A. This means that the fibers will be physically touching inside the connector adapter as they are held under compression. Lack of a PC finish results in an air gap between the fibers and increased attenuation.

There are several polishing methods recommended, which are typically dependent on the ferrule material used. If a ferrule material is very hard, such as ceramic, it is common for the ferrule to be preradiused. Softer ferrule materials, such as composite thermoplastic or glass in ceramic, may be polished flat. These materials wear away at approximately the same rate as the fiber and can be polished aggressively and still maintain a PC finish.

Heat-cured connectors have the advantage of being a cost-effective way to make cable assemblies or to install in a location where a large number of fibers are terminated at one time. Heat-cured connectors typically require more time for the epoxy to harden and skill to polish, and they need epoxies that have limited pot lives. Therefore, they are best used in a controlled environment or where large numbers of connectors are done to take advantage of the low connector costs and special ovens. Connectorized assemblies are almost exclusively heat-cured connectors (Fig. 8.32).

UV-cured connectors can be termed glass-insert connectors. The ferrule features a glass insert surrounded by ceramic. The glass insert allows the installer to do two things. First, the glass insert propagates light. Therefore, a UV-curable adhesive can be used to bond the fiber into the ferrule in a mere 45 s. Second, the glass insert protrudes beyond the ceramic outer sleeve. This means the glass insert is polished along with the fiber. The glass

Fig. 8.32   Heat-cured connector.

Fig. 8.33   Glass-insert connector.

Fig. 8.34   No-cure, no-polish connector.

is approximately the same hardness as the fiber and polishes at the same rate. This results in a flat PC polish (Fig. 8.33).

No-cure, no-polish connectors have the advantage of no epoxy, no polish in the field, no consumables, few tools needed, and minimal setup required. These types of connectors are actually a mini-pigtail housed in a connector body. There is a fiber stub already bonded into the ferrule in the factory, where the end face of the ferrule is polished to a PC finish. The other end of the fiber is cleaved and resides inside the connector. The field fiber is cleaved and inserted into the connector until it "butts up" against the fiber stub. A simple cam actuation process completes the connector with no epoxy or polishing required (Fig. 8.34).

References

1. Englebert, J., S. Hassett, and T. Summers. Optical fiber and cable for telecommunications. In *The electroncis handbook*, ed., J. Whitaker, Chap. 10. Boca Raton, Fla.: CRC Press.
2. IBM. 1996. *FTS direct attach physical and configuration planning,* 2nd ed. New York: IBM.
3. Siecor Corporation. 1995. *Universal transport system design guide, release III.* Hickory, N.C.: Siecor.
4. Siecor Corporation. 1996. *Premises fiber optic products catalog,* 6th ed. Hickory, N.C.: Siecor.

Chapter 9 | Testing Fiber Optic Local Area Networks (LANs)

Jim Hayes

Fotec, Incorporated, Medford, Massachusetts 02155

Greg LeCheminant

Hewlett-Packard, Santa Rosa Systems Division, Santa Rosa, California 95403

With the widespread usage of fiber optics in data communications, the testing and troubleshooting of fiber optic networks has become quite important. Although fiber optic networks have some major physical differences from copper-based networks, testing and troubleshooting them can be easily mastered by technicians with some basic understanding of the technology and training in the procedures.

9.1. Fiber Optic Testing Defined

All the components of a fiber optic network at all levels of integration may require testing. This would include transmitters and receivers in networking equipment, the equipment itself, the fiber optic cable plant, and the entire link. Fiber optic testing primarily consists of tracing or checking continuity and measuring optical power and optical loss and fault location. With the advent of gigabit networks, bandwidth and bit error rate testing has also become important for some networks. Testing is used in various stages of design, installation, and maintenance to ensure the fiber optic link will operate properly.

Tracing and continuity checking simply require using a visible light source to determine if the fibers in the cable are continuous and/or to identify fibers at a connection. Because the biggest problem in installation is connecting fibers correctly, fiber tracers are extremely valuable tools for

HANDBOOK OF FIBER OPTIC
DATA COMMUNICATION
$35.00

BER of Fiber Optic Link

Fig. 9.1   Bit error rate performance of optical link.

installation and upgrades. A tracer may simply be a flashlight that holds a fiber close enough to couple light into the fiber, a visible LED, or a visible laser diode. The sensor is the tester's eyes, looking for visible proof of the continuity or finding the proper fiber.

Optical power is the most basic parameter of fiber optic measurements, equivalent to voltage in electronics. Optical power is measured at the transmitter to determine if the transmitter power output is within specification, and at the receiver to determine if adequate power is available for reliable transmission. The difference between the typical output power of the transmitter coupled into the fiber and sensitivity of the receiver determine the "loss budget" of the link.

Figure 9.1 shows the performance of a typical receiver over variations in input power. As the power increases, the bit-error-rate (BER) decreases, indicating better performance. At higher power levels, the receiver may saturate, with subsequent higher BER. Proper performance means the receiver power must fall within the range shown on the graph. The difference between the transmitted and received power is the amount of loss tolerable in the cable plant connecting transmitter and receiver. An optical power meter adapted for connection to fiber optic cables is used to measure power.

The optical loss budget is determined during the design of the link by the typical power output of the transmitter and sensitivity of the receiver. Because the cable plant is installed before the network equipment is installed, it is tested with a test source similar to the transmitter to simulate loss in the cable plant that approximates what the link will actually see in operation.

Fault location uses high-powered visible diode laser fault locators or optical time domain reflectometers (OTDRs). Visible fault locators use the eye to detect large light losses in cables that are of short length, visible, and readily accessible. OTDRs work like radar to find problems in the cable at longer distances and in locations where the cable is not visible, such as underground runs in conduit.

9.2. Practical Considerations

Most data communications applications of fiber optics use LED (light-emitting diode) sources and multimode fibers. These components allow links to operate at distances up to 4 km and at data rates up to several hundred megabits per second. Beyond these limits, links using laser diode sources and either multimode or single-mode fiber are required. The distance and bandwidth limitation for most links is bandwidth, not power loss. The LED source used with multimode fiber is usually distance limited by the combination of modal and chromatic dispersion of the LED/fiber combination. In gigabit networks, the shorter links are limited by the modal dispersion of the fiber when used with a laser source. Datalinks fall into several basic categories, summarized in Table 9.1.

Most datacom applications are based on multimode fiber, not the single-mode fiber used in telecommunications or CATV networks. The multimode fiber in use is almost always what is called "62.5/125," the dimensions of the core and cladding in microns. This fiber design is over 20 years old and was designed for the last generation of multimode telecom systems of the early

Table 9.1 **Fiber Optic Datalinks**

Description	*Examples*	*Specifications*	*Distance*
Low speed, short distance	10baseF, FOIRL	850 nm LED, multimode fiber	2 km
Medium speed, short distance	FDDI, ESCON, ATM	1300 nm LED, multimode fiber	2–4 km
High speed, short distance	Gigabit Ethernet (GbE), Fibre Channel, 10 GbE	850 or 1310 laser, multimode or single-mode fiber	<500 meters
High speed, long distance	ATM, SONET (SDH), FDDI, ESCON based WAN	1300 nm laser, single-mode fiber	0.5–20 km

1980s. As a result, it was optimized for both high bandwidth and long distance, with the dispersion characteristics optimized at 1300 nm, the lowest loss window feasible at the time. Most 62.5/125 fiber is called "FDDI grade" for its bandwidth specifications designed to meet FDDI network needs.

Datacom links first used 850-nm LEDs and multimode fiber. When datacom links reached higher speeds, 100 MB/s or more with FDDI and ESCON, it was necessary to use 1300-nm sources to properly utilize the bandwidth characteristics of the fiber. Lower speed networks, such as Ethernet and Token Ring, continued to use 850-nm LEDs, because the lower bandwidth and higher attenuation of the fiber worked in concert to limit transmission distances. Laser sources were not considered for datacom, as their higher price and lower reliability was considered undesirable for low-cost networks.

With the introduction of Gigabit Ethernet, which necessitates laser sources to achieve the high modulation rates required, bandwidth of the fiber itself became an important issue. Another multimode fiber was reintroduced, with core/cladding dimensions of 50/125 microns. This fiber was used extensively by the telcos before the introduction of single-mode fiber and was optimized for laser sources. Fiber manufacturers also have introduced new high-bandwidth 62.5/125 fiber that is also more compatible with lasers.

Bandwidth limitations are primarily caused by modal dispersion in the fiber, and in LED-based systems, chromatic dispersion in the fiber. Modal dispersion refers to the dispersion caused by the different modes of light propagating in the fiber, while chromatic dispersion refers to the dispersion caused by the different speed of various wavelengths of light from the source. Chromatic dispersion is a complicated interaction of the wavelength and spectral width of the LED source, combined with the dispersion characteristics of the fiber.

LEDs are made by two different semiconductor processes. Edge-emitting LEDs (ELEDs) are fabricated like semiconductor lasers, emitting light from the edge of the LED chip, but lack the reflective cavity required to stimulate lasing. Surface-emitting LEDs (usually just called LEDs) are made to emit light directly from the surface of the chip. The ELED has a narrower spectral width, which allows it to have lower chromatic dispersion and thereby higher bandwidth than the surface emitter. Coupled with the fact that they have a higher typical power output, ELEDs are preferable for systems that need to go longer distances.

Laser sources have very narrow spectral widths, so modal dispersion is the primary concern for bandwidth of the source/fiber combination.

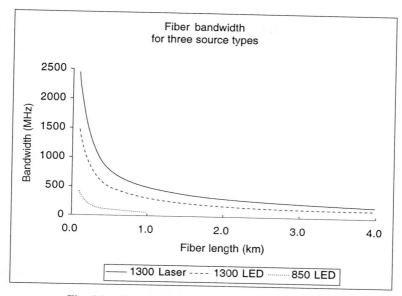

Fig. 9.2 Fiber bandwidth as a function of source type.

Some lasers also are very coherent, so phase shifts among the modes propagated in the fiber can interact at the receiver end to create modal noise. Careful control of the launch conditions with a laser and multimode fiber can minimize the problems.

Figure 9.2 shows the theoretical bandwidth of fiber optic cable as a function of distance for several source types, illustrating the variations in bandwidth capability with source characteristics. It was these kinds of analyses that led to maximum distances being specified for high-speed networks like FDDI and ESCON. These maximum distances are quite "safe," ensuring the networks would be "plug and play," without requiring users to consider any parameters other than cable plant loss. Knowledgeable users often push these limits.

For these fixed limits, the typical cable plant will have much less optical loss than the transmitters and receivers will tolerate. The loss budget is adequate to allow the full fiber length and six or more connections between network equipment with considerable margin for safety.

While multimode links are bandwidth-, not loss-limited, it is generally feasible to test only attenuation on installed networks. The variables needed to test bandwidth are complex and test equipment too expensive for most applications outside the engineering or manufacturing environment. Chromatic dispersion in particular, is very difficult to test. Modal dispersion is more feasible, however, and field test equipment is being developed for this application. For theoretical analysis of link performance, it is possible to simulate bandwidth performance of fibers and sources with computer programs, such as that used to generate the data in Fig. 9.2.

9.3. Component Testing

Fiber optic data links are composed of three components: a transmitter, a receiver, and the interconnecting cable plant. These components must be compatible with the parameters of the intended application. The loss budget must be adequate for the expected loss in the cable plant and the dynamic characteristics must meet the bandwidth requirements set by the network data transmission rate. These two factors are interrelated.

The loss budget is set by the output power of the transmitter and the sensitivity of the receiver. On a static basis, this is determined by the difference in optical power levels, but in reality the issue is a dynamic one, determined by the performance of the components at the data transmission rate of interest. Therefore, one uses a power meter to measure the power levels and a bit error rate (BER) tester to verify the quality of the data transmission. Between these two measurements, a curve similar to Fig. 9.1 will be generated for the data link.

9.3.1. LINK TESTING

The total link performance specifications will include the loss budget, minimum receiver power, and maximum length of fiber optic cable that can be used for the link. When generating bit error rate curves, one generally uses a variable attenuator to simulate the loss in the cable plant. This data does not include the effects of dispersion in the cable plant. One should also perform the tests with lengths of fiber of nominal characteristics to verify the performance under more realistic conditions, right up to the maximum length specified.

9.3.2. CABLE PLANT TESTING

The cable plant is tested after installation to ensure the proper installation of all the components: cables, connectors, splices, patch panels, etc. Even before installation, tracers are used to test the continuity of most fibers in a cable before pulling, to check that the cable has not been damaged during transport.

As the cables are pulled and terminated, each is tested for loss with a test source and power meter to ensure the cable was installed without damage and the terminations have been done properly. After the entire cable plant is installed, end-to-end runs that will connect network equipment are patched in and tested for total loss. This test will determine if the networking equipment will operate properly upon installation.

9.4. Standardization of Testing Procedures

Most fiber optic test procedures have been thoroughly tested and codified as industry standards. These standards are part of EIA/TIA RS455 and are being adapted into IEC standards. Most of these standards refer to tests of individual components under a variety of environmental conditions, so only a few relate to testing the performance of the installed cable plant. With datacom networks, we are primarily concerned with three fiber optic test procedures (FOTPs): FOTP-95 for measurements of optical power, FOTP-171 for testing patchcords, and OFSTP-14 for testing the loss of the installed cable plant. These are described in the following sections.

9.5. Fiber Optic Test Equipment Needed for Testing

The fiber optic power meter is a special light meter that measures how much light is coming out of the end of the fiber optic cable. The power meter needs to be able to measure the light from the fiber optic cable at the proper wavelength and over the appropriate power range. Most power meters used in datacom networks are designed to work at 850 and 1300 nm, the wavelengths used in datacom networks. Power levels are modest, in the range of −15 to −35 dBm for multimode links, 0 to −40 dBm for single-mode links. Power meters generally can be adapted to a number of different connector styles for convenience in testing.

Optical power is usually measured in dBm, or decibels referenced to one milliwatt. This log scale is used because of the large dynamic range of fiber optic links, a range of 1000 or more. Some power levels may be given in microwatts, which many meters measure directly. Power meters measure average optical power, not peak power, so they are sensitive to the duty cycle of the data being transmitted. It is important to specify the test conditions for measuring the optical power of a transmitter or at a receiver in terms of the data transmitted. Most networks have a diagnostic test signal for just this purpose.

Fiber optic power meters, like DMMs, come in a variety of types. The measurement uncertainty of practically all fiber optic power meters is the same, limited by the physical constraints of transferring standards with optical connections. Most meters have an uncertainty of +/− 5% or approximately 0.2 dB, no matter what the resolution of the display may be. Lower cost meters or those for field use usually have a resolution of 0.1 dB, laboratory meters display 0.01 dB, and a resolution of 0.001 dB is available on a few specialized meters.

The appropriate resolution for a measurement should be chosen according to the test. Laboratory measurements of low-loss patchcords, connectors, and splices can be made to 0.01 dB resolution and an uncertainty of 0.05 dB or less if great care is used in controlling the test conditions. Field measurements of absolute power are no better than the absolute calibration uncertainty, but relative power measurements can be made to 0.1 dB. Cable plant loss is affected by not only the meter uncertainty, but also the characteristics of the test source, and may have an uncertainty of 0.5 dB or more.

The test source is a portable version of the source that is used in the communication network attached to the fiber. It simulates the signal in the fiber for loss testing with a power meter, so the test source should match the wavelength and source type of the system source. Generally, multimode fiber is tested with LEDs at both 850 and 1300 nm and single-mode fiber is tested with lasers at 1310 and 1550 nm. The test source will typically be a LED for multimode fiber unless the fiber is being used for Gigabit Ethernet or other high-speed networks that use laser sources. LEDs can be used to test single-mode fibers less than 5 km long, while a laser should be used for long single-mode fibers.

One must always match the source wavelength to the network requirements. While one refers to 850, 1300, or 1550 nm routinely, actual source wavelength may vary. On long lengths of fiber typical of WANs or long-distance networks, the difference becomes important. The fiber attenuation coefficient is a function of wavelength, so source spectral characteristics can be important. When testing long lengths of fiber, one may want to make corrections to nominal source wavelength losses.

In multimode fiber, one may also need to be concerned by "mode power distribution," or how the light is transmitted in the various modes transmitted in multimode fibers. If one uses a test source with a LED typical of the types of LEDs used in actual link transmitters, the differences in mode power distribution will be minimized. Gigabit networks on multimode fiber should be tested with lasers, as noted in the section on testing giga-bit networks.

It is possible to get a source and meter together in a "Test Kit." One can also get an OLTS (optical loss test set), which is a meter and source in one package. But because the source and meter must be at opposite ends of the fiber being tested in the actual installation, an OLTS requires two "sets" of instruments at a higher cost to perform the same measurement.

Reference test cables (sometimes called test jumpers) are needed to test unknown cables. A "launch" cable attaches to your source to set reference conditions for all loss tests. In some tests, you will also need to add a receive cable on the meter after you set your launch reference. These cables must match the fiber type and size in the cable and mate to the connectors on the cables to test. These cables must be tested often to make certain they are good themselves, otherwise all measurements will be in error. While most multimode fibers are the same size (62.5/125 micron), there are now installations that have a mix of 62.5/125 and 50/125 fibers. One cannot mix these two fibers, either in testing or operation, as the mismatch in core sizes can lead to connection losses of up to 4 dB. Most single-mode fiber is the same, simplifying test cable inventory.

Mating adapters are also called bulkhead splices or splice bushings. They allow two cables with connectors to mate. They are also very important to good loss testing as they contribute to the mated connector loss. The ones with plastic alignment inserts are not good for testing as they wear out quickly. Use only those with metal or ceramic alignment sleeves. One can get hybrids too, to mate ST to FC or SC connectors to simplify testing different connector types.

9.6. Measuring Optical Power

Measuring power at the transmitter or receiver requires only a fiber optic power meter, an adapter for the fiber optic connector on the cables used, and the ability to turn on the network electronics. Remember when measuring power, the meter must be set to the proper range (usually dBm, but sometimes microwatts) and the proper wavelength.

When you are ready to measure power, attach the meter to the cable at the receiver to measure receiver power, or to a short test cable (tested and known to be good) that is attached to the source (shown in Fig. 9.3) to measure transmitter power. Turn on the source and note the power the

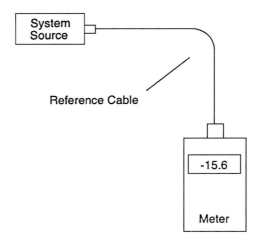

Fig. 9.3   Measuring optical power.

meter measures. Compare it to the specified power for the system and make sure it is in the acceptable range for that system.

9.7.   Testing Loss

Loss testing requires measuring the optical power lost in a cable with a source and power meter. In order to measure loss, it is necessary to create reproducible test conditions for testing fibers and connectors that simulate actual operating conditions. This simulation is created by choosing an appropriate source and mating a known good test cable called a launch reference cable to the source. The reference cable creates controlled test conditions that allow testing other cables mated to a known good connector with a matching fiber type and known source characteristics. (An OTDR or optical time domain reflectometer is used in some situations to measure loss, by analyzing backscattered light in the cable. Because the measurement is indirect and does not simulate actual operating conditions, it should not be used to measure actual link loss. It can allow inspection of intermediate connections and find faults in the cable plant. We'll discuss OTDRs later in this chapter.)

Correct setting of the reference launch power is critical to making accurate loss measurements. The output power of the source and launch cable should be measured by the power meter and recorded for loss calculations

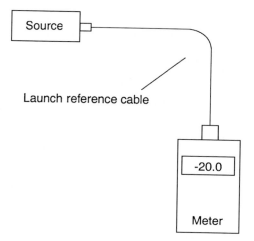

Fig. 9.4   Setting the reference power for loss measurements.

(Fig. 9.4). It is also advisable to periodically recheck this source output to ensure the source has not varied, and it must be rechecked if the launch cable is changed. Some meters will automatically store loss references and calculate loss directly. Even with these meters, it is advisable to record the power level from the launch cable in case the reference value in the meter is lost.

Another important issue in fiber optic testing is cleanliness. Optical fiber is only the size of a human hair and typical airborne particles are almost as big as the light-carrying core of the fiber. Dirt particles on the end of the fiber will create an attenuator when measuring power or when mating connectors to measure loss. It is absolutely mandatory to clean the mating surfaces of the connectors before making measurements (and a good idea to do it again before making cable plant connections!). All connectors and mating splice bushings are supplied with dust caps, which should not be removed except when testing or connecting the components.

No loss tests can be done reliably unless the launch cable is of high quality. Always test reference cables by the single-ended method shown in the next section to make sure they're good before testing other cables. "Reference quality" launch cables are not necessary; only cables that exhibit low loss (less than 0.5 dB max, 0.3 dB or better preferred).

The method previously described for setting the reference for loss measurements is the preferred method in practically all network and cabling standards. Some reference books and manuals show setting the reference power for loss using both a launch and receive cable mated with a splice bushing or even with a launch and receive cable and a "reference system cable." The reference power for loss measurements is supposed to represent the amount of power launched by the source and should be measured directly by a meter from the output of the launch cable. The other methods introduce uncertainties in the measurements, which will adversely affect the measurements.

When using the method of two-cable reference, if either the launch or receive cable is bad, setting the reference with both cables hides that fact. Even if the connectors were so bad the loss of the mated connectors was 3 dB, the reference value would not reflect this, since the loss of the two mated connectors is included in the reference measurement. Therefore one could begin testing with bad launch cables and make all loss measurements inaccurate. This method should never be used unless both reference cables have been tested individually and are known good. Even in the best case, this technique will underestimate connector loss by one connector, compared to how the link loss was specified.

The three-cable reference has the same problems as the two-cable reference, plus the problem that if the system reference cable is bad, it will cause "loss" measurements to show a "gain" with respect to the reference. Fortunately, this method has rarely been recommended.

9.8. Testing Cable Loss

There are two methods that are used to measure loss, which we call "single-ended loss" and "double-ended loss." Single-ended loss uses only the launch cable, while double-ended loss uses a receive cable attached to the meter also.

Single-ended loss (Fig. 9.5) is measured by mating the cable you want to test to the reference launch cable and measuring the power out the far end with the meter. When you do this you measure the loss of the connector mated to the launch cable and the loss of any fiber, splices, or other connectors in the cable you are testing. This method is described in FOTP-171. Because you can reverse the cable to test the connector on the other end, it becomes the best possible method of testing patchcords, since it tests each connector individually.

In a double-ended loss test (Fig. 9.6), you attach the cable to test between two reference cables, one attached to the source and one to the meter.

Fig. 9.5 Single-ended loss.

This way, you measure two connectors' losses, one on each end, plus the loss of all the cable or cables in between. This is the method specified in OFSTP-14, the test for loss in an installed cable plant.

Remember the reference is set the same way for both tests, by measuring the output of the launch cable with the power meter.

Fig. 9.6 Double-ended loss.

9.9. OTDR Testing

The optical time domain reflectometer (OTDR) is both the best known and least understood fiber optic instrument. The OTDR does not measure loss, but instead implies it by looking at the backscatter signature of the fiber. It does not measure cable plant loss that can be correlated to power budgets.

The OTDR (Fig. 9.7) works like radar or sonar, sending out a pulse of light from a very powerful laser, which is scattered by the glass in the core of the fiber. About one millionth of the light is scattered back up the fiber to the OTDR, where it is captured, averaged to improve the signal to noise ratio, and analyzed.

Because the group velocity of light in the fiber can be determined, the round-trip time for the OTDR pulse can be expressed in distance. The pulse is attenuated on the outward path and again on return after backscattering, so the OTDR can calculate the loss of the fiber and any intermediate

Optical Time Domain Reflector (OTDR)

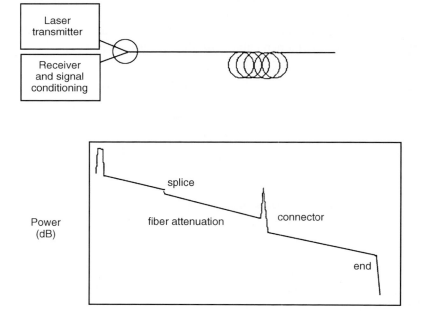

Fig. 9.7 OTDR block diagram.

Fig. 9.8 Information in the OTDR display.

connections from the backscatter signature of the fiber, assuming the backscatter coefficient is constant (Fig. 9.8).

The OTDR suffers from several uncertainties in measurements, and from physical limitations. The measurement uncertainties come primarily from the variations in backscatter of the fiber. The backscatter coefficient is a function of the material properties of the glass in the core and the diameter of the core. Variations of the fiber materials or geometry can cause major changes in the backscattered light, making splice or connector measurements uncertain by as much as $+/- 0.4$ dB. This has often led to confusion by showing a virtual "gain" at a connector, where the fibers involved have different backscatter coefficients. Connector or splice loss must be measured from both directions and averaged to remove this source of error.

In multimode cable, the OTDR laser source does not fill the higher order modes in the fiber, which are the highest loss. Therefore the loss measured underestimates the loss an actual LED source will see in actual use. And the OTDR does not see the connectors on each end of the cables, further underestimating the actual cable plant loss. Although it has proven impossible to correlate OTDR and source/power meter loss techniques, expect the OTDR to underestimate the loss of a link by as much as 3 dB.

While OTDRs can tell where events such as connectors, splices, or breaks are in cables, highly reflective events like connectors can cause "ghosts" that are confusing. The distance is uncertain not only due to instrument limitations, but also the fact that optical cable has about 1% more fiber than the length of the cable, to prevent stress on the fiber during pulling. Thus the actual distance to an event must be corrected for the excess fiber in the cable.

The laser pulse width of the OTDR must be wide in order to induce enough backscattered light to process. The pulse width may be as little as 5 meters or as much as 250 m on long-distance OTDRs. This means the OTDR cannot distinguish most patchcords in a LAN cable plant or any events close to the OTDR itself.

Within these limitations, OTDRs are used primarily to find faults in longer fiber runs, especially when they are remote, buried, or otherwise inaccessible. They should never be used to measure the cable plant loss for power budgeting.

9.10.   Troubleshooting Hints for the Cable Plant

If you have high loss in a cable, the most likely problem is bad connectors. To diagnose the problem cable, reverse it and test in both directions using the single-ended method. Because the single-ended test only tests the connector on one end, you can isolate a bad connector — it's the one at the launch cable end on the test when you measure high loss.

High loss in the double-ended test should be isolated by retesting using the single-ended method, reversing the direction of test to see if the end connector is bad. If the loss is the same, you need to either test each segment separately to isolate the bad segment or, if it is long enough, use an OTDR.

If you see no light through the cable (very high loss — only darkness when tested with your visual tracer), it's probably one of the connectors, and you have few options. The best one is to isolate the problem fiber, cut the connector of one end (flip a coin to choose), and hope it was the bad one. Alternately, if the cable is long enough (>10 meters), it can be tested with an OTDR with long (>100 m) cables on either side to find the bad connector. Of course the OTDR will find breaks or cuts in the cable if it is long enough.

9.11.   Special Test Considerations for Gigabit
Multimode Networks

GBE, as the name says, is Ethernet scaled up to gigabit speeds, providing a migration path from Ethernet at 10 MB/s to Fast Ethernet at 100 MB/s and then on to gigabit speeds for network backbones. It has a data rate of 1 gigabit per second (GB/s, a billion bits per second) and a baud rate of 1.250 Gb/s including the data encoding. Fibre Channel links used for peripheral connections use similar hardware, so we can consider both.

These speeds require laser transmitters, since LEDs are inherently limited to a few hundred MB/s (millions of bits/second). Two fiber optic standards are approved: 1000BASE-SX for short-wavelength lasers over multimode fiber and 1000BASE-LX for long-wavelength lasers over multimode or single-mode fiber. The short wavelength standard will typically be met with a 850-nm VCSEL (vertical cavity surface-emitting laser), a new type of device that offers high performance and very low cost. A CD laser may also be used. Long-wavelength lasers will be standard 1310-nm Fabry-Perot laser single-mode transmitters.

Running GBE over installed multimode fiber cable plants in premises applications (not single-mode fiber like the telco and CATV networks) creates a bandwidth problem. Multimode fiber works well with LEDs, but somewhat unpredictably with coherent laser sources. Coherent sources emit light in phase, and various modes in multimode fiber can interact such that the combination of the two rays interfere. The combination can have more or less intensity, depending on the phase difference between the two rays. In multimode fiber, this can create modal noise, causing bit error rate problems.

Furthermore, LEDs have a wide angle of light output, which must be focused into the fiber, while lasers have a much narrower beam that is more easily coupled to the fiber (Fig. 9.9). But while the LED will fill the higher order modes in the fiber (higher modes = wider angles), lasers will concentrate the light in the center of the fiber, especially the long-wavelength lasers with single-mode pigtails.

While this could theoretically allow higher fiber bandwidth to the laser, the laser can have severe problems with some fibers. The ideal index profile of a multimode fiber is a parabola, but manufacturing realities make this a step function, as different layers of glass are added by discretely changing

Fig. 9.9   Laser and LED launched power into multimode fiber.

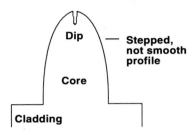

Fig. 9.10   A center dip in the fiber index profile can cause bandwidth problems.

the preform early in the fiber manufacturing process. Furthermore, when the preform is heated to fuse all the layers, there may be incomplete fusion in the center of the preform, causing a slight dip in the index profile (Fig. 9.10).

These properties of the index profile can cause problems with the bandwidth performance of the fiber, especially if the source is a laser. Some tests have shown a correlation of fiber bandwidth to the depth of the index dip. Manufacturers test the bandwidth of fiber with lasers but condition the launch to fill all the modes, minimizing these problems and also simulating performance with a LED.

Another problem that was identified early in the history of fiber optics was mode mixing at splices and connectors. Splices could cause mode mixing that severely affected the bandwidth of a cable plant by changing the mode power distribution following the splice. The bandwidth of the fiber can be affected by this mode mixing.

Fiber manufacturers have begun manufacturing fiber with high bandwidth at 850 nm as well as 1300 nm for these applications. High bandwidth 62.5/125 fiber that will support GBE to lengths much longer than currently proposed in the standard is becoming available. A revival of 50/125 fiber for its greater bandwidth has also been driven by GBE. The new high-bandwidth 62.5/125 fiber will offer even longer distance capability and physical compatibility to today's installed fiber base.

The issue is how far you can run GBE links. For multimode fiber, it's a function of the fiber bandwidth and transceiver design as we'll now discuss in Fig. 9.11. *Note: This is theoretical data calculated for worst case conditions.*

It is not easy to test the fiber already installed for bandwidth. Manufacturers are developing instruments for this purpose and have prototypes in test now and can show preliminary test data.

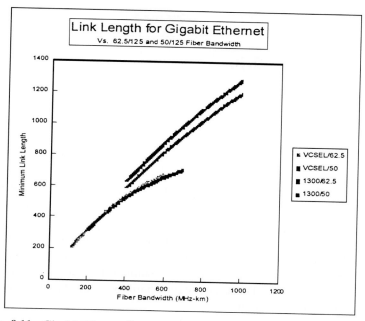

Fig. 9.11   Gigabit Ethernet distance capability as a function of fiber bandwidth.

Bandwidth testing can be done in the frequency domain or time domain. Either method requires a very fast laser transmitter and sensitive receiver. Sophisticated software analyzes the difference between the input and output pulses and calculates the bandwidth of the fiber. Analysis includes looking at the 3 dB bandwidth and if possible, actual waveforms, which like the OTDR trace, have much useful information for the trained user.

Data from a prototype Fotec bandwidth tester is shown in Fig. 9.12. It shows actual input and output pulses and gives data on the end-to-end bandwidth of the fiber.

It is interesting to note the effect of multimode fiber modal dispersion on transmitted pulses. Figure 9.12 shows what happens when bandwidth is inadequate for GBE. The time divisions are 1 ns and the vertical scale is pulse amplitude. Note how the output pulse is wider and misshapen, due to the differential mode delays of the fiber. The pulse not only is widened more than acceptable for GBE, but the distortion aggravates the problem.

Bandwidth testers also allow one to assess the effects of typical components in the cable plant. Using two 100-meter lengths of FDDI-grade fiber with a mid-span connection, one can see the effect of various fiber terminations. To evaluate the effects of termination and splicing, this experiment

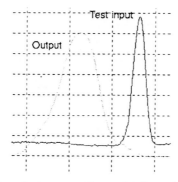

Fig. 9.12   The pulse broadening caused by modal dispersion in a multimode fiber.

used connectors, mechanical splices, and fusion splices to connect the two fibers. The 3-dB bandwidth obtained from multiple tests is shown in Table 9.2. Note the wide variation in bandwidth of the same link with different terminations. These are real variations in the bandwidth of the link caused by the terminations.

9.12.   Cable Plant Loss with Laser Sources

Without dispersion effects, GBE VCSEL transceivers have absolute loss margins of 7.5 dB minimum; about 13 dB maximum. Worst case loss margin, with all the power penalties caused by dispersion and other noise contributions, can be as low as about 3–4 dB.

At 850 nm, the losses in the maximum length cable plant (500 meters) should be within the margin, as long as good connectors are installed. The design of GBE was based around the component specifications for fiber in EIA/TIA 568 standards, which are quite conservative.

However, there are significant differences in cable plant loss when used with 850-nm VCSEL laser transmitters specified in GBE vs 850-nm LEDs

Table 9.2   **Bandwidth Variations with Various Termination Types**

	Connector	*Mechanical splice*	*Fusion splice*
Bandwidth Range	1.28–1.34 GHz	1.09–1.66 GHz	0.83–1.41 GHz

Table 9.3   **Cable Plant Loss with LED and VCSEL Test Sources**

Length (m)	LED loss (dB)	VCSEL loss (dB)	Difference (dB)
10	0.99	0.43	0.56
20 m	1.27	0.85	0.42
250	1.97	1.30	0.67
500 m	2.93	2.30	0.63

used in lower speed networks and most test equipment. The narrower mode fill of the lasers makes the loss of both the fiber and connectors less.

Table 9.3 shows cable plant loss measured per OFSTP-14, with both 850 nm VCSEL and LED sources for six simulated GBE cable plants. Short and long cable plants typical of floor-to-floor and campus cable runs, the minimal and maximum conditions likely encountered in GBE, were tested. Note that the differences do not increase greatly with cable plant length due to the mode mixing occuring at intermediate connections.

The special needs of high-speed multimode fiber optic networks indicates a need for special testing requirements. Loss testing with sources similar to system sources is preferred to minimize rejecting cable plants with adequate performance due to errors caused by source differences. Longer runs on unknown fiber may benefit from bandwidth testing to determine whether legacy fibers can support high-speed networks.

9.13.   Bit Error Ratio Measurements

In any digital communications transmission system, the fundamental measure of the quality of the system is found in the probability that transmitted bits will be correctly received as ones or zeros. The parameter that describes this measure of quality is the bit error ratio (BER). Simply stated, BER is the ratio of the number of bits received incorrectly compared to the number of bits transmitted (for a specified time interval or quantity of bits). Typical acceptable BER levels range from 1E-9 (1 error per 1 billion transmitted bits) to 1E-12.

The test equipment required to perform a BER test includes a pattern generator and an error detector (Fig. 9.13). The pattern generator will produce a known data sequence to the system or device under test. Test patterns can be designed such that they intentionally stress some aspect of the system. For example, a data sequence may be generated that is difficult for a clock

Fig. 9.13 BER test block diagram.

recovery system to synchronize to. Or, data patterns that attempt to mimic real traffic can be produced. A common pattern used is the pseudo-random binary sequence (PRBS). A PRBS has the characteristic of producing all possible combinations of ones and zeros for a given pattern length. For example, a 2^7-1 PRBS pattern will produce all possible combinations of 7-bit binary numbers from 0000001 through 1111111. (The 0000000 sequence is not produced due to limitations of the logic hardware used to generate the pattern.) Pattern generators also typically produce longer PRBS patterns, some as long as $2^{31}-1$, which will generate all combinations of 31 bit binary numbers except the all-zero case.

The error detector is used to determine if the data received matches the transmitted pattern. The error detector will receive the output of the receiver of the system under test. The system receiver under test attempts to determine the logic level of each bit transmitted by the system under test. The output of the receiver decision circuit is then routed to the error detector. In its most basic building blocks, the error detector consists of an internal pattern generator and an exclusive or logic gate. The error detector's internal pattern generator will produce a reference pattern identical to the pattern fed from the pattern generator feeding the test device. The reference pattern is synchronized to the test device output and then compared bit for bit with the exclusive or gate. Determining the BER is then a simple matter of counting the total bits received and the number that were incorrectly received.

In typical digital communications systems, the power from the transmitter is large enough that if it were to arrive unattenuated at the system receiver, there would be error-free communication. System performance in terms of BER is then often characterized in terms of the amount of attenuation between the transmitter or receiver. Similarly, the BER can be characterized in terms of power level at the receiver. Figure 9.14 shows a typical BER characterization of a high-speed system. As the received power is decreased, the signal-to-noise ratio is reduced and the probability of a bit being received in error increases.

Fig. 9.14   BER vs received optical power.

Because BER measurements are used to characterize the effects of system noise, they require analysis from a statistical perspective. Specifically, one must consider how many errors must be detected to render a statistically valid measurement. Ideally, a minimum of 1000 errors should be measured to yield a high confidence level in the accuracy of the BER measurement, although good measurements with as few as 100 errors can be achieved. This then leads to one of the primary difficulties encountered when performing a BER measurement. If a BER of 1E-10 is required, and at least 1000 errors are to be detected, then at least 10 terabits must be transmitted during the measurement. If the transmission rate is 2488 Mb/s, the time required for the measurement will be over an hour. If the data rate is 155 Mb/s, the test time would be almost 18 hours!

As BER levels are made more stringent (some systems require BERs of 1E-13), test times become proportionally longer, measured in days instead of hours. Referring to Fig. 9.14, a common technique for characterizing systems is to make measurements at signals levels that result in high BERs. As many errors are generated, the time required to make the measurements is comparatively low. From these results, conditions for very low BER can be estimated.

BER measurements may also be used to characterize system performance in terms of parameters other than attenuation or signal power. Any parameter that can impair system performance might be considered. Examples include BER vs jitter, sampling instant, and dispersion. Figure 9.15 shows a jitter tolerance measurement. In this measurement, jitter is sequentially increased until the BER is degraded, indicating the system/device under test's ability to perform in the presence of high levels of jitter.

Fig. 9.15   Jitter tolerance measurement.

9.14. Characterizing Digital Communications Waveforms

Complete systems and system receivers are characterized using BER tests. Characterization of transmitters is to a large extent achieved through analysis of the waveforms they produce. The instrument used to view a high-speed digital communications waveform is typically a high-speed digital sampling oscilloscope (DSO). Waveforms can be viewed in one of two ways depending on how the oscilloscope is triggered. A pattern generator, similar to that used in a BERT system, is used to produce a data signal for the transmitter being tested. Pattern generators may produce two types of signals used for timing references (triggers) for the oscilloscope. A pattern trigger is a timing edge that occurs once for every repetition of the data pattern. A clock trigger produces a timing edge for every transmitted bit.

Use of a pattern trigger allows one to view a specific portion of the data pattern. Figure 9.16 shows a data waveform measured with a DSO.

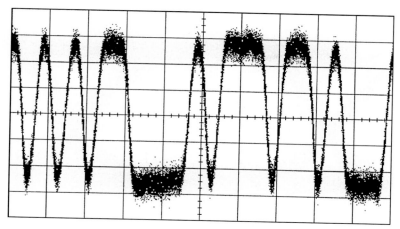

Fig. 9.16   Pulse train from a high-speed laser transmitter.

The measurement shows the waveform for a sequence of several ones and zeros. From this measurement we can see the relative separation between the two logic levels. Rise and fall times, indicating the speed of the transmitter, can also be quantified.

Using a pattern trigger provides a good snapshot view of a small sequence of bits in a transmitted waveform. However, it is important to understand how the transmitter behaves under all operating conditions, and not just a select pattern. This is achieved with the eye-diagram display.

The eye-diagram, shown in Fig. 9.17, is an overlay of many transmitted waveforms. By using a clock signal as the trigger input to the DSO, the transmitted waveform can be sampled over virtually the entire data pattern generated by the transmitter. Thus all the various bit sequences that might be encountered can be sampled to build up the eye-diagram.

For eye-diagram analysis the sample rate of a DSO is typically orders of magnitude slower than the data rate of the signal being characterized. For example, a DSO used to measure a 2488 Mbit/s signal may sample the waveform at a 40 K sample/sec rate. When a point is sampled, the next adjacent point in time sampled will occur thousands of bits later in the transmitted waveform, as shown in Fig. 9.18. It should be noted that the intent of the eye-diagram is not to display individual transmitted bits. Instead, it is constructed from a multitude of waveform samples. It then allows a good visual representation of the overall quality of the waveform.

The eye-diagram includes a significant amount of information about the transmitter output. The relative separation between the two logic levels is

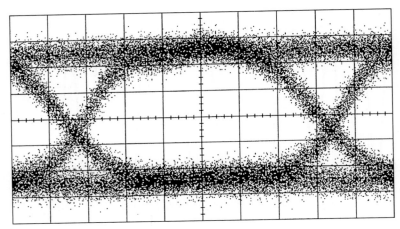

Fig. 9.17   Eye-diagram displayed by a DSO.

easily seen in the vertical "opening" of the eye. Obviously, the more open the eye is, the easier it will be for the receiver to determine the signal logic level. Rise and fall times can be measured on an eye-diagram to determine the transmitter speed. Modern DSOs have the ability to automatically construct histograms to statistically analyze the eye-diagram. Jitter describes

Fig. 9.18   Using histograms to measure jitter.

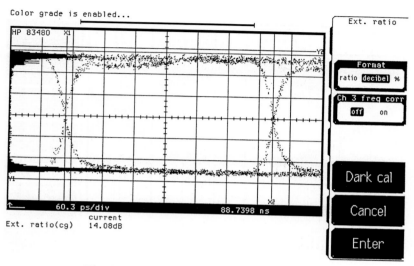

Fig. 9.19 DSO measurement of extinction ratio.

the relative instability of the bit period or data rate of a signal. This can be quantified by measuring the variance in the relative time in which the rising and falling edges occur. By constructing a histogram at the eye-diagram crossing point, jitter is determined, as seen in Fig. 9.19.

Other histogram-based measurements include eye height and width, which describe the vertical and horizontal openings of the eye respectively. A critical measurement for high-speed laser transmitters is extinction ratio. In simple terms, extinction ratio is the ratio of the power level of the logic '1' signal to the power of the logic '0' signal. Extinction ratio is an indicator of how well available laser power is being converted to modulation or signal power. When the extinction ratio is high, virtually all of the transmitted power is used as signal power. Typical extinction ratio values range from 7 (8.5 dB) to 30 (14.8 dB). This too is easily characterized using DSO histogram capabilities.

As extinction ratios approach high levels, measurement error can increase. Fundamentally, a high extinction ratio implies either a very high logic 1 signal power or a very low logic 0 signal power. The latter case is typical. An accurate extinction ratio measurement then requires an accurate measurement of a very small signal. Measurement errors can occur from offset signals or waveform distortion in the instrumentation. Today's modern DSOs are designed to minimize these potential errors.

Fig. 9.20   Eye-diagram mask testing.

Another technique used to characterize transmitted waveforms is to define and measure the shape of the eye-diagram. This is achieved with masks and pulse templates. To define the allowable shape of the eye-diagram, polygons are constructed by the DSO to define regions where the waveform may not exist. Figure 9.20 shows a mask test for a high-speed laser transmitter. Note that the mask consists of three polygons. The polygon in the middle of the eye determines the minimum allowable opening of the eye. The polygons above and below the eye determine allowable overshoot, undershoot, and ringing.

Many communication systems using electrical waveforms will characterize waveshapes by examining individual pulses rather than an eye-diagram. Instead of a mask test, pulses are required to fit within a template, as shown in Fig. 9.21.

Today's DSOs have built-in automatic capabilities to perform mask and template tests. Masks or templates are selected from menus, automatically aligned, and test results logged.

It is important when characterizing the shape of an eye-diagram that the measurement systems have a consistent and well-behaved frequency response. If the measurement channel does not have a flat frequency response, it may introduce waveform peaking or overshoot. Also, a waveform may look much different when measured on a very wide bandwidth system in comparison to a similar measurement made on a lower bandwidth

Fig. 9.21 Pulse template testing.

system. To achieve measurement consistency, standards committees often define the frequency response of the measurement system. For example, the optical receiver used with a DSO for laser transmitter measurements may be followed by a low-pass filter with a 3 dB point such that the test system is set to be 75% of the data rate. All those using such a measurement setup should have comparable results.

Masks and pulse templates are set up and defined by standards bodies to specify equipment to be used in systems such as SONET, SDH, or Fibre Channel. The intent of these tests, coupled with a variety of other specifications and tests, is to guarantee a minimum level of performance in transmitters to allow multiple vendors to supply equipment for a communications system.

9.15. Testing and Troubleshooting Networks

The installed network can be tested quickly and easily with a fiber optic power meter, tracer, and attenuators. If no power meter is available, a fiber tracer or even a flashlight can be used to determine if the cables are transmitting light, to isolate the problem to either the cable plant or the electronics.

Fiber optic link troubleshooting

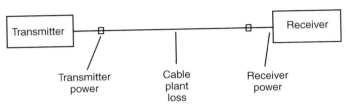

Fig. 9.22   Measurements for troubleshooting fiber optic links.

When using a power meter for diagnostics on a fiber optic link (Fig. 9.22), the network transmitter should be set to transmit a bit stream of known duty cycle. Set the power meter calibration on the proper wavelength and measurement range. To test the received power, the most critical element in the network, merely disconnect the fiber optic cable connector at the receiver, attach the power meter, and measure the power.

If the receiver power is low, the transmitter power should be measured by disconnecting the source jumper cable at the first available connector and measuring the power with the fiber at that point. Alternatively, one can disconnect the cable at the transmitter and use a test jumper of known loss to measure the coupled power. If the output is measured through a short network jumper cable (less than 10 meters), no compensation for fiber loss in the jumper cable is necessary. For longer jumpers, some compensation may be necessary.

If receiver power is low, but transmitter power is high, there is something wrong with the cable plant. They must be tested at every connection to isolate the bad cable(s) and/or connectors. This can be done from either end, but is more convenient starting from the transmitter end. Starting from the transmitter or receiver end, follow the network cables to every patch panel. Disconnect the connector and measure the power at each point. By making measurements in dB, one can easily calculate the loss of the cable network to each point by subtracting successive readings.

When a suspect cable is found, by noting a larger than expected loss in the cable link, the cable needs testing by the appropriate method previously described. If a cable has attenuation that is higher than specifications, but still transmits light, check connectors on a microscope to determine if they have been damaged and should be replaced. If the connectors look good, the best solution may be to replace the cable or switch to a spare. If a visible

Loopback testing

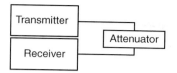

Fig. 9.23   Loopback testing.

laser fault locator is available, it can be used to visually locate breaks in the fiber and find broken connectors. Under some circumstances, such as high loss in long jumper or trunk cables, an OTDR (optical time domain reflectometer) can be used to diagnose cable faults.

9.16.   Transceiver Loopback Testing

Loopback testing is a powerful tool for datalink troubleshooting. It involves looping a signal back across a section of the network to see if it works properly. In a fiber optic link, one wants to test network loopback first from electronics on each end of the datalink, since it includes both transceivers (transmitter/receiver pair) and the cable plant.

The datacom capabilities of the transceiver can be tested with a loopback test also (Fig. 9.23). This test uses an attenuator between the transmitter and receiver and transmits a signal looped back on itself. Thus the transceiver can be tested to see if it can transmit data over the optical loss specified in the loss budget. If transceivers on both ends of a data link pass a unit loopback test but fail a network loopback test, the problem is in the cables, which then need testing by the methods previously described.

9.17.   Conclusion

Testing fiber optic communications requires some specialized knowledge of test and handling procedures but is otherwise straightforward. If networks are properly designed for testability, fiber optic networks are no more difficult to test and troubleshoot than any copper-based network.

Part 3 | The Applications

Chapter 10 | Introduction to Industry Standards

Schelto Van Doorn

Siemens Corporation, Santa Clara, California 95054

10.1. Introduction

This chapter provides some background material and motivation for the processes used to create international data communication standards. There are many standardization bodies that may play a role in this process, such as the IEEE, Internet Engineering Task Force (IETF), Optical Internetworking Forum (OIF), ATM Forum, Infiniband Trade Association, International Telecommunications Union (ITU), and others too numerous to list here. In recent years, the role of industry forums and trade associations has been growing, as the datacom industry attempts to accelerate the delivery of standards to market. Several popular "ad hoc" standards have arisen as a result. For the remainder of this chapter we will concentrate on some more formal examples of the standards process. First, we present an overview of the American National Standards Institute (ANSI), the governing body within the International Standards Organization (ISO), which gave rise to the Fibre Channel Standard (this standard applies to both copper and fiber optic media and uses the term "fibre" to denote both types of physical layers). We will also discuss the Gigabit Ethernet standard, which is similar to Fibre Channel in some respects and was developed with the support of the IEEE as an extension of previous Ethernet standards, as well as under the governance of ANSI. First, however, we will discuss some general features

HANDBOOK OF FIBER OPTIC
DATA COMMUNICATION
$35.00

of the standardization process and the motivation for creating international datacom standards.

The goal of standards is to write a minimal set of requirements to achieve a maximum amount of interoperability. A minimum set of requirements is important to encourage innovation and multiple solutions to a design problem. This in turn gives the end user a choice among multiple products from different suppliers that will interoperate. Standards are documented agreements containing technical specifications or other precise criteria to be used consistently as rules, guidelines, or definitions of characteristics to ensure that materials, products, processes, and services are fit for their purpose. However, it should be noted that a standard usually describes only the minimum, high-level requirements for interoperability and thus it is not the same as a product specification. Indeed, the intent of a standard is to allow for many different design implementations, each of which may have its own specification and may differ in detail that are important to the end user, such as reliability, higher level functionality, etc. We can see that although multiple products may claim compliance with a given standard, they may have component or functional specifications that are different.

Standardization is a condition existing within a particular industrial sector when the large majority of products or services conform to the same standards. It results from consensus agreements reached between all economic players in that industrial sector — suppliers, users, and often governments. They agree on specifications and criteria to be applied consistently in the choice and classification of materials, the manufacture of products, and the provision of services. The aim is to facilitate trade, exchange, and technology transfer through the following:

- Enhanced product quality and reliability at a reasonable price
- Improved health, safety and environmental protection, and reduction of waste
- Greater compatibility and interoperability of goods and services
- Simplification for improved usability
- Reduction in the number of models, and thus reduction in costs
- Increased distribution efficiency and ease of maintenance

Users have more confidence in products and services that conform to international standards. Assurance of conformity can be provided by manufacturers' declarations or by audits carried out by independent bodies.

10.2. Why Are International Standards Needed?

The existence of nonharmonized standards for similar technologies in different countries or regions can contribute to so-called "technical barriers to trade." Export-minded industries have long sensed the need to agree on world standards to help rationalize the international trading process. This was the origin of the establishment of ISO. International standardization is now well established for a wide variety of technologies in such diverse fields as information processing and communications, packaging, distribution of goods, energy production and utilization, shipbuilding, banking, and financial services. It will continue to grow in importance for all sectors of industrial activity for the foreseeable future. The main reasons for this are the following:

- Worldwide progress in trade liberalization: Today's free-market economies increasingly encourage diverse sources of supply and provide opportunities for expanding markets. On the technology side, fair competition needs to be based on identifiable, clearly defined common references that are recognized from one country to the next and from one region to the other. An industrywide standard, internationally recognized, developed by consensus among trading partners, serves as the language of trade.
- Interpenetration of sectors: No industry in today's world can truly claim to be completely independent of components, products, rules of application, etc. that have been developed in other sectors. Bolts are used in aviation and for agricultural machinery; welding plays a role in mechanical and nuclear engineering, and electronic data processing has penetrated all industries. Environmentally friendly products and processes and recyclable or biodegradable packaging are pervasive concerns.
- Worldwide communications systems: The computer industry offers a good example of technology that needs to be standardized quickly and progressively at a global level. ISO's open systems interconnection (OSI) is the best known series of international standards in this area. Full compatibility among open systems fosters healthy competition among producers and offers real options to users because it is a powerful catalyst for innovation, improved productivity, and cost-cutting.

- Global standards are needed for emerging technologies: Standardization programs in completely new fields are being developed. Such fields include advanced materials, the environment, life sciences, urbanization, and construction. In the very early stages of new technology development, applications can be imagined but functional prototypes do not exist. Here, the need for standardization is in defining terminology and accumulating databases of quantitative information.

- Developing countries: Development agencies are increasingly recognizing that a standardization infrastructure is a basic condition for the success of economic policies aimed at achieving sustainable development. Creating such an infrastructure in developing countries is essential for improving productivity, market competitiveness, and export capability.

10.3.   ISO

The technical work of ISO is highly decentralized, carried out in a hierarchy of approximately 2700 technical committees, subcommittees, and working groups. In these committees, qualified representatives of industry, research institutes, government authorities, consumer bodies, and international organizations from all over the world come together as equal partners in the resolution of global standardization problems.

The major responsibility for administrating a standards committee is accepted by one of the national standards bodies that make up the ISO membership — AFNOR, ANSI, BIS, CSBTS, DIN, SIS, etc. The member body holding the secretariat of a standards committee normally appoints one or two persons to do the technical and administrative work. A committee chairperson assists committee members in reaching consensus. Generally, a consensus will mean that a particular solution to the problem at hand is the best possible one for international application at that time.

10.4.   An Introduction to ANSI

ANSI has served in its capacity as administrator and coordinator of the United States private-sector voluntary standardization system for 78 years. Founded in 1918 by five engineering societies and three government agencies, the institute remains a private, nonprofit membership organization supported by a diverse constituency of private- and public-sector organizations. Throughout its history, the ANSI federation has maintained as its primary goal the enhancement of global competitiveness of U.S. business and the American quality of life by promoting and facilitating voluntary consensus standards and conformity assessment systems and promoting their integrity. The institute represents the interests of its nearly 1400 company, organization, government agency, institutional, and international members through its headquarters in New York City and its satellite office in Washington, DC.

ANSI does not itself develop American National Standards (ANSs); rather, it facilitates development by establishing consensus among qualified groups. The institute ensures that its guiding principles — consensus, due process, and openness — are followed by the more than 175 distinct entities currently accredited under one of the federation's three methods of accreditation (organization, committee, or canvass). In 1995, the number of ANSs increased by 10% to a total of 11,500. ANSI-accredited developers are committed to supporting the development of national and, in many cases, international standards, and addressing the critical trends of technological innovation, marketplace globalization, and regulatory reform. ANSI promotes the use of U.S. standards internationally, advocates U.S. policy and technical positions in international and regional standards organizations, and encourages the adoption of international standards as national standards where these meet the needs of the user community. ANSI is the sole U.S. representative and dues-paying member of the two major nontreaty international standards organizations, ISO and, via the U.S. National Committee (USNC), the International Electrotechnical Commission (IEC).

ANSI was a founding member of the ISO and plays an active role in its governance. ANSI is one of five permanent members to the governing ISO council and one of four permanent members of ISO's Technical Management Board. United States participation, through the USNC, is equally strong in the IEC. The USNC is one of 12 members on the IEC's governing Committee of Action and the current president of the IEC is from the United States. Through ANSI, the United States has immediate access to the ISO and IEC standards development processes. ANSI participates in almost the entire technical program of both the ISO (78% of all ISO technical committees) and the IEC (91% of all IEC technical committees) and administers many key committees and subgroups (16% in the ISO and 17% in the IEC). As part of its responsibilities as the U.S. member body to the ISO and the IEC, ANSI accredits U.S. Technical Advisory Groups (U.S. TAGs) or USNC Technical Advisors (TAs). The U.S. TAGs' (or TAs') primary purpose is to develop and transmit, via ANSI, U.S. positions on activities and ballots of the international technical committee. In many instances, U.S. standards are taken forward, through ANSI or USNC, to the ISO or IEC, where they are adopted in whole or in part as international standards. Because the work of international technical committees is carried out by volunteers from industry and government, the success of these efforts often is dependent on the willingness of U.S. industry and the U.S. government to commit the resources required to ensure strong U.S. technical participation in the international standards process.

Conformity assessment, the term used to describe steps taken by both manufacturers and independent third parties to assess conformance to standards, also remains a high priority for the institute. ANSI's program for accrediting third-party product certification has experienced significant growth in recent years, and the institute continues its efforts to obtain worldwide acceptance of product certifications performed in the United States and the promotion of reciprocal agreements between U.S. accreditors and certifiers. One of the best indicators of the strength of the U.S. system is the government's extensive reliance on, and use of, private-sector voluntary standards. Pursuant to Office of Management and Budget (OMB) Circular A119, federal government agencies are required to use voluntary standards for regulatory and procurement purposes when appropriate. State and local governments and agencies have formally adopted thousands of voluntary standards produced by the ANSI federation, and the process appears to be accelerating. In summary, the ANSI federation continues to be fully involved in its support of the goals of U.S. and global standardization and committed to enhancing of the quality of life for all global citizens.

10.4.1. X3 COMMITTEE

The ANSI Accredited Standards Committee X3, more commonly referred to as simply X3, was established in 1961. It is accredited by ANSI to develop voluntary standards. The X3 secretariat is held by Information Technology Industry Council. As a standards developer in information technology, X3 is an active participant in the U.S. TAG to ISO/IEC JTC 1 (JTC1 TAG). X3 provides its 2000 members with countless opportunities to know and influence the key issues, activities, and people that drive national and international information technology standards in dynamic areas of commerce, technology, and society. X3 members have early and ready access to a wealth of technical information affecting the future of information technology on a global basis. Members actively participate in a deliberative consensus-building process that is both national and international in scope. The X3 Operational Management Committee (OMC) is an advisory committee to X3 on new national and international standards development projects, standards requirements, and review of proposed standards. OMC also manages the standardization process within the X3 technical committees (TCs). OMC considers the functional and economic, rather than the detailed technical aspects of standardization. X3's standards are developed by its TCs. A brief description of the TCs relevant to this book follows:

X3T10: I/O interface — Lower level (SCSI)

X3T11: I/O interface — Device level (IPI, HIPPI, FC)
   The current chairman of X3T11:
      Roger Cummings
      Distributed Processing Technology, Inc.
      140 Candace Drive
      Maitland, FL 32751
   Phone: (407) 830-5522 x348
   Fax: (407) 260-5366
   Vice chair:
      Carl Zeitler
      IBM Austin, MS 9250
      11400 Burnet Road
      Austin, TX 78758, USA
      Phone: (512) 838-1797

X3T12: I/O interface — Distributed data (FDDI)

X3T13: I/O interface — AT attachment

You may order completed ANSI publications by calling
   The American National Standards
   11 West 42nd Street
   New York, New York 10036
   Phone: (212) 642-4900 Agents are available from 8:45 AM to 4:45 PM (EST)

For standards in progress:
   Global Engineering Documents
   15 Inverness Way East

Englewood, CO 80112 USA
Phone: 303-397-2715; 800-854-7179 (United States and Canada)

The Fibre Channel Association can be reached on the World Wide Web
at http://www.amdahl.com/ext/CARP/FCA/FCA.html

10.5.   IEEE 802.3z (Gigabit Ethernet)

The set of objectives for IEEE P802.3z (gigabit Ethernet) standard include
the following:

1. Speed of 1000 Mb/s at the MAC/PLS service interface
2. Use 802.3/Ethernet frame format
3. Meet 802 FR, with the possible exception of Hamming distance
4. Simple forwarding between 1000, 100, and 10 Mb/s
5. Preserve min and max frame size of current 802.2 Std
6. Full- and half-duplex operation
7. Support star-wired topologies
8. Use CSMA/CD access method with support for at least one
   repeater/collision domain
9. Support fiber media and, if possible, copper media
10. Use ANSI Fibre Channel FC-1 and FC-0 as basis for work
11. Provide a family of physical layer specifications that support a
    link distance of
    At least 500 m on multimode fiber
    At least 25 m on copper (100 m preferred)
    At least 2 km on single-mode fiber
12. Decide between collision domain diameter of $\geq$50 m or $\geq$200 m
13. Support media selected from ISO/IEC 11801
14. Accommodate proposed P802.3x flow control

There are five criteria motivating the development of gigabit Ethernet
technology; its market potential, compatibility with existing Ethernet standards, its distinguishing features from other standards, and its technical and
economic feasibility. We will review each of these criteria in turn to give
some insight into the forces guiding the standards committee because the
final specification has not yet been written.

10.5.1. BROAD MARKET POTENTIAL

- Broad set(s) of applications
- Multiple vendors, multiple users
- Balance cost, local area network (LAN) vs attached stations

The fast growth of CPU speed is forcing the development of new LANs with higher bandwidth. Applications that can benefit from this capability include backbone, server, and gateway connectivity. Higher bandwidth for multimedia, distributed processing, imaging, medical, CAD/CAM, and other applications is also motivating the aggregation of 100 Mb/s switches and the potentially large upgrade base for installed 10/100 Ethernet. Multiple vendors and users have demonstrated interest by attending the gigabit Ethernet tutorial (more than 200 participants), attending the preliminary study group meeting (more than 120 participants), and enrolling in the higher speed e-mail reflector (more than 210 participants). Eighty-one participants representing at least 54 companies indicate that they plan to participate in the standardization of 1000 Mb/s 802.3. This level of commitment indicates that a standard will be supported by a large group of vendors. This in turn will ensure that there will be a wide variety of equipment to support a multitude of applications. Higher speed 802.3 solutions, which include scaled-up versions of existing 802.3 topologies, have balanced cost. Prior experience with scaling 802.3 across the range of 1 to 100 Mb/s indicates that the cost balance between adapters, cabling, and hubs remains roughly constant, provided that the operating speed can be achieved within the limits of current technology.

10.5.2. COMPATIBILITY WITH IEEE STANDARD 802.3

- Conformance with CSMA/CD media access control (MAC), PLS
- Conformance with 802.2
- Conformance with 802 FR

The proposed standard will conform to the CSMA/CD MAC, with currently authorized extensions appropriately adapted for 1000 Mb/s operation. In a fashion similar to the 100 BASE-T standard, the current physical layers will be replaced with new physical layers (PHY) as appropriate for 1000 Mb/s operation. The proposed standard will conform to the 802.2 LLC interface and the 802 Functional Requirements Document (with the possible exception of Hamming distance). The CSMA/CD access method will not support a 2-km network diameter at this speed while maintaining the current values in the MAC parameter table. This portion of the application space will be addressed at 1000 Mb/s with the full-duplex operating mode of 802.3.

10.5.3. DISTINCT IDENTITY

- Substantially different from other 802.3 specs/solutions
- Unique solution for problem (not two alternatives/problem)
- Easy for document reader to select relevant spec

The proposed standard is an upgrade for 802.3 users, based on the 802.3 CSMA/CD MAC, running at 1000 Mb/s. Maximum compatibility with the installed base of more than 60 million CSMA/CD nodes is maintained by adapting the existing CSMA/CD MAC protocol for use at 1000 Mb/s. Established benefits of CSMA/CD and the 802.3 MAC include the following:

- Optimistic transmit access method
- High efficiency in full-duplex operating mode
- Well-characterized and understood operating behavior
- Broad base of expertise in suppliers and customers
- Straightforward bridging between networks at different data rates

The management information base (MIB) for 1000 Mb/s 802.3 will maintain consistency with the 802.3 MIB for 10/100 Mb/s operation. Therefore, network managers, installers, and administrators will see a consistent management model across all operating speeds. The proposed standard will encompass one physical layer solution for each specific type of network media (e.g., single-mode fiber, multimode fiber, coaxial cable, and balanced pair cable). It will be a supplement to the existing 802.3 standard, formatted as a collection of new clauses, making it easy for the reader to select the relevant specification.

10.5.4. TECHNICAL FEASIBILITY

- Demonstrated feasibility; reports and working models
- Proven technology and reasonable testing
- Confidence in reliability

Technical presentations, given to 802.3, have demonstrated the feasibility of using the CSMA/CD MAC in useful network topologies at a rate of 1000 Mb/s. Technical presentations given to 802.3 from multiple current vendors of full-speed Fibre Channel components have demonstrated the feasibility of physical layer signaling at a rate of 1.06 Gbaud on both fiber optic and copper media. Many of these vendors have expressed support for an increase in the signaling rate to 1.25 Gbaud, which would support a MAC data rate of 1000 Mb/s. The principle of scaling the CSMA/CD MAC to higher speeds has been well established by previous work within 802.3. The 1000 Mb/s work will build on this experience. The principle of building bridging equipment that performs rate adaptation between 802.3 networks operating at different speeds has been amply demonstrated by the broad set of product offerings that bridge between 10 and 100 Mb/s. Vendors of full-speed Fibre Channel components and systems are building reliable products that operate at 1.06 Gbaud and that meet worldwide regulatory and operational requirements.

The FC-1 layer is the encode/decode layer and defines the DC-balanced 8B/10B code, byte synchronization, and character-level error control. In the FCS standard, 8-bit data bytes rate converted to 10-bit transmission characters with an error-checking code; the 8-bit bytes are recovered at the other end of the link. Using 10 bits rather than 8 provides for 1024 possible encoded values rather than only 256. Not all these values are used; to maintain the DC balance of the code, only characters that contain 4 zeros and 6 ones, 6 zeros and 4 ones, or 5 zeros and 5 ones are used. Some of the extra characters are reserved for lwo-level link control, and one special character called a comma or K28.5 is used for byte synchronization.

The FC-2 layer is the framing protocol layer; it defines the signaling protocol including frame and byte structure. This level includes framing protocol, flow control, 32-bit CRC generation, and various classes of service definitions. A frame is the smallest unit of data that can be transmitted on a Fibre Channel link. Frame size depends on the implementation and is variable up to a maximum of 2148 bytes. Each frame contains a 4-byte start of frame delimiter, a 24-byte header, up to 2112 bytes of FC-4 payload, a 4-byte CRC, and a 4-byte end of frame delimiter. The FC-4 payload consists of 0–64 bytes of optional headers and 0–2048 bytes of data. Related frames are grouped together to form a sequence, and multiple sequences can be grouped to form an exchange [comparable to a SCSI input/output (I/O) process]. If an error is detected that requires a recovery operation, then the sequence containing that frame may be retransmitted. The FC-2 layer also defines three classes of service. Class 1 establishes a dedicated connection between two devices (a circuit-switched connection), guarantees delivery of frames in the order in which they were transmitted, and provides confirmation of delivery. Class 2 is a multiplexed, connectionless, frame-switched link; delivery is guaranteed with acknowledgment of receipt but not necessarily in the order of transmission. Class 3 does not guarantee either delivery or order of receipt. If no fabric is present, class 2 and 3 become special cases of point-to-point interconnection.

The FC-3 or common services layer defines a set of services common across multiple ports of a node. This layer is still under discussion and no functions have been formally defined at this time.

The FC-4 or protocol mappings layer is the topmost layer and defines mapping of the upper level protocol (ULP) layers to Fibre Channel. Each ULP supported by FCS is specified in a separate FC-4 document; for example, the Fibre Channel protocol for SCSI is called FCP; it supports SCSI without requiring an SCSI bus because the upper layer protocols are not tied to a particular physical medium or interface.

10.5.5. ECONOMIC FEASIBILITY

- Cost factors known and reliable data
- Reasonable cost for performance expected
- Total installation costs considered

Cost factors are derived from the current full-speed Fibre Channel component supplier base. A reasonable cost increase (three times that of 100BASE-FX) with a 10-fold increase in available bandwidth in the full-duplex operating mode will result in an improvement in the cost/performance ratio by a factor of 3.33 for multimode fiber applications. The provision for a half-duplex operating mode using the 802.3 CSMA/CD MAC will permit the construction of very inexpensive repeating hubs. Customers will in many cases be able to re-use their existing fiber that has been installed in accordance with ISO/IEC 11801. Installation costs for new fiber runs based on established standards are well known and reasonable. Costs for coaxial-based short-run copper links are well established for full-speed Fibre Channel. Although the cost model for the horizontal copper cabling is well established, the cost model for 1000 Mb/s physical layers that will operate on horizontal copper cabling has not yet been firmly established. Presentations have been given to the high speed study group (HSSG), that suggest a cost multiple of two times relative to 100BASE-T2.

Chapter 11 | Intramachine Communications

John D. Crow

IBM Corporation, Yorktown Heights, New York 10598

Alan F. Benner

IBM Corporation, Poughkeepsie, New York 12601

11.1. Introduction

Despite many years of predictions of the imminent arrival of optical computers and optically interconnected computers, there has been relatively little penetration of optoelectronics in the computer interconnect field. There are incidents, but the impressive and continuous growth in the capability of integrated circuit (IC) chip and packaging technology has combined to shrink both the size of computer system building blocks to well within the capabilities of copper wire and electrical connections and the price of computing power to the point that it is difficult for any new technology to be cost-effective. Nevertheless, there has been a recent paradigm shift in the interconnect architecture of computing systems that may provide opportunities for optical interconnect technology. This chapter describes these opportunities and their requirements.

Historically, the data processing subsystem of a computer has been confined to a single equipment frame with highly integrated ICs and dense packaging. In this configuration, wiring density is essential, and the aggregate bandwidth of electronic interconnects per unit area has been greater than that which optical interconnects could achieve for distances within a machine. High aggregate bandwidth for electronic interconnects is achieved by tightly packing wires in two-dimensional formats [e.g., hundreds of pins per cable connector and >700 pin ball grid array (BGA) chip carriers].

HANDBOOK OF FIBER OPTIC
DATA COMMUNICATION
$35.00

Optical interconnects have until recently been almost exclusively single-fiber connections. Because optical links are ultimately driven by electronic circuits, and electronic circuits cannot be made to operate hundreds of times faster, regardless of only having to drive within a module, there has been no reason to use optics.

Over the past few years, it has become increasingly difficult to define "intramachine" connection due to the advent of distributed computing systems. When, for example, an ocean/atmosphere weather simulation is run, with the ocean model being executed on an Intel Touchstone Delta and the atmosphere model executed on a Cray YMP — with interface data being exchanged constantly during the simulation [1] — it is somewhat difficult to say whether one machine or two are being used. Even more difficult is describing the operation of a distributed memory parallel processing machine such as the IBM RS/6000 SP-2, in which each node has its own separate operating system, memory, disk, and input/output (I/O) adapters, but specialized high-performance hardware and software lets the machine be programmed as a single computing resource. On such a machine, parallel computing jobs are started, debugged, checkpointed, run, and stopped on one, several, or all nodes simultaneously. Here, the question of whether one machine or many are being used at any particular time is more a statement of system management and system software rather than cabling and addressability.

There is an increasing trend in the high-performance computing world toward aggregating computing resources and performing computing over physically distributed processing engines. This trend naturally leads toward increasing requirements on the computer interconnects to allow close coupling between computing facilities and allow software to view them as a single distributed computing resource.

In this chapter, we describe the general characteristics of computer networks that couple distinct classic computers [central processing unit (CPU) + memory] into a single closely coupled distributed computing resource. Because the requirements on such a network are fairly well established and unique, we introduce the notion of a system area network (SAN) and enumerate characteristics that distinguish it from a local area network (LAN) and from the internal wiring of a single computer. As the performance requirements for LANs increase to the level currently required for SANs, much of the physical layer of cables, and signaling, will be extended from SANs to LANs.

We also enumerate the characteristics of optical components and optical data links that would be required to satisfy the requirements of a system area network and show that the requirements (particularly aggregate bandwidth) can be most cost-effectively addressed using parallel optical interconnects (POIs). Parallel optical interconnects are a relatively new technology, consisting fundamentally of an array of lasers in a single package connected by a single connector to a ribbon of optical fibers and to an array of receivers. We describe the current state of the art in parallel optical interconnects and directions for the technology.

11.2. Current Intramachine Optics Applications

Before discussing system area networks and parallel optical interconnect hardware, we will discuss various intramachine computer interconnect applications in which optical components have already been applied. We concentrate here on the applicability of these technologies to computer systems and on the reasons why, despite their higher cost [2], they have been able to replace electronic components in those applications.

11.2.1. OPTICAL CLOCK DISTRIBUTION

A large computing resource such as a parallel supercomputer will generally consist of a number of distinct processors (CPU + memory) that may be physically separated into different racks. All of the digital logic will require clocking signals, and the clock frequencies at all processors will generally be at least approximately equal.

There are two common strategies for generating such clock signals. Either each processing element may have its own oscillator and clock distribution system or the whole system may be clocked from a single oscillator with a large clock distribution system. With the first method, the clock distribution problem is easier (a clock signal may not have to propagate outside a single rack), but the machine will require asynchronous buffers between clock domains and idle cycles over the communication links between clock domains to compensate for slightly different clock frequencies in different domains. A single distributed clock signal over the entire machine can significantly increase performance for low-overhead operations compared to multiclock designs. For machines constructed using a single oscillator, a clock distribution system must be constructed that takes a single high-frequency oscillator signal and distributes it over the full complex of digital circuitry with low jitter and without radiating electromagnetic interference (EMI) from the computing complex. For such a system, the high fan-out capabilities and low EMI of an optical clock distribution system can be attractive, particularly at high (>500 MHz) clock frequencies. Several commercial Cray machines have been built with optical clock distribution. In such a system, a single oscillator drives several lasers (for redundancy), and the optical clock signal is fanned out and distributed to each portion of the parallel computer, where a receiver converts the signal to electronic format and an electronic network further distributes the clock signal over the digital logic.

11.2.2.   I/O BUS EXTENSION

The other main area in which optical components have been successful has been in the extension of I/O buses, such as a personal computer interface (PCI) bus, out to longer distances than can be supported using electronic cabling. There are two main applications for optical bus extenders: LAN communications and long-distance disk I/O. Optical LAN cables have become increasingly prevalent as multimode fiber cabling inside a building has become competitive in price with copper cabling [3].

A significant fact in the usage of optics for disk I/O has been the decreasing size of disk drives. Disk drives platters have recently become prevalent with platter diameters approximately half the size of a parallel small computer system interface (SCSI) connector. This has driven the advent of serial disk interfaces [e.g., Fibre Channel or storage system architecture (SSA)], which use a single signal line and can be either optical or electronic. The simplicity of converting between electronic and optical cabling has encouraged the flexible placement of disk drives at relatively long distances from the processors to which they are attached. The latency of the physical cabling medium (~5 ns/m) is relatively insignificant relative to the access latency of the disks, which is measured in milliseconds.

The primary reason for optical bus extenders has generally been the desire to separate processing components to different parts of a single machine room or to different rooms or buildings. Some examples are optical channel extenders for high-performance parallel interface (HIPPI) optical implementations of Fibre Channel and serial HIPPI, and, at the high end, optical extenders for the Cray GigaRing.

A second primary reason for optical bus extenders has been the need to simplify cabling for multichannel installations. Mainframe installations, such as the IBM S/390, often need to interconnect multiple separate processing engines with multiple sets of disk drives and multiple external connections. Wiring of such installations can be made much simpler and more reliable if intermachine connections are converted to serial optical format and ganged together in multifiber ribbons. As many as 144 1-Gb serial optical lines can be ganged together in a single connector housing, significantly simplifying wiring, configuration, and reconfiguration of large systems [4].

Parallel optical interconnects are currently used in I/O extenders to serve a line concentration function when many I/O channel source and sink in the same physical location (see Chapter 13). There is little or no synchronism between the fiber lines, and typical direct access storage device (DASD) I/O channel data rates are used. A similar use for POI is found in large telecommunications multiframe switching systems.

11.3.   System Area Networks

We introduce the concept of a system area network here to distinguish the types of systems in which parallel fiber interconnects are most likely to be used first.

The 1990s have seen the widespread adoption of paralleling processing modalities for large systems. In these systems, multiple processors are coupled together to solve parts of a large-scale problem. Because each processor can work semi-independently, and because the group of processors can be programmed together, a group of processors can solve problems faster than a single processor. In the best cases, adding processors to a multiprocessor system can linearly speed up the problem-solving speed. (There are even some cases in which there is better than linear speed-up because each processor's limited caching space is better utilized when operating over a smaller portion of the full job.)

The evolution of parallel processing machinery has seen a steady evolution toward a system model that might be termed "clustered multiprocessors." In this architecture, a cluster of shared memory processors (SMPs) operate as a single computing resource. Each SMP contains between 2 and 16 processors, sharing (logically) a single system address bus, and having a single multithreaded operating system. Building SMPs larger than 16-way becomes increasingly difficult and less rewarding due to the difficulty of scaling the operating system and increased competition for the system bus.

To further increase scalability, the SMPs are clustered together over a high-speed network. The high-speed network is termed a system area network. The characteristics of this network are intermediate in performance

Fig. 11.1 Diagram of a system area network, illustrating how it interfaces to other data communications networks.

and operation between those of a system bus and a LAN. The performance of the entire parallel processing machine is dependent, aside from the CPU performance and parallelizability of the algorithm being executed, on the performance of this network.

This kind of network is given the name system area network to emphasize that this network operates within the bounds of a set of processors that manage as a single system but shares architectural similarities with a classical network rather than a bus.

Figure 11.1 illustrates how a SAN might be implemented functionally and how it relates to data storage and data communications networks. In this figure, building blocks of microprocessor + local memory are interconnected to form a "single-image system." The single-image concept arises because a user (application) connected to any one node of the SAN does not know which or how many of the nodes are processing his or her job. The system architecture and software determines how to distribute his or her job to optimize resource. To maintain this transparency of the system nodes and interconnects to the user, the performance demands on the interconnect hardware can be large. Conversely, the connections with the storage system and the global data communications network, shown in Fig. 11.1, are isolated or buffered from the computing operation by interface adapters, that contain both hardware and software. The user/application definitely knows when data are sent over these interconnects because of the communication protocols that must be added to the data and also because of (usually) a serious degradation in system performance.

11.3.1. SAN DISTINGUISHING CHARACTERISTICS

The following are the characteristics that distinguish a system area network. Performance values are current as of early 1997. Clearly, the absolute numbers will improve, but it appears that the distinction between system bus, SAN, and LAN interconnects as distinct technologies will remain.

11.3.1.1. Bandwidth

The bandwidth of a SAN is higher than that of LAN, ideally as close as possible to the bandwidth of a system bus. Practically, this will mean a bandwidth higher than 100 MBytes/s.

11.3.1.2. Latency

Because communications over the SAN are intended to approach the performance over a system bus, where latency is measured in hundreds of ns, the latency across a SAN (including communication protocol) must be $<100\ \mu$s, preferably $<10\ \mu$s.

11.3.1.3. Scalability

A SAN would be expected to support economic scalability from a few nodes up to thousands of nodes. It does not imply direct connectivity to tens of thousands of nodes, given the difficulty of programming applications scalable to this range, although some SIMD machines could be considered to be in this range.

11.3.1.4. Packet-Switched Network Topology

Given the requirement of connecting up to thousands of nodes simultaneously, a SAN cannot depend on a shared medium. Also, the low latency requirement drives toward implementation as a packet-switched network. Some SANs will incorporate support for circuit switching as well, and advanced implementations will incorporate support for multistage networks for economical scalability over the full range of supported system sizes.

11.3.1.5.   Flow-Controlled Transmission
for Congestion Control

At a general level, congestion in a network caused by multiple sources sending to the same destination or over the same link or switch can be handled by either (i) dropping or rerouting low-priority packets for retransmission and later delivery or (ii) inserting link-level flow control so that the sender of data is prevented from sending it until it is assured of being receivable. There will also generally be a mechanism for end-to-end flow control.

For long-distance communications or communications in which it is not vital that every bit be received correctly (e.g., voice or video traffic), flow control is generally used less frequently because link-level flow control requires the transmitter to maintain data buffering of a size equal to at least the link bandwidth times the round-trip time to maintain full transmission bandwidth. For SANs, however, the link round-trip times are short enough that the cost of the buffering is less detrimental than the cost of the protocol processing required for retransmitting dropped packets. A SAN will generally not handle congestion control by dropping packets but will use link-level and end-to-end flow control.

11.3.1.6.   Controlled Topology

Because all nodes on the SAN network are known and managed as a single system, the topology of the network can be determined, fixed, and set in the source routing tables. This is generally the case with a LAN as well. The point is emphasized here to contrast it with the case of internetworked LANs, in which (i) routing information is not made available between LANs and bridges or routers must add source and destination identifiers to routing information, at a cost in latency of many microseconds; and (ii) the source-level protocol must assume that intermediate nodes may or may not be available to route messages to a destination node.

11.3.1.7.   Source Routing

Source routing means that the originator of a message can specify the entire routing through the network at the same time that it is doing the rest of the protocol processing, eliminating the need for any routing look-ups in network switches. This characteristic is in contrast to asynchronous transfer mode (ATM) networks, which require the network switches to convert virtual path identifier and virtual circuit identifier information into routing information and back at each switch stage. Because ATM networks are largely targeted at longer distance networks, where the propagation time may be hundreds of μs, the extra latency of a routing look-up is negligible. In a SAN, the low latency and fast routing requirement dictate that the source be able to specify routing through the entire network. In some instances, it is possible to achieve adequate performance if routing information can be calculated at each switching stage by an extremely fast and simple conversion from the destination identifier.

11.3.1.8. Addressability

Some system area networks may also manage their addressing and routing by maintaining a single physical address space across the network so that a physical address presented on one SMP bus would be responded to with data residing in random access memory chips on another bus. Some may maintain cache coherence over the network in very tightly coupled systems.

11.3.1.9. Optimized User-Space Communication Protocol

The latency and bandwidth capabilities supported by the hardware cannot be utilized with a traditional protocol stack driven by the operating system. For example, the latency for TCP/IP or UDP communications is several hundred microseconds, and the maximum bandwidth supported is below 50 Mb/s for most machines. The protocol-processing performance is improving, of course, but not as quickly as the performance of the network. The ratio between latency and bandwidth performance supported by the network and that supported by the traditional protocol stack is at least a factor of 2 and growing.

To circumvent this problem, a SAN will generally use a highly optimized communication protocol. In a SAN, as opposed to a LAN, some optimizations are possible by (i) utilizing limited addressing, for communicating with a limited and relatively stable set of communication partners, and (ii) minimizing the number of copies for data and control information by collapsing the number of protocol stack layers.

The biggest breakthrough in this area has been the advent of "user-space" communication protocols. These protocols minimize protocol processing overhead and latency by operating entirely in the user's process space through library calls and allowing the user-level applications to directly address network interface resources, eliminating the need for a lengthy processor context switch into kernel-level protocol processing code.

The general picture of a system area network is a multiprocessor computing resource in which similar control mechanisms, bandwidth requirements, and latency requirements to those of a computer's system bus are implemented over a distributed network.

Most important, this requires protocol processing mechanisms that are different (faster and more controlled) than those in a LAN interconnection. The TCP/IP protocol is a sample case in which the processing power of a Cray YMP is only capable of processing the TCP/IP protocol fast enough to send data at 80 Mb/s, and yet the system bus speed for such a system (and for a large-scale SMP of 1996 and beyond) will be more than an order of magnitude faster.

11.3.2. SAN PROTOTYPE MACHINES

There are a large number of commercially available server machines that have internal connection networks fitting the description of a system area network. Table 11.1 lists typical examples with some of their interconnect

Table 11.1   **Sample of Multiprocessor Computers with Integrated Networks**[a]

System	Type	Interconnect performance	Status
DEC 7000 Open VMS	Message passing	STAR coupled 70 MBytes/s	Product 1982 25K+ Installations
IBM ES9000 Sysplex	Message passing	Switched 12 MBytes/s	Product 1994
SUN SPARC cluster 2000	Message passing	Ethernet	Product 1994
Hewlett–Packard Exemplar SPP 1200	Shared memory	Switched SCI, 128 CPUs	Product June 1995
IBM RS/6000 SP-2	Message passing	150 MBytes/s, 512 CPUs	Product March 1996
Tandem Servernet	Message passing	50 MBytes/s	Product 1996
Sequent NUMA-Q	Shared memory	1 GByte/s IQ-Link SCI, 32 CPUs	Product December 1996
Data General NUMALiNE	Shared memory	1 GByte/s SCI, 32 CPUs	Product 1997

[a]The newer machines have networks with SAN-like characteristics.

features. This is by no means an exhaustive list. The description essentially fits a description of all parallel processing and distributed memory machines. All the systems in Table 11.1 are (or can) be wired for short distances using parallel copper wire technology, but most could also use parallel interconnect technology if it becomes cost-effective.

Clearly, the performance and node count trend is favorable for optical implementation of the interconnect network, but it cannot be overemphasized that without cost-effectiveness and the demonstration of reliable and flexible operation in these systems, the designers will continue to rely on copper and will design assuming copper's limitations.

11.4. SAN Physical Layer Technology Requirements

To date, there has been very little penetration of optical components into the system area network market, but this situation is expected to change quickly with improving optical technology and increasing requirements on computer interconnects. The greatest barrier to entry has been the inability to inexpensively run multiple optical fibers between transceivers. Where copper interconnects cables may have as many as 50 signal lines per direction, fiber optic links have been serial, using discrete device components. Supporting equivalent bandwidth requires supporting much higher modulation bandwidth (which is difficult, because copper and fiber links both use the same generation of circuit technology) and/or multiple independent transceivers (greatly increasing the cost). The first technology that has shown true promise in cost-effectively addressing the problem of optical computer connect at the SAN level is the parallel optical interconnect technology, which is described in Section 11.6.

In this section, we enumerate some of the other characteristics of the physical layer technology that are required to support the higher level capabilities of a SAN-level interconnect. These characteristics serve both as a current requirements list and as a guide for future development of optical link technology.

11.4.1. LATENCY

The ability to transport memory data blocks in a very few processor clock cycles is a primary requirement for the interconnect network. Although much of latency optimization can only be achieved in the software layers of the protocol stack (i.e., the optimized user-space protocols described previously), it is important that latency be minimized at the hardware layers as well.

Although some hardware delay is caused by speed-of-light propagation over the link cables, which cannot be improved, a greater amount can be caused by media-dependent operations, such as data encoding and decoding, buffering, clock recovery, and link control at the transceivers. Although some amount of data encoding and buffering may be required, depending on link requirements, it is important at the system level to minimize these latencies as much as is practicable.

A further improvement in latency can be gained by incorporating as much of the protocol as possible into hardware (chips) rather than into software (programming). Because chip design is complex, difficult, and expensive, the simpler this can be made, the better. For this reason, a physical layer link module that, for example, supports higher level media access functions as well as media-dependent protocol functions, will likely see more acceptance from system designers than one that supports only electron–photon conversion.

11.4.2. BANDWIDTH

The bandwidth requirements for optical interconnects for SANs are strongly driven by the capabilities of the system buses and network switches and interface adapters that interface with them. These requirements increase approximately at the rate of the increase in processor performance — approximately a factor of 3 every 2 years. The exact bandwidths required are somewhat dependent on implementation but are currently in the range of 100–1000 MBytes/s.

11.4.3. DISTANCE

Distance for a SAN is quite a bit shorter than would be required for LAN networks. Two major factors limited the required distance. First, the latency over a long link limits overall system performance — the speed-of-light round-trip delay over a 100-m link can be several hundred processor clock cycles. Second, and more important, because the link-level flow control is generally used for congestion management, longer distances either imply lower link speeds or large flow control buffers, increasing overall system cost.

In practice, most SANs will be implemented within a single machine room, with the large majority of SAN links being less than approximately 20 m long. Some links will have to be longer, to handle widely distributed systems spanning multiple rooms or multiple buildings, in which case links of 100–400 m would be useful.

11.4.4. PACKAGING DENSITY

The density of lines in and out of the modules and in the cables is high to conserve real estate and keep the cables manageable in a machine room or lab environment. This is an increasingly difficult challenge for copper wiring because line size fundamentally must increase with line speed and line distance required. Some signal processing can be applied to data on the copper lines, but this adds to the protocol complexity and the overall latency of the data transfer.

11.4.5. RELIABILITY, AVAILABILITY,
AND SERVICEABILITY

The SAN is the "center" of the entire computing system, and failures in any way that stop the system cannot be permitted. This must be true 24 h a day for the life of the system, typically about 5 years. Therefore, the link components must be very reliable compared to typical data communications links. Fault tolerance is also designed into the network so that alternate paths can be found around any failed part.

Network failure also refers to the network corrupting data. The system generally recovers from link error by a detect and retransmit operation. This should occur less than a few times in a month, implying that raw bit error ratio on a line be better than 10^{-15}. All systems employ error detection, usually at the packet level (e.g., cyclic redundancy-check), and some systems are considering error-correction schemes.

In general, SANs will be installed, serviced, and maintained by users who are unfamiliar with optical component technology and who are accustomed to the reliability, availability, and service ability characteristics of complementary metal oxide semiconductor (CMOS)-based computer technology. In practice, this means that SAN designers and users are unlikely to use optical interconnect components unless they meet the same order of magnitude performance as CMOS-based components in terms of fault tolerance, hot plugability, servicing and reconfiguration, and component lifetimes.

11.4.6.   FLEXIBILITY

As SAN applications expand from the supercomputer arena to commercial applications, and from a few large installations to many smaller installations, other technology drivers also become important. Commercial
customers require the capability to grow their system in a modular way,
adding processing and/or memory power (i.e., by adding network nodes)
as needed or affordable. This approach is also desired for a smooth upgrade
path.

This flexibility demand affects the interconnect hardware. Standard PC
and workstation cards and boards are more cost-effective (compared to
custom high-density multiprocessor cards, which must be paid for at the
time of initial installation); this implies that more interconnect distance
is needed between the boards and frames. Hot pluggability for the interconnect cards is required to add nodes to the network while it is running
and upgrade or replace computing nodes. Hot pluggable cards also make
it easier for the user to install and reconfigure the system. This flexibility
in interconnect configuration also implies that there will be more direct
customer involvement in installation and reconfiguration.

11.4.7.   SAFETY

Designers and users of SAN equipment (and other computer equipment)
will strongly resist optical component technology that requires any special
handling for safety that is more difficult than is required for CMOS-based
components. For optical implementations this translates, in practice, into
a requirement for "class 1 laser" eye safety. The specific requirements of
class 1 laser safety are quite detailed, but in general they imply that a class
1-compliant component cannot cause eye damage even if someone looks
directly into the optical cable.

Class 1 laser safety can be implemented either by ensuring (i) that the
laser transmitter power is low enough to be eye safe, or (ii) that there is an
open fiber control safety interlock, such as described in the Fibre Channel
standard [5], which disables both transmitters of a bidirectional link if the
link is opened. Using current open fiber control circuitry places restrictions
on the topologies that can be built because it requires that each link be
bidirectional, whereas the strategy of lowering transmitter power places
fairly strict requirements on the receiver sensitivity.

11.4.8. COST

The cost of the interconnect is directly borne by the computer system owner. This makes the network customer much more cost sensitive than the typical customer in "shared-media" applications [e.g., metropolitan area network (MAN) or wide-area network (WAN)]. Historically, the user does not ammortize his or her interconnect network over many product cycles and generally does not care about standard solutions (unless standard implies how cost through multiple sources). For the commercial environment, in which low-cost microcomputers are used as the nodes, the interconnect network cost must be even lower, on the order of 10–20% of the cost of the microcomputer (as in the office PC communications card environment).

Today, with copper wire interframe connections, this implies a physical layer module cost in the range of $10–15 per Gigabit/second aggregate bandwidth and cable costs of appropximately $20 per meter (2 bytes, full duplex).

11.5. Copper vs Optical: Technology Trade-Offs

Computer communications can be viewed as a hierarchy of interconnects:

1. Intrachip connections between transistors, gates, and functional block
2. Interchip connections, ranging from direct chip to chip on a single circuit board to bused connections between multiple chips on multiple circuit boards
3. Rack-to-rack connections, for closely packed racks in large machines within a single machine room
4. Buildingwide or campuswide connections over a LAN
5. Intercampus connections across a metropolitan area network (MAN)
6. WAN machine connections with a large city or between cities

The general characteristics of these types of interconnects are shown in Table 11.2. The actual numbers are approximate and will not of course be equivalent for a personal computer and a supercomputer; however, with the increasing convergence between long-end and high-end computer technology, the values are correct over a perhaps surprisingly wide range of computing systems. The basic electronics that drive each of these levels is

Table 11.2   **Hierarchy of Computer System Interconnect Technologies**

Interconnect level	Technology options	Bit rate per line	Signal lines per connection	Line distance
Intrachip	Metal on CMOS	50–1000 Mb/s	64–256	<25 mm
Chip to chip	Metal on circuit board	30–1000 Mb/s	32–128	1–100 cm
Rack to rack	Twisted pair, coax, multimode fiber	20–200 Mb/s	16–54	1–10 m
LAN	Twisted pair, coax, multimode fiber	10–100 Mb/s	1–10	10–1000 m
MAN	Single-mode fiber, multimode fiber	56 kbps– 1.544 Mbps	1–10	1–10 km
WAN	Single-mode fiber	2488 Mbps	1	>10 km

approximately the same. To within a close approximation, the speed of the transistor that drives a wire to the other side of a very large-scale integration (VLSI) chip is the same as the speed of the transistor that drives a coaxial line or a laser driver.

A number of factors determine whether a given level of interconnect is implemented in electronic or optical transmission technology. These include the following:

1. Optical data transmission technology is, in general, more expensive than electronic data transmission because of the greater costs incurred in electronic/optical/electronic conversion. For the longest distances (multi-km), the cabling and repeater costs of copper outweigh the optical module costs, and optical technology becomes cost-effective.

2. Independent of technology, the cost of an interconnect increases dramatically with distance — for example, a single interchip wire costs less than a penny, whereas a LAN cable costs tens to hundreds of dollars and a WAN line costs thousands of dollars. Copper wire is more cost-sensitive to a data rate × distance figure of merit.

3. Signals are degraded over increasing distance by attenuation, distortion, and cross-talk. These factors are more significant for electronic data transmission than they are for optical.

4. The density of interconnect wires for intrachip and chip-to-chip interconnects is much higher than that for the other levels, driven by the small size of CMOS chips. Electronic interconnect densities can be higher than optical interconnect densities at shorter distances but are lower at longer distances due to cross-talk penalties.

This combination of factors has historically led to optical data transmission technology being used at the higher levels of interconnect (WAN, MAN, and, increasingly, LAN), at which the distance × bandwidth product of the signal wires favored the cost of optical technology. Electronic data transmission has been used at more tightly integrated levels, where the distance · bandwidth requirements have not been quite so high.

With increasing clock speeds in CMOS, and increasingly distributed computing algorithms requiring high communication-to-computation ratios, the bandwidth · distance product for lower levels interconnects is increasing, which increasingly favors optical data transmission technology at lower levels of computer system interconnects.

11.5.1. PACKING DENSITY

Fiber technology in larger distances has been driven by the increase in packing density relative to copper cable interconnects. This has two types: density of fiber connectors vs multiline copper connectors and density of fiber cables vs copper cables.

The HIPPI network, which has been the de facto standard for supercomputer interconnection, uses a 50-pair shielded copper cable for distances up to 25 m. Although adequate for wiring within a machine room, this cable cannot be carried through a building. Fiber extenders have recently been demonstrated that run the same data rate over fiber cabling.

11.5.2. DISPERSION

Pulse dispersion is a basic advantage of optic over electronic data transmission. On copper interconnect, whether on circuit board, twisted pair, or coaxial cable, the propagation speeds of different frequency components of a data pulse are different. A data pulse, which contains a range of frequency components, will then spread out as it passes down a transmission line. In optical transmission, there is still frequency dispersion (in fiber cables, it is referred to as chromatic dispersion), but the amount is lower by several orders of magnitude.

11.5.3. POWER

At extremely short distances (intraboard), the power requirements of a
copper line driver and a fiber line driver are approximately equivalent —
both drive about 10 mA at 3.3 or 5 V.

With higher distance · bandwidth product interconnect requirements, the
effect of the much higher frequency-dependent loss in electronic cables vs
optical increases, greatly increasing the difference in drive current required
for electronic vs optical data links. Also, for electronic data transmission,
distortion-correcting circuitry needs to be added to link interface chips,
further increasing electrical driver power consumption. With low-threshold
lasers for optical line drivers, the optical drive current can be less than
10 mA.

11.5.4. SKEW

Skew is a factor in multifiber vs multiwire ribbons. Without special expense
and selection, the skew in a copper cable can be as much as three 2-ns bit
periods over 20 m, implying across-line skew of 6 ns/20 m, or 300 ps/m.

In fiber ribbons, on the other hand, the average skew for even unselected
fibers is well under 10 ps/m, and selected fiber ribbons are available with
skew of 1 or 2 ps/m. This factor means the difference between requiring
and not requiring skew-compensation circuitry.

Aside from the intrinsic fiber skew, there is also skew in the parallel
fiber interconnects due to several other factors. Chromatic dispersion in the
fibers, together with lasers in the array operating on a different wavelength,
can result in bit skew at the receiver. Skew in the transmitter can be due to
variations in laser threshold across the array coupled with a constant bias
point from the driver.

11.5.5. DELAY PREDICTABILITY

In copper cables, the delay predictability is relatively low. As evident from
the skew variations described previously, in copper the delay variability is
up to 6 ns/(20 m · 4 ns/m) = 7.5%. This means that the speed of propagation
of a signal through twisted pair wire can vary unpredictably by 7.5%. In
optical fiber, the variation in signal arrival time is between 30 and 200 times
better. For some applications, this can make a significant difference.

11.5.6. COST

For the distances and bandwidth requirements discussed in Sections 11.3 and 11.4, it is clear that it is technically possible to build copper cables that will satisfy most current needs.

For interconnect distances within a frame, copper wiring is mature and the costs are very low by current fiber optics standards. It is not expected that optical interconnects will be cost-effective in this environment for approximately 10 years.

For interframe connections, the copper solutions meeting the performance and reliability requirements are not inexpensive — cables may cost several hundred dollars — but they are not beyond the bounds of practicality. There will be some applications in which copper interconnect solutions will be inadequate due to power or space requirements, but these will generally be of limited applicability.

Figure 11.2 represents a "snap-shot" of the relative costs of a copper bus and an optical bus of 500 Mb/s per line and 20 lines wide (nominally 2 bytes wide) as a function of the distance. It illustrates that as the distance increases, the copper version will have to use more expensive cabling, eventually employing discrete passives into the cable to compensate for distortion, and the copper driver and receiver modules will have to employ signal processing to further correct for distortion. The optical modules

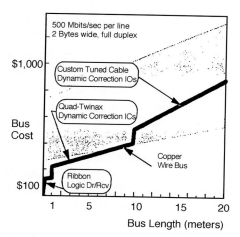

Fig. 11.2 Comparison of copper and fiber bus cost vs distance for 500 Mb/s per line, 20 lines wide, including modules and cables.

will have the higher module costs associated with optoelectronic signal conversions, but the cables can be lower in cost with distance extendability.

At some distance to the right of Fig. 11.2 (at a length of <100 m), the copper wire bus cannot support the 500 Mb/s line speed, and a signal regenerator will be required to redrive the copper bus at a significant increase in price.

The reader who is also a user is forewarned that the costs and performance of both copper and optical technology are dynamic, even though copper is a relatively mature technology compared to optical. Historically, the cost of copper components have dropped approximately 20–40% per year. The reader who is also an optical technology developer is also forewarned: Optical interconnects for SANs must compete with copper solution primarily on a cost basis. The historical trail of optical interconnects for computer applications is littered with the bones of exotic, overdesigned, and overpriced optical solutions waiting for the "killer app" to give them life.

11.6.   Parallel Optical Interconnect Hardware

POIs seek to exploit the advantages of optical technology for the interframe and intraframe applications. The fundamental advantages of parallel connections are (i) the potential of higher line density for distances of 1–500 m and lower cable bulk, (ii) higher per line bandwidth, (iii) freedom from susceptibility to electromagnetic noise or generation of electromagnetic noise (e.g., through frame openings for connectors), (iv) DC electrical isolation between frames, and (v) simpler electronic interfaces to the interconnect (due to lower power requirements and less need to correct signal for the distortion, skew, and cross-talk associated with electrical cables). Disadvantages compared to parallel electrical interconnect are (i) extra optoelectronic conversion chips are required (more card space consumed and more card cost); (ii) the technology is new and has not demonstrated the reliability and robustness of electrical interconnect; and (iii) many implementations of the technology are significantly more expensive than electrical options, both in parts and installation costs. The main differences in the optoelectronic technology for POI compared to those for the duplex links covered in other chapters of this handbook are the greater requirement for high-density chips and packages and the lack of a requirement for long-distance (approximately >300 m) lines. Whereas single or duplex line optical links are built from discrete OE devices and low lead count packages, POI technology is fundamentally an integrated technology [6]. It thus represents a new, advanced generation in optical interconnect technology.

POI technology has been pursued in research labs for more than a decade because the challenges of integrating with optoelectronics and optics are formidable. Lately, there have been chip and package technology, break-throughs leading to a rapid introduction of POI into the marketplace and the potential of becoming cost competitive with high-speed parallel copper interconnects, while offering 10 times smaller and lighter interframe cables and 10 times longer link distances than copper can support. As was emphasized in Section 11.5, cost competitiveness is essential to the use of POI and should be a major consideration in making the performance trade-offs discussed in this section.

This section will review the relevant component design issues and the status of the components. The section will focus on component proper-ties relating to integration, arrays, and uniqueness to parallel applications. The reader can learn general component characteristics from Part I of this handbook.

Figure 11.3 illustrates schematically the elements of a parallel opti-cal interconnect technology. The module may have transmitter or receiver functions or both. There are many parallel electrical data and control lines

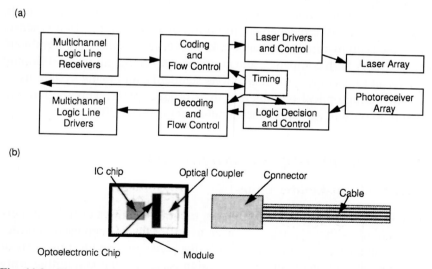

Fig. 11.3 The components of parallel optical interconnect technology. (a) The logical functions that might be contained in a POI module. (b) The physical components of one end of the interconnect.

on and off the module. The lines can be bit synchronized to a clock line; this interface is called synchronous and is usually measured in multiples of bytes (0.5–4 bytes wide have been demonstrated in various prototypes). The SAN application would use synchronous lines. The lines can also be asynchronous (no timing relation between the lines). However, all lines generally operate at the same data rate and with the same protocols. The functions in the module may be simple optical to/from electrical conversions with associated device drivers/receivers, but most users want more functions to condition the interface and make it easier to use. Thus, there can also be functions for logic level receivers and drivers, adding transmission protocols to enhance the reliability of data transmission, data flow control functions, and maintenance, test, and safety functions.

The multiple fibers in the interconnect link are generally captured into one cable and connector structure and hence all lines must terminate at the same physical location. The technology can support associated fan-in and fan-out jumpers to support limited physical spreading of sources and sinks. The cabling technology also supports mid-span connectors, bulkhead connectors, and installation tooling and support.

The reader should be forewarned that the technology of POI is still new and immature. The devices, packaging, and tools continue to evolve at a rapid pace. There are no standards for the components on such fundamental issues as fiber type, connector type, cable pitch, light signaling wavelength, package electrical or mechanical footprint, or power supply. In this sense, this topic is premature for a "handbook." Nevertheless, there are significant resources (by both companies and governments) being put into developing POI for intramachine communications standards, and many system users are integrating optical options into their product plans. This section will attempt to present an "up-to-the-minute" status (early 2001).

11.6.1.   OPTOELECTRONIC AND IC CHIPS

11.6.1.1.   Light-Emitter Chips

For cost-effective parallel applications, monolithic arrays of light emitters are required. Although arrays of discrete devices have been used in prototype to enhance the yield of arrays [7], the packaging and testing costs of this approach are unattractive. The design issues for the monolithic array are to achieve the required device speed, optical power output, efficiency, and noise characteristics consistent with high water-level functional yield of good die, power supply constraints, driver circuit constraints, and packaging constraints. High yield is achieved through device design (i) with large tolerance for fabrication process error, (ii) for uniformity of functional properties, and (iii) for simplicity in wafer fabrication processes. The laser diode device yield on a wafer must be very high to achieve a practical die yield for the wafer. For example, to achieve greater than an 80% yield for a 2-byte-wide (e.g., 20-device array) die, the individual laser device yield would have to be more than 99% on the wafer.

The emitter device could be either an injection laser diode (LD) or a light-emitting diode (LED). Because the modulation speed of lasers is at least 10 times higher than that of LEDs, and the electrical to optical power conversion efficiency is at least 10 times greater, the LD is the preferred device to fabricate into monolithic arrays, despite the added complexity in LD device processing compared to the LED.

Both edge-emitting and surface-emitting laser arrays have been used in POI prototypes and products. The advantages and disadvantages of each laser type for array applications are listed in Table 11.3.

The lowest cost light-emitter today is the edge-emitting audio disk laser (at less than $1 per packaged device), but unfortunately it has not demonstrated the ability to yield monolithic arrays, multigigabit/s speeds, or high-power conversion efficiency. The volume application for this low-cost emitter is moving to visible (red and eventually blue) wavelengths with only a modest increase in speed. Unfortunately, this is incompatible with low-loss fibers.

The most significant recent development in communications emitters has been the emergence of the vertical cavity surface-emitting laser (VCSEL) from research into production. Although complex in fabrication, the high functional array yields demonstrated, efficient light conversion, wafer-level testability, and symmetric beam have generated considerable enthusiasm in POI suppliers, and products have been introduced using monolithic arrays of this laser.

In synchronous bus operation, the emitters in the array are driven by a bit-synchronized data pattern, which is commonly resynchronized at the receiver using a common clock. Thus, skew between the optical outputs of the array is important. The skew depends on uniformity of lasing threshold, laser turn-on characteristics, and laser bias conditions and is also dependent on the data patterns applied, the temperature, and applied voltage variations. The link skew also depends on the uniformity of laser wavelength across the array because this leads to propagation delay variations. Thus, the laser

Table 11.3   **Comparison of Edge- and Surface-Emitting Lasers for POI Applications**

Property	Edge-emitter array	Surface-emitter array
Wavelength (nm)	Made in many material systems, at all communications wavelengths; $\lambda = 650$, 780, 850, 1300, and 1500 nm, but with low to moderate yield	Primarily a GaAs device; 650 nm (low yield), 850 nm (high yield), and 1300 nm (very low yield)
Speed	Tens of GHz	Comparable
Optical beam	Asymmetric and divergent, difficult to couple to fibers; in the plane of the chip, good for low-profile package	Symmetric and low divergence; easy to couple to fibers; normal to chip surface, needs beam redirection for low-profile package
Electrical to optical power conversion efficiency	Moderate ($<10\%$)	High ($<40\%$)
Modal characteristics	Single or multimode	Comparable
Processing	Complex heteroepitaxy growth, regrowths in long wavelength devices; simple or complex postprocessing depending on device; processing on die edge required for passivation	Very complex heteroepitaxy growth, with atomic dimension thickness control required for many layers; simple postprocessing
Temperature sensitivity	Moderate, $\lambda = 780$–850 nm; high, $\lambda = 1300$ nm	Low, $\lambda = 850$ nm
Power consumed	20–100 mW per device	2–10 mW per device
Optical coupling to fiber	Optical lensing component required	Can be butt-coupled
Testing	Must be cleaved into bars to be tested	Automated wafer-level testing demonstrated

Table 11.3   continued

Property	Edge-emitter array	Surface-emitter array
Wafer yield	Low to moderate depending on material and structure	Greater than 99.5% in early results at 850 nm
Reliability	High reliability (MTTF > 10^6 h) demonstrated	Insufficient reliability data presented, but early results are encouraging for MTTF > 10^6 h
Cost	Low, $\lambda = 780$ nm; moderate, $\lambda = 850$ nm; high, $\lambda = 1300$ nm	Data not available; should be low

choice for a specific transmitter is not straightforward. The reader is referred to Ref. [8] for further details about the design trade-offs.

The emitter module for a parallel optical interconnect must satisfy laser safety as defined by the OSHA and International Electrotechnical Community (IEC) class 1 criteria for eye safety, as discussed in Section 11.4. However, the entire optical output of all emitters must be considered. The maximum permissible power depends on the safety standard definition and the emitter wavelength, and the reader is referred to the standards documents themselves for exact definitions [9]. In the case of arrays, the light source is considered "extended" (i.e., not a point source), which leads to special considerations for safe power levels. In general, the total array power must be less than approximately 600 μW for 850-nm wavelengths and 1000 μW for 1300-nm wavelengths.

11.6.1.2.   Light-Receiver Chips

As was discussed for light emitters, the photodetectors used in the POI receiver are generally made in monolithic arrays. Two types of photodetectors are used: the positive–intrinsic–negative (PIN) photodiode and the metal–semiconductor–metal (MSM) photodiode.

The MSM photodiode has some unique advantages for monolithic arrays. Because the capacitance per unit area of a transverse structure MSM detector is low, a fast photodiode with large area can be made. This is an important feature for the MSM arrays, in which it is important to have large tolerance to misalignment of the optics from the chip to keep module assembly cost down.

Because the optical and thus photoelectric signals are low (measured in μA of photocurrent for μW of light power), the data rate is high, and channels are on a small pitch, there can be considerable cross-talk between the channels in a parallel array. This makes it advantageous to integrate an amplifier on the chip with the photodetector. This is called an integrated photoreceiver. Other advantages of integrated receivers in the array include the ability to do wafer-level testing and screening and the uniformity of channel-to-channel performance.

A comparison of a typical PIN photodiode array and a typical MSM photoreceiver array is made in Table 11.4. In serial Gb/s links, the integrated receiver has a lower cost than a 1300-nm photodiode but a more expensive cost than a Si photodiode. However, a significant portion of the cost of a hybrid (photodiode + amplifier) module is in the assembly and test of the two chips.

11.6.1.3. Receiver and Transmitter ICs

The support chips used in transmitter and receiver must supply both digital and analog functions at high speed. The transmitter chip must receiver multichannel logic-level signals from off module — data, control, and clock(s). It provides the bias and analog modulation circuits for each laser channel. It may add coding, to improve transmission integrity, and additional link control functions. The receiver chip amplifies the photocurrent; makes a logic decision using a timing signal that has either been extracted from a data channel or provided on a separate line; and drives logic-level signal off the module.

As mentioned in Section 11.4, it is desirable from both a system latency viewpoint and an ease-of-use viewpoint to provide media-dependent drive and control functions and reliability functions on the chips in the POI modules. Most current prototypes and products do not provide functions beyond optoelectronic signal conversion and some amplification. This is done for the convenience of the POI module supplier, who can then make one module fit many applications. The system integrator then must develop or purchase support modules to provide the link interface functions and develop cards to handle high-speed analog lines and provide special power supplies. The trade-off is design time and cost savings for the POI

Table 11.4   **Comparison of PIN and MSM Photodetectors in Monolithic Arrays**

	PIN photodiode array	*MSM photoreceiver array*
Bandwidth	Tens of GHz	Same
Photoresponse	Fraction of a μA of current output for each μW of light input (depends on wavelength)	Logic voltage swing output for μW of light input
Wavelength	650–850 nm using Si or GaAs; 1300 nm using InGaAs	780–850 nm using GaAs
Interchannel cross-talk	Dominated by mm-long wirebond cross-talk	Dominated by chip substrate isolation
Device structure	Vertical pn junction device; mature device for data link products	Horizontal Schottky barrier device; compatible with MESFET IC processes
Testing	Screened at wafer, functional at module	Functional at wafer
Cost	Moderate for InGaAs (1300 nm) arrays; low for Si (780–850 nm) arrays	Low for GaAs-integrated receiver arrays (780–850 nm)

module supplier versus design time and cost savings for the SAN subsystem user.

As mentioned previously, there are no standards for the functions provided in a POI interconnect, and variations of all the previously described functions are to be found in prototypes and products. Si bipolar and CMOS and GaAs MESFET IC technologies have all been used. General issues under consideration today include the following:

1. Should the links be AC- or DC-coupled optically? DC coupling allows unconstrained codes to be transmitted and requires simpler circuits. AC is more reliable when there are long run-length data patterns, bias control variations, and adverse environmental effects.

2. Can CMOS ICs provide all electronic functions or are there advantages to bipolar ICs?

3. Should the functions be designed as custom or application-specific integrated circuit (ASIC)? To keep the cost of the chip (both in design and in fabrication) competitive, it is best to choose among ASIC technologies, although many prototype POI modules still use custom designs.

4. An important consideration is the supporting design and test tools for the chip technology chosen, both the accuracy of the device models (especially OE device models) and the ability of the tools to handle both digital and analog simulation. The state-of-the-art in optoelectronic device and circuit simulation tools is far behind the electronics industry expectations at this point in time.

The previous topics are discussed in detail in Ref. [6].

11.6.2.   OPTICAL MODULE PACKAGING

11.6.2.1.   Optical Components in the Module

The transmitter and receiver modules of a parallel optical interconnect have optical components inside to couple the light emitter and detector chips to the fibers of the cable. The optical component should be multielement to couple all channels with one alignment operation and keep package assembly costs low and module size small. The optical coupling mechanism can be reflective, refractive, or light guiding, and all have been utilized in prototypes. The optical coupler is often used as part of an environmental barrier (e.g., hermetic seal) between the inside and outside of the module.

The fiber optics industry has historically focused on developing single-channel couplers based on miniature optical elements of high precision and high-coupling efficiency and with little regard to cost. Multichannel optics compatible with IC module packaging is only just emerging. Likewise, IC module technology has developed materials and processes without regard for optically transparent materials or optical alignment tolerances. Thus, packaging optical elements into IC packaging will likely be the most expensive part of the POI technology.

The issues that influence cost and performance for the optical coupling elements include the following:

1. *Alignment tolerances:*   As discussed in the other chapters, single-mode fiber optics required submicrometer tolerances for package alignment, multimode fiber optics requires tolerances of a few micrometers, and "large-core" multimode fiber optics requires tolerances of tens of micrometers. When multichannel couplers are required, these tolerances must be maintained over dimensions of tens of millimeters. This creates a challenge for the fabrication and stability of the monolithic optical coupling element. The lowest cost approach is to use a coupler with greatest tolerance to misalignment consistent with the coupling requirements.

2. *Coupler material and fabrication processes:*   Glass microoptics (lens or reflector) has good optical and mechanical properties but it is difficult to fabricate in multielement arrays and expensive to finish. These couplers usually must be assembled into other structures in order to align them to the OE chip. This adds cost. Glass-filled plastics can be used in a transfer-molded process to make monolithic couplers (also with single-mode precision). The lowest cost approach is to use injection-molded transparent plastic to make an array of coupling elements in a monolithic structure containing the mechanical alignment features. There is no assembly of this coupler. Injection-molded plastic parts are commonly used for consumer and PC applications when low cost is the objective. The challenge is to find transparent materials with correct mechanical and thermal properties for easy molding and durability.

3. *Coupler assembly into module:*   The coupler must be aligned to the emitter and detector chips and to the fiber array (either in the pigtail or in the connector ferrule). The most precise approach is to use active alignment; that is, manipulating the coupler into alignment with an OE device while it is operating and then fixing it into the optimum position. This is generally employed for single-mode systems, but it is time-consuming and expensive. In multimode systems, the coupler can be passively aligned to the chip. Though generally less expensive, this can also be time-consuming due to the number of optical elements to simultaneously align and the number of degrees of freedom in the alignment (both rotational and positional). In some modules, solder bump reflow is used to passively align the chip to the optical component (see Chapter 18). With "relaxed-alignment tolerance" optical components, self-alignment

(or snap-together alignment) of components has been demonstrated as the lowest cost approach.

4. *Light bending in small dimensions:*   For multichannel couplers it is desirable to keep channels on a small pitch to conserve chip real estate, but small dimensions and tight pitch are more expensive to implement. To keep the packages profile low in modules in which the OE chip is mounted on the carrier along with IC chips, a 90° light turning must be made in a few millimeters, but this implies reflective optics [extra coatings or high numerical aperture waveguides], which can adversely affect fiber bandwidth and coupler loss. If the coupler does not bend the light, i.e., end fire coupling between the OE device and the fiber, it is generally simpler to fabricate and has lower coupling loss. Unfortunately, for surface-emitting/detecting devices, this implies that the OE chip must be mounted perpendicular to the main IC module and the high-speed electrical traces must be carried around package corners (EMI, cross-talk, and reliability exposure). This also adds another package layer to the module.

5. *Compatibility with IC chip carriers:*   If the coupler can be aligned with the OE chips using standard IC chip carrier features, then a special OE submount does not have to be developed, and added assembly and testing steps can be avoided.

Most optical couplers are derivatives of the three types described and illustrated below: type 1, an array of discrete microoptic elements (e.g., fibers) affixed into a precision V groove carrier; type 2, a monolithic array of plastic elements (e.g., lenses or reflectors); and type 3, a multielement planar lightguided structure.

Type 1:   The V groove with microoptics coupler consists of a piece of silicon wafer with precision-etched grooves on one surface. Fibers are laid in the grooves, thus aligning them with each other, and epoxied. The fibers + silicon subassembly is then end-polished at a 45° angle so that light can be reflected into or out of the fiber from an OE device above the subassembly. The end face is coated with metal. Single-mode tolerances can be held over array dimensions up to approximately 5 mm; coupling loss of less than 3 dB can be obtained; but the assembly costs for this coupler are high. This is illustrated in Fig. 11.4(b).

(a)

(b)

Fig. 11.4 Photographs of two optical couplers for parallel optical interconnect modules: (a) Jitney coupler and (b) OETC coupler.

Type 2: The injection-molded coupler is fabricated by injecting plastic into a mold that has the desired optical and mechanical features — an array of microlens dimples, an array of reflectors, or an array of lightguides. This can be the lowest cost approach to making optical elements (by 10 times), but for the small dimenions and pitch of the array considered for POI, there is a challenge in making the mold with optical precision over distances up to 1 cm and keeping the mold clean. The part cost in this process can be very low (<$1) as long as assembly into a submount is not required.

Type 3: The lightguide coupler can be made by molding but can also be made by laminating or depositing a blanket of the plastic or glass lightguide material on a substance (sometimes containing mechanical alignment features) and photolithographically processing the lightguides.

Figure 11.4 shows two of the optical coupler types — type 1 and type 2. The Jitney coupler is 22 reflector vanes (on a 500-μm pitch) of metallized, transparent plastic. Each vane preserves VCSEL divergence while bending the light path by 90°. Note the "ears" on the end for passive registration into the package.

11.6.2.2. Modules for Array Optoelectronic and IC Chips

The chip carrier modules used for POI have generally been derivatives of IC chip carriers. Lead-frame and pin grid array carriers were used in early prototypes, but the ball grid array is the direction many suppliers are headed. The IC industry is increasingly using plastic packages for low cost, but the mechanical dimensional control and rigidity are concerns for adaptation to optoelectronic array packages. Many optical chip carriers still use more expensive ceramic or metal materials. The technical issues for a good optoelectronic chip carrier are the following:

1. *Module electrical lead bandwidth:*   The applications demand data rates per line from hundreds to thousands of megabits/s, leading to low inductance, impedance control, and high electrical isolation requirements for the leads between module and card. Generally, differential signals are utilized, with generous use of grounding and power leads. Surface-mount techniques are best for high-speed electrical signal integrity in leads on and off the module. Through-hole packages (e.g., pin grid arrays) can have cross-talk arising from the lack of close proximity return paths for the signals. Multiple levels of signal and power planes are advantageous for signal integrity in the package but add to module cost.

2. *Optical alignment capability:*   If optical components permitting tens of micrometers of misalignment are used in the module, IC packages can have mechanical alignment features (pins and molded surfaces) included in module fabrication. However, for alignments of submicrometer or a few micrometers, a separate level of package — the optical subassembly (OSA) — is needed. In this OSA subpackage the precision alignments are made, and the OSA is electrically attached to the module. OSAs use grooved Si wafer "optical benches" or small leadframes with embedded lightguides. The advantage of this approch is that it partitions the electrical and the

optical/mechanical functions. The disadvantage is that it increases package cost by adding many more parts, assembly processes, and testing steps.

3. *I/O count for electrical interface:* The POI module must support enough I/O pins for multibyte data transfers. This excludes discrete device packages (e.g., TO can) or common OE packages [e.g., 1 × 9 fiber distributed data interface (FDDI) carrier]. Peripheral leaded packages have 20–150 leads and can support up to approximately 2- to 3-interfaces of approximately 500–700 Mb/s per line. Arial I/O packages (e.g., BGA) have 100–1000 leads and can support wider interfaces or full duplex interfaces, and at speeds greater than 1 Gb/s/line. All these options must be balanced against the higher cost of high lead count packages.

4. *Mechanical strength and rigidity:* The rigidity of the package for ensuring optical component alignment was discussed previously. With an OSA the rigidity is put into the submount, whereas direct OE chip attach approaches use metal heat slugs or rely on ceramic chip carriers. The trends in low-cost IC packages are to use plastic chip mounts, however. Although some POI modules have been made and installed with optical fiber "pigtails," the approach adds greatly to the expense of card assembly and handling and card robustness (due to the dangling short optical leads). Many parallel link users will not accept such an approach for their applications (especially because all other IC modules on the card employ surface mount for electrical connection and a card-edge connector socket for cables). Thus, the POI module should also include an optical socket for a cable. This adds the mechanical requirement that the module must strain-relieve the cable attached to the module when it is on the card. Some applications require rear-mounted connection of the card to the cable or board (blind connect) and this influences the mechanical module design.

5. *Cooling:* Virtually all applications are in an air-cooled environment (0–200 linear ft per minute) with forced air of 0–65°C ambient temperature.

6. *Footprint:* The applications are generally in a dense card part of the system, requiring the optical module to be compatible in size to an IC module or an electrical card-edge connector (i.e., approximately 10 × 40 mm for a 1-byte-wide, full-duplex connection).

11.6.3. FIBER OPTIC RIBBON CABLES
AND CONNECTORS

11.6.3.1. Cables

The cables for POI are generally intraframe harnesses, jumper cables between frames, or installed cables in a raised-floor or horizontal building environment. The installations are dynamic in the sense that there is continual rearrangement (30% per year) of the interconnect system as nodes are added or equipment is upgraded. The downtime tolerated for installation or upgrading of the interconnect network is measured in hours. Thus, a flexible and easy to install cabling solution is a requirement. For raised-floor installations, sufficient durability to withstand the stress of power cables laying on top, impact from floor tiles, etc. must be considered.

The cables are generally connectorized at a factory, although the ability to field-install connectors must be supported. The cost of the cables is especially important because they are generally not amortized over many system upgrades or many users. They must be competitive with the copper cables used in today's systems, even though they may offer more performance.

The parallel optical cables and connectors used in early prototypes were taken from the telecommunications industry's trunking applications. Although the performance and durability was good, the cost has been very high (5–10 times that of equivalent copper solutions). This has caused a number of system developers to reject optical interconnect solutions in favor of developing more advanced copper cabling systems (including signal processing at the driver and receiver to enhance data integrity). Fortunately, there are now cost-effective parallel interconnect cables and connectors emerging from development that focus on the data communications installation environment, limited distance requirements, and pricing structure (both parts and installation costs).

Three fiber types have been used in POI cables: the large-core step and graded-index fiber, the data communications standard multimode fiber, and the telecommunications single-mode fiber. The cable characteristics are compared in Table 11.5. Plastic optical fiber is also a good candidate for low-cost, jumper cable-length connections, but to date there have been no ribbon cables developed.

Note that the bandwidth and loss characteristics in Table 11.5 are for full-mode excitation or equilibrium light propagation conditions. In short-distance links (the definition of "short" length depends on cable type — generally <100 m, but sometimes more than 1 km) under laser excitation

Table 11.5   **Comparison of Fibers Used in Ribbon Cables**

Property	Large-core multimode	Datacomm multimode	Single mode
Core/cladding dimension (μm)	200/230	62.5/125	9/125
Alignment tolerance (μm) (1 dB added connector loss)	80	5	1
Fiber bandwidth per line (MHz/km at λ)	SI, 50 ($\lambda = 850$) GI; 350 ($\lambda = 850$)	SI, 165 ($\lambda = 850$) GI, 500 ($\lambda = 1300$)	SI, 2000 ($\lambda = 1300$)
Fiber loss (dB/km)	4	3 ($\lambda = 850$) 0.8 ($\lambda = 1300$)	Same
Cable pitch (μm)	500	250	250
Cable dimensions (mm)	1 × 14 (20 fibers)	2 × 8 (12 fibers)	2 × 8 (12 fibers)
Maximum cable distance (m)			
500 Mb/s/line	40	300	No data
1 Gb/s/line	30 (VCSEL excitation)	300 (Skew limited)	
Skew	SI, 1.1 ps/m GI, 4–5 ps/m	GI, 4–5 ps/m GI, 2 pm/m (selected fibers)	
Relative cable cost (20-fiber cable)	$7–10/m	$10–15/m	$10–15/m

Note. SI, step-index core; GI, graded-index, full-mode excitation.

(VCSEL or edge-emitting), mode equilibrium is almost never established (even in installed cables) and the effective bandwidth is higher and connector/cable attenuation is less [10].

The large-core fibers have the most relaxed alignment tolerances, permitting plastic connectors and the lowest cost assembly. The step-index fiber bandwidth has been shown to be enhanced with VCSEL laser excitation so that 40-m spans are possible at 500 Mb/s. Graded-index versions of this fiber support 500 Mb/s to more than 300 m [11], but their cost is considerably (more than two times) higher. The datacomm standard multimode fiber is used by most POI links and can cover both jumper cable lengths and building wiring spans (up to 300 m). Table 11.5 indicates current specifications, but there are datacomm multimode fibers coming on the market with enhanced short wavelength ($\lambda = 850$ nm) performance of more than 500 MHz/km. The cost of the ribbon cable types of Table 11.5 are approximately comparable (on a per fiber basis) and dominated by the environmental protection required. For interframe connections, round cables are generally preferred over flat cables, but flat ribbon cables are preferred in the frame.

There are no specific ribbon fiber standards for POI, but it is expected that the environmental requirements will be based on duplex cable requirements. There would be separate requirements for building wiring and for computer room installations. For the cable environmental requirements for building wiring and jumper cable requirements, the Fibre Channel Standard may be consulted [12]. For computer machine room installations, the ESCON cabling guide [4] is representative. However, it should be recalled that many of the SAN applications are designed as an entire system so that a standardized interconnect is not necessary (or may not even be desirable for business reasons). For system performance or cost leverage, ad hoc interconnects may make sense. This does not imply that interconnect robustness, integrity, or flexibility will be compromised, but only that the time-consuming process of standardization is avoided.

11.6.3.2. Connectors

The array connectors that have been developed consist of a ferrule, which maintains the relative position of the fibers with respect to registration surfaces, and a body, which provides the latching mechanisms and strain relief. Ferrules have been made from V-grooved silicon or ceramic pieces, cast-molded silica-filled plastic, or injected-molded pieces. Their characteristics are compared in Table 11.6. Note that for array connectors, dimensional control must be maintained over larger dimensions than for single-fiber cable connectors, and that there are more ferrule dimensions that must be maintained to tight tolerance (i.e., there is not circular symmetry as in the simplex connector ferrule).

For connector body-latching mechanisms, push–pull-type connectors are preferred, with a positive locking insertion. However, it is required that the connectors release prior to module damage or cable damage should they be subjected to strong pulls in field installation or use.

Table 11.6 **Comparison of Multifiber Array Connector Types**

	Connector type			
	MTP	**mini-MAC**	**Jitney**	**Argus**
Material	Cast plastic	Si or cast plastic	Injection-molded plastic	Ceramic
Alignment tolerance	<1 μm	<1 μm	<10 μm	<1 μm
Maximum fiber count demonstrated	12	32	20	18
Release mechanism	Push–pull	Push–pull	Side latch	Push–pull
Dimensions	12 × 7.5 mm	12 × 8.6 mm	24 × 5.5 mm	1.44 × 7.2 mm
Relative cost (including assembly)	10×	10×	1×	3×

A major cost factor in ribbon cables is the cost of preparing the cable for connectorizing and the connector assembly procedure. Cables with extruded plastic jackets are generally more expensive to connectorize because the coatings must be stripped off before connectorizing and reapplied to ensure mechanical integrity after connectorizing. Cables with laminated jackets can be made easier to strip, and it has been shown that ferrules can even be inserted during cable manufacture for very low-cost assembly. For large-core MMF, the ferrule dimensions can be controlled adequately to align to the module body. For datacom MMF or SMF, separate alignment pieces (e.g., pins or balls) are usually fitted to the ferrule to increase the precision of registration. This adds to the connector cost and assembly time.

Figure 11.5 shows two typical plastic connectors for ribbon cables; the MTP connector from US Conec, which must be assembled on the end of an extruded coating cable, and the Jitney connector from 3M, which has been fabricated into the cable during manufacture. The figure illustrates how the large core MMF cable can be connectorized using the ferrule alone as an alignment surface, whereas the datacomm MMF requires the use of steel pins for precision registration.

(a)

(b)

Fig. 11.5 Photographs of two ribbon connectors on their cables; (a) MT connector and (b) Jitney connector.

11.6.4. PARALLEL OPTICAL LINK DESIGN AND TESTING TOOLS

The parallel optical link designer uses optical power budgets to determine the transmitters, receivers, link spans, and number of connectors to be used in implimenting the network, just as does the serial link designer. Effective power loss penalties are added for link noise, data pattern distortion, and environmental effects. Specific penalties to consider for parallel interconnects are the effects of interchannel cross-talk statistical variation of devices across the array, and skew. An example of a typical POI power budget is shown in Table 11.7. Note the wide range in dynamic range required for the optical receiver due, in part, to the inaccuracy in accounting for line

Table 11.7   **Typical POI Optical Power Budget**

Transmitter budget VCSEL link	Best case	Typical	Worst case
Optical output power [dBm] (10 mA peak drive)	3	3	0
Coupling loss to fiber	−1	−3	−3
Convert to average power	−3	−3	−3
Total transmitted optical signal (power)	−1	−3	−6
Link loss [dB]			
Optical connector loss	−0.3	−0.5	−0.7
Number of connectors	0	2	4
Coupling loss to detector	−1	−2	−3
Optical crosstalk		0	
Fiber attenuation [dB] (at 4 dB/km)		0.2	0.2
Total link losses [dB]	−1	−3.2	−6
Total received optical signal power	−2	−8.4	−12
Receiver penalties [dB]			
Extrapolation BER $= 10^{-12}$ to 10^{-15}	1	1	1
Electrical crosstalk	0.5	1	1
DC offset			
Modal noise penalty	0	0.5	0.5
Jitter penalty	0	1	1
Dispersion penalty			
Skew penalty	0	1	1
ISI penalty	0	0	1
Total penalties	1.5	5	6
Receiver sensitivity × [dBm ave.] BER $= 10^{-12}$	−20	−19	−18
Link margin [dB]	16.5	5.5	0.5

noise, distortion, and cross-talk. These power budgets are, at best, a rough (and, it is hoped, conservative) estimate of link field performance.

The datacommunications fiber optic industry suffers, in general, from the lack of system modelling tools to aid in design. For POI, with its array of interacting components, determining the cost-effective component specifications using "power budgets" or prototypes adds considerably to the design time for the network, often forcing designers to build excessive margins (and hence component cost) into the network. This situation

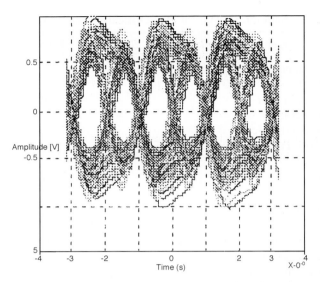

Fig. 11.6   Superimposed eye diagram from 18 separate receiver channels synchronized to a common clock at the transmitter.

must change if computer system network designers are to consider optical interconnects as easy to impliment as are copper wire interconnects.

Though there are no CAD POI simulator products on the market at this time, a number of companies and universities have POI simulation tools [13]. These tools allow much more precise estimations of link performance versus component specification than either power budgets or simple spreadsheets. For example, the IBM or University of Illinois tools include interchannel noise and distortion effects and statistical variations of component parameters (to include manufacturing and environmental effects). Figure 11.6 illustrates how one of these tools can be used to establish the optimum decision time for all 18 channels of a synchronous bus, given the statistical variations of all the components and the cross-talk in all channels of the bus.

For POI components and links to be efficiently characterized and qualified for computer interconnects, automated testing tools and testing methodology still need to be developed and standardized. For example, IBM developed prototypes of such automated testers during the ARPA OETC and NIST-ATP Jitney programs, and tens of thousands of channels of receivers and transmitters were characterized at the wafer and module level [14]. Testing times were reduced to 4–7 s per channel, but this still added $5–20 to the cost of the module. Clearly, rapid screening test for manufacturing (including optical, electrical, and mechanical screens) will be required.

11.7. Examples of Parallel Optical Interconnect Links

11.7.1. INTERFRAME INTERCONNECTS

There are a number of product and prototype parallel optical interconnects reported, and new prototypes are being introduced regularly. Table 11.8 provides an overview comparison of 10 examples current in the literature. The relative cost column is meant only to compare these examples based on technology complexity and infrastructure (an approximation of the cost/performance figure of merit is given); the current prices cover a much broader range and are marketing strategy-sensitive. Table 11.9 compares their functions, and Table 11.10 compares the technology used. There are currently three general design points for the technology.

The first class of POI subsystem is made from the long-wavelength device, single-mode fiber technologies developed by the telecommunications industry for their long-haul applications. These buses are capable of point-to-point spans of kilometers in asynchronous mode (skew would limit their synchronous mode distance) but the module costs are more than $1000 per gigabyte/second. This technology provides a data rate × distance figure of merit that copper interconnects cannot match. The SAN and most LAN applications do not need, and cannot afford, this technology.

By choosing this design point, the telecom industry suppliers hope to leverage their existing serial link technology, expand its markets, and reduce its high costs. Although very expensive (more than $1000 per gigabyte/second), as a method for efficiently packing multiple high-speed lines into compact cabling and module form factors, they will find a niche market. An inherent advantage of using single-mode technology is its coherence, which makes possible the eventual evolution to coherent optical routing and switching devices to form all-optical network. Examples of this technology class in Tables 11.8–11.10 are Nos. 2, 4, 5, and 7.

The second class of POI technology uses short-wavelength OE devices and datacom multimode fibers. The specified spans are for up to 300 m, forcing a requirement for graded-index fibers, and thus active alignment in the packaging. The design point data rates are 400–1250 Mb/s per line. This places their performance (as measured by the data rate × distance figure of merit) more than ten times beyond current copper interconnect capabilities. Unfortunately, their cost-effectiveness figure of merit is between

Table 11.8 Comparison of 10 POIs

Name	Maker	Status (1997)	Comments	Performance	Relative expected cost [1–5 (highest)]
1. Optobus	Motorola	Product	2nd-generation product	400 Mb/s/line, duplex link	2 ($100/Gb/s)
2. MDS2101/ MDR2101	Hitachi	Product prototype	Prototype in field (1993)	250 Mb/s, 100 m, simplex link	5 (>$1K/Gb/s)
3. POLO-2 [15]	HP, DuPont AMP, SDL, USC,	Technology prototype		1 Gb/s/line, <300 m, duplex	4
4. OPTM [16]	Oki	Prototype	ATM switch application	2.48 Gbit/s/line	5
5. Ericsson POI [17]	Ericsson	Prototype	Intraframe applications	4 lines, 400 Mb/s/line	4
6. OETC	AT&T, IBM, Honeywell, Lockheed–Martin	Technology prototype	Samples in field (1995), IBM and LLNL	1 Gb/s/line, 200 m, simplex	3
7. NEC POI [18]	NEC	Technology prototype		1 Gb/s/line, >300 m (estimate), simplex	5
8. Paroli [19]	Siemens	Technology prototype		1.2 Gb/s/line, 300 m, simplex	4
9. Jitney	IBM, 3M, Lexmark	Technology prototype		500 Mb/s/line, 40 m, simplex	1 ($15/Gb/s)
10. P-VixeLink	Vixel Corp.	Product		1.25 Gb/s/line, 100 m, simplex	3 ($200 Gb/s)

$100 and $1000 per gigabyte/second, making them uncompetitive with copper at the distances that copper can reach; thus, it is expected that this approach will be a niche market until after the Year 2000.

Again, both the performance and the cost are beyond copper wire implementations, and the system designer is currently designing for the limits of copper. Suppliers choosing this technology class should try to motivate the system designers to move their system requirements beyond copper capabilities. In any case, it is expected that this second class of design point will eventually be required of POI subsystem for computers. Example Nos. 1, 3, 6, 8, and 10 in Tables 11.8–11.10 are in this category.

The third class of POI technology also uses short-wavelength OE devices but drives even further toward low cost by using large-core fibers, plastic optical components, and passive assembly of the package. The performance design points are limited to data rates of 500–1000 Mb/s per line, and distances up to approximately 30–50 m. Estimates indicate that this class of POI is cost competitive with copper interconnects while offering approximately two times more in the (distance × data rate) figure of merit. Thus, this technology class may displace copper from the volume applications in the near term. However, it is limited in growth to higher data rate at longer distance, and it uses a nondatacom cabling, which may limit its acceptance by some users. A possible motivation of the suppliers of this technology class is to enter the applications market as soon as possible with a cost-effective solution, become a technology driver on the existing application road map, and then improve the technology along that road map's timeframe. Only one example (No. 9) is shown in Tables 11.8–11.10.

This section will not discuss all the examples of Tables 11.8–11.10 (the reader can refer to the references) but only mention some of the design motivations and features of a few significant examples.

Recently, a multisource agreement (MSA) was announced for future parallel optical interconnects in an effort to standardize this interface across different technology suppliers. Initial companies participating in the MSA include W.L. Gore, Agilent, and Mitel; others are expected to join in the coming months.

11.7.1.1. Optobus

The Optobus is the first fully developed product on the market (as opposed to a marketing prototype) backed by a major company in the datacommunications field. It uses common multimode optics technology and common

Table 11.9 Functions of 10 POIs

	Data rate per line	No. of lines (simplex)	Electrical interface	Optical interface	Mechanical interface	Function
1. Optobus	400 Mb/s	10	DC coupled Analog	MT connector 62-μm MMF		OE conversion
2. MDS2101/ MDR2101	250 Mb/s	9 data, 1 clock	CMOS	1-m pigtail, MT or MPO connector single-mode fiber 500-ps skew	14-mm (w) by 9 mm (l) by 6 mm (h) footprint 2.6-W power consumption	OE conversion
3. POLO-2	1 Gb/s	10	ECL	MT connect 62-μm MMF	Ceramic BGA 25 mm (w) × 75 mm (l)	OE conversion
4. Oki OPTM	311 Mb/s	12	ECL, AC-coupled Asynchronous	Pigtail to MPO 50-μm MMF	Ceramic/metal SIP 40 mm (w) × 100 mm (l) × 15 mm (h)	OE conversion Hot plug APC
5. Ericsson	400 Mb/s	4	PECL DC coupled	50-μm MMF or polymer lightguides MT connector	DIP	OE conversion Clocking Line coding

	Speed	Channels	Electrical	Connector/Fiber	Package	Functions
6. OETC	1 Gbaud	32	ECL LVDS DC coupled	MAC connector 62-μm MMF	Ceramic QFP 33 mm (w) × 44 mm (l) × 5 mm (h)	OE conversion Clocking Line coding APC
7. NEC POI	1 Gb/s	8	DC-coupled Analog	Pigtail module ? connector 62.5-μm MMF	Ceramic QFP 18 mm (w) × 25 mm (l) × 5 mm (h)	OE conversion
8. Paroli	1.26 Gb/s	12	ECL-Tx CMS-Rcv. −3.2, 5.2 V	62.5-μm core MMF Pigtail to MT 350-ps skew	5 mm (l) × 10 mm (w) × 4 mm (h) 1.32 W	
9. Jitney	500 Mb/s	20	LVDS	Jitney connector 200-μm core MMF	Plastic QFP 30 mm (w) × 20 mm (l) × 7 mm (h)	OE conversion Cock Control Line coding Self-test
10. PVL-4	1.25 Gb/s	4	PECL differential AC coupled	62-mm MMF Pigtail to MTP	PGA	OE conversion

Table 11.10 Optical Chip and Package Technologies Used in the POIs

Name	Emitter	Receiver	IC	Package	Cable and card mount	Comments
1. Optobus	$\lambda = 850$ nm VCSEL array	GaAs PIN PD array	Bipolar and CMOS	Plastic PGA Plastic lightguide OE submount	62-μm MMF MT connector	
2. MCS2101/ MCR2101	$\lambda = 1310$ nm edge laser array	InGaAs PIN PD array	Si bipolar driver and amplifier	Si V groove submount for optics Active alignment Ceramic on metal stud	SMF MT pigtail	NTT low-skew ribbon
3. POLO-2	$\lambda = 850$ or 980 nm VCSEL array	Si PIN PD array	Bipolar	Ceramic BGA Polymer lightguide couplers	62-μm MMF MT connector	
4. OPTM	$\lambda = 1300$ nm 12-edge LD array	$\lambda = 1300$ nm 12-InGaAs PD array		Ceramic SIP Optical submount Microlens array Active alignment, welded	Pigtail to MPO 50-μm MMF	
5. Ericsson	$\lambda = 1300$ nm 4-edge LD array		BiCMOS	Plastic DIP Si V groove chip carrier	Intraframe harness	Quasi-hermetic Synch and asynch types

	Source	Detector	Electronics	Packaging	Fiber/Connector	Notes
6. OETC	λ = 850 nm VCSEL array	MSM photoreceiver	CMOS	Ceramic QFP Si V groove submount optics Active align	62-μm MMF MAC connector 140-μm pitch	
7. NEC-POI	λ = 1300 nm 4-edge laser array (2 required)	InGaAs PIN PD		Flip-chip mount Si V groove carrier QFP, pigtail	62-μm MMF	No control checks Wide dynamic range design
8. Paroli (see Chapters 5, 6)	λ = 850 nm Edge laser array	InGaAs PIN PD	Si Bipolar for amplifier	Si V groove submount for optics	62-μm fiber MT pigtail Low-skew fiber (1.1 ps/m)	Technology to change to VCSEL
9. Jitney	λ = 850 nm VCSEL array	MSM photoreceiver	Si CMOS	Plastic optical coupler Passive alignment QFP	200-μm MMF Low-skew cable (1.1 ps/m) 3M plastic connector	Wafer-testable chips
10. PVL-4	λ = 850 nm VCSEL array	MSM photodetector		Ceramic OE submount PGA module	62-μm MMF MTP connector	Plastic optics

(Continued)

Table 11.10   continued

Name	Emitter	Receiver	IC	Package	Cable and card mount	Comments
11. LiteBus	$\lambda = 850$ nm VCSEL array	MSM photodetector		Plastic BGA	50-μm MMF MTP compatible duplex	
12. Gore Nlighten	$\lambda = 850$ nm VCSEL array	MSM photodetector			50-μ MMF MTP	
13. Agilent	$\lambda = 850$ nm VCSEL array	MSM photodetector		BGA	50-μ MMF MTP	
14. Molex	$\lambda = 850$ nm VCSEL array	MSM photodetector			50-μ MMF MTP	
15. Mitel	$\lambda = 850$ nm VCSEL array	MSM photodetector			50-μ MMF MTP	

Integrated
Circuits

FR4 Board

VCSEL Array

Photodetector
Array

Waveguide

Optical Fiber
Ribbon Cable

Fig. 11.7 Optobus module.

IC packaging technology within the computer industry generally, although there are notable exceptions. Optobus uses a novel, molded-plastic light-guide technology to couple light to and from the fibers, and a separate submount is required to mount the OE chips perpendicular to the IC package. The electrical interfaces are DC-coupled nonlogic interfaces, so the user is required to supply additional interface chips to make the link compatible with a logic standard bus. The modules provide only optical–electrical conversions. Therefore, all link flow control, clocking, and testing functions must be provided in support chips. This gives the Optobus modules the flexibility to be used in many applications, which, it is hoped, will increase the volume of modules needed. However, they are not expected to be particularly user-friendly to integrate into the SAN system. A photo of the Optobus module is shown in Fig. 11.7.

11.7.1.2. Jitney

The Jitney prototype is unique in that its design is driven by cost competitiveness with copper [11]. It uses modifications of standard datacom OE chips and computer IC packages. Because Jitney uses the large-core, relaxed-tolerance multimode fiber, it can also use standard plastic optic components (from the consumer/printer industry) and QFP packaging (from the consumer/computer industry). Jitney also develops significant manufacturing infrastructure in CAD and wafer-level testable arrays to drive the cost and design time down. It uses a unique approach to cable deployment, using a step-index, large-core fiber cable to achieve cost below copper wiring but limited to distances below 50 m, together with

Fig. 11.8   Jitney module and cable.

a more expensive graded-index large-core fiber cable for distances up to 300 m. Both cable types are mateable to the module. This approach may be attractive if the majority of cabling (e.g., >80%) is short jumpers (this is generally the case in machine room wiring, for example). However, the user is required to support two cable types (similar to the current use of copper). It remains to be seen whether SAN users will accept a cost-optimized cabling strategy for their optical interconnects.

The Jitney modules support logic interfaces and a number of self-test and flow-control functions. The electrical interfaces are DC coupled, but the optical lines employ a run length-limited transmission code to increase the robustness of the links to data pattern effects and component and environment variations. This type of coding has also been adopted for virtually all product serial data links. The Jitney modules and cable are shown in Fig. 11.8.

11.7.1.3.   PVL-4

The Parallel Vixel Link is a product that falls in the second category of POI — using standard multimode fiber components. Its design approach is to maximize the optical line speed and minimize the number of optical lines; it employs only four channels (to keep cost down) at 1.25 Gb/s line speed. It requires 8B/10B line-coded data patterns (run length-limited and DC-balanced transmission code) to be transmitted to simplify its electronics and to improve optical line robustness. PVL-4 provides only light conversion functions. It is designed to be used with a companion set of multiplexer/demultiplexer chips to re-create a byte-wide, NRZ parallel electrical interface. One drawback of this implimentation is that it requires the SAN designer to support eight 1.25 Gb/s lines on his or her card (full duplex).

Fig. 11.9   PVL-4 module.

This is more than twice as fast as that designed for the computer card industry's cost-effective technologies, but perhaps a rapid introduction of Gb/s Ethernet will change this drawback in the near future.

The interesting concept regarding the PVL-4 is the potential follow-up product that integrates the MUX/DEMUX function into the POI module so that the POI module electrical interface is optomized for SAN designer bus width and card speed, whereas the optical interface is optimized to take advantage of the superior optical line driver/receiver speed at a reduced optical I/O count. A PVL-4 is shown in Fig. 11.9.

The conclusion from the sampling shown in Tables 11.8–11.10 is that there is no clear consensus in the specifications from the POI suppliers, partially reflecting a lack of consensus on what customers want and partially reflecting the suppliers' history and marketing strategy. Currently, most suppliers are positioning their products ten times beyond the current copper wire offerings in the data rate × distance performance figure of merit. The functions in the modules are simple optoelectronic conversions because the suppliers attempt to cover as many users as possible with one module type. This also reflects the lack of user consensus in requirements. The cost of this technology will likely be two or three times that of copper. It remains to be seen, once field reliability has been demonstrated, whether SAN system designers will pay such a premium for optical interconnects.

However, some trends are emerging:

1. LVDS logic interfaces, with 3.3-V power supplies
2. 500 Mb/s, 1-byte wide, with a doubling in aggregate data rate approximately every 2 years
3. 300-m span (with the potential of a cost-reduced shorter span)

4. Surface-mount modules (probably BGA for multibyte-wide interfaces or full duplex from one module)

5. Simple OE conversion functions initially, migrating to complete physical layer function integration as standards develop or a volume user appears — this may be supported by a multiple module set initially, but the demand for smaller footprint will likely lead to the added function being integrated into the optical modules

11.7.1.4.   LiteBus

Building on the Jitney technology, IBM has recently introduced the LiteBus line of parallel optical devices. They are among the first implementations of parallel optical transceivers, as opposed to discrete transmitter and receiver modules. The transceiver interface consists of two 1×12 MPO fiber connectors with a common housing. The data rate ranges from 1000 to 2500 Mbaud per line, and supports up to 200 meters distance at the maximum data rate over 50 micron multimode fiber. Through the use of Open Link Shutdown (OLS) logic, the duplex link disables the transmitters in the event of a fiber break or connector opening; in this way, it complies with worldwide IEC class 1 laser eye safety standards. This product is also the only device so far to employ a plastic ball-grid array (PBGA) surface mount technology. This is a low laminate package with a two-dimensional grid of high melt solder balls (Sn/Pb) that use eutectic solder for attachment to standard printed circuit boards. Operating with a single 3.3 V power supply and either low-voltage PECL or LVDS signal levels, this transceiver is intended for high-reliability data communication networks, including Infiniband and the Optical Internetworking Forum Very Short Reach (VSR) OC-192 specifications.

11.7.2.   INTRAFRAME INTERCONNECTS

There have been intraframe versions of POI technology developed (with government sponsorship: United States, ARPA; Europe, RACE). The main additional components are a ribbon wiring harness, which offers smaller connectors and more flexibility than a robbin interframe cable, and an interboard blind-mating card connector.

The OETC project developed a flexible ribbon harness to go with the OETC module (Tables 11.9–11.10, No. 6), called the flexible optical circuit [20]. This harness has fibers embedded in a plastic (Kapton) covering,

Fig. 11.10 Optical wiring harness.

and the fibers are custom wired to connect a set of transmitters and receivers from the array modules into duplex connections at the card edge. A MAC-type connector was modified to be blind mateable along with a series of electrical card-edge connectors. Figure 11.10 illustrates an OETC transmitter and an OETC receiver module connected into one array edge connector.

Ericsson also developed a blind-mateable ribbon harness for interboard connections [17]. The harness uses either embedded optical fibers or polymer lightguides. The connector for this harness was based on the MT ferrule, and it is molded directly on the harness's ribbon to lower cost (similar to the Jitney approach). The harness attaches to the low-profile modules (described in Tables 11.8–11.10, No. 5).

11.8. Conclusions

This chapter has described some key intramachine opportunities for datacom optoelectronics, notably the system area network. The emergence of the SAN, and its increasing impact on computing system architecture, was not enabled by optical interconnects, and the system designers are currently driven by the capabilities of parallel copper wire technology. However, they are also driven by increasing node performance and node count and increasing flexibility in system layout and growth. These drivers all point to the eventual use of parallel optical interconnects. The questions are when and how.

This chapter has also described the parallel optical interconnect technology in products and protypes today. The technology is immature but moving rapidly, and there are no established standards. With few notable exceptions, the suppliers of this technology are hoping to adapt telecom or LAN datacom module and cable components to the intramachine application and maintain the marketing price and service structure established in their earlier markets, in which many users share the optical network in order to defray its high price. These price models do not fit with the pricing and rapid product migration models required in the computer industry, so the SAN developer is still concerned about POI filling more than a niche in the application market. (A similar situation developed in the LAN FDDI market, and copper wire technology took over as the de facto cost-effective standard.) It is significant that there are versions of parallel optical interconnect that can be price competitive due to the arraying of optical devices and component into monolithic structures. The next few years will be critical for both users and suppliers to find the right design point for POI.

References

1. Kleinrock, L., M. Gerla, N. Bambos, J. Cong, E. Gafni, L. Bergman, J. Bannister, S. P. Monacos, T. Buhjewski, P.-C. Hu, B. Kannan, B. Kwan, E. Leonardi, J. Peck, P. Palnati, and S. Walton. 1996 (June). The supercomputer supernet testbed: A WDM-based supercomputer interconnect. *J. Lightwave Tech.* **14**(6):1388–1399.
2. Feehrer, J., J. Sauer, and L. Ramfelt. 1995 (October). Design and implementation of a prototype optical deflection network. *Comput. Commun. Rev.* **24**(4):191–200.
3. Saunders, S. 1996. *The McGraw-Hill high-speed LANS handbook.* New York: McGraw-Hill.
4. ESCON Physical Layer I/O Interface Specification, SA23-0395.
5. Benner, A. F. 1996. *Fibre Channel: Gigabit communications and I/O for computer networks.* New York: McGraw-Hill.
6. Dagenais M., R. Leheney, and J. Crow, ed. 1995. *Integrated optoelectronics.* New York: Academic Press.
7. Hahn, K. H. 1995. POLO—Parallel optic links for gugabyte data communications. *IEEE Electronic Components and Technology Conference* 95, Las Vegas, Nev., May, pp. 368–375.
8. Pepeljugoski, P. K., B. K. Whitlock, D. M. Kuchta, J. D. Crow, and S.-M. Kang. 1995. Modeling and simulation of OETC optical bus. *IEEE Lasers and*

Electro-Optics Society Annual Meeting, San Francisco, Calif., October 30–31, pp. 185–186.

9. "Performance Standards for Laser Products," 21CFR1040, Center for Devices and Radiological Health (HFZ-312), Rockland, Md. Safety of Laser Products, Parts I (IEC 825-1, equipment classification) and II (IEC 835-2, communications systems)," American National Standards Institute, New York, N.Y.

10. Schlager, J. 1997 (June). "Round robin results of the multimode fiber restricted launch group — bandwidth," Minneapolis, Minn.: Telecommunications Industry Association Publication.

11. Crow J. *et al.* 1996 (May). The Jitney parallel optical interconnect. *Proceedings of ECTC'96.* p. 292, ISSN 0569-5503.

12. ANSI Standard X3T9.3. 1992 (June). "Fibre Channel Physical and Signalling Interface, FC-PH," Rev. 3.0.

13. For example, the University of Illinois tool *iFrost* was developed by the DARPA sponsored OETC Consortium of companies, and operates on various workstation platforms. (PC and UNX). See Whitlock, B., *et al.,* "Computer Modeling and Simulation of the OETC Optical Bus," *Proceedings of OFC'96.*

14. Staiwasz, K. *et al.* 1997 (May). "Automated Testing Methodologies for Low Cost, Parallel Optical Bus Components," Proceedings of ECTC'97. pp. 217–224.

15. Hahn, K. *et al.* 1996 (June). Gigabyte/s data communication with the POLO parallel optical link. *IEEE Electronic Components and Technology Conference 96,* p. 301.

16. OKI Electric Industry, Inc. 1994 (September). "Low-Power 11 × 320 Mbps Parallel Transmitter/Receiver LSIs for 2.4 Gbit/s Optical Links," *JETRO,* p. 20.

17. Niburg, M. *et al.* 1996 (September). A complete sub-system of parallel optical interconnects for telecom applications. *Proceedings of ECOC'96,* p. 259.

18. Nagahori, T., *et al.* 1996 (June). 1-Gbyte/sec array transmitter and receiver modules for low-cost optical fiber interconnect. *IEEE Electronic Components and Technology Conference 96,* p. 255.

19. Karstenshen, H. *et al.* 1994 (November). PAROLI — High performance optical bus with integrated components. *Proceedings of LEOS'94.*

20. Wong, Y. M., D. J. Muehlner, C. C. Faudskar, B. D. Buchholz, M. Fusgteyn, J. L. Brandner, W. J. Parzygnat, R. A. Morgan, T. Mullally, R. E. Leibenguth, G. D. Guth, M. W. Focht, K. G. Glogovsky, J. L. Zilko, J. V. Gates, P. J. Anthony, B. H. Tyrone, Jr., T. J. Ireland, D. H. Lewis, Jr., D. F. Smith, S. F. Nati, D. K. Lewis, D. L. Rogers, H. A. Aispain, S. M. Gowda, S. G. Walker, Y. H. Kward, R. J. S. Bates, D. M. Kuchta, and J. D. Crow. 1995 (June). Technology development of a high-density 32-channel 16 Gb/s optical data link for optical interconnection applications for the optoelectronic technology consortium (OETC). *IEEE J. Lightwave Tech.* **13**(6):995–1016.

21. Grimes, G., *et al.* 1996 (November). Applications of parallel optical interconnects. *Proceedings of the IEEE LEOS'96 Annual Meeting,* Vol. 2, pp. 6–7.

22. Crow, J., and F. Tong. (1997). Data processing and data communications networks — The drive for cost effective photonic technology. *Photonic Networks,* New York: Springer, p. 442.

23. Motegi, Y., and A. Takai. 1994 (April). Optical interconnection modules utilizing fiber-optic parallel transmission to enhance information throughput. *Hitachi Review,* pp. 79–82.

24. Lebby, M., *et al.* 1996 (May). Characteristics of VCSEL arrays for parallel optical interconnect. *Proceedings of ECTC'96,* pp. 279–291.

25. Swirhun, S., *et al.* 1996 (May). The P-VixeLink multichannel optical interconnect. *Proceedings of ECTC'96,* pp. 316–320.

Chapter 12 | Applications: Asynchronous Transfer Mode and Synchronous Optical Network

Carl Beckmann

Thayer School of Engineering, Dartmouth College, Hanover, New Hampshire 03755

12.1. Introduction

Early communication networks were driven by telephony and telecommunications requirements. The past 20 years have seen rapid growth in computer networking. Integration of these two applications initially started in wide-area networks, where economic considerations dictated that data networks use existing telephony infrastructures. The increasing use of digital media types promises that integration will reach both the corporate office and home/consumer networks increasingly in the future. This chapter starts with a review of the fundamental characteristics of telecommunications and data networks and the resulting requirements on integrated services networks.

12.1.1. TELEPHONY AND CIRCUIT SWITCHING

In the beginning, there were analog networks for supporting telephony. In the 1960s it became apparent that delivering analog telephone calls using analog frequency-division multiplexing techniques was prone to noise and did not make as effective use of the available bandwidth on copper wires as could digital techniques. Then, long-haul digital networks were installed for delivering long-distance telephone service. (Local service remained

HANDBOOK OF FIBER OPTIC
DATA COMMUNICATION
$35.00

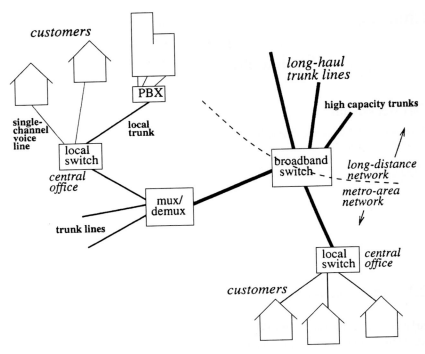

Fig. 12.1 Typical telephone network topology.

primarily analog for some time.) The general organization of telephone networks is illustrated in Fig. 12.1.

For telephony, the basic requirements are to establish a point-to-point connection for a "call" that is typically several minutes in duration. The delay through the network must be small enough not to interfere with the quality of speech and to avoid perceptible echo effects, on the order of milliseconds or less. In North America, voice is digitized to 8 bits precision at 8000 samples per second (to support an analog bandwidth of 300–3000 Hz), for a data rate of 64 kb/s per call. The capacity of a typical copper wire digital trunk link (a T1 line) is 1.5 Mb/s by comparison.

These basic requirements are served well by using reserved connections along all the links from a source to a destination (circuit switching), with time-devision multiplexing (TDM) used to share the bandwidth among multiple calls on the individual links. TDM keeps the latency on each link very low while dividing the available bandwidth evenly between calls, with only a small amount of framing overhead. American and international

Table 12.1   **North American and International
Telecommunications Standards**

Name	Organization	Bit rate	No. voice channels
T1	AT&T	1,544	24
T3	AT&T	44,736	672
DS1	ANSI	1,544	24
DS2	ANSI	6,312	96
DS3	ANSI	44,736	672
DS4	ANSI	274,176	4032
E1	CCITT	2,048	32
E2	CCITT	8,448	128
E3	CCITT	34,368	512
E4	CCITT	139,264	2048

standards bodies have adopted a variety of standard data rates and interoperability specifications, which are summarized in Table 12.1.

12.1.2.   DATA COMMUNICATIONS AND PACKET SWITCHING

Although telecommunications networks can be used to carry data as well, local area data networks can be built much more cheaply and efficiently without resorting to switch-based network architectures.

The requirements for data networks come from the ability to transfer files and other packets of information, such as electronic mail messages, and terminal data interfaces, consisting of typed keyboard strokes from a user and displayed textual and graphic information back to the user. Other traffic comes from remote procedure calls for distributed computing and operating system information and distributed file system transfers. Compared with telephony, a much more heterogeneous mix of traffic exists on data networks.

For most data traffic, there are no hard real-time constraints on its delivery. File transfers should happen quickly to allow for smooth operation and rapid response time to interactive users; slower response time is merely annoying but does not render the service useless.

Table 12.2   **Data Rates on Popular Local Area Network (LAN) Technologies**

Name	Organization	Data rate (Mb/s)	Applications
AppleTalk	Apple Computer	<1	File sharing, printer sharing, electronic mail
Ethernet	ARPA	10	LAN for workstations and mainframes
Fast Ethernet	IETF	100	Buildingwide LAN backbones and servers
FDDI		100	Fast LAN and medium-scale networks (several kilometers)

In early systems, network bandwidth and delay were dictated by near-real-time requirements of character terminal input/output: A fast typist can type 60–100 words per minute. If each word is an average of 6 characters long (including the space), and each character is represented in 8-bit ASCII, then this represents a steady throughput of 80 bits per second. Characters come at an average rate of 10 per second. Terminal equipment was usually connected to the computer (often through an intermediate multiplexer) via a 9600-baud serial data line. Thus, the events of interest (defining the maximal acceptable latency) are on the order of hundreds of milliseconds, and the bandwidth required per connection is on the order of tens of kilobits or less per second. One hundred users could be accommodated with approximately 1 Mb/s worth of usable bandwidth. Table 12.2 lists the data rates in use in several widespread local area network (LAN) schemes.

Due to the highly heterogeneous and unpredictable nature of traffic on data networks, the use of connection-oriented communications, in which bandwidth is reallocated only every few minutes per channel, is not efficient. Moreover, higher latencies are tolerable. This makes connectionless communications based on discrete data packets, each containing its own addressing and format information, much more attractive. The bandwidth of the network can, in effect, be reallocated on a packet-by-packet basis on demand as the packets arrive at the network. To make this scheme efficient, a larger number of data bits should be put in each packet that is required for the packet "header" information. This incurs extra latency but is tolerable.

Finally, LANs can be built cheaply and extensibly by connecting all data equipment to a shared medium (such as coaxial cable) and distributing the

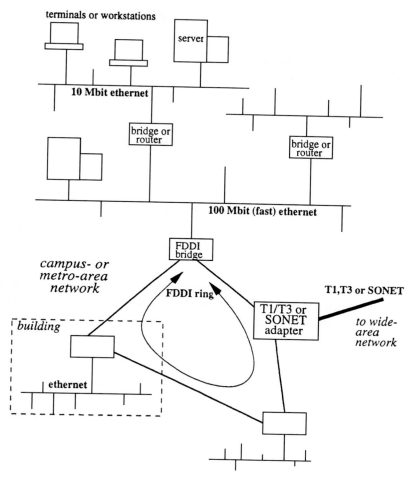

terminals or workstations

server

10 Mbit ethernet

bridge or router

bridge or router

100 Mbit (fast) ethernet

FDDI bridge

campus- or metro-area network

FDDI ring

T1,T3 or SONET

T1/T3 or SONET adapter

to wide-area network

building

ethernet

Fig. 12.2   Topology of a typical LAN.

switch functionality (channel allocation) out to the data nodes. Additional nodes can be added by simply splicing the new equipment into the existing cable. Network interface functionality can be built into each computer system node for a fraction of the cost of the CPU, memory, or hard disk subsystems. Internetworks can be constructed by linking LANs together by bridges, routers, or computing nodes connected to multiple LANs or interconnected via telecom lines. Store-and-forward routing of packets is acceptable because of the longer tolerable latencies for typical datacom applications. The organization of typical local area networks is illustrated in Fig. 12.2.

12.1.3.   BROADBAND INTEGRATED SERVICES DIGITAL NETWORK REQUIREMENTS

The need for integrating data- and telecommunications functions is apparent from several trends:

- Data communication has long made use of digital telephone switching networks for wide-area data transport.
- The use of facsimile and personal computer modems has increased phenomenally during the past several years, using local analog phone lines for digital services.
- New applications in consumer and business fields are emerging that share some of the characteristics of both traditional datacom and telecom applications, such as teleconferencing, advanced television and video-on-demand, interactive multimedia training and education, and telecommuting.

Moreover, as both the bandwidth requirement per user and the number of users connected continue to increase, networks must make increasing use of fiber optics as opposed to electronics at the physical layer.

The requirements on integrated services digital networks (ISDNs) are thus:

- Support for connectionless and packet-oriented data services, in particular TCP/ IP.
- Efficient support for connection-oriented "constant bit rate" (CBR) services such as telephony or (uncompressed) digital video.
- Support for new classes of traffic such as "variable" (but steady) bit rate (VBR) services, such as compressed video or voice; or "bursty" VBR /CBR traffic from interactive multimedia applications.
- Efficient multicasting support, both for media multicasting and for TCP/ IP functionality.

12.1.4.   ASYNCHRONOUS TRANSFER MODE AND SYNCHRONOUS OPTICAL NETWORK OVERVIEW

ATM has been proposed as an enabling network technology to support broadband integrated services. It is not a complete, stand-alone networking standard; rather, ATM defines a common layer of interoperability called the ATM layer, on which various services ranging from telephony and

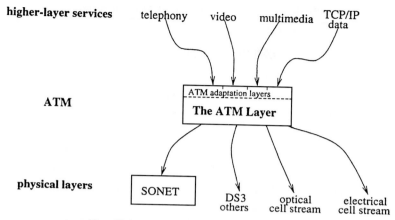

Fig. 12.3 ATM and SONET in perspective.

video conferencing to TCP/ IP data networking and multimedia can be delivered. The ATM layer defines a common format used for switching and multiplexing bit streams from one end of an ATM network to another.

The ATM layer, in turn, uses the hardware facilities of lower layers to deliver the bits across individual links in a network. A variety of such physical layers have been defined, most of which are based on existing standards in order to maximally leverage existing technologies and installed bases. These relationships are summarized in Fig. 12.3.

One family of ATM physical layers is based on SONET. SONET is a synchronous, time-division multiplexing standard based on transmission over optical media (actually, a family of standards at a variety of bit rates). It was primarily designed to support telecommunications and long-haul, broadband services.

12.2. SONET

12.2.1. HISTORICAL PERSPECTIVE

The SONET standards were developed in the mid-1980s to take advantage of low-cost transmission over optical fibers. SONET defines a hierarchy of data rates, formats for framing and multiplexing the payload data, as well as optical signal specifications (wavelength and dispersion), allowing multivendor interoperability. SONET was originally proposed by Bellcore in 1985 and later standardized by ANSI and the CCITT [synchronous digital hierarchy (SDH) is a compatible set of standards in Europe] [1–3].

Table 12.3   **Basic SONET/SDH Data Rates**

SONET			
Electrical	**Optical**	**SDH**	**Rate**
STS-1	OC-1	—	51.840 Mb/s
STS-3	OC-3	STM-1	155.520 Mb/s
STS-9	OC-9	—	466.560 Mb/s
STS-12	OC-12	STM-4	622.080 Mb/s
STS-18	OC-18	—	933.120 Mb/s
STS-24	OC-24	—	1.244160 Gb/s
STS-36	OC-36	—	1.866240 Gb/s
STS-48	OC-48	STM-16	2.488320 Gb/s

SONET is a synchronous, time-division multiplexing scheme designed to support existing telephone network trunk traffic and also designed with broadband ISDN (BISDN) services in mind. Its TDM basis readily supports fixed-rate services such as telephony. Its synchronous nature is designed to accept traffic at fixed multiples of a basic rate, without requiring variable stuff bits or complex rate adaptation. The SONET data transmission format is based on a 125-μs frame consisting of 810 octets, of which 36 are overhead and 774 are payload data. The basic SONET signal, whose electrical and optical versions are referred to as STS-1 and OC-1, respectively, is thus a 51.84 Mb/s data stream that readily accommodates TDM channels in multiples of 8 kb/s.

SONET defines a hierarchy of signals at multiples of the basic STS-1 rate. The SONET rates currently standardized are shown in Table 12.3. SDH is a compatible European counterpart to SONET. Due to compatibility issues with European switching equipment, the basic SDH rate, called STM-1, is three times the STS-1 rate (i.e., STS-3), or 155.52 Mb/s.

12.2.2.   STS DATA RATES AND FRAMING

To efficiently support telephony, SONET bit rates rest fundamentally on voice-quality audio sampling rates, i.e., 8000 samples per second at 8 bits per sample. The SONET data transmission format is therefore based on a 125-μs frame illustrated in Fig. 12.4. This figure shows the basic STS-1 frame. Higher rates are achieved by byte-interleaving multiple STS-1 frames. The 125-μs frame contains 6480 bit periods, or 810 octets (bytes).

column

row	1	2	3	4	...	90
1	*0*	*1*	*2*	*3*	...	*89*
2	*90*	*91*	*92*	*93*	...	*89*
3	*180*	*181*	*182*	*183*	...	*179*
4	section					*etc.*
5	and			Payload		
6	line					
7	overhead					
8						
9	*720*	*721*	*722*	...		*809*

Fig. 12.4   SONET framing format.

This can be viewed as a two-dimensional arrangement of nine rows by 90 columns (of bytes) that is scanned row-wise from the upper left.

Thus, a single voice channel occupies a single octet in each 125-μs frame, and after leaving room for various "overhead" octets (see below), 774 64 kb/s voice channels can be time-division multiplexed into a single STS-1 frame. The bit rate for an OC-N link is thus given by

$$OC\text{-}N \text{ bitrate} = N \cdot 8000 \text{ Hz} \cdot 90 \text{ columns} \cdot 9 \text{ rows} \cdot 8 \text{ bits/octet}$$
$$= 51.84N \text{ Mb/s}, \tag{12.1}$$

and the payload capacity (after accounting for four overhead columns per frame) is

$$OC\text{-}N \text{ capacity} = N \cdot 8000 \text{ Hz} \cdot 86 \cdot 9 \cdot 8 = 49.536 \ N \text{ Mb/s}. \tag{12.2}$$

The first three columns in each frame (i.e., the first three of every 90 octets) are reserved for various overhead bytes. Overhead information is organized into section, line, and path overhead.

SONET can be thought of as following a layered model. At the lowest layer (the physical layer), SONET specifies characteristics of the optical signal, such as maximum dispersion, etc. The lowest level physical link between two pieces of SONET equipment (i.e., an optical fiber pair) is called a section (Fig. 12.5). Multiple sections may be linked together via signal repeaters (regenerators) to form a line. The two ends of a line attach to line termination equipment. At the next level up, a physical line may be used by one or more paths, which are connected on both ends to path termination equipment. It is at the path termination equipment that SONET frames are assembled and disassembled. The layered approach allows the

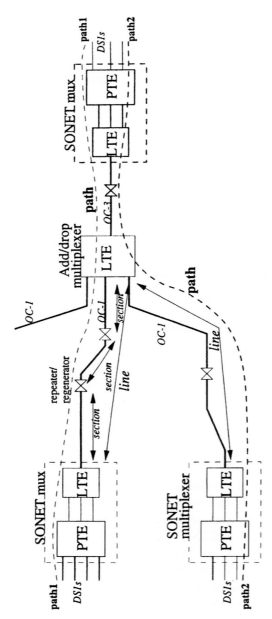

Fig. 12.5 SONET sections, lines, and paths.

Services	DS1, DS3, ATM cells...
Path Layer	Envelope
Line Layer	STS-N blocks
Section Layer	Frame
Physical Layer	Photons

Fig. 12.6   SONET layered architecture.

use of equipment for handling functions related to one or the other layer individually, keeping costs down by not requiring all layers to be handled at once. Figure 12.6 illustrates the SONET layers.

The first three columns of each frame contain the section and line overhead bytes. The first three rows of this are for the section overhead, and the last six rows are for the line overhead. This is illustrated in Fig. 12.7. The remaining 87 columns contain the synchronous payload envelope (SPE). The SPE contains the actual payload data as well as a single column of path overhead bytes.

Note that the SPE need not be exactly aligned in the payload frame. In fact, the first byte of the SPE may reside (and usually does) anywhere within the frame; hence, the path overhead is not always in column 4. Overhead octets H1 and H2 form a pointer to the location of the first SPE octet. This feature is useful in connecting two lines whose bit clocks differ slightly, as they do in practice. This allows the SPE to "slip" slightly with respect to

		1	2	3	4		90
section	1	A1	A2	C1			
overhead	2	B1	E1	F1			
	3	D1	D2	D3			
line	4	H1	H2	H3	path overhead		
overhead	5	B2	K1	K2	J1		
	6	D4	D5	D6	B3		
	7	D7	D8	D9	C2		
	8	D10	D11	D12	G1		
	9	Z1	Z2	E2	F2	payload	
next					H4		
frame					Z3		
					Z4		
					Z5		

Fig. 12.7   SONET frame overhead bytes.

Table 12.4 **SONET Section Overhead Octets**

Symbol	Bits	Name	Description
A1, A2	16	Framing	F628 Hex (1111011000101000 binary); provided in all STS-1 signals within an STS-N signal
C1	8	STS-1 identification	Unique number assigned just prior to interleaving that stays with STS-1 until deinterleaving
B1	8	Section BIP-8	Allocated in each STS-1 for a section error monitoring function
E1	8	Orderwire	Used as a local orderwire channel; reserved for communications between regenerators, hubs, and remote terminal locations
F1	8	Section user channel	This byte is set aside for the network provider's purpose; it is passed from one section level entity to another and is terminated at all section-level equipment
D1–D3	24	Section datacomm	A 192 kb/s channel for alarms, maintenance, control, etc. between section terminating equipment

the frame. A stuff byte is provided in H3 to make up the bandwidth deficit in the case in which the signal to transmit is faster than the line clock. This scheme separates the synchronization of data payload frames from the generation of the framing signals, which can be done from a transmitter's local clock.

12.2.2.1. Section Overhead Octets

The first three rows of the first three columns in each frame are used for section-related functions. The functions of these bytes, which include framing, identification, section error monitoring, and auxiliary data channels, are summarized in Table 12.4.

12.2.2.2. Line Overhead Octets

The last six rows in the first three columns of each frame are used for line-related functions, as summarized in Table 12.5.

Table 12.5 **SONET Line Overhead Octets**

Symbol	Bits	Name	Description
H1, H2	16	Pointer	Indicates the offset in bytes between the pointer and the first byte in the STS SPE
H3	8	Pointer action	Stuff byte for downstream frame advancement
B2	8	Line BIP-8	Allocated in each STS-1 for a line error monitoring function; used as a local orderwire channel; reserved for communications between regenerators, hubs, and remote terminal locations
K1, K2	16	APS channel	Allocated for APS signaling between two line-level entities; also carries other management signals
D4–D12	72	Line datacomm	Nine bytes (576 kb/s) allocated for line data communication for alarms, maintenance, control, etc.
Z1, Z2	16	Growth	Future expansion
E2	8	Orderwire	Express orderwire between line entities

12.2.2.3. Path Overhead Octets

The first column in the SPE of an STS-1 signal is used for various path-related functions, as summarized in Table 12.6. In an OC-N signal, which carries N byte-interleaved STS-1 SPEs, the first column in each STS-1 is used for path-related overhead. By contrast, in a "concatenated" OC-Nc signal, there is only a single column of path overhead, with the remaining $87N-1$ columns available for payload data.

12.2.3. PAYLOAD ENVELOPE POINTER

The SPE of a SONET frame need not be perfectly aligned with the framing overhead. Pointer octets H1 and H2 are used to locate the SPE within the frame. The lower 10 bits of H1 and H2 are an offset to the beginning of the SPE, i.e., the number of octets between H3 and J1, the first octet in the SPE. This feature makes it easier to synchronize multiple signals and multiple pieces of equipment, while allowing each signal source to generate its own framing structure based on a local clock.

Table 12.6   SONET Path Overhead Octets

Symbol	Bits	Name	Description
J1	8	STS path trace	Used by path-terminating equipment to verify its connection to the source, which continuously sends a fixed 64-byte pattern
B3	8	Path BIP-8	Path error monitoring
C2	8	STS path signal label	Indication of valid construction of SPE
G1	8	Path status	Path-terminating status and performance, back to an originating path
F2	8	Path user channel	For network provider
H4	8	Multiframe	A 192 kb/s channel for alarms, maintenance, control, etc. between section terminating equipment
Z3–Z5	24	Growth	Future expansion

The upper 4 bits of H1 and H2 are used to signal changes in the pointer value: A value of 0110 signals an increment or decrement by 1, and a value of 1001 signals some larger change. In the frame in which the pointer is incremented by 1, the lower 10 (H1,H2) bits do not contain the new pointer value but rather the old pointer value, with all the even bits (including the LSB) inverted; on a decrement by 1, the odd bits are inverted. Once the pointer stabilizes the true new value is used in the lower 10 (H1,H2) bits.

Because the frequency deviation imposed by the standard is small, pointer adjustments will take place infrequently in practice. If an upstream clock is too slow, the downstream equipment will have to periodically increment its pointer and delay outgoing SPEs. When eventually the pointer overflows the maximum value of 809, an entire frame will be skipped. If the upstream clock is too fast, the pointer will have to be decremented periodically. When this happens, the missing byte is put in the H3 octet to compensate. Essentially, the H3 stuff byte provides the extra bandwidth needed for slow-running clocks to keep up with the required data rate.

12.2.4.   MULTIPLEXING

Higher speed transmission than STS-1 rates is achieved by byte-interleaving N STS-1 signals to obtain an STS-N signal (which is then converted to an

Table 12.7   **Virtual Tributaries**

Name	Service	Data rate	No. columns
VT1.5	DS1	1.544	3
VT2	CEPT	2.048	4
VT3	DS1C	3.088	6
VT6	DS2	6.176	12

optical OC-N signal). This allows, for example, several STS-1 signals to be multiplexed for tranmission over an OC-3 (or higher) link.

Alternately, higher speed channels can be obtained using concatenated STS-1s to achieve a single channel with N times the capacity of an STS-1. In this case, N STS-1 frames are again byte-interleaved to obtain the STS-Nc framing structure. In the STS-Nc frame, there are $3N$ columns for transport (section and line) overhead, with $87N$ columns remaining for the payload. However, this payload is multiplexed, switched, and transported through the network as a single entity. Hence, only a single column of path overhead is needed (leaving slightly more bandwidth available for data capacity compared to noncontatenated STS-N).

12.2.5. VIRTUAL TRIBUTARIES

In order to directly support services with lower bandwidth requirements than the basic STS-1 payload, several standard "virtual tributary" formats have been defined for SONET. These are summarized in Table 12.7. The VT1.5, for example, allows a DS1 or T1 signal to be carried end to end on a SONET path without having to remultiplex the 24 DS0 (voice) channels contained therein.

Each virtual tributary format is defined as some integral number of columns of the SONET SPE, which includes room for the carried signal as well as any VT-related overhead octets.

12.2.6. INTERNATIONAL INTEROPERABILITY

SONET is compatible with an international set of standards called the SDH. SDH was developed based on SONET, but with the additional goal of providing compatibility between North American and European telecom carriers. Whereas SONET starts with a 51.84 Mb/s signal consisting of nine rows by 90 columns every 125 μs (STS-1), SDH starts with a 9×270 frame every 125 μs, or a 155.52 Mb/s signal.

The basic 155.52 Mb/s SDH signal, called STM-1, is similar and can be made compatible with SONET STS-3. There are some differences in the usage of section and line overhead octets between SONET and SDH. For a more detailed discussion of the differences, the reader is referred to Minoli [3]. See also Table 12.3 for SDH data rates.

12.2.7. SONET PHYSICAL SPECIFICATIONS

Specifications for the transmitter, receiver, and optical signal path characteristics for various SONET line rates are given in Table 12.8 [4].

12.3. ATM

12.3.1. CELL VS PACKET SWITCHING

ATM is designed for high-speed transport of a variety of traffic types. Due to its high-speed nature, it is believed that using fixed-size cells will allow efficient hardware implementations of various multiplexing and routing functions. Unlike LAN environments using Ethernet, fast Ethernet, or fiber distributed data interface (FDDI), and capable of tens to hundreds of megabits of throughput on variable-sized packet traffic, ATM is designed to work into the gigabit per second range [2, 3, 5–7]. Moreover, ATM is designed to be able to support both switched- and packet-oriented applications. Finally, ATM allows the quality of service to be specified within a range of parameters during call setup time.

The use of small, fixed-sized cells has several advantages over larger variable-sized packets as used in Ethernet or FDDI:

- Cell boundaries can be easily recognized at high speed in hardware, should loss of framing occur.
- Individual packets cannot monopolize the bandwidth of the channel.
- Cell-handling decisions (e.g., during congestion or for traffic policing of individual connections) can be easily made based solely on the number of cells, without having to examine their headers for packet size information.
- Cell-buffering hardware in switches and other equipment is simplified.
- Circuit-like switching of cells replaces store-and-forward routing of packets, with much lower latency over multihop paths.

Table 12.8 SONET Physical Layer Optical Specifications

Parameter	Units	OC-1	OC-3	OC-9	OC-12	OC-18	OC-24	OC-36	OC-48
Data rate									
Bit rate	Mb/s	51.84	155.52	466.56	622.08	933.12	1244.16	1866.24	2488.32
Tolerance	ppm			100					
Transmitter									
Type		MLM/LED	MLM/LED	MLM/LED	MLM/LED	MLM	MLM	MLM	MLM
$\lambda_{W_{min}}$	nm	1260	1260	1260	1260	1260	1260	1260	1265
$\lambda_{W_{max}}$	nm	1360	1360	1360	1360	1360	1360	1360	1360
$\Delta\lambda_{max}$	nm	80	40/80	19/45	14.5/35	9.5	7	4.8	4
$P_{T_{max}}$	dBm	-14	-8	-8	-8	-8	-5	-3	-3
$P_{T_{min}}$	dBm	-23	-15	-15	-15	-15	-12	-10	-10
$r_{e_{min}}$	dB	8.2	8.2	8.2	8.2	8.2	8.2	8.2	8.2
Optical path System									
ORI_{max}	dB	na	na	na	na	20	20	24	24
$D_{SR_{max}}$	ps/nm	na	na	31/na	13/na	13	13	13	12
Max sndr. to rcvr. reflectance	dB	na	na	na	na	-25	-25	-27	-27
Receiver									
$P_{R_{max}}$	dBm	-14	-8	-8	-8	-8	-5	-3	-3
$P_{R_{min}}$	dBm	-31	-23	-23	-23	-23	-20	-18	-18
P_0	dBm	1	1	1	1	1	1	1	1

a) Cell Format

b) Header Format

Fig. 12.8   ATM cell format.

The disadvantages are that header information may consume a larger fraction of available bandwidth than for large packets, and that sending very small amounts of information is less efficient than it is for small packets (although both are inefficient).

The structure of ATM cells is shown in Fig. 12.8 [8]. It consists of a 5-byte header followed by 48 bytes of payload data. The header contains the following fields: Generic flow control (GFC), virtual path identifier (VPI), virtual channel identifier (VCI), payload type (PT), a cell loss priority bit (CLP), and header error correction (HEC). A brief description of each of these fields is given in Table 12.9.

12.3.2.   CELL VS CIRCUIT SWITCHING

Another key feature of ATM is its ability to transport "constant bit rate" data such as (uncompressed) telephony or video over virtual circuits with guaranteed bandwidth and latency characteristics. In other words, ATM provides a service that mimics a point-to-point, synchronous connection normally provided by a TDM network. Features of ATM that enable this include the following:

- Cell size is kept small, because cell size directly affects latency at the source and destination associated with packing and unpacking a bit stream into cells and, to some extent, affects latency of cell handling at network-switching elements.
- Keeping cell size fixed makes it feasible to allocate link bandwidth to individual connections, and reducing the cell size increases the bandwidth resolution at which this can be done.
- Fixed-size cells make scheduling of periodic or pseudoperiodic traffic at switching elements feasible in principle.

Table 12.9   **ATM Cell Fields**

Field	Bits	Name	Description
GFC	4	Generic flow control	
VPI	8	Virtual path identifier	Identifies 1 of 256 possible paths out of the current switch or device. Used with VCI to distinguish and locally route different cell streams
VCI	16	Virtual channel identifier	Identifies 1 of 65,536 possible channels in the given path out of the current switch or device. Used with VPI to distinguish and locally route different cell streams
PT	3	Payload type	Differentiates control vs data cells, etc.
CLP	1	Cell loss priority	Used to mark low-priority cells that may be discarded if network traffic is high
HEC	8	Header error correction	A CRC checksum on the first four header octets, using the generator polynomial $x^8 + x^2 + x + 1$. The resulting code is also XORed with 01010101 to get the HEC bits
Data	48×8	Payload	User data and headers/trailers from higher network layers

However, the main justification for ATM is its ability to mix synchronous with other types of traffic such as variable bit rate or connectionless and "bursty" data: By using cells, rather than TDM time slots, the channel bandwidth can be reallocated to different "virtual connections" on a cell-by-cell basis instead of requiring a TDM time slot to be allocated (requiring an end-to-end call setup) as short-lived connections come and go or as the bandwidth requirement of a single channel waxes and wanes. This makes efficient statistical multiplexing possible, where a large number of variable-bandwidth connections can be supported over a broadband channel with capacity for the sum of the connections' average bandwidth requirement, even though the sum of maximum instantaneous bandwidth requirements may exceed the channel's capacity.

Paradoxically, one of the physical layers used for ATM is the SONET. Here, SONET frames are simply used to transport ATM cells across a SONET path. The payload carried by the ATM cells need not be synchronous, however, because from frame to frame (and from cell to cell), the payload carried by the ATM cells can come from completely different ATM connections. The ATM cells in the SONET payload are opaque to the SONET layer. Allocation of the ATM cell bandwidth to CBR, VBR, and connectionless data channels is handled entirely at the ATM layer and above.

12.3.3. ATM LAYERED ARCHITECTURE

ATM is based on a layered architecture (Fig. 12.9). The major layers are the physical layer, the ATM layer, and the ATM adaptation layer (AAL). Above the AAL reside the data source layers, corresponding approximately to open systems interconnection (OSI) layers 3–7. The physical layer is further divided into a lower physical medium-dependent sublayer (PMD) and the transmission convergence sublayer (TC). The adaptation layer is also divided into the segmentation and reassembly sublayer (SAR) and the convergence sublayer (CS).

ATM layers do not correspond to the standard seven-layer OSI model, although an approximate correspondence is shown in Fig. 12.9. In most applications, AAL, ATM, and the TC sublayer of PHY can be thought of as providing the functionality of the OSI data link layer, i.e., the error-free transmission of bits from one end of a link to another. Although this

ATM model			OSI model
			Transport
	source layers (3-7)		Network
AAL	Adaptation Layer	*Segmentation and Reassembly* (SAR)	Data
		Convergence Sublayer (CS)	
ATM		ATM Layer	Link
PHY	Physical Layer	*Transmission Convergence* (TC)	
		Physical Medium Dependent (PMD)	Physical

Fig. 12.9 ATM reference model.

Table 12.10   **ATM Service Classes**

| | CBR (class 1) | VBR (class 2) | ABR | |
			Class 3	Class 4
Timing	Synchronous	Synchronous	Asynchronous	Asynchronous
Bit rate	Constant	Variable	Variable	Variable
Connection mode	Connection oriented	Connection oriented	Connection oriented	Connectionless

may involve the traversal of several switches (which in turn uses routing information in cells' VPI and VCI fields), the actual network layer function of establishing these routes on call setup is left to higher layers.

The ATM layer is fixed, but a variety of adaptation layers and physical layers have been defined. The services provided by the adaptation layer depend on the traffic type being supported. Traffic types vary in their data rate characteristics (constant data rate versus variable or bursty data traffic), connectionless (datagram) versus connection-orientedness, allowing ATM to support the spectrum of services including voice, video, and computer data services and interactive multimedia. Four basic traffic classes have been defined as shown in Table 12.10, and an adaptation layer has been defined for each (AAL-1–AAL-4). (The adaptation layers for class 3 and class 4 available bit rate traffic has been combined into a single layer, AAL-3/4.)

A fifth adaptation layer, AAL-5 (originally called SEAL, the simple and efficient adaptation layer), has also been defined to serve as a convenient application programmer interface (API) for computer applications to build directly on top of ATM services.

12.3.4. ATM PHYSICAL LAYERS

ATM is a switching and multiplexing scheme for BISDN, but it is not necessarily tied to a particular physical layer. Fiber optic as well as electronic physical layers are possible at a variety of data rates [8–11]. At the time of this writing, the ATM Forum Technical Committee has standardized the following physical layers for ATM: 155 and 622 Mb/s fiber optic layers based on SONET; 100 and 155 Mb/s cell-stream fiber optic layers; 155 and 25 Mb/s layers for twisted-pair connections; and a DS1 (1.5 Mb/s) layer based on T1. The physical layers standardized for ATM are summarized in Table 12.11 [12].

Table 12.11   **Standardized ATM Physical Layers**

Rate (Mb/s)	Media	Framing	UNI specification
1.554	Twisted pair	DS1	Public
2.048	Twisted pair, coax	E1	Public
6.312	Coax	J2	Public
25.6	UTP-3	Cell stream, 32 Mbaud	Private
34.368	Coax	E3	Public
44.736	Coax	DS3	Public
51.84	UTP-3	SONET STS-1	Private
100	MMF	Cell stream, 125 Mbaud	Private
155.52	SMF	SONET OC-3c	Public/Private
155.52	UTP-3, coax	SONET STS-3c	Private
155.52	MMF	Cell stream, 194.4 Mbaud	Private
155.52	STP	Cell stream, 194.4 Mbaud	Private
622.08	SMF, MMF	SONET OC-12	Private

The ATM user–network interface (UNI) specification [8] includes two kinds of interfaces; public and private. Public ATM service providers and any equipment connecting to public ATM networks must adhere to the public UNI specification, whereas the less stringent private UNI specification is suitable for use in local-area networking equipment. The private UNI does not need the operation and maintenance complexity or the link distance provided by the public UNI for telecom lines.

12.3.4.1. SONET/SDH

SONET-based fiber optic physical layers for ATM have been defined at 155 Mb/s (OC-3) and 622 Mb/s (OC-12) rates. In both these cases, the PMD sublayer is essentially identical to the corresponding SONET–SDH specification. The TC sublayer makes use of SONET framing by encapsulating ATM cells into the SONET SPE. The relationship between the ATM physical layer and the SONET layered architecture is shown in Fig. 12.10.

The SONET payload envelope presents a bandwidth resource that is used by the TC sublayer to carry ATM cells. However, because the ATM cell size (53 octets) does not evenly divide the size of either the STS-3c or STS-12c payload envelopes, no synchronization between ATM cells and

Fig. 12.10   OC-3 and OC-12 SONET-based ATM physical layer.

the SONET framing structure is implied (i.e., cells may cross SONET frame boundaries). In the STS-3c UNI, the available capacity for ATM cells is nine rows by 260 columns (the payload envelope minus one column of path overhead), or 149.760 Mb/s. In the 622 Mb/s interface, there are three fixed stuff columns following the path overhead, so the available cell carrying capacity is $9 \times (1044 - 4)/125 \ \mu s = 599.04$ Mb/s. In both cases, the available capacity is packed with ATM cells, and any rate decoupling between the ATM and PHY layers is accomplished by inserting empty cells into the stream.

Because of the asynchrony, the TC sublayer is also responsible for cell delineation. This is accomplished via the HEC bits in the cell headers. If cell synchronization is lost, the TC sublayer receiver continuously scans the SONET payload, testing whether each new octet starts a 5-octet ATM header with a valid HEC field. If so, it enters a presynch state, and if several valid cells are detected in a row, the synch state is entered and cell synchronization is assumed. HEC checking continues during normal transmission to verify that cell synchronization is not lost.

As long as cell synchronization is maintained in steady state, the HEC field is also used to correct any single-bit errors found in individual cell headers. The HEC field uses a polynomial code as indicated in Table 12.9 to perform single-bit correction and multiple-bit error detection on the header portion of each cell.

Finally, prior to insertion in the cell stream (and after removal on the receive side), the TC sublayer scrambles the payload portion of ATM cells to avoid any problems with DC levels or repeated bit patterns in the SONET payload envelope. This uses a self-synchronizing scrambler polynomial described in ITU recommendation I.432 [8].

12.3.4.2. Cell Stream

Alternately, cells may be sent directly over optical media, without using SONET framing. Several such physical layers have been defined for ATM at

100 and 155.52 Mb/s data rates. In order to maintain DC levels on the line and aid in synchronization, a 4B/5B (100 Mb/s) or 8B/10B (155 Mb/s) line code is used; hence, the actual required baud rate on the line is 125 or 194.4 Mb/s, respectively. Similarly, an electronic cell-stream physical layer has also been defined for twisted-pair cable, with a data rate of 25.6 Mb/s (32 Mbaud, due to 4B/5B coding) [10]. The 100 Mb/s PHY layer uses the same symbol coding and physical fiber and signal specifications as the FDDI LAN standard.

In the cell-stream interfaces, the TC sublayer is responsible for the functions of cell delineation and HEC verification and for the 155 Mb/s UNI, 125-μs clock recovery. In both of these private UNIs, the HEC is used for detection of errored cells only and not for correction because the use of a 4B/5B (or 8B/10B) code means that any line bit errors result in multiple data bit errors. ATM cells are simply discarded from the stream sent to the ATM layer.

12.3.4.2.1.   100 Mb/s TC Sublayer

In the 100 Mb/s interface (also called TAXI) there is no framing structure; therefore, when no cells are being transmitted, a special 8-bit symbol "JK" (Sync) is continuously sent (not the FDDI "idle line" code). Three special 8-bit symbols are used by the ATM PHY layer, each of which is a pair of FDDI line symbols (5-line bits each), as shown below:

FDDI code pair	Symbol/Function
JK (Sync)	Idle
TT	Start of cell
QQ	Loss of signal

The start of a new cell is indicated by the "TT" byte. This is followed immediately by the 53 bytes of the cell.

Thus, although ATM cells are available from the ATM layer, they are transmitted on the line continuously as 54 FDDI symbol pairs each (a start of cell followed by 53 data bytes). However, the JK Sync code is also used to gain byte alignment; therefore, as a minimum, 1 JK symbol pair must be inserted by the TC sublayer on the line every 0.5 s.

12.3.4.2.2.   155-Mb/s TC Sublayer

The 155 Mb/s interface, on the other hand, does have a framing structure consisting of 1 ATM cell used as a physical layer overhead unit (PLOU) followed by 26 cells of data. All 27 cells consist of 53 bytes each, and each byte is coded on the line as a single 8B/10B symbol as specified in the Fibre Channel physical layer document. Synchronization is maintained by using special codes as the PLOU cell's header (four "K28.5" symbols followed by one "K28.7" symbol).

Unlike the 100 Mb/s interface, cell rate decoupling is performed by inserting idle cells rather than some idle line symbols. Thus, as in the case of the SONET-based PHYs, the available bandwidth is packed with a contiguous stream of whole cells (some of which may be "empty").

An additional function performed by the 155 Mb/s TC sublayer is the delivery of a 125-μs clock across the link. This is done by using another special line code (K28.2), which can be inserted anywhere within the symbol stream. On the receive end, this symbol is used as a 125-μs strobe and removed from the symbol stream prior to extracting ATM cells.

12.3.4.3. Physical Media Requirements

The optical ATM physical layers include those based on SONET–SDH and direct cell-stream physical layers. Currently, all public UNI specifications for fiber optic transmission are based on SONET–SDH; hence, they are suitable for long-distance links. The optical specifications can be found in Table 12.8.

The non-SONET cell-stream PHYs have been approved only for the private UNI. These are intended for shorter distance links (up to 2 km) in LANs. The 100 Mb/s layer is based on the physical specifications for FDDI (see Chapter 15). The optical specifications for the cell-stream ATM PHYs are summarized in Table 12.12 [8].

12.3.4.4. Payload Capacity Comparison

It is instructive to compare the delivered payload capacity of various SONET and ATM formats. Table 12.13 gives the baud rate (line data rate), the bit rate (nominal symbol rate), and delivered payload capacity of raw SONET and several ATM PHY layers in the nominal 100–155 Mb/s range. In the ATM case, the payload capacity listed is the total data payload (no cell headers or HEC) presented to the ATM layer. In the case of SONET, it is the synchronous payload envelope minus any path overhead bytes.

In terms of overall efficiency, raw SONET requires only approximately 4% overhead but delivers only synchronous data. ATM incurs an additional 10% of overhead (5 cell header bytes per 48 bytes of data) — the price for the added flexibility of cell versus synchronous TDM switching.

Table 12.12 **ATM Optical Specifications for the Cell-Stream Private UNIs**

Parameter	Units	100 Mb/s TAXI	155 Mb/s
Data rate			
Data rate	Mb/s	100	155.52
Link rate	Mb/s	125	194.4
Tolerance	ppm		100
Transmitter			
$\lambda_{W_{min}}$	nm		1270
$\lambda_{W_{min}}$	nm		1380
Special width (FWHM)	nm		<200
$P_{T_{max}}$	dBm		−14
$P_{T_{min}}$	dBm		−20
Optical path			
Type		Multimode fiber	Multimode fiber, graded index
Diameter	μm	62.5	62.5/125 or 50
Model bandwidth	MHz/km		500
Zero dispersion Wavelength, λ_0	nm		1295–1365
Max dispersion Slope, s_0	ps/nm^2 – km		0.110
Max distance	km	2	2
Receiver			
$P_{R_{max}}$	dBm		−14
$P_{R_{min}}$	dBm		−29

Table 12.13 **Effective Payload Capacity Comparison**

	SONET		ATM		
	OC-3	OC-3c	OC-3c	155 Mb/s	100 Mb/s TAXI
Baud rate	155.52	155.52	155.52	194.4	125
Bit rate	155.52	155.52	155.52	155.52	100
Payload capacity	148.608	149.760	135.6317	135.6317	88.889
Total efficiency	95.56%	96.30%	87.21%	69.77%	71.11%

The ATM-cell-stream formats lose 20% in overhead due to the 4B/5B or 8B/10B line symbol coding (compared to 4% for some of the same functionality provided by SONET). Note that the SONET OC-3c and 155 Mb/s cell-stream ATM PHYs have the same effective payload capacity by design, although the forms of overhead are different (SONET framing versus one PLOU per 26 cells).

12.3.5.   ATM LAYER

As discussed previously, ATM can be implemented atop a variety of physical layers. On the other hand, ATM supports a variety of different services and traffic classes by providing different adaptation layers to higher network levels. The ability to do so efficiently over a shared infrastructure is made possible by a common middle layer, called the ATM layer.

ATM does not use the standard OSI seven-layer reference model, but the ATM layer performs many of the functions of OSI level 2, the data link layer. For example, the ATM layer and AAL-5 together provide data link layer functionality similar to OSI layer 2, i.e., the ability to transmit error-free frames of size up to 64,000 bytes from a source to a destination ATM entity [5].

The ATM layer is responsible for switching and multiplexing of ATM cells. Because ATM is based on switched point-to-point links, as opposed to a broadcast medium, ATM functionality is basically connection oriented (although connectionless services are supported through an adaptation layer). Although the ATM layer performs the basic operations that transmit cells along multilink paths from source to destination, the establishment of these paths [i.e., network-layer routing using, for example, an Internet protocol address] is left to higher layers. The ATM layer is designed for simplicity and for ease of high-speed hardware implementation.

12.3.5.1.   Virtual Channels and Paths

A virtual channel is a contiguous stream of cells transmitted between two points in an ATM network (e.g., a single user's data stream). To reach the destination from the source, this data stream must traverse a set of ATM switches, going out a particular port on each switch to reach the next switch. This constitutes a virtual path, and many virtual channels may share the same virtual path. Virtual channels can be thought of as being contained inside virtual paths (Fig. 12.11).

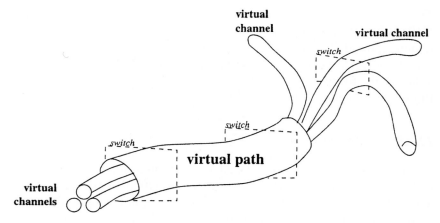

Fig. 12.11   Virtual channels and virtual paths.

Each ATM cell header has 24 bits for identifying the VC and VP that a cell belongs to, the VCI and VPI fields, respectively. This information ultimately is used to route the cell to the correct output port on each switch in its intended path. The generic local routing procedure at each switch is as follows: When a cell enters an input port, its header is examined and the VCI and/or VPI field is extracted. This information is used to index a look-up table, which (i) identifies the output port to send the cell to and (ii) provides new values to be placed into the outgoing cell's VCI and/or VPI fields.

Hence, the VCI and VPI fields are not constant, but rather change on each hop through the network. A virtual path (channel) is a string of VPI (VCI) values stored as values in the switch look-up tables, forming a linked list of table entries along the path (Fig. 12.12). A connection-establishment procedure is responsible for setting up the proper look-up table values whenever a new channel or path is initiated.

Depending on the kind of switch, either just the VPI information is used in routing or both VPI and VCI fields are used. Because logically VCs are viewed as being contained inside VPs, VP-only switches are not allowed to substitute VCI fields of outgoing cells; VP-only routing is considered a lower sublayer than VC routing [3].

Note also that cells entering a particular switch destined for different output ports must have distinct VPI values. Within a given VP, different VCs are distinguished by differing VCI values. However, different channels may share the same VCI value if their VPI fields differ (they belong to a different virtual path).

Fig. 12.12   Virtual path establishment.

References

1. Hac, A., and H. B. Mutlu. 1989 (November). Synchronous optical network and broadband ISDN protocols. *Computer* **22**(11):26–34.
2. Stallings, W. 1992. *ISDN and broadband ISDN*, 2nd ed. New York: Macmillan.
3. Minoli, D. 1993. *Enterprise networking: Fractional T1 to SONET, frame relay to BISDN*. Boston: Artech House.
4. DeCusatis, C. 1995. Data processing systems for optoelectronics. In *Optoelectronics for data communication*, ed. R. Lasky, U. Österberg, and D. Stigliani, chap. 6. New York: Academic Press.
5. Miller, A. 1994 (June). From here to ATM. *IEEE Spectrum* **31**(6):20–24.
6. Jungkok Bae, J., and T. Suda. 1991 (February). Survey of traffic control schemes and protocols in ATM networks. *Proc. IEEE* **79**(2):170–189.
7. Roohalamini, R., V. Cherkassky, and M. Garver. 1994 (April). Finding the right ATM switch for the market. *Computer* **27**(4):16–28.
8. ATM Forum Inc. 1995. *ATM user network interface (UNI) specification Version 3.1, 1/e*. New York: Prentice Hall.
9. The ATM Forum Technical Committee. 1994. DS1 physical layer specification. Technical Report AF-PHY-0016.000, The ATM Forum, September.
10. The ATM Forum Technical Committee. 1995. Physical interface specification for 25.6 Mb/s over twisted pair cable. Technical Report AF-PHY-0040.000, The ATM Forum, November 7.
11. The ATM Forum Technical Committee. 1996. 622.08 Mb/s physical layer specification. Technical Report AF-PHY-0046.000, The ATM Forum, January.
12. Klessig, B. 1995 (July). Status of ATM specifications. http://www.3com.com/0files/mktg/pubs/3tech/795atmst.html.

Chapter 13 | Fibre Channel Standard

Alan F. Benner

IBM Corporation, Poughkeepsie, New York 12601

13.1. Introduction

A large and increasing problem in data communications, particularly optical data communications, is the increasing difference in performance between link speed and protocol-processing speed. The actions of generating and receiving data packets require multiple memory accesses, for storing or retrieving data packets and information describing how those data packets are to be handled. Memory access latency has improved at a much slower rate than either processor clock rate performance or link bit rate performance — in modern high-performance servers with multiple caching levels, the effective memory latency for non-cacheable data has actually steadily gotten worse.

The result of this continuing trend is that protocol-processing performance, rather than link performance, is the primary bottleneck for high-speed data communications. At this writing, it is straightforward and relatively inexpensive to build bidirectional optical links operating at (10+10) Gb/s on a fiber pair, and at (24+24) Gb/s on a 12-fiber ribbon pair. However, doing the protocol processing for a TCP/IP protocol stack with short packets may occupy 100% of the CPU cycles for a high-performance system running over a NIC operating at only (1+1) Gbps. This issue is getting to be so important that a whole new class of processors which are

HANDBOOK OF FIBER OPTIC
DATA COMMUNICATION
$35.00

highly optimized for processing communications protocols, termed "network processors," is being developed by various companies.

A more productive way to address the issue, which is used in the Fibre Channel ANSI standard, is to architecture the network such that the protocol processing work is distributed away from the host processor and onto hardware engines in the network interface and in the network. Effective use of this can decrease the host CPU loading from near 100% to below 5%, while achieving the same transmission rate.

Considering that link speed is continuing to increase dramatically faster than memory access speeds, these methods will continue in the future to grow in importance. The subject of network architectures that can continue to effectively use the quickly increasing bandwidth of optical technologies to deliver real data communications performance to applications is in a state of intense research and rapid development. Fibre Channel is a particularly illustrative example of a network architecture which has been very successful in many areas in achieving these goals.

This chapter gives an overview of Fibre Channel at a fairly detailed level. It also describes the current main application area for Fibre Channel, in networking together servers with their shared storage devices to make "Storage Area Networks," or SANs. Finally, this chapter gives an overview of how Fibre Channel, in contrast with other network architectures, leverages the advantages of high-speed, high-reliability optical technology to provide high overall data communications performance.

13.2. Overview

A Fibre Channel network is logically made up of one or more bidirectional point-to-point serial data channels, structured for high-performance capability. The basic data rate over the links is just over 1 Gbps, providing > 100 MBps data transmission bandwidth, with half-, quarter-, eighth-, double-, and quadruple-speed links defined and 10 Gbps (1,000 MB/s) links under development. Although the Fibre Channel protocol is configured to match the transmission and technological characteristics of single- and multimode optical fibers, the physical medium used for transmission can also be copper twisted pair or coaxial cable.

Physically, a Fibre Channel network can be set up as (1) a single point-to-point link between two communication Ports, called "N_Ports," (2) a network of multiple N_Ports, each linked through an "F_Port" into a switching

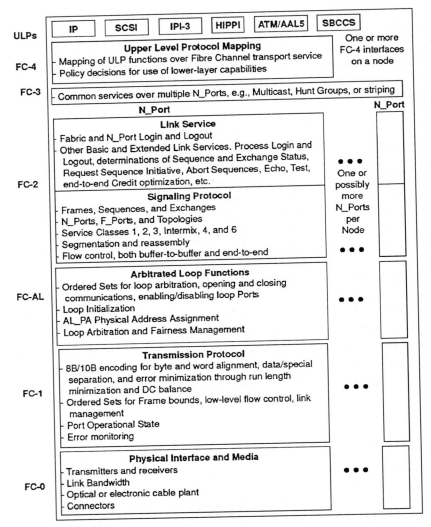

Fig. 13.1 Fibre Channel structural hierarchy.

network, called a Fabric, or (3) a ring topology termed an "Arbitrated Loop," allowing multiple N_Port interconnection without switch elements. Each N_Port resides on a hardware entity such as a computer or disk drive, termed a "Node." Nodes incorporating multiple N_Ports can be interconnected in more complex topologies, such as rings of point-to-point links or dual independent redundant Fabrics.

Logically, Fibre Channel is structured as a set of hierarchical functions, as illustrated in Fig. 13.1. Interfaces between the levels are defined, but vendors are not limited to specific interfaces between levels if multiple levels are implemented together. A single Fibre Channel Node implementing one or more N_Ports provides a bidirectional link and FC-0 through FC-2 or FC-4 services through each N_Port.

- The FC-0 level describes the physical interface, including transmission media, transmitters and receivers, and their interfaces. The FC-0 level specifies a variety of media and associated drivers and receivers that can operate at various speeds.
- The FC-1 level describes the 8B/10B transmission code that is used to provide DC balance of the transmitted bit stream, to separate transmitted control bytes from data bytes and to simplify bit, byte, and word alignment. In addition, the coding provides a mechanism for detection of some transmission and reception errors.
- The FC-2 level is the signaling protocol level, specifying the rules and mechanisms needed to transfer blocks of data. At the protocol level, the FC-2 level is the most complex level, providing different classes of service, packetization and sequencing, error detection, segmentation and reassembly of transmitted data, and Login services for coordinating communication between Ports with different capabilities.
- The FC-3 level provides a set of services that are common across multiple N_Ports of a Fibre Channel Node. This level is not yet well defined, due to limited necessity for it, but the capability is provided for future expansion of the architecture.
- The FC-4 level provides mapping of Fibre Channel capabilities to preexisting Upper Level Protocols, such as the Internet Protocol (IP) or SCSI (Small Computer Systems Interface), or Single-Byte Command Code Sets, which are the command codes used for ESCON (Enterprise System Connection).

13.2.1. *FC-0: PHYSICAL INTERFACE AND MEDIA*

The FC-0 level specifies the link between two Ports. Essentially, this consists of a pair of either optical fiber or electrical cables along with transmitter and receiver circuitry which work together to convert a stream of bits at one end of the link to a stream of bits at the other end. Each fiber is attached to

468   Alan F. Benner

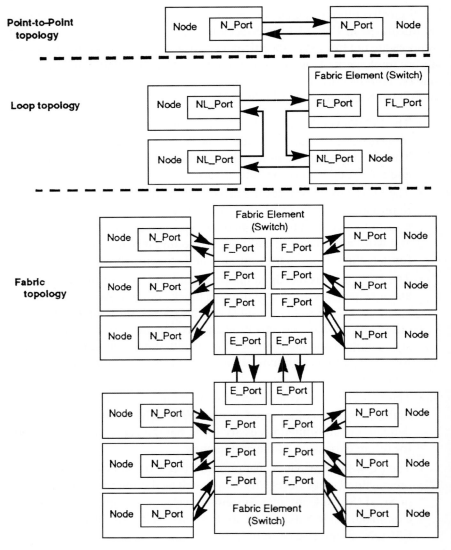

Fig. 13.2 Examples of Point-to-point, Arbitrated Loop, and Fabric topologies.

a transmitter of a Port at one end and a receiver of another Port at the other end. The simplest configuration is a bidirectional pair of links, as shown in Fig. 13.2. A loop topology allows multiple Ports to be connected together without a routing switch, and a full switched fabric, with possibly multiple stages of switches, allow large numbers of Ports to be interconnected.

The FC-0 level describes the various kinds of media allowed, including single-mode and multi-mode optical fibers, as well as coaxial and twisted pair electrical cables for shorter distance links, as shown in Fig. 13.3. The FC-0 level is designed for maximum flexibility and allows the use of a wide variety of media to meet a wide range of system requirements. Fig. 13.4 and Fig. 13.5 show the cable, transmitter, and receiver characteristics for single-mode and multi-mode links, respectively. Similar specifications, which are not listed here, are defined for electrical links over twisted pair and coaxial cable.

For optical fiber, Fibre Channel uses the SC duplex connector. This connector is specially keyed to prevent insertion of a multimode fiber into a single-mode receptacle, which helps to ensure Class 1 laser eye safety. Another feature to ensure Class 1 laser eye safety is the implementation of "Open Fiber Control" (OFC), which allows operation at Class 1 over bidirectional links with laser powers which would normally not meet Class 1 safety standards. Transceivers with Open Fiber Control circuitry will turn off data transmission and only transmit light at a very low duty cycle whenever they are not receiving light, so that an interrupted duplex link is quickly shut down at both ends.

A multi-link communication path between two N_Ports may be made up of links of different technologies. For example, it may have copper coaxial cable links attached to end Ports for short-distance links, with single-mode optical fibers for longer-distance links between switches separated by longer distances.

13.2.2. FC-1: TRANSMISSION PROTOCOL

In a Fibre Channel network, information is transmitted using an 8B/10B data encoding. This coding has a number of characteristics that simplify design of inexpensive transmitter and receiver circuitry that can operate at the 10^{-12} bit error rate required. It bounds the maximum run length, assuring that there are never more than 5 identical bits in a row except at synchronization points. It maintains overall DC balance, ensuring that the signals transmitted over the links contain an equal number of 1s and 0s. It minimizes the low-frequency content of the transmitted signals. Also, it allows straightforward separation of control information from the transmitted data, and simplifies byte and word alignment.

The encoding and decoding processes result in the conversion between 8-bit bytes with a separate single-bit "data/special" flag indication and 10-bit

470   Alan F. Benner

Optical			
Single-mode (9 μm)	Multi-mode (62.5 μm)	Multi-mode (50 μm)	Multi-mode (50 μm) w/o OFC
400-SM-LL-L	100-M6-SL-I*	200-M5-SL-I	400-M5-SN-I
200-SM-LL-L	50-M6-SL-I*	100-M5-SL-I	200-M5-SN-I
100-SM-LL-L	25-M6-SL-I*	50-M5-SL-I	100-M5-SN-I
100-SM-LL-I	25-M6-LE-I*	25-M5-SL-I	
50-SM-LL-I	12-M6-LE-I	25-M5-LE-I*	
25-SM-LL-L		12-M5-LE-I*	

Electrical				
Long Video	Video Coax	Twin Axial	Shielded Twisted Pair	Miniature Coax
100-LV-EL-S	100-TV-EL-S	100-TW-EL-S	100-TP-EL-S	100-MI-EL-S
50-LV-EL-S	50-TV-EL-S	50-TW-EL-S	50-TP-EL-S	50-MI-EL-S
25-LV-EL-S	25-TV-EL-S	25-TW-EL-S	25-TP-EL-S	25-MI-EL-S
12-LV-EL-S	12-TV-EL-S	12-TW-EL-S	12-TP-EL-S	12-MI-EL-S

Fig. 13.3   FC-0 Cable Plant Technology options and terminology.

Parameter	Units	400-SM-LL-L	200-SM-LL-L	100-SM-LL-L	100-SM-LL-I	50-SM-LL-L	25-SM-LL-L	25-SM-LL-I
FC-0 and Cable Plant								
data rate	MBps	400	200	100	100	50	25	25
nominal bit rate/1062.5	Mbaud	4	2	1	1	1/2	1/4	1/4
operating range (typical)	m	2-2k	2-2k	2-10k	2-2k	2-10k	2-10k	2-2k
loss budget	dB	6	6	14	6	14	14	6
dispersion related penalty	dB	1	1	1	1	1	1	1
reflection related penalty	dB	1	1	1	1	1	1	1
cable plant dispersion	ps/nm·km	12	12	35	12	60	60	12
Transmitter (S)								
min. center wavelength	nm	1270	1270	1270	1270	1270	1270	1270
max. center wavelength	nm	1355	1355	1355	1355	1355	1355	1355
RMS spectral width	nm (max.)	6	6	3	6	3	6	30
launched power, max.	dBm (ave.)	−3	−3	−3	−3	−3	−3	−3
launched power, min.	dBm (ave.)	−12	−12	−9	−12	0	−9	−12
extinction ratio	dB (min.)	9	9	9	9	3	6	6
RIN_{12} (maximum)	dB/Hz	−116	−116	−116	−116	−114	−112	−112
eye opening @ BER=10^{-12}	% (min.)	57	57	57	57	61	63	63
deterministic jitter	% (p-p)	20	20	20	20	20	20	20
Receiver (R)								
received power, min.	dBm (ave.)	−20	−20	−25	−20	−25	−25	−20
received power, max.	dBm (ave.)	−3	−3	−3	−3	−3	−3	−3
optical path power penalty	dB (max.)	2	2	2	2	2	2	2
return loss of receiver	dB (min.)	12	12	12	12	12	12	12

Fig. 13.4 FC-0 Specifications for Single-Mode Physical Links.

"Data Characters" and "Special Characters." Data Characters and Special Characters are collectively termed "Transmission Characters."

Certain combinations of Transmission Characters, called Ordered Sets, are designated to have special meanings. Ordered Sets, which always contain four Transmission Characters, are used to identify Frame boundaries, to transmit low-level status and command information, to enable simple hardware processing to achieve byte and word synchronization, and to maintain proper link activity during periods when no data are being sent.

There are three kinds of Ordered Sets. Frame delimiters mark the beginning and end of Frames, identify the Frame's Class of Service, indicate the Frame's location relative to other Frames in the Sequence, and indicate data validity within the Frame. Primitive Signals include Idles, which are transmitted to maintain link activity while no other data can be transmitted, and the R_RDY Ordered Set, which operates as a low-level acknowledgment

Parameter	Units	200-M5-SL-I	100-M5-SL-I	50-M5-SL-I	25-M5-SL-I	25-M6-LE-I	25-M5-LE-I
FC-0 and Cable Plant							
data rate	MBps	400	200	100	100	25	25
nominal bit rate/1062.5	Mbaud	2	1	1/2	1/4	1/4	1/4
operating range (typical)	m	2-300	2-500	2-1k	2-2k	2-1.5k	2-1.5k
fiber core diameter	μm	50	50	50	50	62.5	50
loss budget	dB	6	6	8	12	6	5.5
multi-mode fiber bandwidth							
@ 850 nm	MHz·km	500	500	160	500	160	500
@ 1,300 nm	MHz·km	500	500	500	500	500	500
numerical aperture	(unitless)	0.20	0.20	0.20	0.20	0.275	0.275
Transmitter (S)							
type		laser	laser	laser	laser	LED	LED
min. center wavelength	nm	770	770	770	770	1280	1280
max. center wavelength	nm	850	850	850	850	1380	1380
RMS spectral width	nm(max.)	4	4	4	4	(see	(see
FWHM spectral width	nm(max.)					text)	text)
launched power, max.	dBm(ave)	1.3	1.3	1.3	0	−14	−17
launched power, min.	dBm(ave)	−7	−7	−7	−5	−20	−23.5
extinction ratio	dB (min.)	6	6	6	6	9	9
RIN_{12} (max.)	dB/Hz	−116	−114	−114	−112		
eye opening @BER=10^{-12}	% (min.)	57	61	61	63		
deterministic jitter	% (p-p)	20	20	20	20	16	16
random jitter	% (p-p)					9	9
optical rise/fall time	ns (max.)					2/2.2	2/2.2
Receiver (R)							
received power, min.	dBm(ave)	−13	−13	−15	−17	−26	−29
received power, max.	dBm(ave)	1.3	1.3	1.3	0	−14	−14
return loss of receiver	dB (min.)	12	12	12	12		
deterministic jitter	% (p-p)					19	19
random jitter	% (p-p)					9	9
optical rise/fall time	ns (max.)					2.5	2.5

Fig. 13.5   FC-0 Specifications for Multi-Mode Physical Links.

for buffer-to-buffer flow control. Primitive Sequences are used in Primitive Sequence protocols for performing link initialization and link-level recovery and are transmitted continuously until a response is received.

In addition to the 8B/10B coding and Ordered Set definition, the FC-1 level includes definitions for "transmitters" and "receivers." These are blocks which monitor the signal traversing the link and determining the integrity of the data received. Transmitter and receiver behavior is specified by a set of states and their interrelationships. These states are divided into "Operational" and "Not Operational" types. FC-1 also specifies monitoring

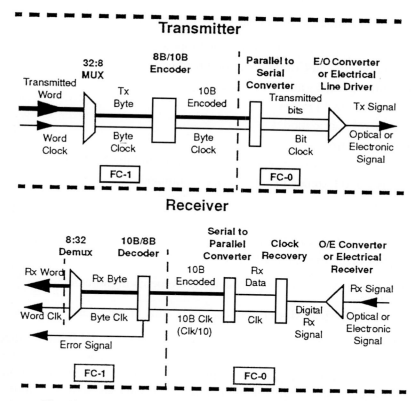

Fig. 13.6 Transmitter and receiver FC-1 and FC-0 data flow stages.

capabilities and special operation modes for transmitters and receivers. Example block diagrams of a transmitter and a receiver are shown in Fig. 13.6. The serial and serial/parallel converter sections are part of FC-0, while the FC-1 level contains the 8B/10B coding operations and the multiplexing and demultiplexing between bytes and 4-byte words, as well as the monitoring and error detection functionality.

13.2.3. FC-2: SIGNALLING PROTOCOL AND LINK SERVICES

The FC-2 level is the most complex part of Fibre Channel and includes most of the Fibre Channel-specific constructs, procedures, and operations. The basic parts of the FC-2 level are described in overview in the following sections, with full description left to later chapters. The elements of the

FC-2 level include the following:

- Physical Model: Nodes, Ports, and topologies
- Bandwidth and Communication Overhead
- Building blocks and their hierarchy
- Link Control Frames
- General Fabric model
- Flow control
- Classes of service provided by the Fabric and the N_Ports
- Basic and Extended Link Service Commands
- Protocols
- Arbitrated Loop functions
- Segmentation and reassembly
- Error detection and recovery

The following sections describe these elements in more detail.

13.2.3.1. Physical Model: Nodes, Ports, and Topologies

The basic source and destination of communications under Fibre Channel would be a computer, a controller for a disk drive or array of disk drives, a router, a terminal, or any other equipment engaged in communications. These sources and destinations of transmitted data are termed "Nodes." Each Node maintains one or possibly more than one facility capable of receiving and transmitting data under the Fibre Channel protocol. These facilities are termed "N_Ports." Fibre Channel also defines a number of other types of "Ports," which can transmit and receive Fibre Channel data, including "NL_Ports," "F_Ports," "E_Ports," etc., which are described below. Each Port supports a pair of "fibres" (which may physically be either optical fibers or electrical cables) — one for outbound transmission, and the other for inbound reception. The inbound and outbound fibre pair is termed a "link." Each N_Port only needs to maintain a single pair of fibres, without regard to what other N_Ports or switch elements are present in the network. Each N_Port is identified by a 3-byte "Port identifier," which is used for qualifying Frames and for assuring correct routing of Frames through a loop or a Fabric.

Nodes containing a single N_Port with a fibre pair link can be interconnected in one of three different topologies. Each topology supports bidirectional flow between source and destination N_Ports. The three basic types of topologies include:

Point-to-point: The simplest topology directly connecting two N_Ports is termed "Point-to-point," and it has the obvious connectivity as a single link between two N_Ports.

Fabric: More than two N_Ports can be interconnected using a "Fabric," which consists of a network of one or more "switch elements" or "switches." A switch contains two or more facilities for receiving and transmitting data under the protocol, termed "F_Ports." The switches receive data over the F_Ports and, based on the destination N_Port address, route it to the proper F_Port (possibly through another switch, in a multistage network), for delivery to a destination N_Port. Switches are fairly complex units, containing facilities for maintaining routing to all N_Ports on the Fabric, handling flow control, and satisfying the requirements of the different Classes of service supported.

Arbitrated Loop: Multiple N_Ports can also be connected together without benefit of a Fabric by attaching the incoming and outgoing fibers to different Ports to make a loop configuration. A Node Port which incorporates the small amount of extra function required for operation in this topology is termed an "NL_Port." This is a blocking topology — a single NL_Port arbitrates for access to the entire loop and prevents access by any other NL_Ports while it is communicating. However, it provides connectivity between multiple Ports while eliminating the expense of incorporating a switch element.

It is also possible to mix the Fabric and Arbitrated Loop topologies, where a switch Fabric Port can participate on the Loop, and data can go through the switch and around the loop. A Fabric Port capable of operating on a loop is termed an "FL_Port."

Most Fibre Channel functions and operations are topology-independent, although routing of data and control of link access will naturally depend on what other Ports may access a link. A series of "Login" procedures performed after a reset allow an N_Port to determine the topology of the network to which it is connected, as well as other characteristics of the other attached N_Port, NL_Ports, or switch elements. The Login procedures are described further below.

13.2.3.2. Bandwidth and Communication Overhead

The maximum data transfer bandwidth over a link depends both on physical parameters, such as clock rate and maximum baud rate, and on protocol parameters, such as signaling overhead and control overhead. The data transfer bandwidth can also depend on the communication model, which describes the amount of data being sent in each direction at any particular time.

The primary factor affecting communications bandwidth is the clock rate of data transfer. The base clock rate for data transfer under Fibre Channel is 1.0625 GHz, with 1 bit transmitted every clock cycle. For lower bandwidth, less expensive links, half-, quarter-, and eighth-speed clock rates are defined. Double- and quadruple-speed links have been defined for implementation in the near future as well. The most commonly used data rates will likely be the full-speed and quarter-speed initially, with double- and quadruple-speed components becoming available as the technology and market demand permit.

Figure 13.7 shows a sample communication model, for calculating the achievable data transfer bandwidth over a full speed link. The figure shows a single Fibre Channel Frame, with a payload size of 2048 bytes. To transfer this payload, along with an acknowledgment for data traveling in the reverse direction on a separate fiber for bidirectional traffic, the following overhead elements are required:

SOF: Start of Frame delimiter, for marking the beginning of the Frame (4 bytes),

Frame Header: Frame header, indicating source, destination, sequence number, and other Frame information (24 bytes),

CRC: Cyclic Redundancy Code word, for detecting transmission errors (4 bytes),

EOF: End of Frame delimiter, for marking the end of the Frame (4 bytes),

Fig. 13.7   Sample Data Frame + ACK Frame transmission, for bandwidth calculation.

Idles: Inter-Frame space for error detection, synchronization, and insertion of low-level acknowledgments (24 bytes),

ACK: Acknowledgment for a Frame from the opposite Port, needed for bidirectional transmission (36 bytes), and

Idles: Inter-Frame space between the ACK and the following Frame (24 bytes).

The sum of overhead bytes in this bidirectional transmission case is 120 bytes, yielding an effective data transfer rate of 100.369 MBps:

$$1.0625[Gbps] \times \frac{2048[payload]}{2168[payload + overhead]} \times \frac{1[byte]}{10[codebits]} = 100.369$$

Thus, the full-speed link provides better than 100 MBps data transport bandwidth, even with signaling overhead and acknowledgments. The achieved bandwidth during unidirectional communication would be slightly higher, since no ACK frame with following Idles would be required. Beyond this, data transfer bandwidth scales directly with transmission clock speed, so that, for example, the data transfer rate over a half-speed link would be $100.369/2 = 50.185$ MBps.

13.2.3.3. Building Blocks and Their Hierarchy

The set of building blocks defined in FC-2 are:

Frame: A series of encoded transmission words, marked by Start of Frame and End of Frame delimiters, with Frame Header, Payload, and possibly an optional Header field, used for transferring Upper Level Protocol data.

Sequence: A unidirectional series of one or more Frames flowing from the Sequence Initiator to the Sequence Recipient.

Exchange: A series of one or more non-concurrent Sequences flowing either unidirectionally from Exchange Originator to the Exchange Responder or bidirectionally, following transfer of Sequence Initiative between Exchange Originator and Responder.

Protocol: A set of Frames, which may be sent in one or more Exchanges, transmitted for a specific purpose, such as Fabric or N_Port Login, Aborting Exchanges or Sequences, or determining remote N_Port status.

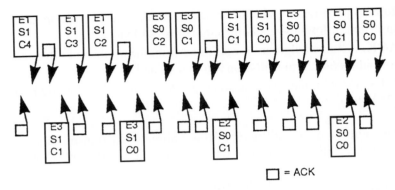

Fig. 13.8 Building blocks for the FC-2 Frame/Sequence/Exchange hierarchy.

An example of the association of multiple Frames into Sequences and multiple Sequences into Exchanges is shown in Fig. 13.8. The figure shows four Sequences, which are associated into two unidirectional and one bidirectional Exchange. Further details on these constructs follow.

13.2.3.4. Frames

Frames contain a Frame header in a well-defined format, and may contain a Frame payload. Frames are broadly categorized under the following classifications:

- Data Frames, including
  - Link Data Frames
  - Device Data Frames
  - Video Data Frames
- Link Control Frames, including
- Acknowledge (ACK) Frames, acknowledging successful reception of 1 (ACK_1), N (ACK_N), or all (ACK_0) Frames of a Sequence
- Link Response ("Busy" (P_BSY, F_BSY) and "Reject" (P_RJT, F_RJT) Frames, indicating unsuccessful reception of a Frame
- Link Command Frames, including only Link Credit Reset (LCR), used for resetting flow control credit values

Frames operate in Fibre Channel as the fundamental block of data transfer. As stated above, each Frame is marked by Start of Frame and End of Frame delimiters. In addition to the transmission error detection capability provided by the 8B/10B code, error detection is provided by a 4-byte

CRC value, which is calculated over the Frame Header, optional Header (if included), and payload. The 24-byte Frame Header identifies a Frame uniquely and indicates the processing required for it. The Frame Header includes fields denoting the Frame's source N_Port ID, destination N_Port ID, Sequence ID, Originator and Responder Exchange IDs, routing, Frame count within the Sequence, and control bits.

Every Frame must be part of a Sequence and an Exchange. Within a Sequence, the Frames are uniquely identified by a 2-byte counter field termed SEQ_CNT in the Frame header. No two Frames in the same Sequence with the same SEQ_CNT value can be active at the same time, to ensure uniqueness.

When a Data Frame is transmitted, several different things can happen to it. It may be delivered intact to the destination, it may be delivered corrupted, it may arrive at a busy Port, or it may arrive at a Port which does not know how to handle it. The delivery status of the Frame will be returned to the source N_Port using Link Control Frames if possible. A Link Control Frame associated with a Data Frame is sent back to the Data Frame's source from the final Port that the Frame reaches, unless no response is required, or a transmission error prevents accurate knowledge of the Frame Header fields.

13.2.3.5. Sequences

A Sequence is a set of one or more related Data Frames transmitted unidirectionally from one N_Port to another N_Port, with corresponding Link Control Frames, if applicable, returned in response. The N_Port that transmits a Sequence is referred to as the "Sequence Initiator" and the N_Port that receives the Sequence is referred to as the "Sequence Recipient."

Each Sequence is uniquely specified by a Sequence Identifier (SEQ_ID), which is assigned by the Sequence Initiator. The Sequence Recipient uses the same SEQ_ID value in its response Frames. Each Port operating as Sequence Initiator assigns SEQ_ID values independent of all other Ports, and uniqueness of a SEQ_ID is only assured within the set of Sequences initiated by the same N_Port.

The SEQ_CNT value, which uniquely identifies Frames within a Sequence, is started either at zero in the first Frame of the Sequence or at 1 more than the value in the last Frame of the previous Sequence of the same Exchange. The SEQ_CNT value is incremented by 1 each subsequent Frame. This assures uniqueness of each Frame header active on the network.

The status of each Sequence is tracked, while it is open, using a logical construct called a Sequence Status Block. Normally separate Sequence Status Blocks are maintained internally at the Sequence Initiator and at the Sequence Recipient. A mechanism does exist for one N_Port to read the Sequence Status Block of the opposite N_Port, to assist in recovery operations, and to assure agreement on Sequence state.

There are limits to the maximum number of simultaneous Sequences an N_Port can support per Class, per Exchange, and over the entire N_Port. These values are established between N_Ports before communication begins through an N_Port Login procedure.

Error recovery is performed on Sequence boundaries, at the discretion of a protocol level higher than FC-2. Dependencies between the different Sequences of an Exchange are indicated by the Exchange Error Policy, as described following.

13.2.3.6. Exchanges

An Exchange is composed of one or more non-concurrent related Sequences, associated into some higher level operation. An Exchange may be unidirectional, with Frames transmitted from the "Exchange Originator" to the "Exchange Responder," or bidirectional, when the Sequences within the Exchange are initiated by both N_Ports (nonconcurrently). The Exchange Originator, in originating the Exchange, requests the directionality. In either case, the Sequences of the Exchange are nonconcurrent, i.e., each Sequence must be completed before the next is initiated.

Each Exchange is identified by an "Originator Exchange ID," denoted as OX_ID in the Frame Headers, and possibly by a "Responder Exchange ID," denoted as RX_ID. The OX_ID is assigned by the Originator, and is included in the first Frame transmitted. When the Responder returns an acknowledgment or a Sequence in the opposite direction, it may include an RX_ID in the Frame Header to let it uniquely distinguish Frames in the Exchange from other Exchanges. Both the Originator and Responder must be able to uniquely identify Frames based on the OX_ID and RX_ID values, the source and destination N_Port IDs, SEQ_ID, and the SEQ_CNT. The OX_ID and RX_ID fields may be set to the "unassigned" value of x'FFFF' if the other fields can uniquely identify Frames. If an OX_ID or RX_ID is assigned, all subsequent Frames of the Sequence, including both Data and Link Control Frames, must contain the Exchange ID(s) assigned.

An Originator may initiate multiple concurrent Exchanges, even to the same destination N_Port, as long as each uses a unique OX_ID. Exchanges may not cross between multiple N_Ports, even multiple N_Ports on a single Node.

Large-scale systems may support up to thousands of potential Exchanges, across several N_Ports, even if only a few Exchanges (e.g., tens) may be active at any one time within an N_Port. In these cases, Exchange resources may be locally allocated within the N_Port on an "as needed" basis. An "Association Header" construct, transmitted as an optional header of a Data Frame, provides a means for an N_Port to invalidate and reassign an X_ID (OX_ID or RX_ID) during an Exchange. An X_ID may be invalidated when the associated resources in the N_Port for the Exchange are not needed for a period of time. This could happen, for example, when a file subsystem is disconnecting from the link while it loads its cache with the requested data. When resources within the N_Port are subsequently required, the Association Header is used to locate the "suspended" Exchange, and an X_ID is reassigned to the Exchange so that operation can resume. X_ID support and requirements are established between N_Ports before communication begins through an N_Port Login procedure.

Fibre Channel defines four different Exchange Error Policies. Error policies describe the behavior following an error, and the relationship between Sequences within the same Exchange. The four Exchange Error policies include:

Abort, discard multiple Sequences: Sequences are interdependent and must be delivered to an upper level in the order transmitted. An error in one Frame will cause that Frame's Sequence and all later Sequences in the Exchange to be undeliverable.

Abort, discard a single Sequence: Sequences are not interdependent. Sequences may be delivered to an upper level in the order that they are received complete, and an error in one Sequence does not cause rejection of subsequent Sequences.

Process with infinite buffering: Deliverability of Sequences does not depend on all the Frames of the Sequence being intact. This policy is intended for applications such as video data where retransmission is unnecessary (and possibly detrimental). As long as the first and last Frame of the Sequence are received, the Sequence can be delivered to the upper level.

Discard multiple Sequences with immediate retransmission: This is a special case of the "Abort, discard multiple Sequences" Exchange Error Policy, where the Sequence Recipient can use a Link Control Frame to request that a corrupted Sequence be retransmitted immediately. This Exchange Error Policy can only apply to Class 1 transmission.

The Error Policy is determined at the beginning of the Exchange by the Exchange Originator and cannot change during the Exchange. There is no dependency between different Exchanges on error recovery, except that errors serious enough to disturb the basic integrity of the link will affect all active Exchanges simultaneously.

The status of each Exchange is tracked, while it is open, using a logical construct called an Exchange Status Block. Normally separate Exchange Status Blocks are maintained internally at the Exchange Originator and at the Exchange Responder. A mechanism does exist for one N_Port to read the Exchange Status Block of the opposite N_Port of an Exchange, to assist in recovery operations, and to assure agreement on Exchange status. These Exchange Status Blocks maintain connection to the Sequence Status Blocks for all Sequences in the Exchange while the Exchange is open.

13.2.3.7.   Link Control Frames

Link Control Frames are used to indicate successful or unsuccessful reception of each Data Frame. Link Control Frames are only used for Class 1 and Class 2 Frames — all link control for Class 3 Frames is handled above the Fibre Channel level. Every Data Frame should generate a returning Link Control Frame (although a single ACK_N or ACK_0 can cover more than one Data Frame). If a P_BSY or F_BSY is returned, the Frame may be retransmitted, up to some limited and vendor-specific number of times. If a P_RJT or F_RJT is returned, or if no Link Control Frame is returned, recovery processing happens at the Sequence level or higher; there is no facility for retransmitting individual Frames following an error.

13.2.3.8.   General Fabric Model

The Fabric, or switching network, if present, is not directly part of the FC-2 level, since it operates separately from the N_Ports. However, the constructs it operates on are at the same level, so they are included in the FC-2 discussion.

The primary function of the Fabric is to receive Frames from source N_Ports and route them to their correct destination N_Ports. To facilitate this, each N_Port physically attached through a link to the Fabric is characterized by a 3-byte "N_Port Identifier" value. The N_Port Identifier values of all N_Ports attached to the Fabric are uniquely defined in the Fabric's address space. Every Frame header contains S_ID and D_ID fields containing the source and destination N_Port identifier values, respectively, which are used for routing.

To support these functions, a Fabric Element or switch is assumed to provide a set of "F_Ports," which interface over the links with the N_Ports, plus a "Connection-based" and/or "Connectionless" Frame routing functionality. An F_Port is a entity that handles FC-0, FC-1, and FC-2 functions up to the Frame level to transfer data between attached N_Ports. A Connection-based router, or Sub-Fabric, routes Frames between Fabric Ports through Class 1 Dedicated Connections, assuring priority and non-interference from any other network traffic. A Connectionless router, or Sub-Fabric, routes Frames between Fabric Ports on a Frame-by-Frame basis, allowing multiplexing at Frame boundaries.

Implementation of a Connection-based Sub-Fabric is incorporated for Class 1, Class 4, and Class 6 service, while a Connectionless Sub-Fabric is incorporated for supporting Class 2 and 3 service. Although the term "Sub-Fabric" implies that separate networks are used for the two types of routing, this is not necessary. An implementation may support the functionality of Connection-based and Connectionless Sub-Fabrics either through separate internal hardware or through priority scheduling and routing management operations in a single internal set of hardware. Internal design of a switch element is largely implementation-dependent, as long as the priority and bandwidth requirements are met.

13.2.3.8.1. Fabric Ports

A switch element contains a minimum of two Fabric Ports. There are several different types of Fabric Ports, of which the most important are F_Ports. F_Ports are attached to N_Ports and can transmit and receive Frames, Ordered Sets, and other information in Fibre Channel format. An F_Port may or may not verify the validity of Frames as they pass through the Fabric. Frames are routed to their proper destination N_Port and intervening F_Port based on the destination N_Port identifier (D_ID). The mechanism used for doing this is implementation dependent, although address translation and routing mechanisms within the Fabric are being addressed in current Fibre Channel development work.

In addition to F_Ports, which attach directly to N_Ports in a switched Fabric topology, several other types of Fabric Ports are defined. In a multi-layer network, switches are connected to other switches through "E_Ports" (Expansion Ports), which may use standard media, interface, and signaling protocols or may use other implementation-dependent protocols. A Fabric Port that incorporates the extra Port states, operations, and Ordered Set recognition to allow it to connect to an Arbitrated Loop, is termed an "FL_Port." A "G_Port" has the capability to operate as either an E_Port or an F_Port, depending on how it is connected, and a "GL_Port" can operate as an F_Port, as an E_Port, or as an FL_Port. Since implementation of these types of Ports is implementation-dependent, the discussion in this book will concentrate on F_Ports, with clear requirements for extension to other types of Fabric Ports.

Each F_Port may contain receive buffers for storing Frames as they pass through the Fabric. The size of these buffers may be different for Frames in different Classes of service. The maximum Frame size capabilities of the Fabric for the various Classes of service are indicated for the attached N_Ports during the "Fabric Login" procedure, as the N_Ports are determining network characteristics.

13.2.3.8.2. *Connection-Based Routing*

The Connection-based Sub-Fabric function provides support for Dedicated Connections between F_Ports and the N_Ports attached to these F_Ports for Class 1, Class 4, or Class 6 service. Such Dedicated Connections may be either bidirectional or unidirectional and may support the full transmission rate concurrently in each direction, or some lower transmission rate. Class 1 Dedicated Connection is described here. Class 4 and Class 6 are straightforward modifications of Class 1.

On receiving a Class 1 connect-request Frame from an N_Port, the Fabric begins establishing a Dedicated Connection to the destination N_Port through the connection-based Sub-Fabric. The Dedicated Connection is pending until the connect-request is forwarded to the destination N_Port. If the destination N_Port can accept the Dedicated Connection, it returns an acknowledgment. In passing the acknowledgment back to the source N_Port, the Fabric finishes establishing the Dedicated Connection. The exact mechanisms used by the Fabric to establish the Connection are vendor-dependent. If either the Fabric or the destination Port are unable to establish a Dedicated Connection, they return a "BSY" (busy) or "RJT" (reject) Frame with a reason code to the source N_Port, explaining the reason for not establishing the Connection.

Once the Dedicated Connection is established, it appears to the two communicating N_Ports as if a dedicated circuit has been established between them. Delivery of Class 1 Frames between the two N_Ports cannot be degraded by Fabric traffic between other N_Ports or by attempts by other N_Ports to communicate with either of the two. All flow control is managed using end-to-end flow control between the two communicating N_Ports.

A Dedicated Connection is retained until either a removal request is received from one of the two N_Ports or an exception condition occurs that causes the Fabric to remove the Connection.

A Class 1 N_Port and the Fabric may support "stacked connect-requests." This function allows an N_Port to simultaneously request multiple Dedicated Connections to multiple destinations and allows the Fabric to service them in any order. This allows the Fabric to queue connect-requests and to establish the Connections as the destination N_Ports become available. While the N_Port is connected to one destination, the Fabric can begin processing another connect-request to minimize the connect latency. If stacked connect-requests are not supported, connect-requests received by the Fabric for either N_Port in a Dedicated Connection will be replied to with a "BSY" (busy) indication to the requesting N_Port, regardless of Intermix support.

If a Class 2 Frame destined to one of the N_Ports established in a Dedicated Connection is received, and the Fabric or the destination N_Port doesn't support Intermix, the Class 2 Frame may be busied and the transmitting N_Port is notified. In the case of a Class 3 Frame, the Frame is discarded and no notification is sent. The destination F_Port may be able to hold the Frame for a period of time before discarding the Frame or returning a busy Link Response. If Intermix is supported and the Fabric receives a Class 2 or Class 3 Frame destined to one of the N_Ports established in a Dedicated Connection, the Fabric may allow delivery with or without a delay, as long as the delivery does not interfere with the transmission and reception of Class 1 Frames.

Class 4 Dedicated Connections are similar to Class 1 connections, but they allow each connection to occupy a fraction of the source and destination N_Port link bandwidths, to allow finer control on the granularity of Quality of Service guarantees for transmission across the Fabric. The connect-request for a Class 4 dedicated connection specifies the requested bandwidth, and maximum end-to-end latency, for connection, in each direction, and the acceptance of connection by the Fabric commits it to honor those Quality of Service parameters during the life of the connection.

Class 6 is a Uni-Directional Dedicated Connection service allowing an acknowledged multicast connection, which is useful for efficient data replication in systems providing high availability. In Class 6 service, each Frame transmitted by the source of the Dedicated Connection is replicated by the Fabric and delivered to each of a set of destination N_Ports. The destination N_Ports then return acknowledgements indicating correct and complete delivery of the Frames, and the Fabric aggregates the acknowledgments into a single response which is returned to the source N_Port.

13.2.3.8.3. *Connectionless Routing*

A Connectionless Sub-Fabric is characterized by the absence of Dedicated Connections. The connectionless Sub-Fabric multiplexes Frames at Frame boundaries between multiple source and destination N_Ports through their attached F_Ports.

In a multiplexed environment, with contention of Frames for F_Port resources, flow control for connectionless routing is more complex than in the Dedicated Connection circuit-switched transmission. For this reason, flow control is handled at a finer granularity, with buffer-to-buffer flow control across each link. Also, a Fabric will typically implement internal buffering to temporarily store Frames that encounter exit Port contention until the congestion eases. Any flow control errors that cause overflow of the buffering mechanisms may cause loss of Frames. Loss of a Frame can clearly be extremely detrimental to data communications in some cases and it will be avoided at the Fabric level if at all possible.

In Class 2, the Fabric will notify the source N_Port with a "BSY" (busy) or a "RJT" (reject) indication if the Frame can't be delivered, with a code explaining the reason. The source N_Port is not notified of non-delivery of a Class 3 Frame, since error recovery is handled at a higher level.

13.2.3.9. Classes of Service

Fibre Channel currently defines five Classes of service, which can be used for transmitting different types of traffic with different delivery requirements. The Classes of service are not mandatory, in that a Fabric or N_Port may not support all Classes. The Classes of service are not topology-dependent. However, topology will affect performance under the different Classes, e.g., performance in a Point-to-point topology will be affected much less by the choice of Class of service than in a Fabric topology.

The five Classes of service are as follows. Class 1 service is intended to duplicate the functions of a dedicated channel or circuit-switched network, guaranteeing dedicated high-speed bandwidth between N_Port pairs for a defined period. Class 2 service is intended to duplicate the functions of a packet-switching network, allowing multiple Nodes to share links by multiplexing data as required. Class 3 service operates as Class 2 service without acknowledgments, allowing Fibre Channel transport with greater flexibility and efficiency than the other Classes under a ULP which does its own flow control, error detection, and recovery. In addition to these three, Fibre Channel Ports and switches may support Intermix, which combines the advantages of Class 1 with Class 2 and 3 service by allowing Class 2 and 3 Frames to be intermixed with Class 1 Frames during Class 1 Dedicated Connections. Class 4 service allows the Fabric to provide quality of service guarantees for bandwidth and latency over a fractional portion of a link bandwidth. Class 6 service operates as an acknowledged multicast, with unidirectional transmission from 1 source to multiple destinations at full channel bandwidth.

13.2.3.9.1. Class 1 Service: Dedicated Connection

Class 1 is a service that establishes Dedicated Connections between N_Ports through the Fabric, if available. A Class 1 Dedicated Connection is established by the transmission of a "Class 1 connect-request" Frame, which sets up the Connection and may or may not contain any message data. Once established, a Dedicated Connection is retained and guaranteed by the Fabric and the destination N_Port until the Connection is removed by some means. This service guarantees maximum transmission bandwidth between the two N_Ports during the established Connection. The Fabric, if present, delivers Frames to the destination N_Port in the same order that they are transmitted by the source N_Port. Flow control and error recovery are handled between the communicating N_Ports, with no Fabric intervention under normal operation.

Management of Class 1 Dedicated Connections is independent of Exchange origination and termination. An Exchange may be performed within one Class 1 Connection or may be continued across multiple Class 1 Connections.

13.2.3.9.2. Class 2 Service: Multiplex

Class 2 is a connectionless service with the Fabric, if present, multiplexing Frames at Frame boundaries. Multiplexing is supported from a single source to multiple destinations and to a single destination from multiple sources. The Fabric may not necessarily guarantee delivery of Data Frames or acknowledgments in the same sequential order in which they were transmitted by the source or destination N_Port. In the absence of link errors, the Fabric guarantees notification of delivery or failure to deliver.

13.2.3.9.3. Class 3 Service: Datagram

Class 3 is a connectionless service with the Fabric, if present, multiplexing Frames at Frame boundaries. Class 3 supports only unacknowledged delivery, where the destination N_Port sends no acknowledgment of successful or unsuccessful Frame delivery. Any acknowledgment of Class 3 service is up to and determined by the ULP utilizing Fibre Channel for data transport. The transmitter sends Class 3 Data Frames in sequential order within a given Sequence, but the Fabric may not necessarily guarantee the order of delivery. In Class 3, the Fabric is expected to make a best effort to deliver the Frame to the intended destination but may discard Frames without notification under high-traffic or error conditions. When a Class 3 Frame is corrupted or discarded, any error recovery or notification is performed at the ULP level. Class 3 can also be used for an unacknowledged multicast service, where the destination ID of the Frames specifies a pre-arranged multicast group ID, and the Frames are replicated without modification and delivered to every N_Port in the group.

13.2.3.9.4. Intermix

A significant problem with Class 1 as described above is that if the source N_Port has no Class 1 data ready for transfer during a Dedicated Connection, the N_Port's transmission bandwidth is unused, even if there might be Class 2 or 3 Frames which could be sent. Similarly, the destination N_Port's available bandwidth is unused, even if the Fabric might have received Frames that could be delivered to it.

Intermix is an option of Class 1 service that solves this efficiency problem by allowing interleaving of Class 2 and Class 3 Frames during an established Class 1 Dedicated Connection. In addition to the possible efficiency improvement described, this function may also provide a mechanism for a sender to transmit high-priority Class 2 or Class 3 messages without the overhead required in tearing down an already-established Class 1 Dedicated Connection.

Support for Intermix is optional, as is support for all other Classes of service. This support is indicated during the Login period, when the N_Ports, and Fabric, if present, are determining the network configuration.

Both N_Ports in a Dedicated Connection as well as the Fabric, if present, must support Intermix, for it to be used.

Fabric support for Intermix requires that the full Class 1 bandwidth during a Dedicated Connection be available, if necessary — insertion of Class 2 or 3 Frames cannot delay delivery of Class 1 Frames. In practice, this means that the Fabric must implement Intermix to the destination N_Port either by waiting for unused bandwidth or by inserting Intermixed Frames "in between" Class 1 Frames, removing Idle transmission words between Class 1 Frames to make up the bandwidth used for the Intermixed Class 2 or 3 Frame. If a Class 1 Frame is generated during transmission of a Class 2 or Class 3 Frame, the Class 2 or Class 3 Frame should be terminated with an End of Frame marker indicating that it is invalid, so that the Class 1 Frame can be transmitted immediately.

13.2.3.9.5. Class 4

A different, but no less significant problem with Class 1 is that it only allows Dedicated Connection from a single source to a single destination, at the full channel bandwidth. In many applications, it is useful to allocate a fraction of the resources between the N_Ports to be used, so that the remaining portion can be allocated to other connections. In Class 4, a bidirectional circuit is established, with one "Virtual Circuit" (VC) in each direction, with negotiated Quality of Service guarantees on bandwidth and latency for transmission in each direction's VC. A source or destination N_Port may support up to 254 simultaneous Class 4 circuits, with a portion of its link bandwidth dedicated to each one. Class 4 does not specify how data is to be multiplexed between the different VCs, or how it is to be implemented in the Fabrics — these functions are determined by the implementation of the Fabric supporting Class 4 traffic.

13.2.3.9.6. Class 6

A primary application area for Fibre Channel technology is in enterprise-class data centers or Internet Service Providers, supporting high-reliability data storage and transport. In these application areas, data replication is a very common requirement, and a high load on the SAN. Class 6 is intended to provide additional efficiency in data transport, by allowing data to be replicated by the Fabric without modification and delivered to each desti-nation N_Port in a multicast group. Class 6 differs from Class 3 multicast in that the full channel bandwidth is guaranteed, and that the destination

N_Ports each generate responses, which are collected by the Fabric and delivered to the source N_Port as a single aggregated response Frame.

13.2.3.10. Basic and Extended Link Service Commands

Beyond the Frames used for transferring data, a number of Frames, Sequences, and Exchanges are used by the Fibre Channel protocol itself, for initializing communications, overseeing the transmission, allowing status notification, and so on. These types of functions are termed "Link Services," and two types of Link Service operations are defined.

"Basic Link Service commands" are implemented as single Frame messages that transfer between N_Ports to handle high priority disruptive operations. These include an Abort Sequence (ABTS) request Frame, which may be used to determine the status of and possibly to abort currently existing Sequences and/or Exchange for error recovery. Aborting (and possibly retransmitting) a Sequence or Exchange is the main method of recovering from Frame- and Sequence-level errors. Acceptance or Rejection of the Abort Sequence (ABTS) command is indicated by return of either a Basic Accept (BA_ACC) or a Basic Reject (BA_RJT) reply. A Remove Connection (RMC) request allows a Class 1 Dedicated Connection to be disruptively terminated, terminating any currently active Sequences. A No Operation (NOP) command contains no data but can implement a number of control functions, such as initiating Class 1 Dedicated Connections, transferring Sequence Initiative, and performing normal Sequence termination, through settings in the Frame Header and the Frame delimiters.

"Extended Link Service commands" implement more complex operations, generally through establishment of a completely new Exchange. These include establishment of initial operating parameters and Fabric or topology configuration through the Fabric Login (FLOGI) and N_Port Login (PLOGI) commands, and the Logout (LOGO) command. The Abort Exchange (ABTX) command allows a currently existing Exchange to be terminated through transmission of the ABTX in a separate Exchange. Several commands can request the status of a particular Connection, Sequence, or Exchange or can read timeout values and link error status from a remote Port, and one command allows for requesting the Sequence Initiative within an already existing Exchange. Several commands are defined to be used as part of a protocol to establish the best end-to-end credit value between two Ports. A number of Extended Link Service commands are defined to manage Login, Logout, and Login state management for "Processes." Implementation of the Process Login and related functions allows targeting of communication to one of multiple independent entities behind a single N_Port. This allows for a multiplexing of operations from multiple Processes, or "images" over a single N_Port, increasing hardware usage efficiency. A set of Extended Link Service commands allow management of Alias_IDs, which allows a single N_Port or group of N_Ports to be known to other N_Ports and by the Fabric by a different ID, allowing different handling of traffic delivered to the same physical destination Port. Finally, a set of Extended Link Service commands allow reporting or querying of the state or the capabilities of a Port in the Fabric.

13.2.4. ARBITRATED LOOP FUNCTIONS

The management of the Arbitrated Loop topology requires some extra operations and communications beyond those required for the point-to-point and Fabric topologies. These include new definitions for Primitive Sequences and Primitive Signals for initialization and arbitration on the loop, an additional initialization scheme for determining addresses on the loop, and an extra state machine controlling access to the loop and transmission and monitoring capabilities.

13.2.5. PROTOCOLS

Protocols are interchanges of specific sets of data for performing certain defined functions. These include operations to manage the operating environment, transfer data, and do handshaking for specific low-level management functions. Fibre Channel defines the following protocols:

Primitive Sequence protocols: Primitive Sequence protocols are based on single-word Primitive Sequence Ordered Sets and do low-level handshaking and synchronization for the Link Failure, Link Initialization, Link Reset, and Online to Offline protocols.

Arbitrated Loop Initialization protocol: In an Arbitrated Loop topology, the assignment of the 127 possible loop address to different Ports attached on the loop is carried out through the transmission of a set of Sequences around the loop, alternately collecting and broadcasting mappings of addresses to Nodes.

Fabric Login protocol: In the Fabric Login protocol, the N_Port interchanges Sequences with the Fabric, if present, to determine the service parameters determining the operating environment. This specifies parameters such as flow control buffer credit, support for different Classes of service, and support for various optional Fibre Channel services. The equivalent of this procedure can be carried out through an "implicit Login" mechanism, whereby an external agent such as a system administrator or preloaded initialization program notifies a Port of what type of environment it is attached to. There is no explicit Fabric Logout since the Fabric has no significant resources dedicated to an N_Port which could be made available. Transmission of the OLS and NOS Primitive Sequences cause an implicit Fabric Logout, requiring a Fabric re-Login before any further communication can occur.

N_Port Login protocol: The N_Port Login protocol performs the same function with a particular destination N_Port that the Fabric Login protocol performs with the Fabric.

N_Port Logout protocol: An N_Port may request removal of its service parameters from another Port by performing an N_Port Logout protocol. This request may be used to free up resources at the other N_Port.

13.2.6.   SEGMENTATION AND REASSEMBLY

Segmentation and reassembly are the FC-2 functions provided to subdivide application data to be transferred into Payloads, embed each Payload in an individual Frame, transfer these Frames over the link(s), and reassemble the application data at the receiving end. Within each Sequence, there may be multiple "Information Categories." The Information Categories serve as markers to separate different blocks of data within a Sequence that may be handled differently at the receiver.

The mapping of application data to Upper Level Protocols (ULPs) is outside the scope of Fibre Channel. ULPs maintain the status of application data transferred. The ULPs at the sending end specify to the FC-2 layer:

- blocks or sub-blocks to be transferred within a Sequence,
- Information Category for each block or sub-block,
- a Relative Offset space starting from zero, representing a ULP-defined origin, for each Information Category, and
- an Initial Relative Offset for each block or sub-block to be transferred.

The "Relative Offset" relationship between the blocks to be transferred in multiple Sequences is defined by an upper level and is transparent to FC-2. Relative Offset is a field transmitted in Data Frame Header used to indicate the displacement of the first data byte of the Frame's Payload into an Information Category block or collection of blocks at the sending end. Relative Offset is not a required function in a Fibre Channel implementation. If Relative Offset is not supported, SEQ_CNT is used to perform the segmentation and reassembly. Since Frame sizes are variable, Frames without Relative Offset cannot be placed into their correct receive block locations before all Frames with lower SEQ_CNT values have been received and placed.

The Sequence Recipient indicates during Login its capability to support Continuously Increasing or Random Relative Offset. If only the former is supported, each Information Category transferred within a Sequence is treated as a block by upper levels. If Random Relative Offset is supported, an Information Category may be specified as sub-blocks by upper levels and the sub-blocks may be transmitted in a random order.

13.2.7. DATA COMPRESSION

Another function included in Fibre Channel is the capability for data compression, for increasing the effective bandwidth of data transmission. ULP data may be compressed on a per Information Category basis within a Sequence, using the Adaptive Lossless Data Compress Lempel Ziv-1 algorithm. When the compression and decompression engines can operate at link speed or greater, the effective rate of data transmission can be multiplied by the inverse of the compression ratio.

13.2.8. ERROR DETECTION AND RECOVERY

In general, detected errors fall into two broad categories: Frame errors and link-level errors. Frame errors result from missing or corrupted Frames. Corrupted Frames are discarded and the resulting error is detected and possibly recovered at the Sequence level. At the Sequence level, a missing Frame is detected at the Recipient due to one or more missing SEQ_CNT values and at the Initiator by a missing or timed-out acknowledgment. Once a Frame error is detected, the Sequence may either be discarded or be retransmitted, depending on the Exchange Error Policy for the Sequence's Exchange. If one of the discard Exchange Error policies is used, the Sequence is aborted at the Sequence level once an error is detected. Sequence errors may also cause Exchange errors which may also cause the Exchange to be aborted. When a retransmission Exchange Error policy is used, error recovery may be performed on the failing Sequence or Exchange with the involvement of the sending ULP. Other properly performing Sequences are unaffected.

Link-level errors result from errors detected at a lower level of granularity than Frames, where the basic signal characteristics are in question. Link-level errors include such errors as Loss of Signal, Loss of Synchronization, and link timeout errors which indicate no Frame activity at all. Recovery from link-level errors is accomplished by transmission and reception of Primitive Sequences in one of the Primitive Sequence protocols. Recovery at the link level disturbs normal Frame flow and may introduce Sequence errors which must be resolved following link level recovery.

The recovery of errors may be described by the following hierarchy, from least to most disruptive:

1. Abort Sequence: Recovery through transmitting Frames of the Abort Sequence protocol;
2. Abort Exchange: Recovery through transmitting Frames of the Abort Exchange protocol;
3. Link Reset: Recovery from link errors such as Sequence timeout for all active Sequences, E_D_TOV timeout without reception of an R_RDY Primitive Signal, or buffer-to-buffer overrun;
4. Link Initialization: Recovery from serious link errors such that a Port needs to go offline or halt bit transmission;
5. Link Failure: Recovery from very serious link errors such as loss of signal, loss of synchronization, or timeout during a Primitive Sequence protocol.

The first two protocols require transmission of Extended Link Service commands between N_Ports. The last three protocols are Primitive Sequence protocols operating at the link level. They require interchange of more fundamental constructs, termed Primitive Sequences, to allow interlocked, clean bring-up when a Port (N_Port or F_Port) may not know the status of the opposite Port on the link.

13.2.9. FC-3: OPERATIONS ACROSS MULTIPLE N_PORTS

The FC-3 level is intended to provide a framing protocol and other services that manage operations over multiple N_Ports on a single Node. This level is under development, since the full requirements for operation over multiple N_Ports on a Node have not become clear. A example function would be "striping," where data could be simultaneously transmitted through multiple N_Ports to increase the effective bandwidth.

A number of FC-3-related functions have been described in the FC-PH-2 and FC-PH-3 updates. These include (1) broadcast to all N_Ports attached to the Fabric, (2) Alias_ID values, for addressing a subset of the Ports by a single alias, (3) multi-cast, for a restricted broadcast to the Ports in an alias group, (4) Hunt groups, for letting any member of a group handle requests directed to the alias group.

13.2.10. FC-4: MAPPINGS TO UPPER LAYER PROTOCOLS

The FC-4 level defines mappings of Fibre Channel constructs to ULPs. There are currently defined mappings to a number of significant channel, peripheral interface, and network protocols, including:

- SCSI (Small Computer Systems Interface)
- IPI-3 (Intelligent Peripheral Interface-3)
- HIPPI (High Performance Parallel Interface)
- IP (the Internet Protocol) — IEEE 802.2 (TCP/IP) data
- ATM/AAL5 (ATM adaptation layer for computer data)
- SBCCS (Single Byte Command Code Set) or ESCON/FICON/SBCON.

The general picture is of a mapping between messages in the ULP to be transported by the Fibre Channel levels. Each message is termed an "Information Unit," and is mapped as a Fibre Channel Sequence. The FC-4 mapping for each ULP describes what Information Category is used for each Information Unit, and how Information Unit Sequences are associated into Exchanges.

The following sections give general overviews of the FC-4 ULP mapping over Fibre Channel for the IP, SCSI, and FICON protocols, which are three of the most important communication and I/O protocols for high-performance modern computers.

13.2.10.1. IP over Fibre Channel

Establishment of IP communications with a remote Node over Fibre Channel is accomplished by establishing an Exchange. Each Exchange established for IP is unidirectional. If a pair of Nodes wish to interchange IP packets, a separate Exchange must be established for each direction. This improves bidirectional performance, since Sequences are non-concurrent under each Exchange, while IP allows concurrent bidirectional communication.

A set of IP packets to be transmitted is handled at the Fibre Channel level as a Sequence. The maximum transmission unit, or maximum IP packet size, is 65,280 (x'FF00') bytes, to allow an IP packet to fit in a 64-kbyte buffer with up to 255 bytes of overhead.

IP traffic over Fibre Channel can use any of the Classes of service, but in a networked environment, Class 2 most closely matches the characteristics expected by the IP protocol.

The Exchange Error Policy used by default is "Abort, discard a single Sequence," so that on a Frame error, the Sequence is discarded with no retransmission, and subsequent Sequences are not affected. The IP and TCP levels will handle data retransmission, if required, transparent to the Fibre Channel levels, and will handle ordering of Sequences. Some implementations may specify that ordering and retransmission on error be handled at the Fibre Channel level by using different Abort Sequence Condition policies.

An Address Resolution Protocol (ARP) server must be implemented to provide mapping between 4-byte IP addresses and 3-byte Fibre Channel address identifiers. Generally, this ARP server will be implemented at the Fabric level and will be addressed using the address identifier x'FF FFFC.'

13.2.10.2. SCSI over Fibre Channel

The general picture is of the Fibre Channel levels acting as a data transport mechanism for transmitting control blocks and data blocks in the SCSI format. A Fibre Channel N_Port can operate as a SCSI source or target, generating or accepting and servicing SCSI commands received over the Fibre Channel link. The Fibre Channel Fabric topology is more flexible than the SCSI bus topology, since multiple operations can occur simultaneously. Most SCSI implementation will be over an Arbitrated Loop topology, for minimal cost in connecting multiple Ports.

Each SCSI-3 operation is mapped over Fibre Channel as a bidirectional Exchange. A SCSI-3 operation requires several Sequences. A read command, for example, requires (1) a command from the source to the target, (2) possibly a message from the target to the source indicating that it's ready for the transfer, (3) a "data phase" set of data flowing from the target to the source, and (4) a status Sequence, indicating the completion status of the command. Under Fibre Channel, each of these messages of the SCSI-3 operation is a Sequence of the bidirectional Exchange.

Multiple disk drives or other SCSI targets or initiators can be handled behind a single N_Port through a mechanism called the "Entity Address." The Entity Address allows commands, data, and responses to be routed to or from the correct SCSI target/initiator behind the N_Port. The SCSI operating environment is established through a procedure called "Process Login," which determines operating environment such as usage of certain non-required parameters.

13.2.10.3. FICON, or ESCON over Fibre Channel

ESCON (Enterprise Systems Connection) has been the standard mechanism for attaching storage control units on IBM's zSeries eServer (previously known as S/390) mainframe systems since the early 1990s. ESCON channels were the first commercially significant storage networking infrastructure, allowing multiple host systems to access peripherals such as storage control units across long-distance, switched fabrics. In 1998, IBM introduced ESCON over Fibre Channel, termed FICON (Fibre Connection), which preserves the functionality of ESCON, but uses the higher performance and capability levels allowed by Fibre Channel network technology.

At the physical layer, FICON is essentially the same as Fibre Channel, using the same transmitters and receivers, at the same bit rates, but there are some small differences. FICON uses the LC duplex connector rather than the standard SC, but the LC connector may be standardized in the future as well. FICON also supports optical mode conditioners, that let single-mode transmitters operate with both single-mode and multimode fiber. This feature, which is also incorporated into Gigabit Ethernet, is not natively defined for Fibre Channel. FICON links at 1.06 Gbps and 2.1 Gbps are available, allowing time-division multiplexing of up to 8 ESCON channels over a single FICON channel.

At the protocol level, FICON is conceptually quite similar to SCSI over Fibre Channel, with a set of command and data Information Units transmitted as payloads of Fibre Channel Sequences. However, the FICON control blocks for the I/O requests, termed CCWs (Channel Control Words), are more complex and sophisticated than the SCSI command and data blocks, to accommodate the different format and the higher throughput, reliability, and robustness requirements for data storage on these systems.

In addition, FICON is optimized, in terms of overhead and link proto-col, to support longer distance links, including links using DWDM (Dense Wavelength-Division Multiplexing), which allow transmission without per-formance droop out to 100 km.

13.3. Storage Area Networks

The main use of Fibre Channel currently is in Storage Area Networks, or SANs. In a SAN, a network sits between the server and its storage devices, i.e., disks or tapes. An example system topology is shown in Fig. 13.9, with a layer of clients connecting to a layer of fewer high-performance, and a few very large storage systems holding data for the whole network. A ser-vice provider system could look very similar, with thin servers replacing the clients and with more connectivity to Wide-Area Networks.

This system is in contrast to more traditional systems without storage networks, where each storage device is directly associated with a server. In the more traditional systems, servers can only access the data associated with other servers by requesting that the associated server fetch it.

In systems using Storage Area Networks, the packets going between the servers and the storage devices contain disk I/O block requests. Because the rate of traffic may be very high, and the disk data may be simultaneously accessed by several servers, it's very important to get high performance.

This type of topology also allows "LAN-free backup," where backup traffic from disks to tapes doesn't cross the LAN connecting the clients and servers and also "server-free backup," where backup traffic goes directly between intelligent disk to tape controllers without crossing through and loading a server.

13.4. How Fibre Channel Leverages Optical
Data Communications

Fibre Channel has a number of characteristics that work well for optical data communications networks. These characteristics arise from the following observations:

- Memory access time has not improved substantially, especially for the non-sequential memory accesses encountered often in protocol processing. Typical DRAM access time best-case is 30 nsec, which means that, for example, on a link transmitting packets with a minimum size of 40B, a 100 Mbps link allows time for 106 memory

Fig. 13.9 Example of an Enterprise or Service Provider SAN + LAN Topology.

accesses per packet for protocol processing, while a 10 Gbps link barely allows time for 1. This trend will continue to get worse — even very advanced memory technologies such as on-chip SRAM and embedded DRAM only improve access time by 3X to 10X, not 100X.

- The number of transistors per chip is growing steadily at 2X every 18 months. This trend, popularly known as Moore's law, is expected to continue for the foreseeable future.

- Power conservation will continue to be an increasingly critical factor in network and link design. That is, Gbps per Watt will be an important measure of network performance, because Gbps per Watt is the critical factor in determining how tightly integrated function can be, which in turn is a critical factor in determining system price.

13.4.1.   PHYSICAL LAYER AND CODING

In comparison to copper data links, optical data links have the characteristics that (1) the received optical power is variable over the life of the link, and (2) only single-ended, rather than differential signalling, is required.

These characteristics imply that the transmitted data must be coded to ensure transitions, and to move the transmitted spectrum up in frequency (i.e., away from DC). Fibre Channel uses the 8b/10b code to accomplish this. Although the 25% overhead of this code is relatively high, this code has the advantage of providing guarantees on the transmission characteristics (minimum transition density, DC balance, etc.) which makes high-sensitivity receiver circuitry easier to build. Lower-overhead codes, such as data scrambling, don't provide as robust guarantees, and make the analog circuitry more difficult to build.

As the bit rates increase and CMOS circuitry gets more dense, digital coding to allow simpler analog circuitry will become an increasingly cost-effective tradeoff.

13.4.2.   ORDERED, RELIABLE DELIVERY
BY THE HARDWARE

Applications generally incorporate the expectation that data either arrives intact, in the order transmitted, or doesn't arrive at all. If the network does not guarantee this feature, it has to be guaranteed by protocol processing layers between the network and the application layer. This is the function delivered by the TCP layer in a TCP/IP protocol stack.

Since this type of processing requires numerous memory accesses, to check for correct ordering and intact delivery, it is much better to eliminate it by having the network assure reliable, ordered delivery to the endnodes.

13.4.3. SIMPLE ADDRESSING

A major factor in network performance is the switching function, since it's not feasible to provide direct links between any 2 machines that might wish to communicate. Ideally, the packets should be self-routing, or should only require a single switch lookup to determine the packet routing.

In Fibre Channel, routing is determined by short 3-byte Node ID, which can be assigned by the switch to allow simple routing. This allows much higher-performance for the routing function than, for example, Ethernet, which requires a hash table or CAM to route by the 48-bit MAC addresses assigned to the various network interface cards.

13.4.4. DATA TRANSFER

Because the per-packet processing is such a large part of the overall processing required (vs. per-byte), it's very important to allow for large data transfers as single units.

In Fibre Channel, the basic unit of transfer between the hardware and the software is the Sequence, which may be up to 128 MB. Sequences are actually transmitted through the network as Frames, but since the segmentation and reassembly of Frames is clearly architected, it is straight-forward to build engines to do this segmentation and reassembly in hardware.

13.4.5. REQUEST/RESPONSE COUPLING

In all communication, there is a combination of at least one request with at least one response, even if the response is just an acknowledgment that the request was received — more complex protocols may require multiple coupled requests and responses. If these are explicitly coupled together, it simplifies the processing.

In Fibre Channel, Exchanges tie together Sequences in both directions, so a single I/O operation can be more easily managed in hardware, without resorting to higher layers for request-response coupling.

13.4.6.   FLOW CONTROL

A high-performance network should have very good flow control support, such that data is not sent unless there is guaranteed space to receive it at every step along the way and at the destination.

In Fibre Channel, there is defined flow control both at the link level (Buffer-to-buffer credit) and between N_Ports (End-to-end credit). When a Source N_Port has available both Buffer-to-buffer credit and End-to-end credit, it can send a Frame, being assured that the neighboring port can buffer the Frame, and that the Destination N_Port will be able to receive it. This guarantee assures that the same protocol processing work won't have to be done multiple times for the same packet.

13.4.7.   MANAGEMENT SERVICES

An important part of high-performance networking is that the performance is not only high, but predictable. In Fibre Channel, the network management is centralized by the switch fabric, such that the N_Ports can request information from the Fabric, rather than having to calculate things itself or request information from all the other N_Ports on the Fabric. This feature moves network management from the hosts onto the Fabric, letting them concentrate on applications, rather than on the network.

13.4.8.   FUTURE, AND COMPARISON
WITH OTHER NETWORKS

Clearly, link and switch speeds will increase, with WDM and Parallel Fiber Transceivers. Also, Fibre Channel and other networks will soon be expected to operate at 10 Gb/sec., which will accentuate the problems described here. Fibre Channel goes a long way toward addressing these issues, but there are a number of other technologies that attempt to solve similar problems in slightly different ways.

One very interesting technology is the incorporation of TCP offload engines into Ethernet NICs. This is fairly difficult, and requires a lot of processing power on the NICs, but may provide substantial performance benefits, especially for well-defined and predictable traffic patterns, as seen in storage traffic.

Another interesting related technology is InfiniBand, which is an emerging standard aimed at integrating a network between a processor and its

I/O cards, i.e., even more tightly into a server than in Fibre Channel. In-finiBand incorporates many of the features described here, and is expected to operate with links speeds between (2+2) Gbps and (24+24) Gbps.

Fibre Channel will therefore have several competitors in the future, incorporating many of the same mechanisms and additional mechanisms for achieving network performance equal to the link performance that optical data communications make possible.

13.5. Summary

Fibre Channel is a rich and complex architecture, as is necessary for a standard attempting to encompass all protocol levels from physical cabling up through network management, error detection and recovery and interface to upper layer protocols.

The primary characteristics of Fibre Channel with respect to optical data transmission are that, as opposed to previously existing networks such as Ethernet or ATM, Fibre Channel defines many networking functions in detail to be implemented in hardware on host channel adapters, rather than in the processors of the communicating systems. This processor offload capability is of large and growing importance, as the bit rate for optical (and electronic) data links continues to increase faster than memory access speeds.

Fibre Channel has been extremely successful in the very demanding network environment between high-performance servers and their networked storage devices, providing communication at near link speed with very low processor loading. However, in the future, a number of other approaches to addressing this processor offload problem will arise, the most important of which currently appear to be Ethernet NICs incorporating major portions of the TCP/IP protocol stack, and InfiniBand Channel Adapters. The relative value of these different approaches will depend both on technical performance, and on cost-related issues related to volumes and interoperability. It will be interesting to see how and where these different technologies are applied as data communication performance continues to improve.

Web Resources and References

The following web pages provide information on technology related to Fibre Channel, SANs and storage networking, and other high-performance data communication standards.

Hardcopies of the standards documents may be obtained from Global Engineering Documents, An IHS Group Company, at http://global.ihs.com/. Also, electronic versions of most of the approved standards are also available from http://www.ansi.org, and at the ANSI electronic standards store.

Further information on ANSI standards and on both approved and draft international, regional, and foreign standards (ISO, IEC, BSI, JIS, etc.) can be obtained from the ANSI Customer Service Department. References under development can be obtained from NCITS (National Committee for Information Technology Standards), at http://www.x3.org.

The following web sites provide information on technology related to Fibre Channel, SANs and storage networking.

- http://webstore.ansi.org — Web store of the American National Standards Institute Softcopies of the Fibre Channel Standards documents.
- http://global.ihs.com — Global Engineering Documents, An IHS Group Company Hardcopies of the Fibre Channel standards documents.
- http://www.fibrechannel.org — Fibre Channel Industry Association
- http://www.snia.org — Storage Networking Industry Association
- http://www.storageperformance.org — Storage Performance Council Industry organizations related to various aspects of storage networking
- http://www.infinibandta.org — InfiniBand Trade Association — provides information on the objectives, history, and specification of InfiniBand network I/O technology.
- http://www.iol.unh.edu — University of New Hampshire InterOperability Laboratory Tutorials on many different high-performance networking standards.

Benner, Alan. 2001. *Fibre Channel for SANs*. New York: The McGraw-Hill Companies.

Clark, Tom. 1999. *Designing Storage Area Networks: A Practical Reference for Implementing Fibre Channel SANs,* Reading, Mass.: Addison-Wesley Longman, Inc.

Decusatis, C. 1995. Data processing systems for optoelectronics. In *Optoelectronics for data communicaztion*. Eds. R. Lasky, U. Österberg, and D. Stigliani, 219–283. New York: Academic Press.

Farley, Marc. 2000. *Building Storage Networks,* New York: The McGraw-Hill Companies.

The Fibre Channel Association. 1994. *Fibre Channel: Connection to the Future.*

Partridge, Craig. 1994. *Gigabit Networking*: Reading, Mass.: Addison-Wesley.

Primmer, M. 1996 (October). An introduction to Fibre Channel. *Hewlett-Packard J.* 47:94–98.

Tanenbaum, Andrew. 1989. *Computer Networks*. Englewood Cliffs, N.J.: Prentice-Hall.

Widmer, A. X., and P. A. Franaszek. 1983. A DC balanced, partition block 8B/10B transmission code. *IBM J. Res. Dev.* 27:440–451.

Chapter 14 | Enterprise Systems Connection Fiber Optic Link

Daniel J. Stigliani, Jr.

IBM Corporation, Poughkeepsie, New York 12601

14.1. Introduction

The information technology (IT) industry has been evolving at a tremendous rate since the advent of computers in the business world in the early 1950s. This "explosion" of IT (formerly data processing) in the commercial world since the late 1980s has been fueled by the following trends:

- Desire of business to achieve competitive advantage through the use of IT
- Development and exploitation of large-scale integration and complementary metal oxide semiconductor (CMOS) technology
- Desire to have ubiquitous information and processing capability
- Desire to realize efficiencies in the business process and reduce product cost
- Development of new application (e.g., multimedia and real-time video and audio at the desktop) and business communication models (e.g., World Wide Web)

All of these items and more have put tremendous emphasis on dissemination of data from server (host computer) to server, server to client, and client to client. The information transfer is likely to start with a client or server request for data from a large database and end with the storage of

506

HANDBOOK OF FIBER OPTIC
DATA COMMUNICATION
$35.00

new or revised data in a remote database or stored locally, for further processing in the future (e.g., data mining). The key factors in a successful data transfer link or network interconnection are

- High aggregate throughput
- Low response latency
- Easy to use
- Allowance for future growth

This environment has placed tremendous demands on International Business Machines Corporation (IBM) eServer zSeries large servers to improve not only data processing and server capability but also the system interconnection capability. In the early 1990s IBM introduced the first in a series of new large-scale servers that provided a new system structure and architecture (Enterprise Systems Architecture/390) for coupling multiple data processing systems together and Enterprise System Connection (ESCON) architecture to provide high bandwidth interconnection capability for System/390 products and attachments. This was the beginning of the large-server interconnection network evolution into the modern information technology paradigm.

This chapter provides an understanding of the ESCON interconnection from a system perspective and design consideration. Detailed design parameters and implementation requirements of the ESCON fiber optic link are discussed. Section 14.2 discusses the system perspective, interconnection architecture and topology, and link-level protocol. The link design consideration and trade-offs are discussed in Section 14.3. Sections 14.4 and 14.5 present the ESCON physical layer specification for multimode and single-mode links, respectively. Section 14.6 discusses the planning and installation considerations and recommendations of an ESCON link. The link optical loss budget analysis is presented in Section 14.7, whereas link troubleshooting is reviewed in Section 14.8. The final section (14.9) discusses ESCON link eye safety, link qualification, the impending single byte connectivity (SBCON) American National Standards Industry (ANSI) standard, ESCON channel performance, and recent technology updates.

This chapter makes extensive use of illustrations and tables published in documents copyrighted by International Business Machines Corporation, which are listed in the References. The illustrations and figures are reprinted by permission from International Business Machines Corporation.

14.2.   ESCON System Overview

ESCON systems architecture is a total network interconnection system for large server complexes [1, 2]. ESCON encompasses fiber optic technology links, serial data transfer, new link-level protocols, data encoding/decoding, new system transport architecture, and a new topology. The application for ESCON is intended as the backbone network that spans a customer premise. In some cases it may be a machine room, whereas in other cases it could be a large multibuilding campus that may span 20 km or more.

The objective of ESCON is to provide a versatile, robust, expandable, high-bandwidth interconnection system for IBM eServer zSeries servers. Some of the key attributes of ESCON include the following:

- Provide high-speed data transfer capability (200 Mb/s) on a media with the ability to support enhanced performance
- Allow interconnection of ESCON-compatible devices at both short (4 m) and long distance (20 km)
- Allow the addition and deletion of components on the network (including servers) without disruption of the network and its operation
- Provide improved connectivity (multipath) between servers and attached devices
- Reduce the quantity and bulk of interconnection cables
- Ease configuration and installation complexity

14.2.1.   DESIGN CONSIDERATIONS AND ATTRIBUTES

A key element of achieving the attributes described previously was the choice of a link technology that would result in

- High-speed data capability
- Long-distance capability
- Small size and weight
- High reliability
- Significant performance improvement capability

The original copper implementation of the System/390 OEMI channel utilized a parallel data transfer protocol and was severely limited in distance capability by skew on the cable. This would not satisfy the previously mentioned requirements and a serial data transfer approach was adopted. This decision further exacerbated the need for a technology that supports very high-speed data transfer. After significant investigation (both theoretical and experimental) in both copper (e.g., coaxial cable) and fiber optic technology transport capability, it was decided that fiber optics technology could meet all the system requirements and support potential future growth. The inherent property of electromagnetic compatibility for fiber optics was also an important consideration.

In the mid-1980s fiber optic technology was being successfully implemented in the telecom environment but there had been no significant use in the datacom market. IBM System/390 division initiated an exhaustive technology investigation of fiber optic technology to define an application component set that provided the following:

- Long-distance transmission (several kilometers)
- High-speed capability (greater than 100 Mb/s)
- Proven capability for high reliability (both soft and hard failure)
- Application robustness in an uncontrolled building environment

The technology set of 1300-nm long-wavelength light-emitting diode (LW LED), 62.5/125 μm multimode fiber (MMF), and unidirectional transmission per fiber was chosen. To facilitate ease of use and avoidance of fiber misplugging, a push-on duplex optic fiber connection was developed. This is the base ESCON link technology. As the need for distances greater than 2 or 3 km became apparent, the ESCON Extended Distance Feature (XDF) was developed. This extended the ESCON distance to 20 km using a long-wavelength laser source and single-mode fiber (SMF). This capability completed the ESCON fiber technology set. Both of these fiber optic technologies are in use today and will be discussed in detail later in this chapter.

The use of fiber optic technology and long-distance transmission changed the System/390 interconnection paradigm from a parallel asynchronous data connection approach to a high-speed serial data communication link. The properties (e.g., long-distance transmission and extremely small number of random bit errors) of such a link fostered a new architecture. In order to facilitate recovering the data, with minimal latency and error, the architecture was designed such that information "idle characters" are continually transmitted to ensure the receiver is continually synchronized to the transmitter. During a data transfer, the receiver is locked to the incoming data by design and can transmit the data to the processing circuits with minimum delay. The data, which are in byte format, are aggregated into groups of bytes that is called a frame. Also, to enable the receiver to sample the data at the optimum point (usually in the middle of the data pulse) the transmit clock is encoded within the data (in-band signaling). Several techniques are common (e.g., Manchester encoding) in the industry. ESCON uses an 8B/10B code where the 8-bit byte is mapped into a 10-bit character.

The physical topology of ESCON was also a critical factor because it had to ensure connectivity to all devices, maintain high throughput, minimize the number of connections (cables), and offer network growth in a realistic and nondisruptive manner. Several network topologies and infrastructures were investigated (e.g., Ethernet CDMA Ring, Dual Insertion Ring, and Host Star). The topology chosen for ESCON is "switched point-to-point." It offered the highest throughput, excellent connectivity with minimal number of links, and the ability to grow the network in a nondisruptive manner.

14.2.2.   ESCON TOPOLOGY

The switched point-to-point topology utilizes a central switch (director) to direct the network traffic to the various elements of the network [3]. The use of a director allows the connectivity of any unit on the network to any other unit on the network. The physical connections are point-to-point links that are ideally suited to fiber optic technology. This topology enables the ability to isolate links in the network for failure analysis and repair. An n-port nonblocking director can accommodate $\frac{n}{2}$ simultaneous conversations between endpoints in the switched point-to-point network.

This type of network offers some very nice attributes that have been implemented in ESCON [2]:

- Dynamic reconfiguration of data exchanges by terminating one connection and establishing a new connection to a different endpoint
- Multipathing of links between endpoints to ensure 100% availability in all circumstances
- The addition and deletion of physical links without disruption of the existing network
- Dedicated high-priority links between endpoints
- Ability to isolate (fencing) portions of the network
- Cascading of directors for enhanced connectivity

14.2.2.1. Typical Configuration

The application of a switched point-to-point topology in ESCON is straightforward and provides the primary interconnection configuration for System/390 platforms. Most computing complexes today contain multiple servers and devices (any attached unit, e.g., DASD controller or communication element). A typical example is shown in Fig. 14.1. An important availability element of this configuration is that all servers and devices have two paths to each director. This configuration provides not only connectivity between servers and devices but also full redundancy and multipathing. For example, if any one of the links or directors becomes inoperative, there is an alternate path between the system and device. Also, by adding four links (two to each director) a new server can be included in the network nondisruptively with immediate full connectivity.

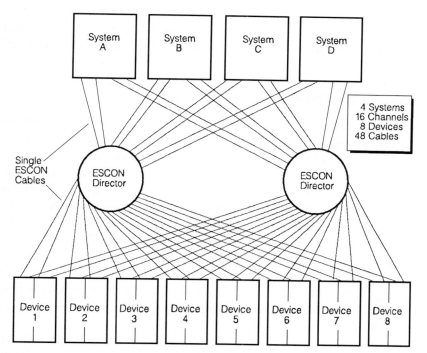

Fig. 14.1 Examples of an IBM System/390 ESCON configuration [2]. (Copyright 1992 by International Business Machines Corporation, reprinted with permission.)

14.2.3.   ESCON ARCHITECTURE AND CHANNEL

The IBM ESCON architecture establishes the rules and syntax used by the server to communicate to attached devices [4]. The architecture was defined to provide efficient transmission of data over long distances via a communication channel with a bit error rate (BER) of 10^{-10} (1 error in 10^{10} bits) or less. The architecture can be divided into two fundamental categories: device level and link level. The device level defines the rules for communication of a large server to an attached device using the facilities of the physical link. It defines data and control messages and the protocol to implement the server input/output (I/O) functions. The link-level architecture defines the actual transmission of information across the physical path. It defines the frame structure, type of frames, link initialization, exchange setup, data and control messages, address structure, and link error recovery. These functions are implemented in the ESCON channel [5] hardware, which interfaces with the fiber optic link hardware.

The ESCON link is a synchronous link with information transferred continually to ensure both transmitter and receive side are in lock step. Data and control information will be sent as required by the server or attached device synchronous to the signal on the line.

14.2.3.1.   Link Protocol

All information on the ESCON link is transferred within a frame structure or a sequence of special characters [4]. The ESCON frame is used to transport control and data information and is structured as shown in Fig. 14.2 [6]. The ESCON frame is delimited by a start of frame (SOF) and end of frame (EOF) ordered set of characters, respectively. The SOF and EOF are unique sets that are also used by the director to establish a connection, continue a connection, or disconnect after completion of the frame transmission. The SOF delimiter is composed of two characters (20 bits). The next 16 bits (before encoding) are reserved for the destination address, the next 16 bits (before encoding) contain the source address, and the next 8 bits (before encoding) are a link control field. The link control field indicates the type and format of the frame. The four fields above are known, as a group, as the link header.

The next field following the header (Fig. 14.2) is the information field, which may contain data or system information and can vary from 0 to 1028 bytes. The link trailer consists of two fields, cyclic-redundancy-check field (CRC) and the EOF field. In order to ensure the data are received correctly,

| SOF | DEST ADDR | SOURCE ADDR | LINK CTL | INFORMATION | CRC | EOF |

| | Header | | | | Trailer | |

SOF :   Two character start-of-frame delimiter.

DEST ADDR :   Two byte destination address of frame.

SOURCE ADDR :   Two byte source address of frame.

LINK CTL :   One byte of link control information.

INFORMATION :   Zero to 1028 bytes of data.

CRC :   Two byte cyclic-redundancy-check information.

EOF :   Three character end-of-frame delimiter.

Fig. 14.2   ESCON frame structure.

a CRC is generated at the transmitter and included in the frame as a 16-bit CRC field. The receiving device uses the CRC to verify the information field. The use of fiber optic technology has ensured that link errors from external stimulus are extremely low and the random bit error rate of the optical link due to receiver noise is less than 10^{-15}. Based on these low error rates, the recovery approach is to retransmit the frame if an error has occurred. Because this happens so seldom, the system performance is not affected by this recovery approach. The EOF field is a three-character (30-bit) field that signifies the end of the frame. The data between the SOF and EOF delimiters are modulo of 8 bits before transmission and encoded into 10-bit characters for transmission on the link.

The architecture also defines an ordered set of sequences that can be transmitted over the link in the presence of a very high error rate condition (in which frames cannot be transmitted correctly). Each sequence contains a continuous repetition of an ordered set to maximize the likelihood that a sequence will be correctly recognized. Some typical sequences are not operational sequence, in which a link-level facility (at the server or device) cannot interpret a received signal, or offline sequence, in which the appropriate link-level facility is indicating that it is off-line with respect to sending any information. These and other sequences are interpreted at a level above the link layer and appropriate action is taken by the server.

An idle character is always sent on the link when no frames or control sequences are being sent. The idle character is a special ordered set of bits (named K28.5) [7]. Also, idles are sent between frames as well. The idle sequence ensures that the receiver is both in bit and character synchronization with the transmitter. If the receiver becomes out of synchronization with the transmitter, the architecture has defined a set of rules and procedures whereby synchronization can be reacquired [7].

14.2.3.2.   Data Encoding/Decoding

High-speed fiber optic receivers perform best over environmental and manufacturing variation when they are AC coupled. The ESCON optical receiver is designed in this manner. In order to prevent DC baseline wander, it is important to ensure that the information on the link is encoded from the normal nonreturn to zero (NRZ) computer code to a DC-balanced code. Several codes (e.g., Manchester and 4B/5B) were investigated and an 8B/10B code was chosen for ESCON. This technique was chosen because it provides the most robust code and a minimum bandwidth overhead (25%). For example, the 8B/10B code contains special control characters that will not degrade into a another valid character with single bit errors.

The 8B/10B encoding transforms a byte (8 bit) of information at a time into a 10-bit transmission character. The 10-bit character is sent serially by bit over the fiber optic link and decoded at the receiver into the original 8-bit byte. Conceptually, the 256-bit combinations of the 8-bit byte are mapped into a subset of the 1024 10-bit characters such that the maximum run length of 1's or 0's is 5. Special control characters and sequences (e.g., idle, SOF, and EOF) are defined that are not derived from the 8-bit original but are meaningful only as architected control and definition characters. For example, the +K28.5 idle character (0011111010) is unique and there is no valid data character with this 10-bit sequence [7]. A single bit error will not result in a valid 10-bit character. Only 536 of the 1024 possible characters are valid. All others will cause an architected error condition.

The running disparity (difference between the number of 1's and 0's in a character) is continually monitored to ensure a DC balance. If disparity exceeds the bounds, an error condition occurs. The 8B/10B code is well behaved with regard to DC balance and the number of transmissions between 0's and 1's is sufficient to ensure that the receiver, retiming, and character recognition circuits can reliably perform the required fuctions.

14.2.3.3. Bit Error Rate Thesholding

The architecture is tolerant of bit errors on the link that may be detected as code violation, sequencing, or CRC errors [8]. A code violation occurs when an invalid transmission character is received. A sequencing error occurs when a sufficient number of consecutive special ordered sets (discussed earlier) cannot be transmitted without error. Finally, a CRC error occurs when the CRC result of the received frame contents is not equal to the expected value.

For the link design, the number of retries due to link errors has a negligible effect on link performance. However, as the rate of retries increases beyond a threshold value the degradation of the link may be noticeable. A report is generated when the specified threshold is reached on a link for further analysis and maintenance. The threshold for ESCON is set at 1 error in 10^{10} bits. At this level the link performance is still tolerable and maintenance can be deferred until a convenient time. Beyond this level, the server will begin to realize degraded performance on that link.

The actual measurement is done by counting the number of code violation events within a specified time. A bit error will likely cause more than one code violation. Consequently the concept of an error burst has been developed. To prevent a single bit error causing multiple error counts, one or multiple code violations within a 1.5-second period are considered as one error burst for the threshold count. Fifteen or more error bursts within a 5-min period will result in a threshold error recorded by the server. The threshold count is reset when the threshold is reached or every 5 min, whichever occurs first. Detailed information is given in Ref. [8].

14.3. ESCON Link Design

The transition of computer interconnection from parallel copper technology to a radically different technology ("serial" fiber optics) generated many questions and concerns. Most of the concerns centered around the reliability of the link in a computer data center environment. Can the technology meet the stringent reliability requirement for both bit errors on the link and hardware failures? The fiber optic link must perform equal to or better than the copper links it replaces. The ESCON link design [9] and component selection were made to achieve both high data rate and reliability.

14.3.1. MULTIMODE DESIGN CONSIDERATIONS

The multimode ESCON link replaced a parallel (8-bit wide) copper coaxial cable link that had proven reliability and performance. Any replacement of the copper link must be easier to use, offer higher data rate and distance performance, be smaller in size, be lighter in weight, and be equal or better in reliability. To achieve those requirements, a technology investigation was initiated to determine the best technology set for a communication link. Fiber optics was the clear choice with both multimode and single-mode fiber. The technology allows for significant future performance enhancements as the need for higher data rates continue. Figure 14.3 depicts the size reduction of the interconnection cable and connector of ESCON compared to the equivalent (two) parallel copper cables and connectors it replaces.

The optical link must extend throughout a campus environment (typically 2 or 3 km) and achieve very reliable data transfer. A optical link BER design of 10^{-15} for the worst-case link (longest length) was chosen.

The optical cable must be easy to handle and install without major complications. The optical cables (jumper cables) that connected to the server units were chosen to be robust, flexible, and contain no metal. The absence of any metal in the cable allowed the link design to take advantage of the inherent electromagnetic compatibility (EMC) of the technology. The optical cable had to have the capability to be pulled under the raised floor,

Fig. 14.3 Parallel copper and ESCON channel cables [9]. (Copyright 1992 by International Business Machines Corporation, reprinted with permission.)

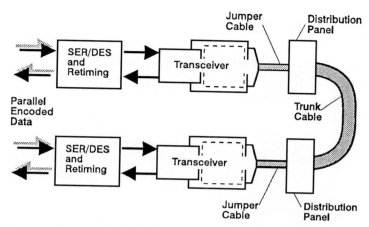

Fig. 14.4   Block diagram of fiber optic link elements [9]. (Copyright 1992 by International Business Machines Corporation, reprinted with permission.)

through ceiling conducts/raceways, etc. and withstand this application environment without any functional or reliability impact. The connector was designed with polarization to ensure the cable could not be misplugged.

14.3.1.1.   Major Components

The major components of the optical link are illustrated in Fig. 14.4. The serializer (typically implemented in CMOS technology) takes the 10 parallel bits of 8B/10B encoded data and serializes the data into a 200 Mb/s rate serial bit stream, whereas the deserializer performs the complementary function. The deserializer also includes the retiming function, which extracts the clock from the serial data. The derived clock is used to latch and reshape the serial data prior to deserialization.

The transceiver contains the electrooptic transducer that converts the electrical signal to an optic signal and vise versa. It also includes the drive and receive electronics that interface to the transducers. The transmitter uses a light-emitting diode (LED) operating at 1300 nm and the receiver uses a positive–intrinsic–negative (PIN) photodiode. Both devices are made of InGaAsP quaternary material. The 1300-nm LED was chosen because this wavelength is at the optimum attenuation and bandwidth of multimode fiber and has excellent reliability and low cost.

For a System/390 CMOS server four channels are implemented on a single printed circuit card as depicted in Fig. 14.5. Individual channels on

Fig. 14.5   An IBM ESCON channel card (courtesy of IBM Corp.).

the card are configured horizontally with the electrooptic transceivers on the right card edge. The CMOS channel logic module is the 25-mm capless module to the left of the transceiver. This module contains the retiming, serializer, deserializer, 8B/10B encoding/decoding, and all channel logic functions associated with the ESCON link.

The jumper cable is a two-fiber (one inbound and one outbound), rugged, yet flexible, cable assembly that uses an aramid fiber strength member. The ESCON connector is a low-profile, polarized, push-on connector that latches into a transmitter receiver subassembly (TRS) or coupler assembly. The fiber used in the jumper is multimode 62.5/125 μm and the ferrules are made of zirconia ceramic material. The ESCON link is designed to be used with either 62.5/125 or 50/125 μm multimode trunk fiber. The use of 62.5/125 μm trunk supports a link length of 3 km, whereas the

50/125 μm trunk fiber supports a 2-km link distance. The difference in distance capability is due to the additional loss associated with connecting a 62.5/125 μm jumper fiber to a 50/125 μm trunk fiber.

14.3.2. SINGLE-MODE DESIGN CONSIDERATIONS

In general, ESCON multimode links satisfy the majority of the distance requirements for a large server complex. However, there has been a significant need to extend the ESCON distance beyond the campus environment. To meet these requirements, it was necessary to develop an alternate technology set that is complementary to the telecommunications infrastructure being established in fiber optics. However, the link signaling, protocol, and architecture remain the same as with the base ESCON link.

The new long-distance ESCON link, called ESCON XDF, uses a long-wavelength laser as the source and single-mode optical fiber (SMF). The single-mode fiber chosen is the same as that used by the telecommunications industry and generally available. This is an important consideration because these long-distances typically will traverse right-of-ways and likely the fiber is owned by another company (e.g., posts, telephone, and telegraphs; local telephone provider; and power company). In general, the computer customer is not interested in fiber optics as an entity but only as a means of efficient communication within his or her network. To ensure ease of use and not require of the customer anything more than the base link requirements, the XDF must be an international class 1 laser safety product. This category allows unrestricted access by uncertified laser personnel because the product conforms with "eye safe" government and industry criteria.

14.3.2.1. Major Components

The single-mode link is very similar in structure to the multimode link illustrated in Fig. 14.4. The major differences include the following:

1. Use of 1300-nm laser source
2. Transceiver uses the Fibre Channel Standard (FCS) duplex SC connector
3. Single-mode fiber jumpers and trunk cables

The single-mode fiber was chosen for its high bandwidth and low loss characteristics. The jumper cables use 9/125 μm fiber, whereas the trunk can use either 9- or 10-μm core fiber. There is no distance penalty associated with the use of 10-μm core trunk fibers.

The XDF feature provides a 20-km link capability at 200 Mb/s without the use of repeaters. The link distance is a function of the optical loss budget and is a trade-off of laser transceiver cost and complexity versus distance. The laser power output is maintained at a low enough power level to ensure compliance with class 1 laser safety standards (see Section 14.9.1).

The laser transceiver discussed in this chapter is a second-generation transceiver that utilizes the single-mode asynchronous transfer mode industry standard module package with the FCS connector. The prior version was a single-mode ESCON connectorized module that is no longer in production. It was designed and produced by IBM because there was no industry product at that time that could meet the requirements of System/390 servers. The new and original laser transceivers are fully compatible and have similar specifications.

14.3.3. OPTICAL LINK DESIGN

The ESCON optical link must perform reliably while allowing maximum user flexibility and performance (data rate and distance). The data rate is fixed at 200 Mb/s and cannot be altered or compromised. The optical link parameters (e.g., transmitter optical coupled power, LED spectral width, and receiver sensitivity) can be optimized to result in the longest possible link using state-of-the-art technology.

The link system design must consider the effects of all the parameters that contribute to an optical loss either directly or indirectly via a bandwidth limitation [10]. The fiber optic cable bandwidth will decrease the achievable link length (for a given data rate) unless this phenomenon is compensated for with an increase in the light signal level. This compensation methodology can be used, generally, for all link parameters provided the individual parameter limitation effects are maintained at modest levels (<3–5 dB). If the effect is very pronounced or extreme, then the use of optical power to compensate for bandwidth effects is very difficult because of the nonlinearity of the correlation function. For example, the slope of the optical signal bandwidth transfer function becomes very steep beyond the classical knee of the curve. At this point an unrealistic power increase is needed to compensate for bandwidth roll-off.

The design of the ESCON link is based on each parameter being represented by an optical loss or an equivalent optical loss (simulation of those parameter effects that are not direct optical loss characterized). Each of the parameters can then be characterized as a statistical parameter element (mean and standard deviation). The individual elements are summed statistically to arrive at a total optical loss. The total optical loss must be less than the optical link power available from the respective transmitter and receiver pair. The statistical summation of the loss elements has the advantage of more precisely representing the link components in the field. It allows the designer to more accurately describe the component elements in the form of distributions and improve focus on cost/performance trade-offs as well as design optimization.

14.3.3.1. Multimode Link Design and Specification

The ESCON link budget elements are grouped into two major categories:

1. Cable plant

    The cable plant loss includes connector loss, fiber attenuation, higher order mode loss, and splices.

2. Available power

    The available power is the resultant optical power available for the link after the optical budget associated with the transmitter and receiver is adjusted for link losses such as

    - Fiber dispersion penalty (modal and spectral)
    - Retiming penalty
    - BER specification conversion from 10^{-12} to 10^{-15}
    - LED end-of-life degradation
    - Transceiver coupling variation
    - Data dependency.

   Link parameters are defined into these categories to allow maximum flexibility over the elements that can be controlled by the user (e.g., fiber attenuation) and incorporate into the available power those elements that are difficult or cannot be controlled (e.g., fiber dispersion) by the user.

   The elements of the available power budget are statistically summed to yield a resultant available power as a distribution with a mean and standard deviation. The following condition must be satisfied for the link to meet its design criteria as follows,

$$U_{\text{av}} - n\sigma_{\text{av}} \geq C_t, \tag{14.1}$$

Table 14.1   **ESCON Maximum Link Loss (at 1300-nm Wavelength)**

Maximum link length (km)	Maximum link loss (dB)	Trunk fiber core size (μm)	Minimum trunk modal bandwidth (MHz/km)
2.0	8.0	62.5	500
2.0	8.0	50.0	800
2.0–3.0	8.0	62.5	800

Note. From Ref. [11]. The maximum link length includes both jumper and trunk cables. The maximum total jumper cable length cannot exceed 244 m when using either 50/125 μm trunk fiber or when a 62.5/125 μm link exceeds 2 km.

where U_{av} is the available power, n is the number of standard deviations, σ_{av} is the standard deviation of the available power, and C_t is the total cable plant optical loss. For an ESCON link $n = 3$ (3 σ design) for the longest link allowed in the configuration at a BER of 10^{-15}. The resultant mean and standard deviation for the available power is determined using a Monte Carlo technique to sum the various elements. This was done because all the parameter distributions are not necessarily Gaussian and in fact the transmitter output power and receiver sensitivity are truncated distributions. The use of a 3 σ design point for the worst-case link (3 km for 62.5-μm trunk and 2 km for 50-μm trunk) ensures that all shorter links are designed conservatively and the risk of an install link budget failure is extremely remote.

Table 14.1 illustrates the resultant specification of the cable plant to ensure the multimode link operates in accordance with the link design requirements. The maximum link loss is established at 8 dB independent of link configuration. The loss budget was maintained at 8 dB by adjusting the fiber bandwidth and in turn the dispersion penalty. The standard 2-km 62.5/125 μm link uses 500 MHz-km fiber, whereas the 2-km 50/125 μm and 3-km 62.5/125 μm link use a higher bandwidth (800 MHz/km) grade of fiber. This allows the customer maximum flexibility to adjust his or her configuration to the environment. The user can trade off number and connector quality with fiber attenuation and length to achieve an optimized installation.

14.3.3.2.   Single-Mode Link Design and Specification

The single-mode link design follows the same approach used for the multimode design. The jumper fiber is 9/125 μm. The XDF link supports both 9- or 10-μm core fiber without any effect on distance. The excess loss

Table 14.2   **ESCON XDF Maximum Link Loss (at 1300-nm Wavelength)**

Maximum link length (km)	Maximum link loss (dB)	Trunk fiber core size (μm)
20.0	14.0	9–10

Note. From Ref. [11]. The maximum link length includes both jumper and trunk cables. The minimum length of a single-mode jumper cable is 4 m. In a single-mode trunk cable, distance between connectors or splices must be sufficient to ensure only the lowest order bound mode propagates.
Single-mode connectors and splices must meet a minimum return loss specification of 28 dB. The minimum return loss of a single-mode link must be 13.7 dB.

(approximately 0.2 dB) associated with the coupling of a 9-μm core jumper to a 10-μm core trunk fiber is included in the available power category and transparent to the overall link budget. The dispersion penalty of the fiber due to spectral width of the laser is small and has also been accounted for in the available power budget along with any effects due to laser mode hopping and relative intensity noise [10]. All these time domain effects are relatively small for a 200 Mb/s single-mode link and are included as a 1.5-dB fixed (no distribution) "AC optical path" penalty. The single-mode link specification is given in Table 14.2. A maximum link length of 20 km can be achieved with a maximum optical cable plant loss budget of 14 dB for the cable plant.

In order to ensure the laser is well behaved under all operating conditions it is important to minimize any optical reflections occurring in the cable plant. This is done by specifying that all connections and splices in the link have a minimum return loss of 28 dB. Mode partition noise in the XDF link is alleviated by specifying that no jumper less than 4 m may be used. The minimum length in conjunction with the specified cutoff wavelength of the fiber ensures only the lowest order bound mode propagates in the jumper. Likewise, the trunk installer must ensure any connectors or splices in the trunk meet the return loss specification and all connectors or splices are placed sufficiently apart to ensure only the lowest order mode is propagating prior to any connectors, splices, or other optical discontinuities.

14.4. Multimode Physical Layer

The multimode physical layer [11] defines a common set of specifications at the device interface that ensures compatibility between the link transmitter and receiver through a defined cable plant. The physical layer provides a design point that allows any server or attached device manufacturer to

Fig. 14.6   Bidirectional fiber optic ESCON link [12].

build an ESCON-compatible link. Transceivers by multiple manufacturers can be intermixed on a link and still conform to the ESCON performance requirements.

The ESCON link is composed of two unidirectional point-to-point links as shown in Fig. 14.6. Information (light pulses) is transferred from the server to a device over one fiber and from the device to the server over the other fiber. Both fibers are encased in a common cable assembly. The link is full duplex — both links may be active simultaneously.

The multimode physical interface specifies the optical and mechanical requirements of the optical output and input interface to the jumper cable and the performance requirements of the cable plant. Data are transmitted over the link at a transmission rate of 200 ± 0.04 Mb/s. The encoding and link protocols were discussed in Section 14.2. A "1" bit corresponds to light on and a "0" bit to light off. All specifications are for worst-case operating conditions, including end-of-life effects.

14.4.1.   MULTIMODE OUTPUT OPTICAL INTERFACE

The optical coupled light specifications required for an ESCON link are given in Table 14.3. The parameters specified will allow the maximum distance requirements and loss budget, as specified in Table 14.1 with a BER of 10^{-15}. The light source is an incoherent light-emitting diode.

14.4.2.   MULTIMODE INPUT OPTICAL INTERFACE

The input optical interface specifications are given in Table 14.4. A loss-of-light function and operation is specified for link failure indication and diagnostic use. The design of the machine receiving this information determines how this state change information is utilized.

Table 14.3   **Multimode Optical Output Interface Specifications**

Parameter	Minimum	Maximum	Unit
Average power [a,b]	−20.5	−15.0	dBm
Center wavelength	1280	1380	nm
Spectral width (FWHM)		175.0	nm
Rise time (t_r) (20–80%)[a,c]		1.7	ns
Fall time (t_f) (80–20%)[a,c]		1.7	ns
Eye window[a]	3.4		ns
Optical output jitter [d]		0.8	ns
Extinction ratio[a,e]	8		dB
t_r, t_f at optical path output[c,f]		2.8	ns

Note. From Ref. [11].

[a] Based on any valid 8B/10B code. The length of jumper cable between the output interface and the instrumentation is 3 m.

[b] The output power shall be greater than −29 dBm through a worst-case link as specified in Table 14.1. Higher order mode loss (HOML) is the difference in link loss measured using the device transmitter compared to the loss measured with a source conditioned to achieve an equilibrium mode distribution in the fiber. The transmitter shall compensate for any excess HOML occurring in the link (e.g., HOML in excess of 1 dB for a 62.5-μm link).

[c] The minimum frequency response bandwidth range of the optical waveform detector shall be 100 kHz to 1 GHz.

[d] The optical output jitter includes both deterministic and random jitter. It is defined as the peak-to-peak time-histogram oscilloscope value (minimum of 3000 samples) using a 2^7-1 pseudo-random pattern or worst-case 8B/10B code pattern. The transmitter output light is coupled to a PIN photodiode O/E converter (e.g., Tektronix P6703A or equivalent) via a 3-m cable and jitter measured with a digital sampling oscilloscope [13].

[e] Measurement shall be made with a DC-coupled optical waveform detector that has a minimum bandwidth of 600 MHz and whose gain flatness and linearity over the range of optical power being measured provide an accurate measurement of the high and low optical power levels.

[f] The maximum rise or fall time (from, e.g., chromatic, modal dispersion, etc.) at the output of a worst-case link as specified in Table 14.1. The 0 and 100% levels are set where the optical signal has at least 10 ns to settle. The spectral width of the transmitter shall be controlled to meet this specification.

14.4.3. *MULTIMODE FIBER OPTIC CABLE SPECIFICATION*

The two optical fibers are assembled into a duplex optical cable assembly for the jumper and assembled into pairs for the trunk. The jumper cable assembly is terminated in the ESCON duplex fiber optic connector. The trunk cable, however, is usually installed in high-count configurations (e.g., 12, 24, 36, 72, and 144 fiber counts) by professionals skilled in the art of fiber optic installation. The planning and installation of the trunk is reviewed in Section 14.6. The two fibers in a jumper cable are assembled as illustrated in the cable cross section (Fig. 14.7). The cable assembly is nonmetallic

Table 14.4   **Multimode Optical Input Interface Specifications**

Parameter	Minimum	Maximum	Unit
Sensitivity[a,b]		−29.0	dBm
Saturation level[a]	−14.0		dBm
Acquisition time[c]		100	ns
LOL threshold[d]	−45	−36	dB
LOL hysteresis[d,e]	0.5		dB
Reaction time for LOL state change	3	500	μs

Note. From Ref. [11].

[a] Based on any valid 8B/10B code pattern measured at, or extrapolated to, 10^{-15} BER measured at center of eye. This specification shall be met with worst-case conditions as specified in Table 14.3 for the output interface and Table 14.1 for the fiber optic link. This value allows for a 0.5-dB retiming penalty.

[b] A minimum receiver output eye opening of 1.4 ns at 10^{-12} should be achieved with a penalty not exceeding 1 dB.

[c] The acquisition time is the time to reach synchronization after the removal of the condition that caused the loss of synchronization. The pattern sent for synchronization is either the idle character or an alternating sequence of idle and data characters.

[d] In direction of decreasing power: If power > −36 dBm, LOL state is inactive; if power < −45 dBm, LOL state is active. In direction of increasing power: If power < −44.5 dBm, LOL state is active; if power > −35.5 dBm, LOL state is inactive.

[e] Required to avoid random transitions between LOL being active and inactive when input power is near threshold level.

and uses aramid fiber as the strength member. All the elements are encased in a flexible polyvinyl chloride (PVC) jacket.

The optical specifications in this section are primarily associated with the fiber and are necessary to ensure the link meets its performance objectives. They also ensure consistency among various ESCON-compatible devices.

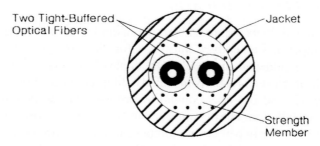

Fig. 14.7   Multimode jumper cable construction [12].

Table 14.5 **Multimode (62.5/125 μm) Jumper Cable Specifications**

Parameter	Specification
Fiber type	Graded index with glass core and cladding
Operating wavelength	1300 nm
Core diameter[a]	62.5 ± 3.0 μm
Cladding diameter[b]	125 ± 3.0 μm
Numerical aperture[c]	0.275 ± 0.015
Minimum modal bandwidth[d]	500 MHz-km
Attenuation	1.75 dB/km at 1300 nm (maximum)

Note. From Ref. [11].
[a] Measured in accordance with EIA 455 FOTP 58, 164, 167, or equivalent.
[b] Measured in accordance with EIA 455 FOTP 27, 45, 48, or equivalent.
[c] Measured in accordance with EIA 455 FOTP 47 or equivalent.
[d] Measured in accordance with EIA 455 FOTP 51 or equivalent.

14.4.3.1. Multimode Jumper Cable Assembly

The MMF jumper cable is only offered in a 62.5/125 μm fiber configuration and the optical specifications are given in Table 14.5. The cable jacket color is orange. All the parameters are specified and measured in accordance with the applicable industry standards as indicated.

14.4.3.2. Multimode Trunk Fiber Specification

Two multimode fiber types are supported for the trunk. The required optical parameters of both trunk fibers are specified in Table 14.6. Both fiber types conform to applicable European and United States industry standards [14–16]. All fiber parameters are specified and measured in accordance with the applicable industry standards as indicated.

14.4.4. *ESCON CONNECTOR (MULTIMODE)*

The ESCON connector (illustrated in Fig. 14.8) is a ruggedized, two-ferrule connector that is polarized to prevent misplugging. The polarization is accomplished by beveling two corners of the connector as shown in Fig. 14.8. The connector is a push-on type that is retained in the receptacle by two latches. The ferrules are protected from handling and dirt by a protective spring-loaded cap that retracts upon insertion in a receptacle, allowing the ferrules to be inserted into alignment sleeves. The ferrules are made of

Table 14.6 **Multimode Trunk Fiber Specifications**

Parameter	Specification
	62.5/125 μm multimode fiber
Fiber type	Graded index with glass core and cladding
Operating wavelength	1300 nm
Core diameter[a]	62.5 ± 3.0 μm
Core noncircularity	6% maximum
Cladding diameter[b]	125 ± 3.0 μm
Cladding noncircularity	2% maximum
Core and cladding offset	3.0 μm maximum
Numerical aperture[c]	0.275 ± 0.015
Minimum modal bandwidth[d]	500 MHz-km at ≤2 km
	800 MHz-km at >2 km and ≤3 km
Attenuation[e]	1.0 dB/km at 1300 nm
	50/125 μm multimode fiber
Fiber type	Graded index with glass core and cladding
Operating wavelength	1300 nm
Core diameter[a]	50 ± 3.0 μm
Core noncircularity	6% maximum
Cladding diameter[b]	125 ± 3.0 μm
Cladding noncircularity	2% maximum
Core and cladding offset	3.0 μm maximum
Numerical aperture[c]	0.200 ± 0.015
Minimum modal bandwidth	800 MHz-km
Attenuation[e]	0.9 dB/km at 1300 nm

Note. From Ref. [11].
[a] Measured in accordance with EIA 455 FOTP 58, 164, 167, or equivalent.
[b] Measured in accordance with EIA 455 FOTP 27, 45, 48, or equivalent.
[c] Measured in accordance with EIA 455 FOTP 47 or equivalent.
[d] Measured in accordance with EIA 455 FOTP 51 or equivalent.
[e] The attenuation is a typical value rather than a specification. Table 14.1 is the specification.

high-precision, stabilized zirconia ceramic. The ferrule ensures accurate alignment of the fiber to an LED, photodiode (PD), or fiber receptacle. The ends of the ceramic are polished into a convex shape to ensure that glass physical contact is always achieved in fiber-to-fiber connections.

Fig. 14.8 ESCON multimode duplex connector [12].

This feature ensures low loss connections (typically less than 0.4 dB) and high return loss from the connection interface.

The color of the multimode connector is black. The connector also has bend radius-limiting boot at the connection of the fiber cable. This ensures that excessive bend stresses cannot be applied to the fiber at the connector.

14.4.4.1.   Multimode Duplex Receptacle Specification

The duplex receptacle for the ESCON multimode connector is specified in detail in Fig. 14.9. All the necessary dimensions and specifications are illustrated to build a transceiver or fiber (coupler) receptacle while maintaining full functional and dimensional compliance with the ESCON multimode connector.

14.5.   Single-Mode Physical Layer

The single-mode physical layer follows the same approach as the multimode physical layer and defines a common set of specifications at the device interface that ensure interoperability [11]. The ESCON XDF link is composed of two simplex, point-to-point counterdirectional links encased in a duplex configuration. Both the optical and mechanical properties of the interface are specified as well as the cable plant.

The physical layer defines a set of specifications that ensure link performance up to a 20-km distance, without retransmission, using dispersion unshifted single-mode fiber. The data rate specification of 200 ± 0.04 Mb/s is the same as multimode. A "1" bit corresponds to a light on condition.

DIMENSIONS ARE IN MILLIMETERS.

Tx AND Rx DENOTE TRANSMITTER AND RECEIVER PORTS RESPECTIVELY.

1. ALIGNMENT BORE TO ACCEPT 2.498/2.499 DIAMETER FERRULE.

2. PLUG FERRULE END SHALL SEAT TO THE OPTICAL REFERENCE PLANE WITH A STATIC FORCE OF 7.1 TO 14 N (1.6 TO 3.1 POUNDS) PER FERRULE.

3. FERRULE MATERIAL SHALL BE ZIRCONIA WITH A MAXIMUM SURFACE RUGHNESS OF 0.05 μm.

4. DATUM TARGETS SHOWN APPLY TO TOP AND BOTTOM SURFACES. THEY ARE TO BE USED IN PAIRS TO ESTABLISH CENTERLINES.

5. THE MODULE BORE MUST BE ABLE TO WITHSTAND A FORCE OF 2.5 N (0.55 POUNDS) APPLIED PERPENDICULAR TO THE OPENING AT THE OUTERMOST CONTACT POINT OF THE SLEEVE.

6. AS SPECIFIED FOR A SOLID MODULE SLEEVE. FOR A SPLIT SLEEVE,

7. BLEND/TAPER BORE ENTRANCE.

Fig. 14.9 Diagram of the multimode duplex receptacle [11].

All specifications are for worst-case operating conditions, including end-of-life conditions.

14.5.1. SINGLE-MODE OUTPUT OPTICAL INTERFACE

The parameter specifications in Table 14.7 define the optical output interface for the light coupled into the single-mode fiber. The parameter specifications ensure the BER does not exceed 10^{-15}. The light source is a 1300-nm semiconductor laser.

Table 14.7   **Single-Mode Optical Output Interface Specifications**

Parameter	Minimum	Maximum	Unit
Average power into SMF[a]	−8.0	−3.0	dBm
Center wavelength[a]	1261	1360	nm
Spectral width (rms)		7.7	nm
Rise time (t_r) (20–80%)[a,b]		1.5	ns
Fall time (t_f) (80–20%)[a,b]		1.5	ns
Eye window[a]	3.5		ns
Optical output jitter[c]		0.8	ns
Extinction ratio[a,d]	8		dB
Relative intensity noise (RIN_{12})[e]		−112.0	dB/Hz
t_r, t_f at optical path output[c,f]		2.8	ns

Note. From Ref. [11].

[a] Based on any valid 8B/10B code pattern, the measurement is made using at 4-m single-mode jumper cable and only includes the power in the lowest order fundamental mode.

[b] The minimum frequency response bandwidth range of the optical waveform detector shall be 100 kHz to 1 GHz.

[c] The optical output jitter includes both deterministic and random jitter. It is defined as the peak-to-peak time-histogram oscilloscope value (minimum of 3000 samples) using a 2^7-1 pseudo-random pattern or worst-case 8B/10B code pattern. The transmitter output light is coupled to a PIN photodiode O/E converter (e.g., Tektronix P6703A or equivalent) via a 3-m cable and jitter is measured with a digital sampling oscilloscope [13].

[d] Measurement shall be made with a DC-coupled optical waveform detector that has a minimum bandwidth of 600 MHz and whose gain flatness and linearity over the range of the high and low optical power levels.

[e] The relative intensity noise is measured with a 12-dB optical return loss into the output interface.

[f] The maximum degradation in input interface sensitivity (from, e.g., jitter, mode hopping, and intersymbol interference) that can occur by using a worst-case link as specified in Table 14.2. The spectral width of the transmitter shall be controlled to meet this specification.

14.5.2.   SINGLE-MODE INPUT OPTICAL INTERFACE

The single-mode optical input interface requirements are given in Table 14.8. A loss-of-light function and operation is specified for link failure indication and diagnostic use. The design of the machine receiving this information determines how this state change information is utilized.

14.5.3.   SINGLE-MODE FIBER OPTIC CABLE SPECIFICATION

The ESCON XDF link is similar to the multimode link. It is composed of two unidirectional links assembled in a common cable and connector

Table 14.8   **Single-Mode Optical Input Interface Specifications**

Parameter	Minimum	Maximum	Unit
Sensitivity[a,b]		−28.0	dBm
Saturation level[a]	−3.0		dBm
Return loss[c]	12.5		dB
Acquisition time[c]		100	ns
LOL threshold[d]	−45	−32	dB
LOL hysteresis[e]	1.5		dB
Reaction time for LOL state change	0.25	5000	μs

Note. From Ref. [11].

[a] Based on any valid 8B/10B code pattern measured at, or extrapolated to, 10^{-15} BER measured at center of eye. This specification shall be met with worst-case conditions as specified in Table 14.7 for the output interface and Table 14.2 for the fiber optic link. This value allows for a 0.5-dB retiming penalty.

[b] A minimum receiver output eye opening of 1.4 ns at 10^{-12} should be achieved with a penalty not exceeding 1 dB.

[c] The measurement is made using a 4-m single-mode jumper cable and only includes the power in the lowest order fundamental mode.

[d] The acquisition time is the time to reach synchronization after the removal of the condition that caused the loss of synchronization. The pattern sent for synchronization is either the idle character or an alternating sequence of idle and data characters.

[e] Required to avoid random transitions between LOL being active and inactive when input power is near threshold level.

housing. The cable is terminated in a polarized duplex connector. The trunk cable is usually installed by professionals in high-count cables and physically distributed in distribution panels (see Section 14.6).

14.5.3.1.   Single-Mode Jumper Cable Assembly

The SMF jumper cable assembly is a second-generation product that conforms to the ANSI Fibre Channel Standard [17]. The cable assembly is nonmetallic and uses aramid fiber as a strength member. The cable construction is illustrated in Fig. 14.10.

Fig. 14.10   Single-mode jumper cable construction [12].

Table 14.9   **Single-Mode Jumper Cable Specifications**

Parameter	Specification
Fiber type	Dispersion unshifted
Operating wavelength	1261–1360 nm
Mode field diameter [a]	$9.0 \pm 1.0 \mu$m
Zero dispersion wavelength [b]	1310 ± 10 nm
Dispersion [b]	6.0 ps/nm-km maximum
Cutoff wavelength (λ_c) [c]	1260 nm maximum
Attenuation [d]	0.8 dB/km maximum

Note. From Ref. [11].
[a] Measured in accordance with EIA 455 FOTP 164, 167, or equivalent.
[b] Measured in accordance with EIA 455 FOTP 168 or equivalent.
[c] Measured in accordance with EIA 455 FOTP 80 or equivalent.
[d] Measured in accordance with EIA 455 FOTP 78 or equivalent.

The single-mode jumper is only offered in 9/125 μm nondispersion-shifted fiber with the optical and mechanical specifications given in Table 14.9. The color of the cable jacket is yellow. All parameters are specified and measured in accordance with the applicable industry standards as indicated.

14.5.3.2. Single-Mode Trunk Fiber Specification

The allowable single-mode trunk fiber optical specifications are given in Table 14.10. The nominal mode field diameter of the fiber can vary from 9 to 10 μm to account for all the standard single-mode fiber types. The specifications are consistent with the applicable European and United States industry standards [14, 18, 19]. All fiber parameters are specified and measured in accordance with the applicable industry standards as indicated.

14.5.4. ESCON XDF DUPLEX CONNECTOR

The single-mode ESCON XDF connector is the same as the FCS SC push-on, polarized (via keying) duplex fiber optic connector (Fig. 14.11) [17]. The ferrules are also made of zirconia ceramic. The ends of the ferrules are polished to ensure physical contact when mated to another connector. For the single-mode version, one key is narrower than the other. It is

Table 14.10   **Single-Mode Trunk Fiber Specifications**

Parameter	Specification
Fiber type	Dispersion unshifted
Operating wavelength	1261–1360 nm
Mode field diameter[a]	9.0–10.0 ± 10%
Core concentricity error[b]	1 μm maximum
Cladding diameter[b]	125 ± 2.0 μm
Cladding noncircularity[b]	2% maximum
Zero dispersion wavelength[c]	1295–1322 nm (1310 nm nominal)
Zero dispersion slope[c]	0.095 ps/nm^2-km maximum
Cutoff wavelength (λ_c)[d]	1280 nm maximum
Cutoff wavelength (λ_{cc})[e]	1260 nm maximum
Attenuation above nominal[f,g]	0.6 dB/km maximum
Attenuation (nominal)[g]	0.5 dB/km at 1310 nm

Note. From Ref. [11].

[a] Measured in accordance with EIA 455 FOTP 164, 167, or equivalent.

[b] Measured in accordance with EIA 455 FOTP 45, 48, or equivalent.

[c] Measured in accordance with EIA 455 FOTP 168 or equivalent.

[d] Measured in accordance with EIA 455 FOTP 80 or equivalent.

[e] Measured in accordance with EIA 455 FOTP 170 or equivalent.

[f] The maximum attenuation for wavelengths from 1261 to 1360 nm shall not exceed the attenuation at 1310 nm by more than 0.06 dB/km. Typically, this specification can be met by fiber with 1383-nm water absorption peaks below 2 dB/km.

[g] The attenuation is a typical value rather than a specification. Table 14.2 is the specification.

Fig. 14.11   ESCON single-mode duplex connector [12].

mechanically retained in the duplex receptacle by latch arms that engage the connector upon plugging. Each fiber has its own subassembly that is compatible with the industry standard SC connector. The mating, external dimension, and interface requirements of the connector conform to the ANSI Fibre Channel Standard [17]. In order to improve usability, the connector housing is gray in color. This helps to differentiate the single-mode link from the multimode link. The connector also uses bend radius-limiting boots to minimize bending stress at the cable/connector interface.

14.5.4.1. Single-Mode Duplex Receptacle Specifications

The duplex receptacle for the XDF single-mode connector is specified in detail in Fig. 14.12. The specification is the same as specified in the Fibre Channel Standard FC-PH document. Any single-mode connector that is compliant with FCS will interoperate with the XDF single-mode receptacle.

14.6. Planning and Installation of an ESCON Link

This section discusses the planning and installation considerations for fiber optic cabling in a large data processing environment [20]. The size, layout, and number of buildings along with the physical location of the equipment to be interconnected determine the cable plant layout. The equipment may be located

- in a single floor within a building;
- or multiple floors or in multiple rooms within a building;
- between two or more buildings within a campus; or
- between two or more campuses.

The distance between ESCON-compatible equipment may vary from 4 m to 20 km and beyond when equipment (e.g., director) is used to retransmit the optical signal.

Fiber optic cable planning is very similar to planning for a copper cable plant. The cables should be installed in troughs or conduits, which may be run in the ceiling or under the floor. For a small installation, where the quantity of equipment is small and the equipment pieces are located close to each other, the cable plant will likely consist of jumper cables only.

DIMENSIONS ARE IN MILLIMETERS

(1) ALIGNMENT BORE TO ACCEPT 2.499/2.4985 DIAMETER FERRULE.

2. PLUG FERRULE END SHALL SEAT TO THE OPTICAL REFERENCE PLANE WITH A STATIC FORCE OF 6.7 TO 13.1 N PER FERRULE.

3. FERRULE MATERIAL SHALL BE ZIRCONIA WITH A MAXIMUM SURFACE ROUGHNESS 0.05 μ m.

Fig. 14.12   Diagram of ESCON XDF duplex receptacle [17].

In a more complex configuration, where equipment is numerous and widely dispersed, the cable plant will consist of a network of distribution panels and trunk cables. It is generally recommended to place the distribution panel at several strategic locations, which minimize the length of the jumper cables. The distribution panels are interconnected with large-count (12, 24, 36, 72, or 144 fibers) trunk cables. This is the more common configuration and requires a significant investment in the procurement and installation of the cable plant. The trunk cables are run throughout the buildings and between buildings. Generally, the planning, purchase, and installation of the cable plant is handled by professionals under contract from the end user. For example, IBM Networking Services will provide a total turnkey design and installation of the cable plant (see Section 14.6.5).

In general, the cable plant is a considerable investment and should be designed with the following considerations:

- Potential future reconfiguration, expansion, and growth
- Future applications
- Local laws and ordinances that would affect the type of cable (e.g., required use of plenum or halogen-free cables in place)
- Adherence to all link specifications with margin for growth (if possible)
- The number and size of the cables required
- Obtaining right-of-way access if cables are to be installed through or over property owned by others
- Potential data rate or dark fiber services that may be available for right-of-way crossings

14.6.1. CABLE PLANT CONFIGURATION CONSIDERATIONS

As mentioned earlier, the physical location of present and future (if possible) equipment is a key factor in determining an optimized cable plant. Establish an equipment location layout of all the equipment to be interconnected. Include any right-of-way requirements in the layout. The layout should also include the number of links to each equipment, to and from locations, and potential future locations. Based-on this layout, a judicious choice of trunk cables, fiber counts, and distribution panel locations can be established.

14.6.1.1. Link Requirements

After determining the optimum link configuration:

1. Use a floor plan to assess the possible link paths, and determine the length and type of each link (for example, jumper only or jumper and trunk, and multimode or single-mode) and identify and locate each connector, splice, and distribution panel.
2. Ensure that link specifications agree with the fiber cable types, calculated link loss is within specifications (see Section 14.7), and links conform to national, state, and local building codes.

3. Establish a labeling/identification technique for ease of installation and future cable management.

14.6.1.2. Installation Planning

Before installing a link, consider the following:

1. Cable routing diagrams
   • Location and length of each link
   • Type, location, and identification of connectors, adapters, and couplers
   • Locations of splices and distribution panels
2. Manufacturer data sheets
   • Cables
   • Bend radius control
   • Strain relief
   • Splices
   • Distribution panels
   • Attached equipment

The link must be compatible with the equipment that attaches to it. Section 14.6.2 contains cable physical specifications required to support ESCON-capable equipment.

14.6.1.3. Trunk Cables

The trunk cable should meet all the requirements in Sections 14.4.3.2 and 14.5.3.2. The connectors used to terminate the trunk should be of high quality, physical contact, and low loss. Industry standard connectors (e.g., ST and FC/PC) meet these requirements and are readily available. Both types are available for multimode and single-mode fiber. Splices are often part of the trunk cable (particularly for lengths greater than 2 km). They should also be high quality, low loss, and high return loss (>28 dB).

Consideration should be given to adding extra fibers in the trunk cable to allow for future growth and expansion and/or additional conduits for the addition of future trunks. In those cases in which the trunk cable communication link is critical to the operation of the data processing complex, and right-of-ways are traversed, an alternate trunk cable route is strongly suggested to ensure uninterrupted service. System/390 servers can be set up to automatically reroute traffic if a trunk cable fails for whatever reason.

14.6.1.4. Distribution Panels

The jumper cable and trunk cable are connected at the distribution panel. In the distribution panel, the trunk cable is split into "two-fiber" groups, connectorized, and connected to an ESCON coupler (e.g., trunk cable terminated with ST connectors and connected to an ESCON/ST coupler). Distribution panels should be located to provide the following [20]:

- An easy entry, exit, and connection point for trunk and jumper cables
- A concentration point for jumper cables to allow manual I/O reconfiguration
- A suitable access point to service an area of processors and equipment

Because of fiber optic channel link distance capabilities, panels can also be located away from the processors and attached equipment, thereby saving floor space.

Distribution panels should contain an adapter and coupler panel, a splice tray, and a storage area for jumper cable slack [20]. The panel should also provide strain relief and radius control for jumper cables:

- Adapter and coupler panels provide the hardware to join identical (coupler) or different (adapter) cable connectors. They can be used separately or in combination within the panel.
- Splice trays should have the same cable capacity as the adapter and coupler panels and should be equipped with hardware to provide strain relief for all cables entering and exiting the tray.
- Jumper cable slack storage provides adequate organizing hardware and storage space for excess jumper cable. The hardware should allow identification and removal of an individual jumper cable, without having to disconnect or remove other jumper cables within the panel.
- Jumper cable strain relief and radius control provides hardware that prevents damage to the jumper cables caused by excessive bending or pulling of the cables.

14.6.2.   CABLE ENVIRONMENTAL REQUIREMENTS

In many instances the fiber optic cables will be routed in troughs in the ceiling or under raised floors along with copper cables. The fiber cables are very durable, rugged, and will withstand significant stress and strain due to rough handling. Nevertheless, it is good practice not to intermingle fiber optic jumper cables and large copper cables (e.g., System/390 bus and tag cables). Trunk cables in general are designed for extremely rugged environments and would be the recommended cable for outdoor, underground, and harsh building environments. When cables are routed properly and the cable environment is organized, no additional fiber cable protection is required.

As a guide to determine the suitability of the jumper cable for the intended environment, the specification, the specifications in Tables 14.11 and 14.12 for multimode and single-mode jumper cables, respectively, govern. Each data processing complex must be carefully analyzed to determine the amount of cable activity or disruption that could occur under the raised floor. If that activity creates a significant exposure to the fiber cables, do any or all of the following:

- Consider installing a direct attached trunk system (e.g., IBM's Fiber Transport System, described in Section 14.6.5).
- Educate personnel about proper installation and handling procedures for fiber cables.
- Bundle the fiber cables. For example, use tie wraps to reduce the possibility of damage.
- Route the fiber cables to minimize the exposure caused by sharp edges, excessive bends, or congestion caused by bus and tag cables.

Also, fiber cables can be installed in conduits, tubings, raceways, troughs, or other protective devices.

14.6.3.   LINK CONTINUITY

Light propagation occurs in a single direction within a fiber. In an ESCON jumper cable (for either multimode or single mode), light enters the "B"-labeled port and exits the "A"-labeled port of the cable assembly (Fig. 14.13). The transceiver output is always coupled to the B port and the transceiver

Table 14.11   **Multimode Jumper Physical and Environmental Specifications**

Parameter	Specification
Physical	
Jacket material	PVC
Jacket outside diameter	4.8 mm
Weight (for information only)[a]	20 g per meter
Installation tensile strength	1000 N maximum
Minimum bend radius (during installation)	4.0 mm; 5 s maximum at 400 N
Minimum installed bend radius	
No load	12 mm
Long-term residual	25 mm at 89 N
Flammability	Underwriters laboratory UL-1666 ONFR (Optical Fiber Nonconductive; Plenum UL-910 is also acceptable)
Crush resistance	500 N/cm maximum
Maximum unsupported vertical rise	100 m
Environmental	
Operating environment	Inside buildings only
Operating temperature	10–60°C
Operating relative humidity	5–95%
Storage and shipping	−40 to 60°C
Lightning protection	None required
Grounding	None required

Note. From Ref. [21].
[a] Cable only; connectors not included.

input is always coupled to the A port. The fiber crossover is designed into the jumper cable assembly and, by virtue of the polarized connector, the transceiver output is always connected to the corresponding transceiver input at the other end of the cable.

The design of the trunk cable plant must be done to account for the fiber crossover. Only an odd number of crossovers will result in a viable link. In general, a link contains two jumper cables, which requires the trunk fiber to be crossed in order to ensure link continuity. This configuration is illustrated in Fig. 14.14.

Table 14.12   **Single-Mode Jumper Physical and Environmental Specifications**

Parameter	Specification
Physical	
Jacket material	PVC
Jacket outside diameter	Zip cord
Weight (for information only)[a]	20 g per meter
Installation tensile strength	1000 N maximum
Minimum bend radius (during installation)	50 mm; 5 s maximum at 100 N
Minimum installed bend radius	
No load	30 mm
Long-term residual	50 mm at 80 N
Flammability	Underwriters laboratory UL-1666 ONFR (Optical Fiber Nonconductive; Plenum UL-910 is also acceptable)
Crush resistance	889 N/cm maximum
Maximum unsupported vertical rise	100 m
Environmental	
Operating environment	Inside buildings only
Operating temperature	0–60°C
Operating relative humidity	8–95%
Storage and shipping	−40 to 60°C
Lightning protection	None required
Grounding	None required

Note. From Ref. [21].

[a] Cable only; connectors not included.

Fig. 14.13   Direction of light propagation in an IBM jumper cable (multimode depicted) [20].

Fig. 14.14   Direction of light propagation in a link with a trunk [20].

14.6.4.   CAMPUS LAYOUT EXAMPLE

The physical location and quantity of equipment are the primary factors affecting the cable layout. The application environment may vary from a single-room configuration with limited equipment to a multicampus environment with multiple buildings and floors within a building.

The example illustrated in Fig. 14.15 is a single campus environment with data processing equipment in two buildings. The point of egress of each building is at a "building interface panel." This panel provides a common access point for each building. The trunk cable (multimode or single mode) between the buildings will be determined by the actual "run" distance of the intended cable. In general, single mode is chosen for distance greater than 3 km. The trunk cable should be a rugged outdoor cable type that can be installed in an underground conduit (preferred) or strung on utility poles. Extra count trunks should be considered for expansion.

Within a building, distribution panels are optimally located. Both building trunks and jumpers are used to interconnect distribution panels. In general, jumpers connecting to distribution panels should be kept as short as possible. Distribution panels should be optimally located with multifiber trunks interconnecting the distribution panels.

The flexibility and congestion relief inherent with distribution panels requires additional termination hardware and increased link loss due to the inclusion of additional connections and splices. This must be considered in the loss budget analysis (Section 14.7) to ensure the loss budget requirements are not violated.

If the two buildings were located on different campuses, then right-of-way access is required from the owner of the land. Leasing of "dark fiber"

Fig. 14.15   Example of a single campus link environment between two buildings [20].

may also be a possibility. The negotiation, however, for right-of-way, leasing, etc. may greatly extend the total planning and installation time. This must be considered in the overall project schedule.

For distance requirements beyond 20 km (XDF feature maximum), channel extenders may be used such as the IBM 9036 ESCON Remote Channel Extender or the IBM 9729 Fiber Optic Multiplexer. The IBM 9036 is a repeater that retransmits the optical signal to achieve an additional 20-km distance. The Fiber Optic Multiplexer utilizes wavelength division multiplexing to transmit up to 10 individual ESCON channels on a single pair of optical fibers. Each channel is converted to a different wavelength and transmitted simultaneously on the fiber. Distances of 50 km can be achieved with the IBM 9729.

14.6.5. DIRECT ATTACH TRUNKS

For large installations, in which the amount of equipment in a data processing complex is large, the use of direct attach trunk cables has become popular [20]. Multifiber ribbon (12 fibers per ribbon) trunk cables are directly attached to an IBM eServer zSeries as illustrated in Fig. 14.16. For large ESCON channel count servers, direct-attached trunks offer the least congested cabling approach.

Fig. 14.16   IBM's direct-attach trunk system in a eServer zSeries [20].

The trunk may contain as many as 12 ribbons (144 fibers total). Each ribbon is termined in a 12-fiber multifiber terminated push-on (MTP) connector. The small size of the connectors easily enables six MTPs to be terminated inside the server frame in a coupler bracket. Several brackets can be installed in a machine. A harness is used inside the machine to convert the 12 fibers (representing six ESCON channels) of a ribbon to six ESCON connectors for insertion into the appropriate transceiver.

Once the harnesses are installed in the machine, all fiber connects and disconnects can be done using the trunk cable MTP connectors. The harnesses remain inside the machine, plugged into the individual ESCON or coupling facility ports, whereas the trunk cables can be quickly removed. The machine can then be relocated within the data complex because the trunk cables are easily rerouted to the new location and replugged into the harnesses. This greatly reduces the time spent unplugging, rerouting, and plugging the individual fiber optic jumper cables back into the machine ports.

MTP-to-MTP multiribbon trunks can be used throughout the complex to directly connect equipment together as well as equipment to distribution panels. For these installations the use of jumper cables is minimized. This cabling approach is used for both multimode and single-mode versions.

IBM Availability Services offers this approach as Fiber Transport System III (FTS-III) and is also available to do the planning, design, and installation of this interconnection system.

14.7.   Loss Budget Analysis

The cable plant loss (C_t) of an ESCON link is a critical parameter and is determined by the link length, fiber attenuation, number of connections and jumper cable loss, jumper/trunk core diameter mismatch splices, etc. [22]. These parameters are greatly influenced by the cable plant layout and quality of components used. As discussed in Section 14.3.3, the required optical cable plant loss is established by the available power as determined by Eq. (14.1). The cable plant optical link loss must be less than or equal to the available power to ensure the link design performance. This is true for both multimode and single-mode links.

14.7.1.   MULTIMODE CABLE PLANT LINK LOSS

The multimode link loss is determined by statistically adding all the individual loss elements in the cable plant. A good approximation is that the individual loss distributions are Gaussian and can be described with a

Table 14.13 **Typical Optical Component Loss Values**

Component	Description	Fiber core size (μm)	Mean loss (dB)	Variance (dB^2)
Connector[a]	Physical contact	62.5–62.5	0.40	0.02
		50.0–50.0	0.40	0.02
		9.0–9.0[b]	0.35	0.02
		62.5–50.0	2.10	0.12
		50.0–62.5	0.00	0.01
Splice	Mechanical	62.5–62.5	0.15	0.01
		50.0–50.0	0.15	0.01
		9.0–9.0[b]	0.15	0.01
Splice	Fusion	62.5–62.5	0.40	0.01
		50.0–50.0	0.40	0.01
		9.0–9.0[b]	0.40	0.01
Cable	IBM MMF jumper	62.5	1.75[c]	NA
	IBM SMF jumper	9.0	0.8[c]	NA
	Trunk	62.5	1.00[c]	NA
	Trunk	50.0	0.90[c]	NA
	Trunk	9.0	0.50[c]	NA

Note. From Ref. [21].

[a] The connector loss value is typical when attaching identical connector types. The loss can vary significantly if attaching different connector types.

[b] Single-mode connectors and splices must meet a minimum return loss of 28 dB.

[c] Actual loss value in dB/km.

mean value and variance. The means and variances of all the individual elements are summed together to yield an overall cable plant mean (M_c) and variance (σ_c^2), where σ_c is the cable plant loss standard deviation. The total cable plant loss is given by

$$C_t = M_c + 3\sigma_c. \qquad (14.2)$$

Representative loss values for the individual elements can be obtained from component manufacturers or from a professional fiber optic network planner. Table 14.13 contains typical loss distributions for the elements in a cable plant. Table 14.13 can be used as a guide if more accurate data are not available; however, it is preferred to use data from the cable plant installer or component manufacturer.

In general, component losses (e.g., fiber attenuation, and connection loss) are measured using an equilibrium mode distribution in the fiber. In order to account for the optical loss of a link associated with mode

Fig. 14.17   Example of a generic ESCON link configuration [22].

redistribution in multimode fibers, the following loss should be included in a link:

- 1.5 dB for 50/125 μm trunk fiber, or
- 1.0 dB for 62.5/125 μm trunk fiber

A typical example of a link is illustrated in Fig. 14.17. As a multimode link, a typical configuration might be two jumper cables with a combined length of 90 m, 1.5 km of 50/125 μm trunk (800 MHz/km bandwidth), two physical contact connectors (one 62.5–50 μm and the other 50–62.5 μm), and six splices. A completed loss worksheet is shown in Fig. 14.18. The element loss values used are from Table 14.13. In this case the total cable plant link loss is 7.3 dB.

14.7.2.   SINGLE-MODE CABLE PLANT LINK LOSS

This section describes the calculation of the loss in a single-mode link. The approach and procedure are the same as those in the multimode link. The individual element losses are statistically added to yield a total link loss mean and variance value. Equation (14.2) is used to calculate the cable plant loss. In the case of a single-mode fiber, no equilibrium mode distribution loss

A. Calculating the Multimode Component Mean Loss

Connection loss multiplied by the number of connections in the link:						
_62.5_ -μm-to- _50.0_ -μm connection:	_2.10_ dB	×	_1_	=	_2.10_ dB	
_50.0_ -μm-to- _62.5_ -μm connection:	_0_ dB	×	_1_	=	_0_ dB	
_____ -μm-to-_____ -μm connection:	_____ dB	×	_____	=	_____ dB	
Splice loss multiplied by the total number of splices in the link:	_0.15_ dB	×	_6_	=	_0.90_ dB	
Jumper cable loss multiplied by the combined length of the jumper cables:	_1.75_ dB/km	×	_0.09_ km	=	_0.16_ dB	
Trunk loss per kilometer multiplied by the total trunk length (in km):	_0.90_ dB/km	×	_1.5_ km	=	_1.35_ dB	

(+)

Total Component Mean Loss   _4.51_ dB

B. Calculating the Multimode Component Variance Loss

Connection variance multiplied by the number of connections in the link:						
_62.5_ -μm-to- _50.0_ -μm connection:	_0.12_ dB²	×	_1_	=	_0.12_ dB²	
_50.0_ -μm-to- _62.5_ -μm connection:	_0.01_ dB²	×	_1_	=	_0.01_ dB²	
_____ -μm-to-_____ -μm connection:	_____ dB²	×	_____	=	_____ dB²	
Splice variance multiplied by the number of splices in the link:	_0.01_ dB²	×	_6_	=	_0.06_ dB²	

(+)

Total Component Variance Loss   _0.19_ dB²

C. Calculating the Total Multimode Link Loss

Total component mean loss:	=	_4.51_ dB
Square root of total component variance loss multiplied by 3:	$\sqrt{\underline{0.19}\ dB^2}$ = _0.44_ dB × 3 =	_1.32_ dB
Higher-order mode loss (ESCON only):	=	_1.5_ dB
50.0-μm trunk = 1.5 dB 62.5-μm trunk = 1.0 dB	(+)	
Calculated Link Loss		_7.3_ dB

Fig. 14.18   Example of a completed multimode link loss worksheet [22].

occurs but there is a small loss associated with mode field diameter mismatches (excess connector loss). This loss adds to a 0.5-dB total (two connections) and is small enough to be included in the loss budget for all configurations (whether present or not). Using the generic configuration in Fig. 14.17, a typical single-mode link could consist of two jumper cables with a combined length of 210 m, 19.76 km of 9/125 μm trunk, two

A. Calculating the Single-Mode Component Mean Loss

Connection loss multiplied by the number of connections in the link: _0.35_ dB   ×   _2_   =   _0.70_ dB

Splice loss multiplied by the total number of splices in the link: _0.15_ dB   ×   _2_   =   _0.30_ dB

Jumper cable loss multiplied by the combined length of the jumper cables: _0.8_ dB/km   ×   _0.21_ km   =   _0.17_ dB

Trunk loss per kilometer multiplied by the total trunk length (in km):   _0.5_ dB/km   ×   _19.76_ km   =   _9.88_ dB

(+)

Total Component Mean Loss   _11.05_ dB

B. Calculating the Single-Mode Component Variance Loss

Connection variance multiplied by the number of connections in the link: _0.06_ dB^2 ×   _2_   =   _0.12_ dB^2

Splice variance multiplied by the number of splices in the link: _0.01_ dB^2 ×   _2_   =   _0.02_ dB^2

(+)

Total Component Variance Loss   _0.14_ dB^2

C. Calculating the Total Single-Mode Link Loss

Total component mean loss: =   _11.05_ dB

Square root of total component variance loss plus jumper assembly variance loss multiplied by 3: $\sqrt{\_0.14\_ + 0.05 \text{ dB}^2}$ =   _0.436_ dB × 3   =   _1.31_ dB

Jumper assembly loss plus excess connector loss: =   0.50 dB

(+)

Calculated Link Loss   _12.86_ dB

Fig. 14.19   Example of a completed single-mode link loss worksheet [22].

physical contact connections, and two mechanical splices. Using the typical values in Table 14.13, Fig. 14.19 illustrates a completed single-mode cable plant loss analysis. In this case the loss is 12.86 dB.

14.7.3. SPECIFICATION VERIFICATION

Upon completion of the link loss analysis, the resultant values should be checked against the requirements given in Tables 14.1 and 14.2 for multimode and single-mode links, respectively. Both of the examples illustrated earlier fulfill this requirement.

14.7.4. CABLE PLANT LINK VERIFICATION

Verification of the cable plant ensures that a link meets IBM specifications before equipment is attached. This is usually done by the cable plant installer prior to delivering the installation to the end user. Verifying a link requires using an optical source to transmit a signal through the link and a power meter to measure this signal at the other end. For multimode links, an optical mode conditioner tool [23], to achieve a equilibrium mode distribution, must be used to provide consistent power loss measurements.

Before you begin link verification, you need to review the link documentation, e.g., link configuration, layout, and distribution panels, to ensure you understand what loss value to expect based on the analysis given earlier in this section. Link verification consists of (i) calibrating the test equipment, (ii) obtaining reference power levels, and (iii) measuring the total loss of the links under verification. The amount of loss introduced by the link depends on the jumper cable lengths, the trunk cable length and specifications, and the number and type of link splices and connectors. This measured loss should be less than the maximum allowable cable plant loss.

14.8. Link Troubleshooting

This section summarizes the link troubleshooting strategy and process if a failure of the link has occurred after initial installation, verification, and operation. A link failure has occurred if the link has failed completely (no intelligible signal can be transmitted over the link) or the bit error rate is higher than the allowable threshold and the processor has declared the link defective. In either case, the processor end of the link will report a malfunction and cause a repair call. Depending on the severity (e.g., one path of multiple paths to the equipment) of the problem at the system level, the channel will be required immediately or at a convenient maintenance interval. The latter is generally the case for System/390 servers. The servers will also tell the system operator or repair technician exactly which link is defective.

The link troubleshooting process consists of verification of the output interface operation, cable loss, and input optical interface, usually in this order [24]. The generic link configurations that may be found in an installation are the following:

- Links consisting of one jumper cable between two devices
- Links consisting of two jumper cable connected together in a distribution panel; In this case, the duplex-to-duplex coupler performs a fiber crossover
- Links consisting of two jumper cables, two distribution panels, and a trunk cable connecting the distribution panels

The various link configuration are treated similarly in the link troubelshooting, fault location process.

The most common cause of link failure, after the link was verified upon installation, is the "no-light" condition on the link. This condition is most likely caused by

1. Reconfiguration of link causing a light continuity error
2. Dust or dirt at the interfaces blocking the light
3. Jumper failure
4. Trunk failure
5. Transmitter operation failure
6. Receiver operation failure

Items 1 and 2 can be investigated by obtaining the link diagram and ensuring that there are an odd number of crossovers in the link (Section 14.6.3), the link connects the correct equipment, and by inspecting and cleaning all connections. The remaining items will be discussed in the following sections.

Fiber optic bandwidth measurements and output optical sensitivity measurements are very difficult to measure in the field. Specially trained technicians or technical people are needed to perform these types of measurements with sophisticated test equipment. These elements are verified at the factory and are not likely to fail in the field. They are rarely measured in the field.

14.8.1.   LINK (MMF OR SMF) ANALYSIS

The link failure analysis consists primarily of optical power measurements at strategic locations throughout the link. The appropriate test tools (e.g., optical power source and optical mode conditioner) will be needed to

Fig. 14.20   Measurement positions in an ESCON link.

perform the measurements. A tool kit is available from IBM or your local cable installer to perform these measurements. Figure 14.20 illustrates the four key measurements points (P_T, P_1, P_2, and P_R) in an ESCON link. The four measurements should be made on both paths of the link or until the problem has been found. A typical procedure follows:

Step 1:   Measure the average output power in dB, P_T, of the transmitter by using an ESCON duplex-to-simplex (ST or Biconic) test jumper and a power meter. The test jumper should be made with 62.5/125 μm fiber for multimode link and 9/125 μm fiber for a single-mode link. Ensure that all connections are clean and the power meter is properly set and calibrated. The average optical power should be within the range specified in Table 14.3 for a multimode link or Table 14.7 for a single-mode link. If the specification is satisfied go to step 2; if not, replace the channel card and retest the new port for average optical power. If the specification is achieved, have the system restart the ESCON channel and verify correct operation.

Step 2:   Measure the average power in dB, P_1, exiting the jumper cable by using a duplex-to-duplex coupler, a large core (100/140 μm) duplex-to-simplex (ST or Biconic) bucket fiber jumper, and the same power meter. Ensure all surfaces are clean. The P_1 reading should be between the saturation and sensitivity levels are specified in Table 14.4 for multimode and Table 14.8 for single-mode. Also, the jumper loss, obtained by subtracting P_1 from P_T, should not exceed 2 dB. If these requirements are satisfied go to step 3, or if this is a single jumper link, repeat steps 1 and 2 for the return path. If any condition is not met, the jumper is defective and should be replaced. Remeasure P_1 with the new jumper and verify the requirements are satisfied. If the specification is achieved have the system restart the ESCON channel and verify correct operation.

Step 3: This measurement is made only if a trunk cable is present. If the link is composed of only two jumper cables, proceed to step 4. Measure the output power in dB, P_2, of the output of the trunk cable at the distribution panel using a 62.5/125 μm duplex-to-simplex (ST or Biconic) fiber jumper and the same power meter. Ensure all services are clean. The P_2 reading should be between the saturation and sensitivity levels as specified in Table 14.4 for a multimode link and Table 14.8 for a single-mode link. Also, the sublink loss, obtained by subtracting P_2 from P_T, should not exceed 8 dB for a multimode link (specified in Table 14.1) or 14 dB for a single-mode link (specified in Table 14.2). If these conditions are satisfied, proceed to step 4. If the requirements are not satisfied, proceed to Section 14.8.2.

Step 4: This procedure measures the amount of light available to the receiver. Measure the output power (dB), P_R, of the total link at the receiver input using a duplex-to-duplex coupler, a large core (100/140 μm) duplex-to-simplex (ST or Biconic) bucket fiber jumper, and the same power meter. Ensure all surfaces are clean. The P_R reading should be between the saturation and sensitivity levels as specified in Table 14.4 for a multimode link and Table 14.8 for a single-mode link. Also, the jumper cable loss, obtained by subtracting measurement P_R from P_2, should be less than 2 dB. If the jumper cable loss requirement is not satisfied, replace the jumper cable and retest the link. If the two conditions are now satisfied, restart the channel and perform a system test. If the average power does not meet the receiver requirements then the trunk cable is defective and proceed to Section 14.8.2. If the specifications in Table 14.4 or Table 14.8 are satisfied and the link still does not function properly, then the receive channel card is likely defective. Replace the receive channel card and restart the channel from the server and perform a system test.

If all the optical conditions on the cable and transmitters are satisfied (e.g., P_T, P_1, P_2, and P_R) and the link still does not function after replacing the receive channel card, then the transmitter is defective (e.g., LED "stuck on") and the transmitter channel card should be replaced. In most cases, the system will be able to identify which path in the channel link is defective. However, in the case in which the defective path of the channel link is not known, repeat the four steps for the return path.

14.8.2. TRUNK CABLE ANALYSIS

If it is determined that the trunk cable or its terminating connection is suspected of being defective, then the trunk cable should be tested with a "conditioned" light source. The trunk loss should be measured from the first distribution panel nearest device 1 of the link to the last distribution panel nearest device 2 (see Fig. 14.17). The trunk cable may be a long link (up to 3 km for multimode and 20 km for single mode) and may interconnect buildings or extend over right-of-ways. The trunk cable test procedure is intended to determine if the total trunk meets requirements and does not attempt to isolate the defective element.

The test equipment required for a multimode trunk include an LED optical source with a mode conditioner, two 62.5/125 μm duplex-to-simplex (ST or Biconic) test jumpers, a large-core (100/140 μm) duplex-to-simplex (ST or Biconic) bucket fiber jumper, a duplex-to-duplex coupler and other couplers as appropriate, and a calibrated power meter. The mode conditioner is placed between the LED optical source and the first test jumper. The purpose of the condition is to ensure the light exiting the "conditioned" optical source is equivalent to an equilibrium mode distribution. This can also be achieved by using 1 km of fiber or wrapping a length of fiber several times about a mandrel.

The test equipment for a single-mode trunk cable is a laser power source, two simplex-to-duplex 9/125 μm test jumpers at least 2 m long, a large-core (100/140 μm) duplex-to-simplex (ST or Biconic) bucket fiber jumper, duplex-to-duplex coupler, and a calibrated power meter. The conditioning tool is not needed because the 2-m length of test jumper ensures single-mode operation.

The same procedure is used for multimode or single-mode trunk:

Step 1:   The intent of this step is to assemble and measure an optical source whose optical output is at the end of a duplex test jumper. Assemble the optical source to the mode conditioner (not required for SMF), and connect the simplex side of the appropriate (MMF or SMF) test jumper to the conditioner (MMF only) output. For single-mode setup connect the SMF test jumper directly to the laser power source.

Measure the output optical power of the assembled source by coupling the large-core bucket jumper to the duplex connector end of the test jumper via the appropriate duplex-to-duplex coupler. Insert the appropriate simplex end into the power meter and measure the average output power, P_1, which is the average optical power inserted into the duplex trunk connector at the first distribution panel.

Step 2: With the optical source connected to the trunk cable (do not disconnect cable after step 1 measurement), connect the appropriate (MMF or SMF) test jumper to the last distribution panel in the link. Insert the appropriate simplex end of the test jumper to the same power meter and measure the average output power, P_2, of the trunk cable. The trunk cable loss is obtain by subtracting P_2 from P_1. This value includes all the losses in the trunk, including the connections at the first and last distribution panel. The measurement should be less than 8 dB for a multimode link and 14 dB for a single-mode link. The measurement should also be compared to the trunk loss measured when the trunk was installed and verified.

If the cable loss criteria in step 2 is not satisfied or the loss is significantly different from the initial installation measurement, the trunk cable installer or a qualified fiber optic professional should be called to reevaluate and repair the trunk. Once the link is corrected, restart the channel from the server and perform a system test.

14.9.   Additional Aspects

This section discusses some additional considerations for the ESCON fiber optical link that are very important but do not conform to the theme of the previous sections. The IBM ESCON link, in conjunction with the ANSI standard discussed in Section 14.9.3, will likely continue to evolve and improve in the future with technology advances. Nevertheless, any evolution of both ESCON and ESCON XDF will be accomplished in a manner such that it is compatible with the existing installed based.

14.9.1.   LASER COMPLIANCE

Both ESCON and ESCON XDF links are designed to be class 1 optical products. Class 1 optical products are not considered hazardous to any untrained personnel who are using or servicing the links. A general requirement imposed on transceiver manufacturers is that class 1 certification (if applicable) must be inherent to the transceiver. No requirements should be placed on the system hardware or software by the transceiver manufacturer to achieve class 1 compliance.

The LED transceiver used in ESCON, by virtue of its low maximum output power (-15 dBm) and incoherence, is inherently compliant to the class 1 criteria used by the Department of Health Services (DHHS) in the United States the International Electrotechnical Community (IEC) in Europe. Certification is not required. For customer awareness, however, optical radiation warning labels should be placed inside the host equipment containing ESCON and class 1 compliance should be discussed in appropriate literature.

The ESCON XDF transceiver contains a 1300-nm semiconductor laser and must be certified by the appropriate agencies to be considered a class 1 laser product. The certification consists of submitting a written report of device operation, maximum conditions, fail-safe mechanisms, manufacturer's quality control, etc. to both U.S. and European certification agencies. The IBM ESCON XDF transceiver modules are certified in the United States to conform to the requirements of DHHS-21 CFR Subchapter J for class 1 laser products. Elsewhere, they are certified to be in compliance with IEC 825 and CENELEC HD 482 S1 as a class 1 laser product. Appropriate labels should be placed inside the host equipment and discussed in the appropriate literature.

14.9.2. QUALIFICATION

The ESCON fiber optic link has undergone an extensive functional and environmental test program to ensure design robustness and that the product meets all the application requirements. The qualification consists of the following major elements:

1. Component functional and environmental testing
2. Link optical and mechanical performance testing over the application environment
3. System-level testing in simulated and actual application environments.

Component testing consists of extensive environmental testing (e.g., temperature, humidity, shock, vibration, and accelerated life testing) and ensuring all functional specification parameters are in conformance during the testing. A measure of reliability is obtained from the data. The components tested include the transceivers, jumper cables, couplers, and channel logic modules.

The link performance testing involves a parametric analysis of link performance over worst-case conditions (functional and environmental) using a representative sample of manufactured components. Link margins are determined and conformance to the link model is assessed utilizing simulated data transmission of all possible combinations. All link configurations are evaluated.

The system-level testing involves utilizing a large sample of links in a real data processing environment. The BER is evaluated with simulated and real workloads, all link configurations and operational environmental extremes. Electromagnetic compatibility (EMC) testing is completed at this time.

14.9.3.   SBCON ANSI STANDARD ESCON

ESCON has proven to be an excellent communication link for the data processing environments. IBM, with the support and encouragement of other companies in the information industry, has requested that ANSI sanction an official work effort to make ESCON an industry standard. The activity was formally approved by ANSI in 1995 and a draft standard SBCON has been established [25]. The SBCON standard was approved in 1997. The information contained in this chapter is consistent with this standard.

14.9.4.   ESCON CHANNEL PERFORMANCE

The ESCON channel provides many benefits in configuration flexibility, performance, and management. The capability of the base ESCON channel architecture and implementation (point-to-point multimode fiber link is limited to 3 km) is designed to support a total channel distance of 9 km or greater between the host and the device, by using intervening directors and/or channel extenders. The distance capability of ESCON is an important attribute of the technology and has been utilized extensively. This is a significant improvement over the prior channel distance of 400 feet, using parallel copper technology. Also, the use of the XDF link, available between Directors, and single-mode fiber extenders can extend the total channel

distance to 60 km. This capability provides improved configuration flexibility to optimize the placement of ESCON-capable data processing equipment with regard to installation and facilities constraints as well as enhanced
disaster recovery. Data processing equipment can now be placed in nearby
buildings, as required by the customer.

Significant consideration was given to performance at longer distances,
in the design of the ESCON channel. The performance on an ESCON channel (particularly to Direct Access Storage Devices, DASD) is dependent
on the host server performance, device performance, propagation delay of
the signal on the fiber, the number of exchanges required to complete an
I/O transaction, etc. The Enterprise System Connection architecture was
designed to greatly reduce the communication required between the host
and the device to process an I/O transaction in order to minimize the effect
of distance on performance. Nevertheless, for long distance of propagation,
the delay (5 us/km) of the optical signal in the fiber will limit the channel
throughput. Figure 14.21 illustrates the performance of an ESCON channel
as a function of distance contributed by the Enterprise System Connection
implementation.

ESCON performance, in general, is relatively constant up to 9 km (de-
grading by 10% at 9 km) and begins to droop substantially thereafter. This
distance capability allows high-bandwidth devices (e.g., DASD) to be
placed anywhere within a "typical" campus environment while maintaining the high performance capability of the ESCON channel. In addition,

Fig. 14.21 Representative ESCON channel performance as a function of distance (courtesy of IBM Corp.).

some low-speed devices (e.g., printers) may be placed as far as 60 km from the host and will perform acceptably.

14.9.5. ESCON TECHNOLOGY UPDATE

The recent announcement and shipment of the IBM eServer zSeries 900 platform included a significant packaging improvement to the prior ESCON card shown in Fig. 14.5. The new ESCON I/O card design is a high-port density card and is illustrated in Fig. 14.22.

The ESCON card density has been improved by 4x with the use of state-of-the-art optical and silicon technologies. The ESCON transceiver has been changed to an industry standard "Small Form Factor" (SFF) transceiver. This upgrade enables the placement of 16 transceivers along the card edge (shown on the right side in Fig. 14.22). The four ASIC modules directly behind the transceivers contain the ESCON channel logic, serializer, and deserializer functions. Each module contains four ESCON

Fig. 14.22   High-density ESCON I/O card (courtesy of IBM Corp.).

Fig. 14.23   MT-RJ ESCON multimode jumper cable photograph (courtesy of IBM Corp.).

independent channels. The remaining components contain support functions. The ESCON optical specifications and link performance have not been changed. The link is functionally fully compatible with existing ESCON links.

The use of the SFF transceiver necessitated a change of the ESCON connector from the original design (Fig. 14.8) to a smaller MT-RJ fiber optic connector. The MT-RJ uses a single plastic molded ferrule instead of two ceramic ferrules. It is an axial push/pull connector with a single latch. The connector conforms to the ANSI/TIA/EIA 604-12 (FOCUS 12) fiber optic connector standard [26]. An MT-RJ fiber optic jumper cable is illustrated in Fig. 14.23.

The smaller connector and cable size improves the handling and flexibility of the jumper cable which, in turn, improves jumper cable installation. The new jumper cable is optically and functionally compatible with existing links and cables. However, a connector conversion is necessary to attach a new transceiver or cable to an existing cable infrastructure. A cable conversion kit from original ESCON to MT-RJ is available from IBM to enable this conversion without changing the existing cable infrastructure.

New installations should design a cable infrastructure compatible to the ESCON MT-RJ cable system.

ACKNOWLEDGMENTS

I wish to thank the International Business Machines Corporation for allowing the extensive use of copyrighted material in this chapter. I also wish to thank the many individuals (too numerous to list here) both within and outside of IBM whose hard work and dedication helped make ESCON a reality. In particular, I wish to thank M. Lewis and J. Quick for the hardware photographs.

References

1. IBM Corp. Introducing Enterprise Systems Connection (IBM Document No. GA23-383). Mechanicsburg, Pa.: IBM Corp.
2. Calta, S. A., deVeer, J. A., Loizides, E., and R. N. Strangwayes. 1992. Enterprise Systems Connection Architecture — System Overview. *IBM J. Res. Dev.* 36:535–551.
3. Georgiou, C. J., Larsen, T. A., Oakhill, P. W., and Salimi, B. 1992. The IBM Enterprise Systems Connection (ESCON) director: A dynamic switch for 200 Mb/s fiber optic links. *IBM J. Res. Dev.* 36:593–616.
4. Elliott, J. C., and Sachs, M. W. 1992. The IBM Enterprise Systems Connection Architecture. *IBM J. Res. Dev.* 36:577–592.
5. Flanagan, J. R., Gregg, T. A., and Casper, D. F., 1992. The IBM Enterprise Systems Connection (ESCON) channel — A versatile building block. 1992. *IBM J. Res. Dev.* 36:617–632.
6. IBM Corp. Enterprise Systems Architecture/390 ESCON I/O interface (IBM Document No. SA22-7202, Chap. 2). Mechanicsburg, Pa.: IBM Corp.
7. IBM Corp. Enterprise Systems Architecture/390 ESCON I/O interface (IBM Document No. SA22-7202, Chap. 8). Mechanicsburg, Pa.: IBM Corp.
8. IBM Corp. Enterprise Systems Architecture/390 ESCON I/O interface (IBM Document No. SA22-7202, Appendix). Mechanicsburg, Pa.: IBM Corp.
9. Aulet, N. R., Boerstler, D. W., DeMario, G., Ferraiolo, F. D., Hayward, C. E., Heath, C. D., Huffman, A. L., Kelley, W. R., Peterson, G. W., and Stigliani, D. J., Jr. 1992. IBM Enterprise Systems multimode fiber optic technology. *IBM J. Res. Dev.* 36:553–577.
10. DeCusatis, C. 1995. Data processing systems and optoelectronics. In *Optoelectronics for Data Communication,* eds. R. C. Lasky, U. L. Österberg, and D. J. Stigliani, Chap. 6. New York: Academic Press.
11. IBM Corp. Enterprise Systems Architecture/390 ESCON I/O interface physical layer (IBM Document No. SA23-0394). Mechanicsburg, Pa.: IBM Corp, courtesy of International Business Machines Corporation.

12. IBM Corp. Planning for fiber optic channel links (IBM Document No. GA23-0367, Chap. 1). Mechanicsburg, Pa.: IBM Corp, courtesy of International Business Machines Corporation.
13. Gregurick, V. 1992. Fiber optic transmitter measurement procedure, IBM Engineering Specification 49G3489. East Fishkill, N.Y.: IBM Corp.
14. Electronics Industry Association/Telecommunications Industry Association Commercial Building Telecommunications Cabling Standard (EIA/TIA-568-A).
15. Electronics Industry Association. Detail specification for 62.5 μm core diameter/125 μm cladding diameter class 1a multimode. Graded index optical waveguide fibers (EIA/TIA-492AAAA). New York: Electronics Industry Association.
16. International Telecommunications Union. Characteristics of a 50/125 μm multimode graded index optical fibre cable (CCITT Recommendation G.651).
17. Fiber Channel Standard. Physical layer and signaling specification. American National Standards Institute X3T9/91-062, X3T9.3/92-001 (rev. 4.2).
18. Electronics Industry Association Detail specification for class IVA dispersion unshifted single-mode optical wavelength fibers used in communications systems (EIA/TIA-492BAAA). New York: Electronics Industry Association.
19. International Telecommunications Union. Characteristics of single-mode optical fibre cable (CCITT Recommendation G.652).
20. IBM Corp. Planning for fiber optic channel links (IBM Document No. GA23-0367, Appendix A). Mechanicsburg, Pa.: IBM Corp, courtesy of International Business Machines Corporation.
21. IBM Corp. Planning for fiber optic channel links (IBM Document No. GA23-0367, Chap. 2). Mechanicsburg, Pa.: IBM Corp, courtesy of International Business Machines Corporation.
22. IBM Corp. Planning for fiber optic channel links (IBM Document No. GA23-0367, Chaps. 3 and 4). Mechanicsburg, Pa.: IBM Corp, courtesy of International Business Machines Corporation.
23. Light launch conditions for long-length graded-index optical fiber spectral attenuation measurements (EIA/TIA-455-50A), available from Electronics Industry Association, New York.
24. IBM Corp. Maintenance information for enterprise systems connection links (IBM Document No. SY27-2597). Mechanicsburg, Pa.: IBM Corp.
25. American National Standards Institute. Single-byte command code sets CONnection architecture (SBCON). Draft ANSI Standard X3T11/95-469 (rev. 2.2).
26. American National Standards Institute. ANSI/TIA/EIA 604-12 (FOCUS 12) Fiber Optic Connector Intermateability Standard, Type MT-RJ.

Chapter 15 | Local Area Networks, Fiber Distributed Data Interface, Ethernet, and Token Ring

Rakesh Thapar

Marconi, Warrendale, Pennsylvania 15086

15.1. Introduction

Local area networks (LANs) are commonly used in many applications and environments. As the technology is changing so are the LANs. Tomorrow's LANs are promising much higher speeds and reach. Unfortunately, the competing technologies being developed are based on different philosophies, thereby making them incompatible. This chapter introduces the modern fiber-based LAN technologies that either exist today or are expected in the market during the next few years. In particular, the chapter discusses Fast Ethernet, gigabit Ethernet, VG-AnyLan, Internet protocol (IP) over asynchronous transfer mode (ATM), ATM LAN emulation, and fiber distributed data interface (FDDI).

Several proposed technologies promise better performance for tomorrow's LANs. The promises create difficult choices for LAN owners who need better performance today but do not want to buy equipment that will be nonstandard or obsolete when tomorrow's technologies are standardized. In this chapter, I give an overview of the current LAN technologies, some of which are still undergoing an active standardization process.

HANDBOOK OF FIBER OPTIC
DATA COMMUNICATION
$35.00

15.2. Fast Ethernet

Fast Ethernet [1–3] or the 100BASE-T, is a 100 Mb/s networking technology based on IEEE 802.3u standard. It uses the same media access control (MAC) protocol, the carrier sense multiple access/collision detection (CSMA/CD) method (running 10 times faster) that is used in the existing Ethernet networks [International Standards Organization/International Electrotechnical Community (ISO/IEC) 8802-3] such as 10BASE-T, connected through a media-independent interface (MII) to the physical layer device (PHY) running at 100 Mb/s. The supported PHY sublayers are 100BASE-T4, 100BASE-TX, and 100BASE-FX. The general name of the PHY group is 100BASE-T. One of the most important advantages of Fast Ethernet is that it uses the standard 802.3 MAC. This allows data to be interchanged between a 10BASE-T and 100BASE-T nodes without protocol translation, thereby allowing the possibility of a phased introduction of 100BASE-T in the existing 10BASE-T networks and a cost-effective migration path that maximizes use of existing cabling and network management systems. Autonegotiation is another key element of the 100BASE-T Fast Ethernet IEEE 802.3u standard. Autonegotiation allows an adapter, hub, or switch capable of data transfer at both 10 and 100 Mb/s rates to automatically use the fastest rate supported by the device at the other end. Autonegotiation signals the capabilities it has available, detects the technology that exist in the device it is being connected to, and automatically configures to the highest common performance mode of operation. The relationship of Fast Ethernet model to the open systems interconnection model is shown in Fig. 15.1.

15.2.1. THE PHYSICAL LAYER

15.2.1.1. 100BASE-T4

This physical layer defines the specification for 100BASE-T Ethernet over four pairs of category 3, 4, or 5 unshielded twisted-pair (UTP) wire. This is aimed at those users who want to retain the use of voice-grade twisted-pair cable. Additionally, it does not transmit a continuous signal between packets, which makes it useful in battery-powered applications.

With this signaling method, one pair is used for carrier detection and collision detection in each direction and the other two are bidirectional. This allows for a half-duplex communication using three pairs for

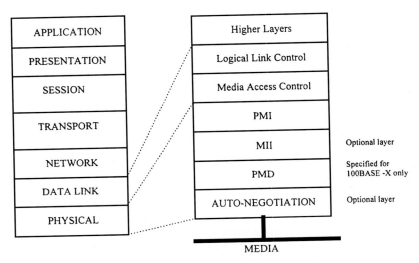

Fig. 15.1   Fast Ethernet (IEEE 802.3u) model.

data transmission. The unidirectional pairs are the same ones used in 10BASE-T (it uses only two pairs) for consistency and interoperability. Because three pairs are used for transmission, to achieve an overall 100 Mb/s each pair needs only to transmit at 33.33 Mb/s. If Manchester encoding was to be used at the physical layer, as in 10BASE-T, the 30-MHz limit for the category 3 (cat 3) cable would be exceeded. To reduce the rate, a 8B6T block code is used that converts a block of 8 bits into six ternary symbols, which are then transmitted out to three independent channels (pairs). The effective data rate per pair thus becomes $(6/8) \cdot 33.33 = 25$ MHz, which is well within the cat 3 specifications of 30 MHz.

15.2.1.2.   100BASE-X

Two physical layer implementations, 100BASE-TX and 100BASE-FX, are collectively called 100BASE-X when referring to issues common to both. 100BASE-X uses the physical layer standard of FDDI by using its physical media-dependent sublayer (PMD) and medium-dependent interfaces (MDI). The 125 Mb/s full-duplex signaling system for a twisted pair, defined in FDDI, forms the basis for 100BASE-TX and the system defined for transmission on optical fiber forms the basis for 100BASE-FX. Basically, 100BASE-X maps the characteristics of FDDI PMD and MDI to the services expected by the CSMA/CD MAC. It defines a full-duplex signaling

Table 15.1   **Fast Ethernet Copper Connector Pinouts**

Contact	100BASE-T4		100BASE-TX	
	PHY without internal crossover	PHY with internal crossover	PHY without internal crossover	PHY with internal crossover
1	Transmit +	Receive +	Transmit +	Receive +
2	Transmit −	Receive −	Transmit −	Receive −
3	Receive +	Transmit +	Receive +	Transmit +
4	Bidirectional-1 +	Bidirectional-2 +	Not used	Not used
5	Bidirectional-1 −	Bidirectional-2 −	Not used	Not used
6	Receive −	Transmit −	Receive −	Transmit −
7	Bidirectional-2 +	Bidirectional-1 +	Not used	Not used
8	Bidirectional-2 −	Bidirectional-1 −	Not used	Not used

standard of 125 Mb/s for multimode fiber, shielded twisted-pair and unshielded twisted-pair wiring. The physical sublayer maps 4 bits from MII into 5-bit code blocks and vice versa using the 4B/5B encoding scheme (same as FDDI).

15.2.1.3. 100BASE-TX

This physical layer defines the specifications for 100BASE-T Ethernet over two pairs of category 5 UTP wire or two pairs of shielded twisted-pair (STP) wire. With one pair for transmit and the other for receive, the wiring scheme is identical to that used for 10BASE-T Ethernet (see Table 15.1 for the connector details).

15.2.1.4. 100BASE-FX

This physical layer defines the specification for 100BASE-T Ethernet over two strands of multimode (62.5/125 μm) fiber cable. One strand is used for transmit, whereas the other is used for receive.

100BASE-T Fast Ethernet and 10BASE-T Ethernet differ in their topology rules. 100BASE-T preserves 10BASE-T's 100-m maximum UTP cable runs from hub to desktop. The basic rules revolve around two factors: the network diameter and the class of the repeater (or hub). The network diameter is defined as the distance between two end stations in the same collision domain. Fast Ethernet specifications limit the network diameter to approximately 205 m (using UTP cabling), whereas traditional Ethernet could have a diameter up to 500 m.

Repeaters are the means used to connect segments of a network medium together in a single collision domain. Different physical signaling systems can be joined into a common collision domain using a repeater. Bridges can also be used to connect different signaling systems; however, each system connected to the bridge will have a different collision domain.

In traditional Ethernet, all hubs are considered to be functionally identical. In Fast Ethernet, however, there are two classes of repeaters: class I and class II. Class I repeaters perform translations when transmitting, enabling different types of physical signaling systems to be connected to the same collision domain. Because they have internal delays, only one class I repeater can be used within a single collision domain when maximum cable lengths are used, i.e., this type of repeater cannot be cascaded and the maximum network diameter is 200 m. On the other hand, class II repeaters simply repeat the incoming signals with no translations, i.e., provide ports for only one physical signaling system type. Class II repeaters have smaller internal delays and can be cascaded using a 5-m cable with a maximum of two repeaters in a single collision domain if maximum cable lengths are used. Cable lengths can always be sacrificed to get additional repeaters in a collision domain. This is summarized in Table 15.2.

With traditional 10BASE-T Ethernet, networks are designed using three basic guidelines: maximum UTP cable runs of 100 m, four repeaters in a single collision domain, and a maximum network diameter of 500 m. With 100BASE-T Fast Ethernet, the maximum UTP cable length remains 100 m. However, the repeater count drops to two and the network diameter

Table 15.2 **Maximum Collision Domain Diameter**

	All UTP cable (m)	UTP (T4) and fiber (m)	UTP (TX) and fiber (m)	All fiber (m)
Point-to-point (no repeater)	100	N/A	N/A	412
One class I repeater	200	231	260	272
One class II repeater	200	304	308	320
Two class II repeaters	205	236	216	228

drops to 205 m. On the surface, the 100BASE-T Fast Ethernet rules seem restrictive, but with use of repeaters, bridges, and switches, Fast Ethernet can be easily implemented in a network.

15.2.2. AUTONEGOTIATION

Autonegotiation is the process that enables two devices, sharing a link segment, to communicate necessary information with one another in order to interoperate taking maximum advantage of their abilities. The basis of autonegotiation is a modified 10BASE-T link integrity pulse sequence as defined in Section 28 of the Institution of Electrical and Electronic Engineers (IEEE) 802.3u standard. The technologies currently supported by autonegotiation are 10Base-T half duplex, 10Base-T full duplex, 100Base-TX half duplex, 100Base-TX full duplex, and 100Base-T4.

The foundation for all of autonegotiation's functionality is the fast link pulse (FLP) burst, which is simply a sequence of 10Base-T normal link pulses, also known as link test pulses (in 10Base-T technology) that come together to form a "word" (link code word) that identifies supported operational modes. On power-on or command from management entity, a device capable of autonegotiation issues a FLP burst.

Each FLP is composed of 33 pulse positions, with the 17 odd-numbered positions corresponding to clock pulses and the 16 even-numbered positions corresponding to data pulses. The time between pulse positions is 62.5 ± 7 μs, and therefore 125 ± 14 μs between each clock pulse. All clock positions are required to contain a link pulse. On the other hand, if there is a link pulse present in a data position, it is representative of a logic 1, whereas the lack of a link pulse is representative of a logic 0. The amount of time between FLP bursts is 16 ± 8 ms, which corresponds to the time between consecutive link test pulses produced by a 10Base-T device to allow interoperation with fixed-speed 10BASE-T devices.

The 16 data positions in an FLP burst come together to form a link code word that is typically encoded as shown in Fig. 15.2.

The 5-bit selector field contains 32 possible combinations, only 2 of which are currently defined and allowed to be sent (Table 15.3). The next 8 bits, which make up the technology ability field, are used by a device to advertise its abilities to support various IEEE 802.3 technologies. The abilities are advertised in parallel, i.e., a single selector field value will advertise all the supported technologies. A logic 1 in any of these positions symbolizes that the device is capable of a particular IEEE 802.3 technology

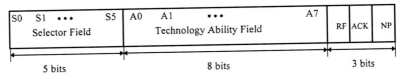

(as shown in Table 15.4). The device should advertise only the technologies that it supports.

The devices at the two ends of the link segment may have an ability to support multiple technologies; the highest common denominator ability is always chosen, i.e., the technology with the highest priority that both sides can support. The relative priorities of the technologies supported by the IEEE 802.3 selector field value are as follows (from highest to lowest):

1. 100BASE-TX full duplex
2. 100BASE-T4
3. 100BASE-TX
4. 10BASE-T full duplex
5. 10BASE-T

Remote fault: This bit can be set to inform a station that a remote fault has occurred.

Acknowledge: This bit is set to confirm the receipt of at least three complete, consecutive, and consistent FLP bursts from a station. This functionality will be discussed in detail later.

Next page: This is a means by which devices can transmit additional information beyond their link code words. This bit simply indicates

Table 15.3   **Selector Field Values in Link Word**

	Bit position				
S0	**S1**	**S2**	**S3**	**S4**	**Description**
1	0	0	0	0	IEEE 802.3
0	1	0	0	0	IEEE 802.9
Other combinations					Reserved for future use

Table 15.4   **Ability Field Values in**
Link Word

Bit	Technology represented
A0	10BASE-T
A1	10BASE-T full duplex
A2	100BASE-TX
A3	100BASE-TX full duplex
A4	100BASE-T4
A5	Reserved
A6	Reserved
A7	Reserved

whether or not there are more Next Pages to come. When set to logic 0, it indicates that additional pages will follow, whereas a logic 1 indicates that there are no remaining pages.

15.2.3. BASIC OPERATION

When a Fast Ethernet device is brought up, it performs autonegotiation if capable. The objective, as described previously is to agree to operate in the highest common denominator mode. Once the operating parameters are determined, the operations are exactly like the traditional Ethernet or IEEE 802.3 standard MAC operation. The MAC frame formats, back-off algorithms, and the collision detection scheme work exactly as in traditional CSMA/CD (in IEEE 802.3 standard).

15.3. Gigabit Ethernet

The purpose of the gigabit Ethernet is to extend the 802.3 protocol to an operating speed of 1000 Mb/s in order to provide a significant increase in bandwidth while maintaining maximum compatibility with the installed base of 10/100 CSMA/CD nodes, research and development, network operation, and management. It is being studied by an IEEE working group aiming to standardize this technology as IEEE 802.3z [4].

The following objectives have been adopted by the IEEE 802.3 High-Speed Study Group for this technology:

1. Speed of 1000 Mb/s at the MAC–PHY service interface
2. Use 802.3/Ethernet frame format
3. Meet 802 FR, with the possible exception of Hamming distance
4. Simple forwarding between 1000, 100, and 10
5. Preserve minimum and maximum frame size of the current 802.3 standard
6. Full- and half-duplex operation
7. Support star-wired topologies
8. Use CSMA/CD access method with support for at least one repeater per collision domain
9. Support fiber media and, if possible, copper media
10. Use American National Standards Institute (ANSI) Fibre Channel FC-1 and FC-0 as basis for work
11. Provide a family of physical layer specifications that support
    - A link distance of at least 2 km on single-mode fiber
    - A link distance of at least 500 m on multimode fiber
    - A link distance of at least 25 m on copper (100 m preferred)
12. Decide between collision domain diameter of ≥ 50 m or ≥ 200 m
13. Support media selected from ISO/IEC 11801
14. Accommodate proposed P802.3x flow control

15.4. VG-Any LAN

The demand priority protocol (IEEE 802.12) [2, 5, 6], commonly referred to as 100VG-Any LAN [5, 6], is a local area network standard that seeks to provide a high-speed, shared-media LAN utilizing the existing copper media. The fundamental components of VG-Any LAN are end nodes and repeaters (Fig. 15.3).

Typically, the end nodes are user workstations, bridges, routers, switches, or file servers. They are connected to the network through an interface card in the end node. End nodes have two modes of operation: private and promiscuous. Promiscuous nodes receive all frames on the network, whereas private nodes receive only the messages specifically sent to them.

Repeaters are devices that interconnect end nodes and other components in demand priority. They manage the network access by performing

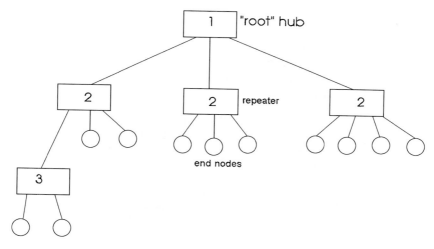

Fig. 15.3 A general VG-Any LAN topology.

"round-robin" scans of network port requests while keeping track of addresses and making sure that the link is up and the end station functional. Each repeater has a special uplink port through which another repeater can be cascaded. Each port on the repeater can be configured to operate in normal mode or promiscuous mode. In the normal configuration, a port receives only packets addressed to the attached node. In the promiscuous mode, the port receives all packets. The repeater may be configured to handle IEEE 802.3 CSMA/CD or IEEE 802.5 Token Ring frame formats. However, all repeaters in the same segment must use the same format.

A demand priority network employs a physical star-wired tree topology. A minimum network of this type will consist of one repeater, two end nodes, and two network links. All the end nodes in the network are connected to the repeaters. A usual small network consists of between 6 and 32 end nodes connected to a single repeater. A larger network can be constructed by cascading several repeaters using the uplink. A maximum of five levels of cascading are allowed by the protocol because of the timing issues. The layered model for demand priority is shown in Fig. 15.4.

15.4.1. PHYSICAL LAYER

The physical layer is made of the physical media-independent (PMI) sublayer, the MII sublayer, and the physical media-dependent (PMD) sublayer. The network links are made of either copper or fiber media (see Table 15.5).

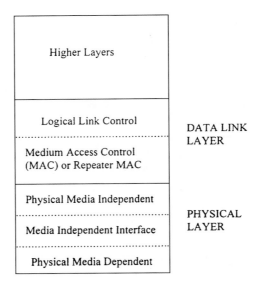

Fig. 15.4 Demand priority model.

The bundled cable (25 pair) is allowed in end node-to-repeater links provided that the repeater supports privacy mode and store-and-forward capability. Bundled cable is not allowed in repeater-to-repeater, repeater-to-bridge, or repeater-to-promiscuous end node connections. No loops or branches are allowed in the topology. Flat cable is prohibited. The maximum number of nodes in one segment is limited to 1024.

PMI sublayer, which is link independent, is responsible for taking a frame from the MAC layer and dividing it into 5-bit data streams (called quintets) and then sequentially distributing them among four channels. If four-pair UTP is used, then channel 0 is transmitted on leads 1 and 2, channel 1 uses leads 3 and 6, channel 2 uses leads 4 and 5, and channel 3 uses leads 7 and 8. When fiber optic or the two-pair STP cable is used,

Table 15.5   **Cable Type vs Distance for Demand Priority**

Cable type	Category	Maximum distance (m)
4 pair UTP	3,4	100
2 pair STP	3,4	100
4 pair UTP	5	200
Fiber	Single or multimode	2000

the PMD sublayer uses a multiplexing scheme to convert the four channels into one or two channels. A data scrambling algorithm is used to randomize bit patterns on each transmission pair to minimize interference and cross-talk. After scrambling, the quintets are converted to 6-bit sextets using the 5B/6B encoding scheme. Finally, a preamble, start delimiter, and end frame delimiter are added to each data channel along with padding, if needed. A reverse role is played when PMI receives a frame to be handed to MAC or repeater MAC (RMAC) layers.

The MII is an interface between the PMI and the PMD. PMD service primitives are passed across this interface and provide for interchange-ability of PMDs, easy reconfiguration of end nodes, and ports to support multiple link media.

The PMD sublayer is link dependent and provides control signal generation and recognition, data stream signal conditioning, clock recovery, and channel multiplexing appropriate to the link medium. The transmission over the physical medium is half duplex. The actual signaling of data uses non-return-to-zero (NRZ) encoding scheme. The clock rate for transmitting is 30 MHz for each wire pair; therefore, 30 Mb/s are transmitted on each pair, with a total data rate of 120 Mb/s. Because every 5 data bits are encoded as 6 bits, the actual data transmission rate is 100 Mb/s.

15.4.2. MEDIA ACCESS CONTROL

The MAC layer is the control center for data transfer for a specific node. It is responsible for handling data transfer requests to the network and encapsulation of data into frames, on receipt removing data from frames and error checking.

The end nodes are continually polled by the repeater for any data it needs to send. The transmitted data can be either normal priority or high priority. The repeater first takes the high-priority requests and then the normal priority data while arbitrating between the ports in round-robin. Starvation of normal data by high-priority data is avoided by changing normal priority data to high priority if it has been waiting for longer than the threshold. Data packet transmission consists of a series of handshaking sequences in which the sending side of the end node–repeater local port on a point-to-point connection makes a request and the other side acknowledges the request. The sequence of sending a data packet transmission is requested by an end node and controlled by the repeater. The sequence of events proceeds as follows:

1. If an end node has a data packet ready to send, it transmits either a *Request_Normal* or *Request_high* control signal. Otherwise, the end node transmits the *Idle_Up* control signal.

2. The repeater polls all local ports to determine which end nodes are requesting to send a data packet and at what priority level (normal or high).

3. The repeater selects the next end node with a high-priority request pending. Ports are selected in port order. If no high-priority requests are pending, then the next normal-priority port is selected (in port order). This selection causes the selected port to receive the *Grant* signal. Packet transmission begins when the end node detects the *Grant* signal.

4. The repeater then sends the *Incoming* signal to all other end nodes, alerting them to the possibility of an incoming packet. The repeater decodes the destination address from the frame being transmitted as it is being received.

5. When an end node receives the *Incoming* control signal, it prepares to receive a packet by stopping the transmission of requests and listening on the media for the data packet.

6. Once the repeater has decoded the destination address, the packet is delivered to the addressed end node or end nodes and to any promiscuous nodes. Those nodes not receiving the data packet receive the *Idle_Down* signal from the repeater.

7. When the end node(s) receives the data packet, it returns to its state prior to the reception of the data packet, either sending an *Idle_Up* signal or making a request to send a data packet.

This process is used throughout the demand priority protocol to allow end nodes to transmit data packets to other end nodes.

Demand priority is designed to operate in compatibility modes for both Ethernet and Token Ring frame formats. This means that software and protocols above the logical link control (LLC) need only to know that they are operating on either an Ethernet or Token Ring network with regard to frame formats. The LLC provides the MAC with the primitives containing the information used to construct the Ethernet or Token Ring frame. The MAC then constructs a demand priority frame containing the proper elements. The supported demand priority frame formats are as follows:

- *802.12 training frames:* These frames are used during link training to establish validity of a link. They are initially constructed at the lower end of the link. These frames are forwarded to all network repeaters.
- *802.12 void frames:* These frames are generated when a repeater detects that the acknowledged device did not send a packet within an expected time span or because the packet has errors.
- *802.3 frame:* This frame format allows the demand priority network to communicate with existing Ethernet networks.
- *802.5 frame:* This frame format enables the demand priority network to communicate with existing Token Ring networks.

15.4.3. REPEATER MAC

An RMAC is to a repeater what a MAC is to an end node. The RMAC accepts transmit requests from the end nodes, arbitrates transfer orders, and routes the incoming packets to the proper outbound ports. The two major tasks of RMAC are round-robin polling and link training.

In order to provide each node fair access to demand priority network, repeaters perform round-robin polling. Each repeater maintains two requested pointers to keep track of the next port to be serviced: one for the next normal-priority port and one for the next high-priority port. All repeater ports are polled at least once per packet transmission to determine which have requests pending. High-priority requests exist so that time-sensitive transmissions, such as video, sound, etc., will not be caught in normal traffic.

High-priority requests are serviced before normal-priority requests. No requests are interrupted once data are being transmitted. To avoid starvation, any normal-priority traffic that has been waiting to transmit for 200–300 ms is automatically elevated to high-priority status and is serviced in port order by the high-priority pointer.

The repeater–end node link is initialized using link training. The actual training involves the transmission of training frames in each direction between the lower entity and the repeater. Such frames are always considered as normal-priority frames. In order to join the network, the lower entity must send and receive 24 consecutive, uncorrupted training frames. This ensures the cable validity. Within the training frame is the field that indicates the "requested configuration" of the lower entity. When the repeater responds, the "allowed configuration" field instructs the lower entity of the allowed configuration. The lower entities' address must be acknowledged by the upper repeater.

Link training is used for many purposes, such as verification of cable quality for data transmission, allowing the receiver to adapt to the link, establishing the MAC address of the end node, determining the link node type — 802.3 or 802.5, etc. The link training is performed each time a link is established, such as power-up, or at cable connections or during certain error conditions and is always initiated by the lower entity device desiring to connect to the network, which can be either an end node or a repeater.

15.4.4. CLASSICAL IP OVER ATM

Classical IP over ATM (RFC 1577) [7, 8] had been proposed by the Internet Engineering Task Force (IETF) as a way of connecting IP-based work-stations on ATM. RFC 1577 basically emulates the IP layer (network layer) over ATM to provide end-to-end connectivity to the higher layers. In this approach, the IP end stations connected to the ATM cloud are divided into logical IP subnets (LISs). The subnets are administered in the same manner as the conventional subnets. Hosts connected to the same subnetwork can communicate directly; however, communication between two hosts on different LISs is only possible through an IP router, regardless of whether direct ATM connectivity is possible between the two hosts.

Implementation of classical IP over ATM requires a mapping between IP and ATM addresses. IP addresses are resolved to ATM addresses using the ATM address resolution protocol (ATMARP) and vice versa using the inverse ATMARP (InATMARP) within a subnet. ATMARP is used for finding the ATM address of a device given the IP address. It is analogous to the IP–ARP associated with IP protocol. Just like conventional ARP, it has a quintuple associated with it: source IP address, source ATM address, destination IP address, and destination ATM address. On the other hand, InATMARP is used to find the IP address of a station given the ATM address [almost equivalent to conventional reverse address resolution protocol (RARP)]. Typical use of InATMARP is by the ATMARP server to find out the IP address of the station connected to the other end of an open virtual circuit. This information is used to update database entries and to refresh the entry on time-outs.

Every end station is configured statically with the address of the ATMARP server. On initialization, the end station opens a virtual channel connection (VCC) to the ATMARP server. The ATMARP server, on detecting a new VCC, performs an InATMARP on it to find the IP address of the end station connected at the other end of the VCC. This information is stored in the tables of the ATMARP server for further use. Each end station maintains a local ARP cache that acts as the primary cache and the APR server acts as a secondary cache. If the ATM end station wants to contact another station, it will query its local cache first for the ATM address for a given IP address. If that fails, it queries the ARP server for the ATM address. Once it has the ATM address of the destination, the end station proceeds to open a direct VCC to the destination.

The basic drawback to this approach is that this work only for IP-type traffic. It does not support multicast or broadcast. It requires static assignment of ARP server address and the ARP server becomes the single point of failure. The IETF has recently removed some of these drawbacks by introducing a new concept to enhance RFC 1577, i.e., multicast address resolution server. Work on it has been going on since late 1994.

15.4.5. ATM LAN EMULATION

LAN emulation (LANE) [9–11] has been proposed by ATM Forum and has been widely accepted by the ATM industry as a way to emulate conventional LANs. The necessity of defining LANE arose because most of the existing customer premises networks use LANs such as IEEE 802.3/802.5 (Ethernet and Token Ring) and customers expect to keep using existing LAN applications as they migrate toward ATM. To use the vast repertoire of LAN application software, it became necessary to define a service on ATM that could emulate LANs. The idea is that the traditional end-system applications should interact as if they are connected to traditional LANs. This service should also allow the traditional (legacy) LANs to interconnect to ATM networks using today's bridging methods.

LANE has been defined as a MAC service emulation, including encapsulation of MAC frames (user data frames). This approach, as per ATM Forum, provides support for maximum number of existing applications. This is not easy because there are some key differences between legacy LANs and ATM networks (Table 15.6).

The main objective of LAN emulation service is to enable existing applications to access an ATM network via protocol stacks, such as NetBIOS,

Table 15.6   **LAN vs ATM**

LANs	*ATM*
Connectionless	Connection oriented
Broadcast and multicast are easily accomplished through shared medium of LAN	ATM does not support broadcast or multicast
LAN MAC addresses based on manufacturers organizationally unique identifiers and serial numbers are independent of topology	ATM addresses depend on topology, i.e., where the station is plugged in (i.e., network prefix)

IPX, IP, AppleTalk, etc., as if they were running over traditional LANs. In many cases, there is a need to configure multiple separate domains within a single network. This objective is fulfilled by defining an emulated LAN (ELAN) that comprises a group of ATM-attached devices. It appears as a group of stations connected to a IEEE 802.3 or 802.5 LAN segment. Several ELANs could be configured and membership in an ELAN is independent of the physical location of the end station. An end station could belong to multiple ELANs.

15.4.5.1. Components

There are four basic components in LANE: LAN emulation client (LEC), LAN emulation configuration server (LECS), LAN emulation server (LES), and broadcast and unknown server (BUS).

LEC is an entity in the ATM workstation or ATM bridges that performs data forwarding, address resolution, and other control functions. This provides a MAC-level emulated Ethernet/IEEE 802.3 or IEEE 802.5 service interface to applications running on top. It implements the LANE user–network interface (LUNI) when communicating with other entities within the emulated LAN.

The LES implements the control coordinating function for the ELAN. It provides a facility for registering and resolving MAC addresses or route descriptors to ATM addresses. LECs register the LAN destinations they represent with the LES. A client can also query the LES when the client wishes to resolve a MAC address to an ATM address. A LES will either respond directly or forward the query to other clients so they may respond.

BUS handles data sent by an LEC to the broadcast MAC address ("FFFFFFFFFFFF" hex), all multicast traffic, and, as an option, some initial unicast frames sent before the target ATM address is resolved.

15.4.5.2. LEC Initialization Phases

The basic states that a LEC goes through before it is operational are shown in Fig. 15.5 and described as following:

Initial state: In this state LES and LEC know certain parameters (such as address, ELAN name, maximum frame size, etc.) about themselves.

LECS connect phase: LEC sets up a call to LECS. The VCC that is opened is referred to as configuration-direct VCC. At the end of configuration, this VCC may be closed by the LEC.

Fig. 15.5   Initialization phases of LANE [9]. Copyright 1995 The ATM Forum.

Configuration phase: LEC discovers LES in preparation for join phase.

Join phase: During the join phase, LEC establishes its control connections to the LES. Once this phase is complete, the LEC has been assigned a unique LEC identifier (LECID), knows the emulated LAN's maximum frame size and its LAN type, and has established the control VCC with the LES.

Initial registration: After joining, an LEC may register any number of MAC addresses in addition to the one registered during the join phase.

BUS connect: In this phase a connection is set up to the BUS. The address of the BUS is found by issuing an *LE_ARP* for an ATM address with all 1's. The BUS then establishes a multicast-forward VCC to the LEC.

15.4.5.3. Connections

A LEC has separate VCCs for control traffic and for data traffic. Each VCC carries traffic for only one ELAN. The VCCs form a mesh of connections between the LECs and other LANE components such as LECS, LES, and BUS.

15.4.5.4. Control Connections

A control VCC links the LEC to the LECS and LEC to the LES. The control VCCs never carry data frames and are set up as a part of the LEC initialization phase. The control connection terminology is listed below:

Configuration-direct VCC is a bidirectional point-to-point VCC set up by a LEC as part of the LECS connect phase and is used to obtain configuration information, including the address of LES. This connection may be closed after this phase is over.

Control-direct VCC is a bidirectional point-to-point VCC to the LES set up by a LEC for sending control traffic. This is set up during the initialization process. Because LES has the option of using the return path to send control data to the LEC, this requires the LEC to accept control traffic on this VCC. This VCC must be maintained open by both LES and LEC while participating in the ELAN.

Control-distribute VCC is a unidirectional point-to-multipoint or point-to-point VCC from LES to the LEC to distribute control traffic.

This is optional and LES, at its discretion, may or may not set this up. This VCC is also set up during the initialization phase. This VCC, if set up, must be maintained while participating in the ELAN.

15.4.5.5. Data Connections

Data VCCs connect the LECs to each other and to the BUS. These carry either Ethernet or Token Ring data frames and under special conditions a flush message (optional). Apart from flush messages, data VCCs never carry control traffic:

Data-direct VCC is a bidirectional point-to-point VCC established between LECs that want to exchange unicast data traffic.

Multicast-send VCC is a bidirectional point-to-point VCC from LEC to BUS. It is used for sending multicast data to the BUS and for sending initial unicast data. The BUS may use the return path on this VCC to send data to the LEC, so this requires the LEC to accept traffic from this VCC. The LEC must maintain this VCC while participating in the ELAN.

Multicast-forward VCC is either a point-to-multipoint VCC or a unidirectional point-to-point VCC from the BUS to the LEC after the LEC sets up a multicast-send VCC. It is used for distributing data from the bus. The LEC must attempt to maintain this VCC while participating in the ELAN.

Figure 15.6 shows all type of connections. LEC1 maintains its configuration directly though LEC2 and LEC3 has torn down that connection after becoming operational. Here, all three LECS are operational.

15.4.5.6. Operation

To get to the operational state, i.e., the state at which the LEC can start exchanging information with other LECs, it has to go through an initialization process that consists of several phases. First, if required, it must contact the LECS. This phase is optional and may not exist if a preconfigured switched virtual circuit or permanent virtual circuit (PVC) to LES is used. The LEC will locate the LECS by using the following mechanisms to be tried in the following order: (i) Get the LECS address via interim local management interface (ILMI) using the ILMI Get or ILMI Get Next to obtain the ATM

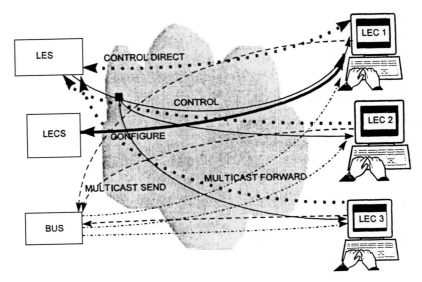

Fig. 15.6   Various lane connections.

address of the LECS for the UNI; (ii) using the well-known LECS address: If LECS address cannot be obtained via ILMI or if LEC is unable to establish a configuration direct VCC to that address, then an ATM Forum specified well-known address "47.00.79.00.00.00.00.00.00.00.00.00-00.A0.3E. 00.00.01-00" hex must be used to open a configuration direct VCC; (iii) using a well-known PVC: If VCC could not be established to the well-known address in the previous step then the well-known PVC of virtual path identifier = 0 and virtual channel identifier = 17 (decimal) must be used.

The configuration phase prepares the LEC for join phase by providing the necessary operating parameters for the emulated LAN that the LEC will join. Once the LECS is found then LEC sends a *LE_Configure_Request* and waits for a *LE_Configure_Response*, which is a part of the LE configuration protocol. All control frames have the structure shown in Fig. 15.7.

Marker is always a fixed 2-byte value "FF00" hex. The op-code determines the type of control frame, e.g., "0001" for *LE_Configure_Request* "0101" for *LE_Configure_Response*, etc. Status is used in the responses to inform about reasons of denial for the requests or to indicate success. Type length values are used to exchange specific information in the control frames such as timer values, retry counts, etc.

Byte Offset

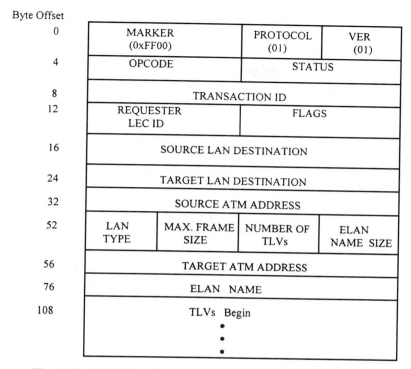

Fig. 15.7   LANE frame format [9]. Copyright 1995 The ATM Forum.

During the configuration, the LECS provides the LEC with the ATM address of LES and also provides all kinds of timers values, time-out periods, and retry counts.

Armed with this information, LEC enters the join phase. Here, the LEC establishes its connection with the LES and determines the operating parameters of the emulated LAN. The LEC implicity registers one MAC address with the LES as a part of the joining process. LEC must initiate the UNI signaling to establish a control-direct VCC (or use a control-direct PVC) and then send a *LE_JOIN_Request* over this VCC to the LES. The LES, may optionally establish a control-distribute VCC back to the LEC. After that the LES will send back a *LE_JOIN_Response* that may be sent on either control direct or control distribute (if created). To each LEC that joins, the LES assigns a unique LECID.

If the join phase is successful then the LEC is allowed to register additional MAC addresses, which it represents with the LES. This is called the registration phase. However, this can happen any time and is not restricted to this phase. However, additional registrations cannot be done before joining the ELAN.

This is followed by the BUS connect phase in which LEC has to establish connection to the BUS. For this purpose, the LEC needs to find out the address of the BUS. This is accomplished by the ARP. In this procedure whenever a LEC is presented with a frame for transmission whose LAN destination is unknown to the client, it must issue LANE ARP (*LE_ARP*) request frames to the LES over its control-direct VCC. The LES may issue an *LE_ARP* reply on behalf of a LEC that had registered the LAN destination earlier with the LES or alternatively can forward the request to the appropriate client(s) using the control-distribute VCC or one or more control-direct VCCs and then the *LE_ARP_Reply* from the appropriate LEC will be relayed back over the control VCCs to the original requester. Each LEC also maintains a local cache of addresses.

For connecting to the BUS, LEC first issues an *LE_ARP_Request* to the LES for the broadcast MAC address, i.e., all 1's ("FFFFFFFFFFFF" hex). The *LE_ARP_Response* gives the ATM address of the BUS. The LEC then proceeds to set up a multicast-send VCC to the BUS, which then immediately opens a multicast-forward VCC back to the LEC. At this point the LEC is considered operational.

Now, if the LEC wants to exchange information with another LEC, it can use the address resolution procedure to get its address and then set up a data-direct VCC to the other LEC and transfer information. However, to save time, if the target LEC's address is not known, then the originating LEC issues a *LE_ARP_Request* and starts sending frames through the BUS. Once the *LE_ARP_Reply* is received the LEC is required to stop using the BUS and open a data-direct VCC. Despite all this, ATM guarantees in-order delivery and therefore a flush message is sent to BUS that ensures that no frames are transmitted on the data-direct VCC until all the previous ones routed through the BUS are delivered. Flush request message is a way to inform the other side that following that request, data will be transmitted on a different channel, e.g., switching from multicast-send to data-direct VCC. The flush request needs to be responded by flush response so that the side issuing the flush request understands that all the previously sent messages have been delivered on the old channel and it is safe to switch channels and still maintain in-order delivery of messages.

15.5. FDDI

15.5.1. STANDARD

FDDI is an ANSI standard and has an ISO approval. The basic FDDI standard [12–14] can be categorized as shown in Fig. 15.8.

It is based on a dual-ring topology composed of two counterrotating rings that operate at a data rate of 100 Mb/s. It uses dual counterrotating rings to enhance reliability: the primary ring and the secondary ring. Secondary ring is used in case of failures. Multiple failures, however, result in multiple rings. Multimode fiber connects each station together and the total ring can be 100 km in length. A maximum of 500 stations can be connected in the ring and hence it forms an ideal backbone network.

FDDI consists of a set of stations serially connected by a transmission medium to form a closed loop. Information is transmitted as symbols (equivalent of 4 information bits) from one active station to the next. Each station, including the destination, simply regenerates each symbol on the ring in the direction of rotation. The destination, however, makes a copy as the information passes it. The frame is removed by the station that originated the transmission.

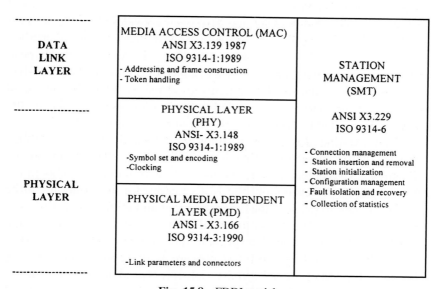

Fig. 15.8 FDDI model.

15.5.2. STATION TYPES

There are two types of FDDI stations: a dual-attach station (DAS), which is connected to both rings, and a single-attached station (SAS), which is attached only to the primary ring. A dual-attached station has at least two ports — an A port, where the primary ring comes in and the secondary ring goes out, and a B port, where the secondary ring comes in and the primary goes out. A station may also have a number of M ports, which are attachments for single-attached stations. Stations with at least one M port are called concentrators. SASs have S ports, which can be connected to the FDDI ring only through concentrators using M ports. A dual ring of DAS devices cannot be readily moved, changed, or added into. A single user disconnecting a dual-attached workstation causes a break in the ring. Dual-attach stations require twice the number of connectors and cables.

15.5.3. PHYSICAL LAYER SPECIFICATIONS

FDDI supports two basic media types — copper and fiber. With copper it works with STP cable and allows data at 100 Mbs for up to 100 m. When using STP cabling, we get FDDI over copper commonly known as copper distributed data interface. Copper cables will be used to connect work-stations to hub only. UTP has also been considered by the X3T9.5 committee. When using fiber, there are three alternatives: single-mode fiber, multimode fiber, and the plastic or low-cost fiber. The single-mode fiber can be used to work at FDDI speed over a distance of 50 km but is expensive. Multimode fiber is the most commonly used media. Plastic or low-cost fiber were on the market for a while but, due to the nature of plastic, the distance was limited to 100 m and interest in them was short lived.

An FDDI station attaches to the fiber optic cable by a media interface connector (MIC) whose main function is to mechanically align the fibers. MIC plugs have mechanical latches that mate with the latch points in the MIC receptacle. As an option, the MIC receptacles of a station can be keyed to prevent improper attachment of input and output. Keys are used to determine the PMD entries contained in the FDDI node. Four keys are defined: type A or red key/label (primary in/secondary out) and type B or blue key/label (secondary in/primary out) are used to connect DAS into an FDDI ring. Type M or green key/label (master) is used on the concentrator side to connect a SAS to a concentrator. Type S or no key/white label (slave) is used on the station side to connect a SAS to the concentrator

Table 15.7   **MIC Connector Key Usage**

FDDI connection type	Key type
Workstation to wall	S/S
Distribution panel to concentrator M port	M/M
Distribution panel to concentrator A port	A/A
Distribution panel to concentrator B port	B/B
Concentrator A to concentrator M port	A/M
Concentrator B to concentrator M port	B/M
Concentrator 1A to concentrator 2B port	A/B
Concentrator 1B to concentrator 2A port	B/A
Workstation to concentration M port	S/M

(i.e., to connect to a port M). Table 15.7 summarizes the general MIC connector key usage and Table 15.8 gives the FDDI interface power levels.

15.5.3.1.   Coding and Symbol Set

FDDI uses 4B/5B encoding, i.e., each 4 bits are encoded into a 5-bit symbol. FDDI communicates all of its information using symbols. This 5-bit encoding provides 16 data symbols (normal hex range: 0–F), eight control symbols (Q, H, I, J, K, T, R, and S), and eight code violation symbols (V) as shown in Table 15.9. After the data are 4B/5B encoded, they are further encoded using non-return-to-zero invert coding before the bits are let out on the media.

15.5.3.2.   Line States

Line states are used by the physical layer to communicate information. A line state is a continuous stream of a certain symbol(s) sent by transmitter (physical layer) that, upon receipt of another station, will uniquely identify

Table 15.8   **FDDI Interface Power Levels**

	Output interface		Input interface	
	Min	**Max**	**Min**	**Max**
Center wavelength	1270 nm	1380 nm	1270 nm	1380 nm
Average power	−20.0 dBm	−14.0 dBm	−31.0 dBm	−14.0 dBm

Table 15.9   **FDDI Symbol Set**

Data symbols			Control symbols		
Symbol	Binary	Bit stream (4B/5B encoded)	Symbol		Bit stream (4B/5B encoded)
0	0000	11110	Q	QUIET	00000
1	0001	01001	I	IDLE	11111
2	0010	10100	H	HALT	00100
3	0011	10101	J	Used in Starting Del.	11000
4	0100	01010	K	Used in Starting Del.	10001
5	0101	01011	T	Used in Ending Del.	01101
6	0110	01110	R	RESET (logical 0)	00111
7	0111	01111	S	SET (logical 1)	11001
8	1000	10010		Code violations	
9	1001	10011	V or H	To be taken as HALT	00001
A	1010	10110	V or H	To be taken as HALT	00010
B	1011	10111	V		00011
C	1100	11010	V		00101
D	1101	11011	V		00110
E	1110	11100	V or H	To be taken as HALT	01000
F	1111	11101	V		01100
			V or H	To be taken as HALT	10000

the state of the communication line. The following are the commonly encountered line states:

- Quiet line state: A continuous stream of "Q" symbols. This line state is used to break a link and restart a connection.
- Halt line state: A continuous stream of "H" symbols. This line state is used during connection management signaling.
- Master line state: An alternating series of "H" and "Q" symbols. This line state is used to indicate a trace and also used during physical connection management (PCM) signaling.

- Idle line state: A continuous stream of "I" symbols. This line state is used to separate information bits and to provide clock synchronization.

- Active line state: This line state is asserted by a "JK" pair and continues to be asserted until it receives "R," "S," or "T" symbols. It is used by MAC layer to transmit data frames.

- Link state unknown: This is an indication by the physical layer if the receiver is unable to determine the line state.

- Noise line state: This is an indication by the physical layer if the receiver has received enough symbols but is unable to uniquely recognize the line state (i.e., the receiver is getting garbled information on the line).

15.5.4. MAC SPECIFICATIONS

FDDI is a controlled-access network. The access to transmission is through a token. The token is a particular unique combination of symbols that circulates on the ring in one direction from station to station. A station receives the token in its entirety on its input port and then regenerates the token on the output port to continue circulation. In order to transmit, a station waits for the token and on receipt of the token removes it from the ring, thereby capturing the token. It then puts out the frame(s) it wants to transmit and then at the end introduces a new token. A station can hold on to a token for a finite time defined as the token-holding time. The station that transmits a frame is also responsible for removing the frame from the ring. The basic two protocol data units that are used by the FDDI MAC are the token and the frames.

15.5.4.1. Token

The general format of a token is shown in Fig. 15.9. The preamble (PA) is a minimum of 16 symbol of idles. Because of ring latency and timing constraints, subsequent stations may see a smaller length of preamble. Tokens are recognized when received with a preamble of zero or greater length. The start delimiter is made up of two symbols, J and K, which will not be seen anywhere except to mark the beginning of a new token or a frame. Similarly, the end delimiter is made of two symbols, which are both a T. For a token there are two possible values (as shown previously) for the

Fig. 15.9   FDDI token format.

frame control field depending on whether it is a nonrestricted or restricted token. The nonrestricted mode is the normal mode of operation in which the bandwidth is shared equally among all requesters. The restricted mode is used to dedicate bandwidth to a single extended dialog between specific requesters (these are described later in this section).

15.5.4.2.   Frame

FDDI defines three basic types of frames: (i) the MAC frames, which carry MAC control data; (ii) logical link control frames (LLC), which carry user–user information; and (iii) Station management frames (SMT), which carry management information between stations and layers. MAC frames include frames used in ring initialization and beacon frames used in the process of ring fault isolation. These frames never leave the FDDI ring (i.e., they never go across a bridge or a router) because they are specific for a ring only. LCC frames carry user information between nodes on a network. These frames do cross bridges or routers. SMT frames, like MAC frames, do not cross bridges or routers because they are used to control the operation of a ring only.

Figure 15.10 shows the general format used to transmit the LLC, MAC, and SMT frames. The SMT frames are discussed later and this section

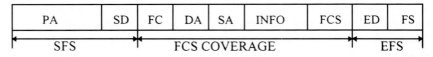

Fig. 15.10   FDDI LLC and MAC frame format. SFS, start of frame sequence; SD, start delimiter (2 symbols); PA, preamble (≥4 symbols); FC, frame control (2 symbols); DA, destination address (4 or 12 symbols); SA, source address (4 or 12 symbols); INFO, information; FCS, frame check sequence (8 symbols); EFS, end of frame sequence; ED, ending delimiter (1 or 2 symbols); and FS, frame status (3 symbols).

will focus on MAC and LLC frames: The preamble, start delimiter, and the end delimiter are the same as in a token. The frame control is used to distinguish between a token, MAC frame, LLC frame, or SMT frame. The Fibre Channel Standard (FCS) is a 32-bit polynomial checksum.

The frame status consists of three indicators that may have one or two values. The indicators can either be set (S) or reset (R). Every frame is originally transmitted with all the indicators set to R. The indicators can be set by intermediate stations when they retransmit the frame. The three indicators are error, address recognized, and copy. Error is set when a station determines that the frame is in error. This may be due to a format error or failure of cyclic redundancy checksum (CRC). The address recognized bit is set when a station receives the frame and determines that the destination address is its own MAC address (or if it was broadcast, then the first recipient sets it). Copy indicator is set when the destination station is able to copy the contents into its buffers.

15.5.4.3. Ring Operation

Access to the ring is controlled by passing a token around the ring. The token gives the downstream station (relative to the one passing the token) an opportunity to transmit a frame or a sequence of frames. If a station wants to transmit, it must first capture the token, i.e., strip the token off the ring and then begin transmitting its eligible queued frames. After finishing transmission, the station immediately issues a new token for use by downstream stations, without waiting for frames to return on the ring. Optionally, only for various station management functions, the station may wait to see one or more or all the transmitted frames return before it issues the token. If a station has nothing to transmit, it acts as a simple repeater, i.e., it repeats the incoming symbol stream. While repeating, the station must determine if the information is destined for it. This is done by matching the destination address (DA) to its own address or a relevant group address. If a match occurs, then the subsequent symbols up to FCS are processed by the MAC (for MAC frames) or sent to LLC (for LLC frames).

Each transmitting station is responsible for stripping its own transmitted frames from the ring. This is done by stripping the remainder of each frame whose source address (SA) matches the station's address and replacing it with IDLE symbols. This process leaves remnants of frames, consisting of PA, start delimiter (SD), frame control (FC), DA, and SA fields, followed by IDLE symbols, because the decision to remove the frame is based on matching of the SA field and that cannot occur until after the initial part of the frame is repeated. Remnants are easily distinguishable because they are always followed by IDLE symbols. They are removed from the ring when they encounter a transmitting station.

15.5.4.4.   Ring Scheduling

Transmission of normal data on the ring is controlled by a timed token protocol. This protocol supports two types of transfer: (i) asynchronous — dynamic bandwidth sharing and (ii) synchronous — guaranteed bandwidth and response time. The asynchronous transfer is typically used for those applications that are bursty and whose response time is not critical. The bandwidth is instantaneously allocated from the pool of remaining unallocated or unused bandwidth. On the other hand, synchronous transfer is the choice for those applications that are predictable enough to permit a preallocation of bandwidth [via station management (SMT)].

Each station maintains a token-rotation timer (TRT) to control the ring scheduling. A target token rotation time (TTRT) is negotiated during ring initialization via the claim token bidding process. Initially, TRT is set to TTRT. A token arriving before the TRT reaches zero, called an early token, causes the TRT to be reset, and can be used for transmitting both the synchronous and asynchronous traffic. On the other hand, a token arriving after TRT reaches zero, called a late token, can be used only for synchronous transmissions. Each station also maintains a counter called the late counter (LC), which records the number of times the token was late and a token holding timer (THT), which dictates how long a station can hold the token.

This protocol guarantees an average TRT (or average synchronous response time) not greater than TTRT, and a maximum TRT (or maximum synchronous response time) not greater than twice TTRT.

When a station receives an early token, i.e., $LC = 0$, it saves the remaining time of TRT in THT and then resets and enables TRT for countdown, i.e.,

$THT = TRT$

$TRT = TTRT$

Enable TRT countdown

Transmit synchronous frames (if any)

Enable THT countdown

Transmit asynchronous frames as long as $THT > 0$

If the received token is a late token, i.e., LC > 0, then

LC = 0

TRT not changed and continues to run

Transmit only synchronous frames

Each station has a known allocation of synchronous bandwidth, i.e., the maximum time that the station may hold the token without THT being enabled. Synchronous bandwidth is expressed as a percentage of TTRT, i.e., a station would require a 100% allocation to transmit for a time equal to TTRT before issuing a token. The sum of current allocations of all stations should not exceed the maximum useable synchronous bandwidth of the ring (like successive expiration of TRT with late counter in the MAC or invalid frames on the ring or a logical/physical break in the ring).

15.5.4.5.   Claim Token Process

Any station detecting a requirement for ring initialization shall initiate a claim token process. In this process one or more stations bid (TTRT) for the right to initialize the ring by continuously transmitting claim frames. Each station also looks for incoming claim frames and compares the received bid with the station's own bid. Any station receiving a lower bid shall enter or reenter the bidding, whereas any station receiving a higher bid shall yield. The claim token process is completed when one station receives its own claim frames after the frames have passed around the ring. At this point all stations around the ring have yielded to this station's claim frame. The winning station proceeds to initialize the ring.

15.5.4.6.   Beacon Process

When a station detects that claim token process has failed or upon a request from SMT, it initiates the beacon process. This happens when the ring has, most probably, been physically interrupted or reconfigured. The purpose of the beacon process is to signal to all remaining stations that a significant logical break has occurred and to provide diagnostic or other assistance to the restoration process. In this process, the station begins continuous transmission of beacon frames. If a station receives a beacon frame from its upstream neighbor, it repeats that beacon and stops its own. If it receives its own beacon, it assumes that the logical ring is restored and it stops beaconing and starts the claim token process. If a station receives no beacon and it has been continuously beaconing for at least the time indicated by stuck beacon timer, then ring management begins the stuck beacon recovery procedure. The recovery procedure starts with a directed beacon and eventually traces the stations around the fault, at which time all stations around the fault remove themselves from the ring, perform a self-test, and can rejoin the ring if they pass those tests.

15.5.5. STATION MANAGEMENT

Station management provides the control necessary at the station (node) level to manage the processes under way in the various FDDI layers such that a station may work cooperatively as a part of an FDDI network. SMT provides services such as connection management, station insertion and removal, station initialization, configuration management, fault isolation and recovery, scheduling policies, and collection of statistics. A variety of internal node configurations are possible. However, a node shall have one, and only one, SMT entity. It may, however, have multiple instances of MACs, PHYs, and PMDs. For a SMT frame, the information field in the general frame format is occupied by a SMT header and a SMT information portion. The SMT header is the protocol header for all SMT frames. SMT information is the information indicated by the SMT header. An SMT frame has subfields as shown in Fig. 15.11.

15.5.5.1. SMT Header

SMT frames are identified by their frame class, e.g., neighbor information frame [NIF(01 hex)], configuration status information frame [SIF-Cfg (02 hex)], operation status information frame [SIF-Oper (03 hex)],

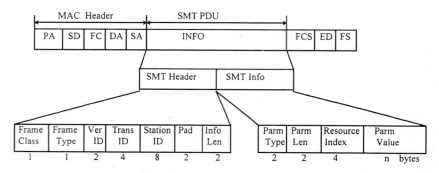

Fig. 15.11   SMT frame format.

Echo frame [ECF (04 hex)], and so on. The frame type is an indicator of whether this frame is an announcement (01 hex), a request (02 hex), or a response (03 hex). The *Version_ID* field indicates the structure of SMT into field. At the moment only two possible values are acceptable — 0001 hex for stations using a version lower than 7.x and 0002 hex for stations using version 7.x of SMT. To ensure backward compatibility, NIF, SIF, and ECF frames will have a constant *Version_ID* of 0001 hex. *Transaction_ID*'s sole purpose is to match the request and response frames. The *Station_ID* is the unique identifier of the FDDI station transmitting the SMT frame. The pad field is two bytes of all zeros. This is used to make the header an even 32 bytes. The length of the information field is reported in the info length field. This value does not include the length of the SMT header or this field. Its value can be between 0 and 4458 bytes.

15.5.5.2. SMT Information

The SMT info field consists of a list of parameters. If more than one parameter is present in the frame, they will be listed one after another.

NIFs are used by a station for periodic announcement of its MAC address and basic station description. NIFs are used in the neighbor notification protocol that allows a MAC to determine its logical upstream neighbor address and logical downstream neighbor address. The protocol also detects duplicated MAC addresses on an operational ring. NIFs can be used by a monitoring station to build a ring map.

SIFs are used, in conjunction with the SIF protocol, to request and provide in response to a station's configuration and operation information. There are two classes of SIFs to provide this function: the SIF configuration and the SIF operation request and response frames. Potential uses are fault isolation and statistics monitoring.

ECFs are defined for SMT-to-SMT loopback testing on an FDDI ring using the ECHO protocol. The ECH frames may contain any amount of echo data (up to the maximum frame size). Potential uses include confirming that a station's port, MAC, and SMT are operational.

Resource allocation frames are defined to support a variety of network policies such as allocation of synchronous bandwidth.

Request-denied frames are used to notify a requester that its request has been denied because of version or protocol problems, e.g., if a station receives a request with a *Version_ID* that is not supported or if the request frame has a length error.

Status report frames are used by a station to announce the station status, which may be of interest to the manager of the ring. These are sent when certain conditions become (in)active. The conditions that are reported include frame errors, link errors, duplicate address, peer wrap condition, and not-copied condition.

Parameter management frames (PMFs) provide the means for remote access to station attributes via the parameter management protocol. They operate on all SMT MIB attributes. There are two types of PMFs: PMF-Get frames to look up a value and PMF-Set frames to change a value. Not all stations support the PMF-Set protocol.

15.5.5.3.   Physical Connection Management

Within each FDDI station there is one SMT entity per port, called the physical connection management (PCM) entity. The number of PCM entities within a station is exactly equal to the number of ports of that station. PCM entities are a part of SMT that control the operations of the ports. In order to make a connection, two ports must be physically connected to each other by means of fiber optic or copper cable. When this happens, the PCMs that are responsible for those ports can recognize each other's existence and begin communicating. They do this by sending line states out of the port onto the cable. The PCM at the other end of the connection will recognize the line state and respond accordingly.

When the PCM sees another PCM on the other end of the connection, they will synchronize and communicate with each other. During this communication, the following important things happen: (i) the PCMs figure out the type of each port and determine if they are compatible and (ii) the PCMs perform a link confidence test (LCT). The LCT determines if the quality of the link is good enough to establish a connection. If not, then a connection will not be made (allowed); otherwise, PCM will establish a connection and place the ports on a token path that goes through that station. At this point the ports become a part of the network and data can be sent through these ports.

15.6.   FDDI-II

FDDI-II is an enhancement to the original FDDI being developed by the X3T9.5 committee. Both FDDI and FDDI-II run at 100 Mb/s on the fiber. FDDI can transport both asynchronous and synchronous types of frames. FDDI-II has a new mode of operation called hybrid mode. Hybrid mode uses a 125-μs cycle structure to transport isochronous traffic in addition to synchronous/asynchronous frames. FDDI-II supports integrated voice, video, and data capabilities and therefore expands the range of applications of FDDI. FDDI and FDDI-II stations can be operated in the same ring only in the basic mode.

FDDI-II has the ability to carry the isochronous (constant bit rate) and multimedia data, which is sometimes not possible in the original FDDI due to variation in delay that can be obtained with the token access control method.

FDDI-II nodes can run in basic mode. If all nodes on the ring are FDDI-II nodes, then the ring can switch to the hybrid mode in which isochronous service is provided in addition to basic mode services. In the basic mode on FDDI-II, synchronous and asynchronous traffic is transmitted in a manner identical to that on FDDI. In the basic mode, FDDI-II behaves in the same way as FDDI, i.e., all ring accesses for transmission are controlled by a token and the available bandwidth is time shared.

In the alternative mode, also called the hybrid mode, the bandwidth is divided into channels using time-division multiplexing, whereas some of the channels can be reserved for isochronous data. The establishment of multiple channels is done by a station known as the "cycle master," which also controls the utilization of the channels. To provide isochronous service, FDDI-II deploys a special frame called a cycle. A cycle is generated every 125 μs. At 100 Mbs, 1262.5 bytes can be transmitted in 125 μs. The bytes of the cycles are preallocated to various channels on the ring. The 1560 bytes of the cycle are divided into 16 wideband channels (WBCs) that can carry either circuit-switched or packet-switched traffic. All packet-switched WBCs are concatenated together to form a single channel, which is operated with the FDDI-timed token protocol.

Despite the strengths and the characteristics of FDDI-II, it never caught on in the market. Very few vendors have products supporting this standard and the interest seems to be dying.

References

1. IEEE Std 802.3. Local and metropolitan area networks, supplement—Media access control (MAC) parameters, physical layer, medium attachment and repeater for 100 Mb/s operation, type 100BASE-T (clauses 21–30). Copyright ©1995 by the Institute of Electrical and Electronics Engineers, Inc. The IEEE disclaims any responsibility or liability resulting from the placement and use

in the described manner. Information is reprinted with the permission of the IEEE.

2. Molle, M., and G. Watson. 1996. 100Base-T/IEEE 802.12/packet switching. *IEEE Commun. Mag.* **34**(8):64–73.

3. F. Halsall. 1995. *Data communications, computer networks and open systems.* 4th ed. Reading, Mass.: Addison-Wesley.

4. IEEE 802.3z, http://stdbbs.ieee.org/pub/802_main/802.3/gigabit.

5. IEEE Std 802.12. Demand-priority access method, physical layer and repeater specifications for 100Mb/s operation. Institute of Electrical and Electronic Engineers, 345 East 47th Street, New York, N.Y. 10017.

6. Watson, G., A. Albrecht, J. Curcio, D. Dove, S. Goody, J. Grinham, M. P. Spratt, and P. A. Thaler. 1995. The demand priority MAC protocol. *IEEE Network* **9**(1):28–34.

7. RFC 1577. Classical IP over ATM, request for comments. Internet Engineering Task Force.

8. D. E. Comer. 1995. *Internetworking with TCP/IP — Volume 1,* 3rd ed. Englewood Cliffs, N.J.: Prentice Hall.

9. ATM Forum. 1995. *LAN Emulation over ATM Version 1.0,* af-lane-0021.000. ATM Forum, 303 Vintage Park Drive, Foster City, Calif. 94404-1138.

10. Siu, K., and R. Jain. 1995. A brief overview of ATM: Protocol layers, LAN emulation and traffic management. *ACM SIGCOMM* Comput. *Commun. Rev.* **25**(2):6–20.

11. Finn, N., and T. Mason. 1996. ATM LAN emulation. *IEEE Commun. Mag.* **34**(6):96–100.

12. ANSI X3T9.5. FDDI — Physical layer medium dependent (PMD). American National Standards Institute, 1430 Broadway, New York, N.Y. 10018.

13. ANSI X3.148. FDDI — Token ring physical layer protocol. American National Standards Institute, 1430 Broadway, New York, N.Y. 10018.

14. ANSI X3.139. FDDI — Token ring media access control (MAC). American National Standards Institute, 1430 Broadway, New York, N.Y. 10018.

Part 4 | The Manufacturing Technology

Chapter 16 | Semiconductor Laser and Light-Emitting Diode Fabrication

Wenbin Jiang
Michael S. Lebby

Phoenix Applied Research Center, Motorola, Incorporated,
Tempe, Arizona 85284

16.1. Introduction

In Chapter 2, we briefly described the technology behind light-emitting diodes (LEDs), edge-emitting lasers, and vertical cavity surface-emitting laser (VCSEL) sources. We will now focus on manufacturing issues for these optical sources, such as material growth and device fabrication. We will concentrate on the fabrication of laser sources because manufacturing LEDs uses most of the same principles as those in the fabrication of edge-emitting semiconductor laser diodes, albeit with simpler manufacturing in general. VCSEL manufacturing will be described in detail because this is an emerging field that will potentially have a great impact on the data communication industry.

16.2. LED Fabrication

Figure 16.1 shows a cross-sectional and top view of an LED. In manufacturing an LED, a substrate such as GaAs or GaP is grown as an ingot, using a seed crystal of appropriate crystal orientation. During the crystal growing process, the ingot is doped n type and possibly with other dopants such as sulfur (S) consistent with the bandgap associated with the desired

603

Fig. 16.1   Cross-sectional and top view of an LED.

wavelength of emitted light, as described in Chapter 2. The ingot is sawed to make wafers or starting substrates.

Epitaxial (epi) layers are deposited using a chemical vapor deposition process, with the composition of the epi layers gradually changed to maintain a perfect crystal lattice. Appropriate dopants, again consistent with the desired bandgap and n-type doping, are incorporated during the epitaxial deposition process. The wafers are then coated with a layer of a dielectric, such as silicon nitride, using chemical vapor deposition.

The silicon nitride is patterned using photolithography, and the silicon nitride is selectively etched. A p-type dopant such as zinc is introduced to the surface of the wafer, with the silicon nitride acting as a barrier against the diffusion of p-type dopant wherever it remains after the selective etch process. The diffusion of zinc into the opening forms the pn junction for the light-emitting diode.

Metal such as AuBe or aluminum is deposited using a process such as sputter deposition and patterned using photolithography and metal etch processing so that it contacts the p-type diffused area and also extends over the silicon nitride-capped region, providing a bonding pad.

The back of the wafer is exposed, and a metal such as AuGe or AuSn is deposited by sputter or evaporative deposition onto the back of the wafer, forming the backside contact.

16.3. Material Growth — Lasers

Edge-emitting semiconductor lasers and vertical cavity surface-emitting lasers are based on the same material system but with very different epitaxy growth control requirements. VCSEL technology is in essence an epitaxy growth technology because a VCSEL structure may consist of more than 100 epitaxy layers. The thickness and the doping level of each layer must be precisely controlled. Any slight deviation to the designed parameters may affect the VCSEL performance, which will end up affecting the final manufacturing yield. Fortunately, modern epitaxy techniques have made the VCSEL growth possible after many years of development. In contrast, edge-emitting semiconductor laser epitaxy growth involves much fewer steps and is more tolerant to the parameter variation.

16.3.1. EDGE-EMITTING LASERS

Three major epitaxial growth techniques are used for edge-emitting laser diodes: liquid phase epitaxy (LPE), metal organic chemical vapor deposition (MOCVD), and molecular beam epitaxy (MBE). LPE was the dominant growth technique in the early stage of the semiconductor laser development [1] but has gradually been replaced by MBE and MOCVD techniques due to the requirements for high quality, high uniformity, and high reproducibility in heteroepitaxial material growth, especially in the growth of quantum well structures. Today, the vast majority of the commercial semiconductor lasers are grown by either MBE or MOCVD.

Molecular beam epitaxy offers an advantage to grow very thin active layers down to only one monolayer because of its slow growth rate. It is therefore an excellent technique to grow single-quantum well or multiquantum well lasers. In practice, MBE has been used for the growth of commercial GaAs/AlGaAs laser structures. In GaAsP/InP and InGaAlP/InGaP structures have also been demonstrated with MBE [2, 3]. Figure 16.2 is a schematic diagram of a MBE system. The MBE growth is conducted in an ultrahigh vacuum (UHV) of approximately 10^{-10} torr with heated effusion cells filled with Ga, Al, As, and doping sources to produce an adequate beam flux onto the substrate. The effusion cells are enclosed in liquid nitrogen-cooled shrouds. The Al content of $Al_x Ga_{1-x} As$ is controlled by varying the partial gas pressures of Al and Ga. The typical doping sources are Si or Sn for n type and Mg or Be for p type. During the growth, the substrate can be rotated to achieve the thickness uniformity across the whole wafer.

Fig. 16.2   Schematic diagram of a MBE system.

Reflection high-energy electron diffraction (RHEED) is usually used for *in situ* growth monitoring. The detailed description of MBE growth can be found in Ref. [4]. Recently, MBE techniques have also been used for GaN-based material growth [5, 6], but the crystal quality has not reached that grown by MOCVD techniques.

MOCVD, also known as metal organic vapor-phase epitaxy, offers excellent growth uniformity across a wafer and run-to-run reproducibility. Because it has the advantages of high growth rate and multiwafer growth capability, MOCVD has become the dominant technique in the growth of commercial semiconductor lasers.

Figure 16.3 is a schematic diagram of a typical MOCVD system. The growth sources used for MOCVD are arsine (AsH_3) or phosphine (PH_3) for group V elements and trimethylgallium (TMGa), trimethylaluminum (TMAl), and trimethylindium (TMIn) for group III elements. Some reactors also use triethylgallium (TEGa), triethylaluminum (TEAl), and triethylindium (TEIn) sources as group III elements. During the growth,

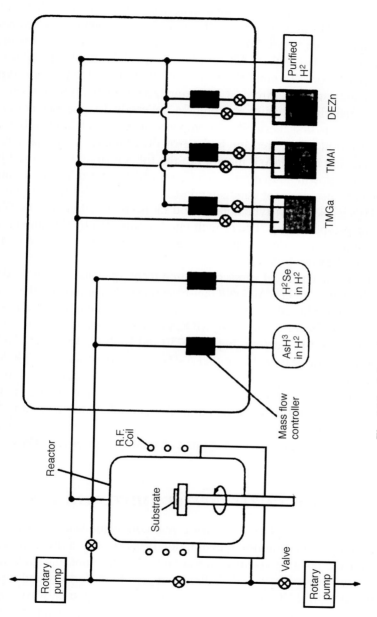

Fig. 16.3 Schematic diagram of a MOCVD system.

liquid-state group V sources having high vapor pressure at room temperature and group III AsH_3 source are transferred to the reactor by bubbling with hydrogen (H_2) as a carrier gas. The sources are decomposed at temperatures ranging from 500 to 800°C in the reactor to form III–V compound crystal structures. Dopants for AlGaAs-based materials are silicon (Si) from disilane (Si_2H_6) or tellurium (Te) from diethyltellerium (DETe) for n-type materials, and carbon (C) from CCl_4 or zinc (Zn) from diethylzinc (DEZn) for p-type materials. Because phosphine is not easily decomposed at room temperature, it is first thermally decomposed in a special furnace before being transferred into the reactor. In the growth of InGaAsP/InP structures, lattice matching of InGaAsP material to the InP substrate is achieved by adjusting the growth source ratio. The group III to group V source ratio adjustment is also used for bandgap control. Selenium (Se) from H_2Se or S from H_2S is used for n-type doping, and Zn from dimethylzinc (DMZn) is used for p-type doping. The growth rate is controlled by the amount of group V sources introduced into the reactor, i.e., by the flow rate of H_2 carrier gas for group V elements.

Metal organic chemical vapor deposition techniques have been used to grow high-quality InGaN/AlGaN/GaN-based crystal structure for blue visible LED. Visible blue edge-emitting laser has also been demonstrated based on the III-nitride materials grown by MOCVD. The material growth temperature in the MOCVD reactor is between 750 and 1050°C. Ammonia (NH_3) is the source of nitride to replace either AsH_3 or PH_3 sources in the MOCVD reactor for the growth. Magnesium (Mg) from *bis*-cyclopentadienyl magnesium (CP_2Mg) is the p-type dopant, and Si from silane (SiH_4) is the n-type dopant. Postgrowth thermal annealing at 700°C for 10 min in N_2 atomosphere [7] or low-energy electron beam irradiation [8] have been effective in activating the p-type dopants and reducing the p-type GaN film resistance.

16.3.2. VERTICAL CAVITY SURFACE-EMITTING LASERS

Although the growth of edge-emitting semiconductor lasers is straightforward and involves minimum calibration because only a few layers are needed for the positive–intrinsic–negative (PIN) structure, the growth of VCSEL structures poses a real challenge in growth uniformity and reproducibility. A typical VCSEL consists of more than 100 quarter-wavelength distributed Bragg reflector (DBR) mirror and multiple quantum well (MQW) active layers. The total variation of each layer thickness across the wafer cannot exceed 1% during the growth. Such a precise thickness control is needed for two reasons. First, the net optical gain in the VCSEL cavity is very small because of the thin active layer. Highly reflective DBR mirrors, typically in excess of 99.5%, are needed to ensure that the laser operates effectively. In order to achieve such a high reflectivity, each repetitive layer in the DBR mirror has to be exactly the same, i.e., one quarter wavelength for each layer, to retain appropriate constructive interferences among those layers. Second, the cavity resonance peaks of a VCSEL are usually spaced widely apart due to the typical one-λ cavity. The requirement that the spectral position of this resonance fall within the optical gain region of the active MQWs places a severe constraint on the optical thicknesses of the layers making up the cavity.

Figure 16.4 shows the simulation results of net reflectivity of a completed 850-nm GaAs VCSEL structure grown on a GaAs substrate that consists of a top DBR, a bottom DBR, and an active region sandwiched between, as shown in Fig. 2.14 in Chapter 2. Figure 16.4a is the reflectivity of an idealy grown structure. A deep Fabry–Perot resonate peak sits in the center of the reflective band (850 nm). Figure 16.4b illustrates the reflectivity of a structure with the top DBR wavelength being longer than the bottom DBR wavelength by 1%. The resonate peak shifts toward the longer wavelength and becomes less pronounced, and a small glitch appears on the long-wavelength end of the high-reflectivity band. Figure 16.4c illustrates the reflectivity of a structure with the top DBR wavelength being shorter than the bottom DBR wavelength by 1%. The resonate peak shifts toward the shorter wavelength and becomes less pronounced, and a small glitch appears on the short-wavelength end of the high-reflectivity band. Figure 16.4d illustrates the laser cavity being 1% longer than the designed wavelength of 850 nm. The center resonate peak moves toward the longer wavelength, and the depth of the resonate peak is reduced. Figure 16.4e illustrates the laser cavity being 1% shorter than the designed wavelength of 850 nm. Accordingly, the center resonate peak moves toward the shorter wavelength, and the depth of the resonate peak is also reduced. In order to achieve better than 1% growth control for the VCSEL structure, not only is the exact knowledge of the refractive index of the required alloy concentration is needed but also the precise growth rate has to be determined and maintained throughout the growth. Typically, the simulation results as shown in Fig. 16.4 are used as references to judge if the growth is successful or how much deviation there is from a designed structure. The information drawn from the comparison will be used to make corrections for the

Fig. 16.4 Net reflectivity simulation results of a complete GaAs VCSEL structure grown on a GaAS substrate for (A) a perfectly grown structure with identical top and bottom DBR mirrors, (B) the top DBR center wavelength being 1% longer than the bottom DBR center wavelength, (C) the top DBR center wavelength being 1% shorter than the bottom DBR center wavelength, (D) the VCSEL cavity being 1% longer than the designed wavelength, and (E) the VCSEL cavity being 1% shorter than the designed wavelength.

subsequent growth runs. A successful growth requires an extremely stable epitaxial growth reactor that has the necessary long-term reproducibility and controllability. Similar to the growth of edge-emitting semiconductor laser materials, two types of state-of-the-art epitaxial techniques are typically used to grow VCSEL structures: MBE and MOCVD.

A modern MBE growth system can usually take a 2-in. or a 3-in. GaAs wafer. The growth process takes 8–12 h to complete the entire VCSEL structure. The standard high-temperature effusion cells provide group III sources, such as Al, Ga, and In, and a group V source such as As. The MBE system is also equipped with cells for n-type and p-type dopants, which are usually Si and Be, respectively. Modifications are usually made to the MBE system to reduce the growth rate transients to less than 1% during the repetitive switching of constituents for the VCSEL multiple layers. When substrate rotation is employed during the growth, 1% lateral thickness uniformity across entire 2-in. wafers can be achieved. This will lead to high yield and high uniformity across the wafer, as is required for manufacturing product.

RHEED oscillation measurements can be used to count the atomic layers being deposited, thus precisely monitoring the growth thickness [9, 10]. However, RHEED can only be used to monitor one particular spot on a wafer. When RHEED is used for *in situ* growth monitoring, the wafer is not allowed to rotate, leading to approximately 5%/cm linear variation in layer thickness across the wafer. Although such a variation may be advantageous in the research environment, it is not acceptable to the commercial application due to wafer growth nonuniformity affecting production yield. Therefore, RHEED oscillation calibrations are usually performed prior to the actual growth on a small test sample centrally mounted on a different sample block. The reevaporation effect of group II species can be a problem that will affect the crystal quality grown by MBE. This effect is strongly substrate temperature dependent. To achieve good crystal quality, the GaAs-AlGaAs-AlAs layers are usually grown at a substrate temperature of 580–630°C and InGaAs at 520–530°C.

The growth rate calibration acquired just before the actual growth is only accurate to approximately 1%. It is not only time consuming but also subject to drift. A modified RHEED *in situ* monitoring scheme is based on discrete substrate rotations [11]. Following the growth of each pair of quarter-wavelength Bragg layers, the growth is paused for a few seconds and the substrate is rotated 180°. Any growth nonuniformity across the wafer will be compensated during the subsequent growth after the substrate rotation. The net effect of the discrete rotation will be equivalent to that of the continuous substrate rotation.

Optical reflectometry measurement is another effective *in situ* monitoring technique in which the bottom DBR mirror and most of the VCSEL cavity, ranging between 94 and 100% of the cavity, is grown first. The partially finished wafer is then taken out to evaluate the reflectivity spectrum.

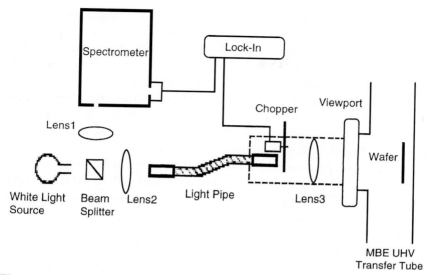

Fig. 16.5 An *in situ* reflectivity measurement system for growing VCSELs by a MBE system. (Reprinted with permission from Ref. [13]. Copyright 1992 American Institute of Physics.)

A comparison with the simulation will reveal information about how to complete the rest of the cavity and the top DBR [12].

A more rigorous *in situ* reflectivity measurement apparatus is shown in Fig. 16.5 [13], in which the measurement is conducted at room temperature through the viewpoint of the MBE transfer tube. The wafer remains in the UHV environment for the measurement so that any potential oxidation-induced defects can be avoided at the interface between the portions grown and the subsequent growth. In order to minimize the weight and bulk of the equipment attached to the viewport, a light pipe is used to direct the white light toward and receive the reflected light from the wafer. A chart recorder synchronized with the scan rate of the spectrometer records the unnormalized reflectivity spectra. During the growth, the reflectivity spectrum is first measured after only a few periods of the bottom DBR mirror have been grown. Any relative error in the center wavelength determined from the measurement will be used to correct the subsequent growth. The second interruption and reflectivity measurement can be made just after the growth of the cavity. However, because the bottom DBR mirror reflectivity is very high and cavity absorption is low, the Fabry–Perot resonant dip is very shallow and sometimes difficult to identify. To enhance the resonant dip visibility, two or three periods of the top DBR mirror can be grown before the second reflectivity spectrum measurement. Any cavity thickness correction can then be made by adjusting the thickness of the immediate one layer in the top DBR mirror when the growth resumes.

Other methods, such as multiwavelength pyrometric interferometry measurements [14] and single-wavelength pyrometric interferometric measurements [15, 16], have also demonstrated that they are capable of *in situ* monitoring the VCSEL growth precisely to within 1%.

Although the predominant number of research publications on near-infrared VCSELs (780–980 nm) are based on MBE, MOCVD has its own advantages in VCSEL applications due to its high growth rate, wide range of host and dopant species, and capability of continuously varying material composition.

In a high-volume manufacturing environment, VCSEL growth time of 8–12 h by a MBE reactor is feasible but may suffer from a high cost of ownership and low-throughput arguments. Conversely, a MOCVD reactor can achieve growth rates approximately three times as high as MBE [17] while maintaining $\pm 1\%$ thickness uniformity across the entire wafer. Reports of good yields have also been demonstrated with multiwafer MOCVD reactors, which can hold up to 15×2-in. wafer loads [18].

In order to maintain sharp doping profiles and prevent the dopant out-diffusion into the active region of the device, dopants with low diffusion coefficients must be used. Zinc is the most commonly used p dopant in MOCVD growth [19] and is the dominant dopant in GaAs-based LEDs. It possesses, however, a high diffusion coefficient and is therefore not recommended for the VCSEL p dopant from device performance consideration [20]. The other commonly used p-doping alternative is carbon (C), which has proven to be a reliable p dopant with very low diffusivity and high solubility [21–24]. The most commonly used carbon precursor for p doping is carbon tetrachloride (CCl_4). The growth rate of C-doped AlGaAs is usually lower than that of undoped or n-doped AlGaAs with the effect of slow etching of AlGaAs growth surface while doping with CCl_4. It should be noted that CCl_4 is an ozone-depleting material and its use has been discouraged from the environmental consideration. Therefore, the search for alternative C-doping method has been ongoing within the research community. One candidate is the use of organometallic As sources such as tertiarybutylarsine (tBAs). Tertiarybutylarsine allows the incorporation of intrinsic carbon into the grown material due to enhancement of background carbon levels from organometallic decomposition and has the advantage of being approximately on an order of magnitude less toxic than arsine. This has encouraged research centers such as universities to install MOCVD reactors in addition to MBE reactors [25–27].

Low series resistance in the p-doped DBR mirror ensures high VCSEL performance, which is usually achieved by grading the heterointerfaces between the alternative DBR mirror pairs, as discussed in Chapter 2. In MBE, the continuous interface grading can be accomplished by varying source cell temperature [28, 29] or by repeatedly switching the cell shutters to create a superlattice grading [30]. Even though the techniques have been demonstrated with high reproducibility, the processes are not easy to implement from the manufacturing standpoint. In comparison, the continuous grading of an arbitrary composition profile can be readily achieved with MOCVD, and it is therefore expected that within the near future MOCVD will dominate the growth process.

The MOCVD growth is typically carried out in a commercially available horizontal quartz tube reactor with graphite susceptor and fast-switching run-vent manifold. Sources used are TMIn, TEGa or TMGa, TMAl, AsH_3, and PH_3. Dopants are Si from Si_2H_6 or Te from DETe for n-type materials, C from CCl_4 for p-type AlGaAs alloys, Zn from DEZn for p-type GaAs cap, and Zn or Mg from bis-cyclopentadienyl magnesium (Cp_2Mg) for p-type AlGaInP. Growth pressure inside the reactor is in the range of 80–200 mbar, and growth temperatures are typically in the range of 725–775°C. With AlGaAs DBR mirrors, the growth rate is approximately 3 μm/h, V to III ratio is 50 : 1, and the interface grading, whether it is linear or parabolic, can be accomplished by simultaneous ramping of TMGa and TMAl flows in very short increments, such as 0.1 s, controlled by the computer. Good yield and uniformity have been reported with 3-in. wafers [31, 32] and with 4-in, wafers [33] by MOCVD.

The growth reproducibility of VCSEL structures is of major concern because the DBR structures are very sensitive to growth fluctuations. The statistics obtained for 150 VCSEL wafers grown by MOCVD and operating at 850 nm are presented in Fig. 16.6, in which the variation of the reflectivity center wavelength as well as bandwidth of the reflectivity spectrum are shown in Fig. 16.6a and 16.6b, respectively. The standard deviation is 11 and 6 nm for the center wavelength and the bandwidth, respectively [34]. The application of *in situ* reflectometry growth monitoring [35–38] has

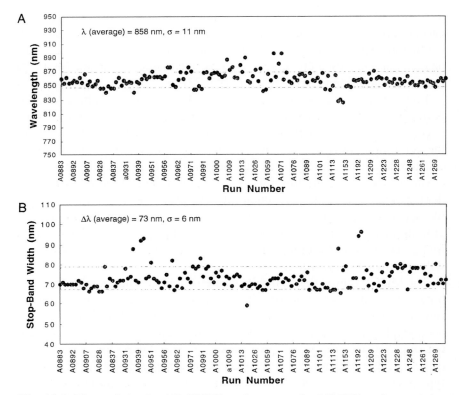

Fig. 16.6 The statistics for 150 VCSEL wafers grown by MOCVD and operating at 850 nm. (A) The variation of the reflectivity center wavelength and (B) the bandwidth of the reflectivity spectrum (after Ref. [34]).

allowed a better run-to-run thickness control across a 3-in. wafer using MOCVD [39].

16.4. Device Processing

Similar to standard silicon device fabrication, semiconductor laser processing includes photolithography, dopant diffusion, ion implantation, oxide deposition, chemical etching, metalization, etc. Unlike standard silicon device fabrication, however, a compound semiconductor wafer that contains epitaxial growth is used to begin the process. Each fabrication step used in the building of lasers may be functionally like that used in the silicon industry but is customized for compound semiconductor material systems.

16.4.1. EDGE-EMITTING LASER FABRICATION

The process procedures of edge-emitting semiconductor lasers vary with device structures. Edge-emitting semiconductor lasers can be catagorized into gain-guided lasers and index-guided lasers based on how current and optical modes are confined. Gain-guided lasers include broad-area lasers and stripe lasers, and index-guided lasers include weak index-guided lasers and strong index-guided lasers.

Broad-area semiconductor lasers do not employ any current confinement scheme, and their fabrication involves only a few process steps. First, p-type contact metal, usually AuZn/Au or Ti/Pt/Au, is deposited onto the epitaxy side of an as-grown wafer. AuZn/Au is an alloy contact and the alloying is carried out at approximately 400–450°C for 30 s. Ti/Pt/Au is a nonalloy contact. After the wafer is thinned down to approximately 100 μm, n-type contact metal, usually Au/Sn or Ni/AuGe, is deposited onto the back side of the substrate. The alloying for the n-type contact metal is carried out at 350–400°C for approximately 30 s. The wafer is then cleaved into bars of appropriate width, which determine the laser cavity length. The bars are finally sawed into individual chips, as shown in Fig. 16.7. Broad-area lasers are frequently used to characterize the laser material quality.

Because the current in broad-area semiconductor lasers is not confined, the broad-area lasers have high threshold current and seldom lase continuous wave (CW) at room temperature. A gain-guided stripe laser has a current restriction stripe built along the junction plane to enhance the carrier confinement and is used to reduce the threshold current. The stripe lasers allow CW operation [40] and exhibit fundamental transverse mode [41]. Three types of typical gain-guided stripe lasers based on AlGaAs/GaAs material systems are shown in Fig. 16.8. The first type is an oxide-stripe laser as shown in Fig. 16.8a. A SiO_2 layer of the p-GaAs contact layer is used to force the injected current flowing into a small region through an opening in the dielectric film [40]. The SiO_2 dielectric film can be deposited by plasma-enhanced chemical vapor deposition (PECVD). Photolithography is used to define a stripe width of 15 μm or less. Either wet chemical etching or reactive ion etching is used to remove the dielectric material within the stripe. After the removal of the photoresist, p-contact metal is deposited onto the entire wafer. The wafer is then thinned to approximately 100 μm and n-contact metal is deposited onto the back side of the substrate. Finally, the wafer is cleaved into individual laser chips. The second type uses proton implantation to define a region of high resistivity that

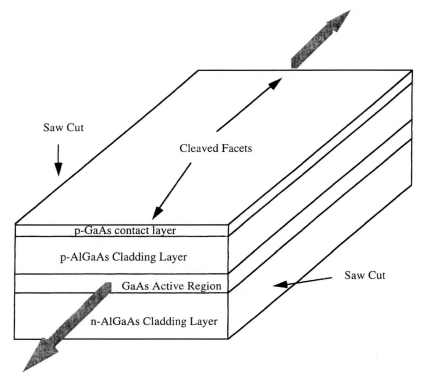

Fig. 16.7   Schematic diagram of an broad-area GaAs semiconductor laser chip.

restricts the current flowing into an opening in the implanted region [41], as shown in Fig. 16.8b. A thick photoresist stripe can be used as a mask for proton implantation to generate high-resistive region outside of the stripe region covered by the photoresist. After the removal of the photoresist, metal contacts are deposited onto both sides of the wafer to finish the device fabrication. The third type is a junction-stripe laser [42], as shown in Fig. 16.8c. Zinc diffusion is used to convert a small region of the top n-type GaAs layer into p type, thus providing a current path. The reverse-biased junction over the remaining region provides current confinement. Typically, a SiO_2 mask is used to define a stripe window on the wafer for the Zn diffusion, which is performed at $700°C$ for 30 min using $ZnAs_2$ as a diffusion source.

Although the current confinement has been improved with a gain-guided stripe laser, the lateral optical mode confinement is poor because of the

Fig. 16.8 Schematic diagram of (A) an oxide stripe GaAS semiconductor laser, (B) a proton implantation stripe GaAs semiconductor laser, and (C) a junction stripe GaAs semiconductor laser with Zn diffusion.

Fig. 16.9 Schematic diagram of (A) a ridge waveguide GaAs semiconductor laser and (B) a rib waveguide InGaAsP semiconductor laser.

carrier-induced antiguiding effect in the active layer. In order to further reduce the laser threshold and increase the differential quantum efficiency, the laser structure is modified to introduce an index step to guide the optical mode.

A weakly index-guided laser relies on layer thickness nonuniformity in the cladding region. The effective index difference between the optical mode guiding region and the region outside of the waveguide is on the order of 10^{-2}, larger than the carrier-induced antiguiding effect. There are a number of weakly index-guided structures that can be grouped into two categories, ridge waveguide devices [43] and rib waveguide devices [44] (Fig. 16.9).

The ridge waveguide laser fabrication (Fig. 16.9a) starts with an epitaxial wafer grown by LPE, MBE, or MOCVD techniques. The wafer is etched to form a ridge 3–5 μm wide. Dry etching is typically used to ensure a better scale control, but wet etching will result in less crystal damage. The etching stops in the p-cladding layer approximately 0.5 μm above the active layer. The etching depth is controlled *in situ* by a reflective laser interferometer [45]. Alternatively, an etch stop layer can be used for precise etching depth control [46]. A layer of SiO_2 dielectric material is then deposited onto the wafer. Photolithography and etching are used to remove the SiO_2 on the ridge for p-type metalization, followed by substrate thinning and n-type metalization on the back side of the substrate. Finally, the wafer is cleaved into laser chips. During the laser operation, a fraction of the optical mode overlaps with the dielectric, which has a considerably lower refractive index than the semiconductor cladding layer. The optical mode overlap introduces an effective lateral index step more significant than that induced by the carrier effect, thus eliminating the antiguiding effect and improving the optical mode confinement.

A channeled substrate planar (CSP) laser [44], one of several basic rib waveguide lasers, is shown in Fig. 16.9b. Epitaxial layers, including n-InGaAsP cladding layer, InGaAsP active layer, p-InP cladding layer, and n-InGaAsP contact layer, are grown by either LPE or MOCVD on a patterned n-InP substrate with etched-stripe structures. Zinc diffusion is used to convert the n-InGaAsP contact layer in the channeled structure region into p type for current injection. Metal contact of p type is then deposited onto the epitaxy side of the wafer. After the wafer thinning, a n-type metal contact is deposited onto the back side of the substrate. The wafer is finally cleaved into laser chips. When this CSP laser is forward biased, current can only flow through the Zn-diffused region because the remainder is reverse biased. A fraction of the optical mode overlaps with the lower index InP substrate on either side of the active region. This overlap introduces a lateral index step for optical mode guiding. The degree of guiding in effect depends on the width of the active region, the thickness of the n-cladding layer, and the depth of the channel stripes.

Strong index-guided semiconductor lasers have active regions buried in the higher bandgap materials (Fig. 16.10). The index differences between the guided region and the region outside is larger than 0.1. These types of

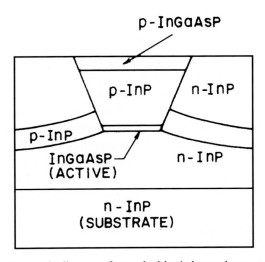

Fig. 16.10   Schematic diagram of an etched-buried mesa heterostructure laser.

lasers are called buried heterostructure (BH) semiconductor lasers. There are a number of different types of BH lasers, such as etched-mesa heterostructure (EMBH) lasers [47], double-channel planar buried heterostructure lasers [48, 49], planar buried heterostructure lasers [50], strip buried heterostructure lasers [51, 52], V-grooved substrate buried heterostructure lasers (which are also known as channel-substrate buried heterostructure lasers [53]), mesa-substrate buried heterostructure lasers [54], and buried-crescent buried heterostructure lasers [55].

Buried heterostructure semiconductor lasers rely generally on high-quality epitaxial regrowth techniques. For example, the fabrication of an EMBH laser (Fig. 16.10) starts with the first growth by LPE or MOCVD of four epitaxial layers, including a n-InP buffer layer of 3 or 4 μm doped at 2×10^{18} cm^{-3}, an undoped InGaAsP active layer of 0.2 μm at a wavelength of 1.3 μm, a p-InP cladding layer of 2 μm doped at 10^{18} cm^{-3}, and a p-InGaAsP contact layer of 0.5 μm doped at 10^{19} cm^{-3} with a bandgap corresponding to a wavelength of 1.1 μm [47]. Mesas of 2 or 3 μm wide are then formed by etching on the as-grown wafer using a solution of 1% Br$_2$ and CH$_3$OH with plasma-deposited SO$_2$ as an etching mask. After the etching, the etching mask SiO$_2$ is removed, and p-InP and n-InP blocking layers are grown using the second LPE or MOCVD growth. After the regrowth, a p-type metal contact is deposited onto the epitaxial side of the wafer. The wafer is then thinned to approximately 100 μm before a n-type metal contact is deposited on the back side of the substrate. Finally, the wafer is cleaved to produce semiconductor lasers. Regrowth techniques are generally used for InP-based material systems, although good-quality GaAs-based regrowth has also been demonstrated [56, 57].

Semiconductor lasers are required to operate at a single longitudinal mode for long-distance optical communications. Distributed feedback (DFB) lasers are the most widely used single longitudinal mode device in commercial applications. A typical long-wavelength DFB laser is shown in Fig. 16.11 [58]. To fabricate the DFB laser, a grating is first formed on a n-GaAs or n-InP substrate with the grating period determined by the required lasing wavelength. The optical length of the grating period should be equal to the half wavelength or its integer multiplication. The first-order grating is used for an edge-emitting laser, and the second-order grating is used for a grating surface-emitting laser that emits light perpendicular to the surface of the substrate [59].

A two-beam holographic exposure method is generally used to fabricate the uniform gratings [60]. A laser beam is first expanded and collimated

Fig. 16.11 Schematic diagram of a long-wavelength buried heterostructure DFB laser.

by a pinhole and a pair of lenses. The collimated beam is then divided by a beam splitter into two beams that meet again in the sample stage. Grating patterns generated by the interference between the two collimated beams are transferred photolithographically onto the surface of a semiconductor substrate that is spin-coated with a photoresist to a thickness of approximately 100 nm. The grating patterns of the developed photoresist are transferred to the substrate by selective wet etching or dry etching. The grating constant is determined by the writing laser wavelength and the angle of the two crossed laser beams. A $\lambda/4$-shifted grating can be fabricated by multilayer photoresist processing [61] or by using a phase mask [62], and a chirped grating can be fabricated by using a spherical beam method that uses divergent beams instead of collimated beams [63]. The $\lambda/4$-shifted grating or the chirped grating will result in more stable single longitudinal mode operation. There are other ways to make gratings, such as a photomask self-interference method [64] or an electron beam lithography technique [65].

After the grating structure is fabricated on the InP substrate, multiple epitaxial layers, including n-InGaAsP lower-guide layer, InGaAsP active layer, p-InGaAsP upper-guide layer, p-InP cladding layer, and p-InGaAsP cap layer, are grown by LPE or MOCVD. Either a wet or dry etching technique is used to form mesa structures approximately 1 μm wide. A small mesa width is used to ensure a fundamental transverse mode. The mesa structures are then embedded epitaxially with p-InP and n-InP blocking layers or with semi-insulating blocking layer. After the wafer thinning and metalization, the wafer is cleaved into bars 200–400 μm wide. The laser emission facets are then coated with a thin dielectric film by either sputtering or electron beam evaporation. Stable operation of a longitudinal single mode relies strongly on the facet reflectivity and is therefore sensitive to the thickness of the coating film. Typically, a $\lambda/4$-shifted DFB laser should have a facet reflectivity of less than 1%. The facet coating also acts as a passivation means to protect the exposed active layer at the facets, improving the device reliability. Finally, the laser bars are separated into individual laser chips.

A DBR laser can also be used to achieve longitudinal single-mode operation [66–68]. The DBR laser grating is formed outside of the active region (Fig. 16.12). Laser oscillation takes place in a resonator formed by two grating mirrors or by one grating mirror and one cleaved facet. The wavelength selectivity of the grating mirrors enables the single-mode oscillation in a DBR laser. Because there are many modes in the laser resonator, temperature variation can cause the DBR laser mode jumping between the adjacent modes. DBR laser fabrication is similar to DFB laser fabrication except that the corrugated grating structure in a DBR laser is in the passive waveguide

a. Active layer, $\lambda_g = 1.59$ μm
b. Antimeltback layer, $\lambda_g = 1.35$ μm
c. External guide layer, $\lambda_g = 1.35$ μm

Fig. 16.12 Schematic diagram of a GaInAsP/InP buried heterostructure DBR laser.

region instead of above or below the active layer, as in an index-coupled DFB laser.

16.4.2. VCSEL FABRICATION

From the emission direction standpoint, there are top-emitting VCSELs and bottom-emitting VCSELs. Any VCSELs based on GaAs substrate will not emit toward the bottom substrate direction if the VCSEL wavelength is shorter than 870 nm because of the substrate absorption, unless the substrate is removed. In this section, we will concentrate on top-emitting GaAs VCSEL processing. The same fabrication technology can easily be extended to bottom-emitting VCSELs, such as strained InGaAs VCSELs at approximately 980 nm.

Similar to edge-emitting laser diodes, VCSELs are either gain guided or index guided. The gain-guided VCSEL has a planar top surface as shown in Fig. 16.13. The current path of the gain-guided VCSEL is surrounded by the proton (H^+) implant buried in the top p-doped DBR mirror, which

Fig. 16.13   Cross-section SEM photo of a plannar GaAs VCSEL structure, where proton (H^+) implantation region is delineated by stain etching.

forms a high-resistance enclosure right above the active region. The optical mode is determined by the active region where the current flows through.

The fabrication of the gain-guided VCSEL is straightforward. Immediately after the photoresist patterning on top of the wafer that defines the VCSEL emission aperture window, BeAu/Ti/Au three-layer p-contact metal is deposited by thermal evaporation [69] for liftoff. The total contact metal thickness is approximately 1500–3000 Å. The emission window diameter typically varies from 5 to 50 μm, depending on the requirements for optical output power and transverse mode design. A thick layer of photoresist is patterned to cover the laser emission window with the diameter slightly larger than that of the emission window as a proton (H^+) implant mask. The photoresist implant mask thickness varies from 6 to 12 μm, depending on the proton implant energy. The implant energy is such that the majority of the protons rest at 0.5 μm above the active region to create a highly resistive layer, leaving the topmost portion of the DBR mirror relatively conductive. For the top DBR mirrors of 18–30 quarter-wavelength stack pairs, the required implant energy varies from 300 to 380 kV. The proton implant dose is approximately 10^{14} or 10^{15} cm^{-2}. The implant straggle width is approximately 0.5 μm. The proton implant defines an annular high-resistance region underneath the p-metal contact and right above the active region in the p-DBR mirror as that the injected current can funnel into the central active region underneath the laser emission aperture window. The diameter of the implant mask is chosen to be larger than the laser emission window defined by the p-metal contact. This helps lower the p-metal contact resistance because of a ring contact with an area unaffected by the implant. The smaller emission window also behaves like a filter that enhances the lowest order transverse mode [69, 70].

After the photoresist mask is stripped off, a new photoresist mask pattern is deposited for a shallow proton implantation to electrically isolate individual VCSELs between each other. The second implantation can be replaced by etched trenches if the requirement for planar surface is not essential [71]. Subsequently, a n-metal contact layer consisting of GeAu/Ni/Au is deposited by either thermal evaporation or e-beam evaporation on the bottom of the GaAs substrate. The wafer is then annealed at 450°C for 30 s. The anneal serves to alloy both contacts and to further shape the resistance profile due to proton implantation by further reducing the surface resistance while retaining a high-resistance layer under the metallized region [72]. A metal interconnect deposition on the top surface completes the device fabrication process. As an option, a layer of dielectric material, such as SiN or SiO_2, can be deposited by PECVD for the purposes of laser device surface passivation.

The gain-guided VCSEL by proton implantation has been proven manufacturable with exceptionally high yield [31, 32]. There are, however, concerns regarding the devices' turn-on delay due to thermal lensing effect [73] and turn-on delay-induced timing jitter [74, 75]. These effects give rise to problems in high-speed fiber optical applications. Adverse effects of the turn-on delay have been alleviated via careful biasing of the laser at or above threshold [76, 77]. To further address this issue, index-guided VCSELs or ridge waveguide VCSELs have been developed and have achieved better performances than the gain-guided VCSELs [32].

A typical index-guided VCSEL structure is shown in Fig. 2.15 of Chapter 2 [78]. The fabrication process starts with the deposition of 4000 Å TiW metal alloy by sputtering and 4000 Å SiN by PECVD. A photoresist mask is used to define the index-guided VCSEL mesa diameter, which usually ranges from 8 to 50 μm. The SiN and TiW outside of the photoresist mask is dry etched by fluorinated chemistry in a reactive ion-etching (RIE) reactor to expose the AlGaAs semiconductor epitaxial layer. After the removal of the photoresist, the remaining SiN serves as a hard mask for the following mesa etching by chlorine chemistry in a RIE reactor. The mesa etching depth is *in situ* monitored using laser interferometry to a level above the active layer. Protons are then implanted into the remaining *p*-DBR mirror to further electrically isolate the mesa region. The implantation energy is 50 keV and the implantation dose is in the range of 10^{14} to 10^{15} cm^{-2}. After the implantation, a layer of 4000 Å SiN is deposited by PECVD onto the entire wafer. A photoresist mask is then spun onto the SiN with the mesa top exposed for dry etching using fluorine chemistry to remove the SiN on the mesa. This etching process is not selective, so some TiW will be etched as well. After the photoresist is removed, 4000 Å TiW will be sputtered onto the entire wafer. A new photoresist mask is then used to define a VCSEL emission aperture on the mesa. The diameter of the emission aperture is smaller than the mesa diameter and fluorine chemistry is used to selectively remove the TiW in the emission aperture. Gold interconnect pads can then be deposited on top of the TiW by a photoresist liftoff process. This completes the device fabrication on the *p*-side. A broad-area *n*-metal contact formed by Ni/AuGe/Au is deposited on the bottom surface of the GaAs substrate. The wafer is then annealed in a rapid thermal annealing chamber for 30 s at 420°C. Finally, a SiN passivation layer can be deposited on the VCSEL surface for protection. The wafer is then sawed into individual chips as opposed to cleaving, which is usually done for edge-emitting lasers. This fabrication process is highly manufacturable and is currently being used in production for Motorola's trademarked Optobus optical interconnect product (Fig. 16.14) [32].

A

B

Fig. 16.14 Motorola Optobus module, a 10-channel data link at 400 MHz/channel.

When the VCSEL is driven at a current well above threshold, multitransverse modes develop partially due to spatial hole burning in the active region. The number of allowable transverse modes of a regular index-guided VCSEL is largely determined by the waveguide, which is composed of etched mesa and the materials surrounding the mesa, such as SiN, metal, or air. As a consequence of the large index difference between the semiconductor material and the surrounding materials, the diameter of the mesa must be very small for the waveguide to support only the fundamental/single transverse mode. Single-mode VCSELs are preferred for an optical interconnect based on single-mode fibers, laser printing, compact disk data storage, bar code scanning, etc. Reducing the mesa diameter also helps reduce the VCSEL threshold. Super low threshold VCSEL designs are appealing for some high-speed applications as well as for some basic physics studies. However, the mesa-type index-guided VCSEL cannot be too small in diameter because of the optical scattering loss from the etched sidewall. The mesa sidewall damage due to exposure to the ion beam [79] may also become nonradiative recombination centers that limit the minimum achievable threshold and cause potential reliability problem.

A selective lateral oxidation defined index-guided VCSEL (shown in Fig. 2.27 in Chapter 2) is a promising alternative to the mesa-defined index-guided VCSEL [80–83]. It has been reported that native oxide of Al_xO_y can be stable enough to be utilized in the device fabrication for current confinement [84–87]. The effective index difference between the semiconductor and its native oxide also provides for an excellent waveguide for optical mode confinement. The diameter of the lateral oxide VCSEL waveguide can be made relatively larger for supporting only the single transverse mode, owing to the relatively moderate effective index difference. In addition, the p metal has a larger area to make contact with for the oxide VCSEL, reducing the series resistance significantly, even for a very small active area. Lower series resistance is effective in reducing the active junction temperature during operating; therefore, it is more desirable from a reliability standpoint.

For example, to fabricate the device structure shown in Fig. 2.27a, a VCSEL epitaxial structure is grown that includes $Al_{0.85}Ga_{0.15}As/Al_{0.16}Ga_{0.84}As$ DBR mirrors and a triple GaAs quantum well active region in a graded AlGaAs cavity, with the aluminum mole fraction of the AlGaAs mirror layer closest to the cavity increased from $x = 0.85$ to $x = 0.98$ on the p side. Devices are formed in mesas created by etching away all the surrounding p-side mirror stacks and part of the cavity materials above the active layer to expose the edge of the cavity layers. A portion of the AlGaAs layers starting from the exposed edge is then converted to an electrically insulating aluminum oxide (Al_xO_y) by wet thermal oxidation at 425°C for a few hours. The aluminum oxide has a refractive index of ~1.55. The lateral oxidation is conducted in a furnace supplied with a flow of N_2 bubbled through deionized water heated to a temperature of 95°C. The $x = 0.98$ layer oxidizes more rapidly [88] and the resultant oxide protrudes further into the mesa to provide electrical and optical confinement about the active region. A back-side substrate contact and an annular contact on the mesa complete the VCSEL [83]. This selective lateral oxidation is the most effective method in achieving record low-threshold VCSELs [89, 90]. Single transverse mode operation of up to 2.7 mW has also been demonstrated [91].

The same oxidation technology described previously can be used to oxidize through the high aluminum mole fraction layer to create DBR mirrors with one layer of semiconductor material and another layer of a native oxide Al_xO_y. The large index difference between the two alternative quarter-wavelength-thick layers enables the demonstration of a high-reflective and wide-bandwidth DBR mirror with only 4 or 5 mirror pairs [92, 93].

Even though the oxide VCSEL has superb reported performances in single-mode power operation [91] and low threshold [89, 90], its long-term reliability, manufacturing uniformity across the wafer, and yield are still unknown. Other techniques for high-power single-mode emission include the leaky mode solution [94], which uses the regrowth technique to deposit higher index materials to surround the etched-mesa pillar [95, 96]. A more simple leaky mode technique involves using a planar proton-implanted VCSEL with smaller current isolation diameter and emission aperture. This technique has allowed the generation of a single-mode output power up to 4.4 mW at 980 nm [97].

In systems that are polarization sensitive, such as some optical data storage systems, the precise knowledge of the VCSEL polarization is required. VCSELs generally polarize along |110> and |1h0> crystalline directions. In order to stabilize the VCSEL polarization along a predetermined direction, certain asymmetries must be introduced into the laser structure. The most straightforward technique is to utilize the anisotropic transverse cavity geometry to select and maintain a dominant polarization state [98, 99]. Typically, the airpost or ridge waveguide mesa structure is elongated into a rectangular, rhombus, or dumbbell shape. This technique works successfully in stabilizing the VCSEL polarization along the elongated mesa direction. Alternatively, the polarization can be stabilized by using a differential loss method obtained from coatings on the sides of the upper DBR [100] or by applying biaxial stress to the active layer [101]. A grating formed on the very top layer of the p-DBR, with metalization on the sidewalls of the grating structure, has also been demonstrated with good polarization extinguishing ratio [102].

Even though the previously mentioned fabrication procedures use GaAs VCSELs for the example, the techniques should apply to InGaAs/GaAs VCSELs at 980 nm and InAlGaP/GaAs VCSELs at red visible wavelengths. Long-wavelength VCSELs at 1.3 and 1.55 μm are of interest to the telecommunication industries, but their fabrication has met some difficulty due to Auger nonradiative recombination-induced loss and the low reflectivity of the monolithic InGaAsP/InP DBR mirrors. Dielectric mirrors with 8.5 pairs of MgO/Si multilayers and Au/Ni/Au on the p side and 6 pairs of SiO2/Si on the n side have been used instead of the semiconductor DBR mirrors [103]. Continuous-wave VCSEL operation at 14°C has therefore been demonstrated at a wavelength of 1.3 μm.

A more rigorous technique in the long-wavelength VCSEL fabrication is to use a wafer-fusion or -bonding process to unite separately made GaAs/AlAs mirrors formed on a GaAs substrate to the InGaAsP double-heterostructure (DH) laser diode [104]. The higher reflectivity of the GaAs system mirrors at either 1.3 or 1.55 μm make them more advantageous than InP system mirrors. The wafer-fused VCSEL fabrication begins by fusing the GaAs/AlAs mirror wafer with the top of the InP DH wafer at 650°C for 30 min in a H_2 ambient environment under a graphite load [105]. After the wafers have been fused together, the InP substrate is removed by a selective wet chemical etchant such as HCL : H_2O (3 : 1). The resulting fused structure, an InP DH active layer with a GaAs/AlAs DBR mirror on a GaAs substrate, is mechanically robust and undergoes typical fabrication procedures without any special handling. The completed mesa structure VCSEL has a Pd/Zn/Pb/Au intracavity ring contact and a Si/SiO2 dielectric mirror on the p side and an AuGe/Ni/Au contact on the n-type GaAs substrate [106]. This device operates pulsed at room temperature and CW at −45°C at a wavelength of 1.3 μm. When both sides of the InP DH wafer are fused with the GaAs/AlAs DBR mirrors, the VCSEL performances are further improved [107]. A double-fused VCSEL operating CW up to 64°C has been demonstrated at 1.55 μm by combining with the lateral DBR mirror oxidation technique [108]. This result represents the best long-wavelength VCSEL performance achieved to date.

16.4.3. CLEAVED-MIRROR FORMATION AND DICING

Before an edge-emitting laser can be probe tested, a resonant cavity must be formed by cleaving the wafer into bars along the crystalline direction, typically |110> or |1h0> (Fig. 16.15). The cleaving facets function like laser cavity mirrors. Usually, the wafer to be cleaved is mounted on a sticky, thin, flexible tape, and small indentations are scribed near one edge of the wafer with a diamond scriber. Slight flexing will then result in cleavage along |110> crystalline perpendicular to the surface. This technique ensures perfectly aligned mirrors. For broad-area lasers, the cleaved bars can then be sawed perpendicular to the mirror faces with a diamond or string saw to separate them into rectangular device chips. The rough sawn sides and a cavity length-to-width ratio of at least 2.5 appear to prevent excitation of internally circulating modes [109]. The natural cleaved facets are usually mirror-like smooth with a reflectivity of approximately 30%, determined by the index difference between the semiconductor material and the air. The laser cavity length between the two cleaved facets usually ranges from 100 μm to 1 mm. When the device is biased above the threshold, the laser will be generated emitting toward both facets with an equal amount of output power. The emission from one of the laser facets can be used to monitor the laser output power from the other facet in applications in which constant output power is needed. When the monitoring photodiode detects

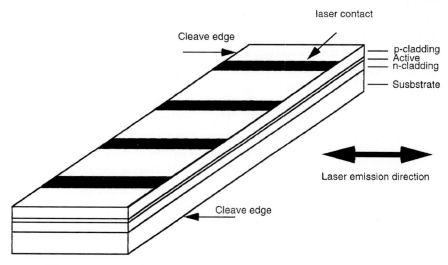

Fig. 16.15 Schematic diagram of a cleaved edge-emitting laser array bar.

UNCOATED
REGION

COATED
REGION

Fig. 16.16 Photo showing the facet erosion in the uncoated region of a partially coated laser diode after 600 h of operation. (Reprinted with permission from Ref. [136]. Copyright 1977 American Institute of Physics.)

the variation in laser output power, it will send a signal back into the driver circuitry to adjust the laser pump current until a predetermined laser output power is reached. This auto power control (APC) technique has been used in substantially all compact disk (CD) pickups and serial optical fiber data links to compensate for the power fluctuation due to ambient temperature change and/or device degradation.

The facets of edge-emitting lasers based on GaAs/AlGaAs are subject to oxidation when exposed to air [110, 111]. The catastrophic failure due to facet degradation is one of the most commonly seen failures occurring with edge-emitting lasers (Fig. 16.16). Facet passivation using SiO_2, Si_xN_y, or Al_2O_3 via deposition by chemical vapor deposition or sputtering is one of the many techniques used to prevent the facets from oxidation and has been proven effective in improving the device reliability.

When higher output power is needed from the laser, one of the laser facets can be antireflectively (AR) coated to reduce the facet reflectivity to only a few percent. More than 90% of the intracavity energy will then be coupled out from this coated facet. The reduced cavity Q by AR coating will cause the laser threshold to increase. A highly reflective (HR) dielectric mirror coating is usually deposited onto the other facet to increase the cavity Q back to or higher than the original level without the AR coating [112]. Some applications require an edge-emitting laser to operate at low threshold. This can be achieved by depositing HR dielectric mirror coatings on both facets, with one of the facets less reflective than the other for the laser to couple out. Edge-emitting lasers with sub-mA thresholds have been demonstrated in this way in the research environment [113].

Both HR coatings and AR coatings are deposited after the wafer cleaving. The individual edge-emitting laser chips, or laser bars, must be mounted one by one for the coatings. This is a high-cost manufacturing process that challenges the device yield and throughput. The high level of stress incorporated during the coating deposition also affects the device reliability. In contrast, the monolithic manufacturing process of VCSELs allows on-wafer probe test with a computer-controlled probe card as per the silicon industry. After testing, the wafer is separated into chips by a diamond saw similar to the dicing of silicon integrated circuits (IC) (Fig. 16.17). Known good dies will then be automatically picked out for packaging using computer control. This highly automated process is compatible with the existing silicon IC infrastructures and therefore preferred for low-cost manufacturing. Furthermore, the VCSEL active layers are not exposed to the air, and no facet-related degradation mode exists. Therefore, the GaAs VCSEL reliability improves dramatically compared to that of conventional GaAs edge-emitting lasers having unprotected facets. In addition, the VCSEL wafer can easily be sawed into chips of two-dimensional (2-D) arrays, which may either be individually addressed or matrix addressed depending on the interconnect metal deposited onto the wafer [114–117], whereas edge-emitting lasers can only be cleaved into one-dimensional array bars.

16.4.4. PACKAGING AND BONDING

The commercial semiconductor laser is usually packaged in either a can type or a dual-in-line pin (butterfly) type package. The can-type package is smaller in design and is currently used in high-volume manufacturing. It is less costly to manufacture and is therefore widely used for CD data storage, bar code scanners, pointers, laser printers, and serial data links. The 14-lead butterfly-type package costs more to manufacture and is typically reserved for long-wavelength semiconductor lasers.

Semiconductor lasers packaged in metal cans (TO cans) dominate today's commercial semiconductor laser market. The majority of earlier TO cans had a diameter of 9 mm, but the trend has been to use a smaller type TO-56 package with a diameter of 5.6 mm (Fig. 16.18). The laser chip is first mounted upside down on a Si submount with Sn solder. Electrical transmission lines can be deposited on the submount for electrical contact. The submount with the laser chip turned on is then mounted sideways with In solder onto a copper heat sink that sits on a stem 5.6 mm in diameter so that the laser emission is vertical to the package. Sometimes the laser chip

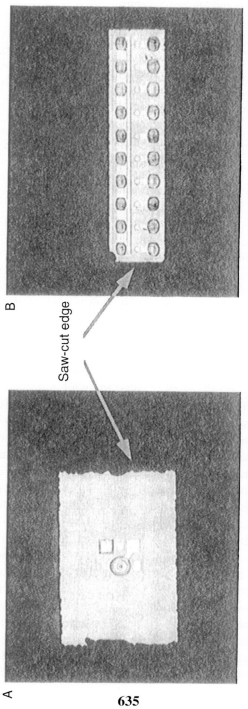

Saw-cut edge

A   B

Fig. 16.17   Photo showing (A) a discrete VCSEL and (B) a 1×10 VCSEL array.

Glass windor
Cap
Stem
Submount
Laser chip
Monitor detector

Header

Fig. 16.18 Schematic diagram of an edge-emitting laser mounted on a TO-56 package with a power monitoring photodiode.

is directly mounted upside down onto the copper heat sink without any Si submount. The stem is based on copper, iron, or nickel, depending on the heat dissipation and cost requirements. Both the heat sink and the stem can be plated with gold for better heat dissipation.

If APC is needed, a silicon PIN detector is directly mounted onto the stem with a Sn or In solder to receive the laser emission from the back facet. The detector on the stem is tilted to avoid any reflected light from coupling directly back into the laser, thus destabilizing the laser operation. The detector can sometimes be made directly on the silicon submount that will detect one-half of the cone-shaped backward laser emission for the laser power monitoring. Gold wire bonding is used to connect electrodes on the semiconductor laser and the detector to the corresponding posts on the stem. Depending on the specific application, the semiconductor laser cathode can be either in common with the detector cathode, as shown in Fig. 16.19a, or in common with the detector anode, as shown in Fig. 16.19b. Finally, a lid is hermetically sealed onto the stem. Hermetic sealing is needed to improve the laser reliability. The glass window on the lid is AR coated to reduce any backward reflection that is detrimental to the laser signal-to-noise ratio. A slanted lid is sometimes used to steer the backward reflection light away from entering the laser cavity (Fig. 16.20).

Dual in-line butterfly pigtail-type laser packages are typically used for optical communications. Laser light is required to couple into a single-mode fiber with a core diameter of approximately 10 μm. Butt-coupling is one of the commonly used active alignment techniques for fiber pigtailing.

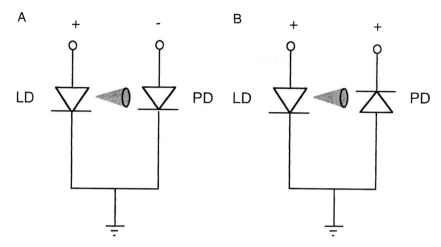

Fig. 16.19 Electrical contact diagram of a laser package with a monitoring photodiode with the laser cathode in common with the photodiode (A) and (B) anode.

In the butt-coupling, a laser chip is first mounted onto a subcarrier and turned on. An optical fiber is brought close to the emission end of the laser to achieve the best coupling efficiency and then secured onto the subcarrier. The butt-coupling usually has a loss of more than 3 dB because of the poor matching in numerical aperture (NA) between the laser emission and the fiber. The coupling efficiency can be improved dramatically by inserting a lens between the laser and the fiber [118]. Many types of lenses can be used, but a gradient-index lens is among the best in performance [119].

Fig. 16.20 A TO-56 package with slanted lid to avoid direct reflected light into the laser.

Fig. 16.21 A four-channel V-groove submount for passive alignment of laser array to fibers (after Ref. [123]).

An integrated ball lens can be made directly using the glass fiber tip and relatively good coupling has been achieved [120]. Many manufacturers use can-type packaged CD lasers at 780 nm because of their commercial availability at a substantially lower price. Typically, a lens between the can and the fiber is used to improve the coupling efficiency [121, 122].

To reduce the fiber pigtailing cost, especially when a laser array is needed to couple into a fiber array, a passive coupling alignment technique is preferred in which the lasers are never turned on during the packaging. This passive coupling alignment can also increase the packaging throughput significantly. Figure 16.21 is an example of a four-channel fiber pigtail package with passive alignment [123]. A Si submount is first made with four V grooves uniformly spaced apart. The spacing is determined by the spacing of the lasers on the laser array chip. A four-channel laser chip is flip-chip mounted adjacent to the end of the V grooves with the position predetermined by the marks on the Si submount. A self-alignment technique can also be used to position the laser chip using solder reflow [124]. Four fiber cores are then placed into the V grooves on the Si submount with their ends leaning against the end of the V grooves. A Si cover carrier with V grooves and a space for the laser chip is then placed back onto the Si submount with optical glue to secure the whole package. No lenses are used between the lasers and the fibers because the laser emission facets are right against the fiber input ends and their relative positions are predetermined with very high placement resolution.

The vertical emission feature of a VCSEL allows it to be packaged at a relatively lower cost. Because the VCSEL has a circular beam shape, its coupling efficiency into the fiber is high. A single VCSEL can be packaged in the same way as a surface-emitting LED on either a TO-type can or a molded leadframe (Fig. 16.22). A molded lens can be integrated to the

Fig. 16.22 Photos of VCSELs packaged with (A) TO-56 cans and (B) plastic molding.

10-channel waveguide

Leadframe

1x10 VCSEL array

Fig. 16.23   Schematic diagram of a VCSEL GUIDECAST package.

leadframe surface mount-type package to adjust the laser beam divergence angle and thus to match the NA of the fiber for better coupling efficiency.

One-dimensional (1-D) or 2-D array capability is the unique advantage of VCSEL technology. The Optobus package is an example how a 1-D VCSEL array is packaged to couple into multimode fibers [32] (Fig. 16.23). The package supports a 1×10 VCSEL array. Leadframes with 10 copper leads for the VCSEL anodes and two copper leads for the VCSEL cathodes are first embedded into a 10-channel polymer waveguide, the Guidecast. The 12 leads are bent toward the input end of the Guidecast for the direct chip attachment of the VCSEL array onto the leads. All the VCSEL anodes and cathodes are on the top side, as shown in Fig. 16.17b. The electrical contacts of the VCSEL array are made of plated Au with the same height above the top emission surface. The VCSEL array is flip-chip mounted onto the Guidecast with a robot. The robot has a placement accuracy of several micrometers — enough to guarantee an accurate alignment of the VCSEL with an emission aperture of 10–14 μm in diameter onto the Guidecast waveguide channels with dimensions on the order of 40×40 μm. The electrical contacts between the VCSELs and the Guidecast leads are connected by conductive epoxy. Underfill material with the same composition as the Guidecast is used to secure a robust attachment of the VCSEL array to the Guidecast. In addition to being electrical connectors, the Guidecast leads also function as thermal heat sinks to dissipate heat from the VCSELs. The output end of the Guidecast is attached to the standard MT-ferrules for light coupling into 10-channel multimode fiber ribbons. The whole concept of

Fig. 16.24 Photo showing a tape-access-bonding (TAB) approach for VCSEL array packaging.

this approach is based on passive coupling alignment and polymer material molding to achieve low-cost packaging while utilizing the array function of VCSELs to achieve high-speed parallel data transmission.

The leadframe type Guidecast works properly for transmitting data at 155 Mb/channel. When a higher data transmission speed is needed, the electrical cross-talk between the leads becomes a limiting factor. To address this issue, a tape access-bonding (TAB) approach is used to make electrical contacts to the VCSEL array [33] (Fig. 16.24). The polymer Guidecast without the copper leads is then attached to the VCSEL array and functions as a waveguide to couple the VCSEL beams into the 10-channel fiber ribbon through the MT-ferrules. This approach has improved the Optobus data link speed to 400 Mb/channel and is expected to work well for the next-generation Optobus data link at 800 Mb/channel.

Vertical cavity surface-emitting laser array packaging for data links has also been demonstrated with direct fiber coupling [125–128] or through the use of flexible Polyguide [129]. Passive coupling alignment with the Polyguide approach has recently been demonstrated [130].

It should be noted that the efficiency to couple light into a fiber is determined by the beam transverse mode and the modes that the fiber can support. Edge-emitting semiconductor lasers, VCSELs, and LEDs have different beam profiles and therefore have different launch conditions for fiber coupling. Good coupling efficiency can be achieved by adjusting the light source beam profile with a lens to match the fiber NA. In telecommunication applications, long-wavelength semiconductor lasers are coupled into a single-mode fiber to achieve the optimum bandwidth–distance performance. Optical fibers are designed for applications at those wavelengths.

The low-cost requirement for high-speed data communications pushes forward the applications of CD lasers operating on multimode fibers, and they have very different launch conditions compared to long-wavelength laser sources due to the difference in optical modes that the multimode fiber can support at the CD wavelength region. Recently, some specifications for gigabit Ethernet were redrafted because the bandwidth and losses were originally calculated based on an incorrect LED launch model.

VCSELs have an advantage of low-cost manufacturing. In addition, GaAs VCSELs at 850 nm have demonstrated superb performance and reliability for applications in data communications. However, most standard graded-index 62.5-μm multimode fibers today are optimized for long-wave applications at either 1.3 or 1.55 μm and have much smaller bandwidth–distance performance at 850 nm. Therefore, today's VCSEL data link market is targeted for short-distance applications up to 300–500 m. As the VCSELs become more important and prevalent, it should be possible to buy fiber with higher bandwidth at shorter wavelengths. For example, a multimode fiber with a minimum of 800 MHz/km at 850 nm has been demonstrated that is achieved by optimizing the fiber manufacturing process at a slightly different point and tightly controlling the process variables. This cable is more expensive today than the standard long-wave fiber, but this trend could change soon if the market demand triggers sufficient volume production.

16.5. Conclusion

The semiconductor laser and LED manufacturing technologies have been discussed in this chapter with the emphasis on the lasers. Edge-emitting laser diodes have been commercialized for decades and are extensively used for CD/DVD data storage, laser printing, bar code scanners, pointers, optical communications, solid-state laser pumping sources, etc. VCSELs have just started to enter the commercial market and are currently used for parallel or serial optical fiber data links. It is expected that VCSELs will also see applications in areas currently dominated by edge-emitting laser diodes. In-plane surface-emitting lasers [59, 131] are another type of lasers that have been extensively studied by many research groups. Circular-grating surface-emitting lasers [132, 133] are considered to be in this category. Manufacturability and manufacturing cost relative to VCSELs are the two factors that have hindered their commercialization.

Edge-emitting lasers are grown by LPE, MBE, or MOCVD techniques, whereas VCSELs are grown by MBE or MOCVD techniques. Growth using

LPE is not accurate enough for monolithic DBR structures of VCSELs. From a manufacturing standpoint, MOCVD offers the best growth through-put and excellent material quality; thus, it is predominantly used by all the commercial VCSEL manufacturers. Many edge-emitting laser manufac-turers are also relying on the MOCVD technology for the high-volume production.

Currently, edge-emitting lasers are dominated by either DH-type or MQW-type structures. Index-guided structures are used to achieve low thresholds and single transverse mode operation. Distributed feedback lasers have been extensively studied to achieve single longitudinal-mode operation for long-distance optical communications. Precise wavelength control is critical for applications in wavelength-division multiplexing [134]. Vertical cavity surface-emitting lasers have only one longitudinal mode because of their extremely short laser cavity. Multitransverse mode operation, however, is common due to the large active area relative to the laser cavity length. There are three basic types of VCSEL structures: proton-implanted VCSELs, etched-mesa VCSELs, and lateral oxidation-confined VCSELs. Many applications require that the laser beam be diffraction lim-ited. Single transverse-mode VCSELs can be achieved by reducing the laser emission aperture, but the single-mode output power becomes limited as well [91, 95–97]. Nevertheless, VCSEL beams have the advantage of be-ing circular, and therefore offer a high coupling efficiency into an optical fiber.

Edge-emitting lasers can be tested only after the wafer is cleaved, and most of the time after being packaged. Vertical cavity surface-emitting lasers can be tested on the wafer, and only known good dies will be packaged after being sawed into chips. From a manufacturing standpoint, VCSELs are more suitable for low-cost mass production.

Edge-emitting lasers at 1.3 and 1.55 μm are still the only choice for long-distance optical communications. Tremendous progress has been made in VCSEL research at 1.3 and 1.55 μm using wafer-bonding techniques [106–108]. The device efficiency and yield sare still low and the manufacturability is unknown. Further development is needed before the long-wavelength VCSELs become practical for commercial applications.

Device reliability is one of the most important considerations for any commercial application. Edge-emitting lasers at 780 nm do not survive long, predominantly due to facet degradation. Facet passivation and hermetically sealed packages are effective in increasing the device lifetime. For applica-tions requiring extreme device reliability, burn-ins and special screenings are necessary. Such procedures certainly affect the manufacturing cost.

Vertical cavity surface-emitting lasers have been proven reliable with extrapolated mean time to failure of up to millions of hours at normal operation condition without any burn-in screening [32, 135]. Hermeticity may
not be necessary for VCSEL packaging. Such an advantage will allow a
VCSEL to be packaged, like a LED, using plastic molding and therefore
substantially reducing the packaging cost.

References

1. Casey, H. C., Jr., and M. B. Panish. 1978. *Heterostructure lasers.* New York:
   Academic Press.
2. Temkin, H., M. B. Panish, R. A. Logan, and J. P. Van der Ziel. 1984. $\lambda \approx$
   1.5 μm InGaAsP ridge lasers grown by gas source molecular beam epitaxy.
   Appl. Phys. Lett. 45:330.
3. Asahi, H., Y. Kawamura, and H. Nagai. 1982. Molecular beam epitaxial
   growth of InGaAlP on (100) GaAs. *J. Appl. Phys.* 53:4928.
4. Herman, M. A. and H. Sitter. 1989. *Molecular beam epitaxy.* Berlin: Springer-
   Verlag.
5. Wang, C. and R. F. Davis. 1993. Deposition of highly resistive, undoped,
   and p-type, magnesium-doped gallium nitride films by modified gas source
   molecular beam epitaxy. *Appl. Phys. Lett.* 63:990.
6. Yung, K., J. Yee, J. Koo, M. Rubin, N. Newman, and J. Ross. 1994. Observation of stimulated emission in the near ultraviolet from a molecular
   beam epitaxy grown GaN film on sapphire in a vertical-cavity, single pass
   configuration. *Appl. Phys. Lett.* 64:1135.
7. Nakamura, S., T. Mukai, M. Senoh, and N. Iwasa. 1992. Thermal annealing
   effects on p-type Mg-dope GaN films. *Jpn. J. Appl. Phys.* 31:L139.
8. Amano, H., M. Kito, K. Hiramatsu, and I. Akasaki. 1989. P-type conduction
   in Mg-doped GaN treated with low-energy electron beam irradiation. *Jpn. J.
   Appl. Phys.* 28:L2112.
9. Neave, J. H., B. A. Joyce, P. J. Dobson, and N. Norton. 1983. Dynamics
   of film growth of GaAs by MBE from Rheed observations. *Appl. Phys. A.*
   31:1.
10. Walker, J. D., D. M. Kuchta, and J. S. Smith. 1991. Vertical-cavity surface-
    emitting laser diodes fabricated by phase-locked epitaxy. *Appl. Phys. Lett.*
    59:2079.
11. Walker, J. D., D. M. Kuchta, and J. S. Smith. 1993. Wafer-scale uniformity
    of vertical-cavity lasers grown by modified phase-locked epitaxy technique.
    Electron. Lett. 29:239.
12. Chalmers, S. A. and K. P. Killeen. 1993. Method for accurate growth of
    vertical-cavity surface-emitting lasers. *Appl. Phys. Lett.* 62:1182.

13. Bacher, K., B. Pezeshki, S. M. Lord, and J. S. Harris. 1992. Molecular beam epitaxy growth of vertical cavity optical devices with *in situ* corrections. *Appl. Phys. Lett.* 61:1388.

14. Grothe, H., and F. G. Boebel. 1993. *In situ* control of Ga(Al)As MBE layers by pyrometric interferometry. *J. Cryst. Growth* 127:1010.

15. Houng, Y. M., M. R. T. Tan, B. W. Liang, S. Y. Wang, L. Yang, and D. E. Mars. 1994. InGaAs(0.98 μm)/GaAs vertical cavity surface emitting laser grown by gas-source molecular beam epitaxy. *J. Cryst. Growth* 136:216.

16. Houng, Y. M., M. R. T. Tan, B. W. Liang, S. Y. Wang, and D. E. Mars. 1994. *In situ* thickness monitoring and control for highly reproducible growth of distributed Bragg reflectors. *J. Vacuum Sci. Tech.* B12:1221.

17. Lear, K. L., R. P. Schneider, K. D. Choquette, S. P. Kilcoyne, J. J. Figiel, and J. C. Zolper. 1994. Vertical cavity surface emitting lasers with 21-percent efficiency by metalorganic vapor phase epitaxy. *IEEE Photon. Tech. Lett.* 6:1053.

18. Hibbs-Brenner, M. K., R. P. Schneider, R. A. Morgan, R. A. Walterson, J. A. Lehman, E. L. Kalweit, J. A. Lott, K. L. Lear, K. D. Choquette, and H. Juergensen. 1994. Metalorganic vapour-phase epitaxial growth of red and infrared vertical-cavity surface-emitting laser diodes. *Microelectron. J.* 25:747.

19. Kawakami, T., Y. Kadota, Y. Kohama, and T. Tadokoro. 1992. Low-threshold current low-voltage vertical-cavity surface-emitting lasers with low-Al-content p-type mirrors grown by MOCVD. *IEEE Photon. Tech. Lett.* 4:1325.

20. Yoon, S. F. 1992. Observation of nonbiased degradation recovery in GaInAsP/InP laser diodes. *IEEE J. Lightwave Tech.* 10:194.

21. Cunningham, B. T., M. A. Haase, M. J. McCollum, J. E. Baker, and G. E. Stillman. 1989. Heavy carbon doping of metalorganic chemical vapor deposition grown GaAs using carbon tetrachloride. *Appl. Phys. Lett.* 54:1905.

22. Cunningham, B. T., J. E. Baker, and G. E. Stillman. 1990. Carbon tetrachloride doped $Al_xGa_{1-x}As$ grown by metalorganic chemical vapor deposition. *Appl. Phys. Lett.* 56:836.

23. de Lyon, T. J., N. I. Buchan, P. D. Kirchner, J. M. Woodall, G. J. Scilla, and F. Cardone. 1991. High carbon doping efficiency of bromomethanes in gas source molecular beam epitaxial growth of GaAs. *Appl. Phys. Lett.* 58:517.

24. Buchan, N. I., T. F. Kuech, G. Scilla, F. Cardone, and R. Potemski. 1989. Carbon incorporation in metal-organic vapor phase epitaxy grown GaAs from CH_xI_{4-x}, HI, and I_2. *J. Electron. Mater.* 19:277.

25. Stringfellow. 1989. Non-hydride group V sources for OMVPM. *Mater. Res. Soc. Symp. Proc.* 145:III–V Heterostructures for Electronic/Photonic Devices, 171.

26. Lum, R. M., J. K. Klingert, and F. A. Stevie. 1990. Controlled doping of GaAs films grown with tertiary-butylarsine. *J. Appl. Phys.* 67:6507.

27. Kuech, T. F., D. J. Wolford, E. Veuhoff, V. Deline, P. M. Mooney, R. Potemski, and J. Bradley. 1987. Properties of high-purity $Al_xGa_{1-x}As$ grown by the metal-organic vapor-phase-epitaxy technique using methyl precursors. *J. Appl. Phys.* 62:632.

28. Chalmers, S. A., K. L. Lear, and K. P. Killeen. 1993. Low resistance wavelength-reproducible p-type (Al, Ga)As distributed Bragg reflectors grown by molecular beam epitaxy. *Appl. Phys. Lett.* 62:1585.

29. Lear, K. L., S. A. Chalmers, and K. P. Killeen. 1993. Low threshold voltage vertical cavity surface-emitting laser. *Electron. Lett.* 29:584.

30. Peters, M. G., B. J. Thibeault, D. B. Young, J. W. Scott, F. H. Peters, A. C. Gossard, and L. A. Coldren. 1993. Band-gap engineered digital alloy interfaces for lower resistance vertical-cavity surface-emitting lasers. *Appl. Phys. Lett.* 63:3411.

31. Hibbs-Brenner, M. K., R. A. Morgan, R. A. Walterson, J. A. Lehman, E. L. Kalweit, S. Bounnak, T. Marta, and R. Gieske. 1996. Performance, uniformity, and yield of 850-nm VCSEL's deposited by MOVPE. *IEEE Photon. Tech. Lett.* 8:7.

32. Lebby, M., C. A. Gaw, W. B. Jiang, P. A. Kiely, C. L. Shieh, P. R. Claisse, J. Ramdani, D. H. Hartman, D. B. Schwartz, and J. Grula. 1996. Use of VCSEL arrays for parallel optical interconnects. *Proc. SPIE* 2683:81.

33. Lebby, M. L., C. A. Gaw, W. B. Jiang, P. A. Kiely, P. R. Claisse, and J. Grula. 1996 (November). Key challenges and results of VCSELs in data links. LEOS '96, WV2, Boston.

34. Grodzinski, P., S. P. Denbaars, and H. C. Lee. 1995. From research to manufacture — The evolution of MOCVD. *JOM — J. Miner. Metal. Mater. Soc.* 47:25.

35. Kawai, H., S. Imanaga, K. Kaneko, and N. Watanabe. 1987. Complex refractive indices of AlGaAs at high temperature measured by *in situ* reflectometry during growth by metalorganic chemical vapor deposition. *J. Appl. Phys.* 61:328.

36. Breiland, W. G. and K. P. Killeen. 1995. A virtual interface method for extracting growth rates and high temperature optical constants from thin semiconductor films using *in situ* normal incidence reflectance. *J. Appl. Phys.* 78:6726.

37. Frateschi, N. C., S. G. Hummel, and P. D. Dapkus. 1991. *In situ* laser reflectometry applied to the growth of $Al_xGa_{1-x}As$ Bragg reflectors by metalorganic chemical vapour deposition. *Electron. Lett.* 27:155.

38. Azoulay, R., Y. Raffle, R. Kuszelewicz, G. Leroux, L. Dugrand, and J. C. Michel. 1994. *In situ* control of the growth of GaAs GaAlAs structures in a metalorganic vapour phase epitaxy reactor by laser reflectometry. *J. Cryst. Growth* 145:61.

39. Hou, H. Q., H. C. Chui, K. D. Choquette, B. E. Hammons, W. G. Breiland, and K. M. Geib. 1996. Highly uniform and reproducible vertical-cavity surface-emitting lasers grown by metalorganic vapor phase epitaxy with *in situ* reflectometry. *IEEE Photon. Tech. Lett.* 8:1285.

40. Ripper, J. E., J. C. Dyment, L. A. D'Asaro, and T. L. Paoli. 1971. Stripe-geometry double heterostructure junction lasers: Mode structure and cw operation above room temperature. *Appl. Phys. Lett.* 18:155.

41. Dyment, J. C., L. A. D'Asaro, J. C. North, B. I. Miller, and J. E. Ripper. 1972. Proton-bombardment formation of stripe-geometry heterostructure lasers for 300 K CW operation. *Proc. IEEE* 60:726.

42. Yonezu, H., I. Sakuma, K. Kobayashi, T. Kamejima, M. Ueno, and Y. Nannichi. 1973. A GaAs-Al$_x$Ga$_{1-x}$As double heterostructure planar stripe laser. *Jpn. J. Appl. Phys.* 12:1585.

43. Kaminow, I. P., R. E. Nahory, M. A. Pollack, L. W. Stulz, and J. C. DeWinter. 1979. Single-mode CW ridge-waveguide laser emitting at 1.55 μm. *Electron. Lett.* 15:763. [Erratum, *Electron. Lett.* 16:75 (1980)].

44. Ueno, M., I. Sakuma, T. Furuse, Y. Matsumoto, H. Kawano, Y. Ide, and S. Matsumoto. 1981. Transverse mode stabilized InGaAsP/InP (λ = 1.3 μm) plano-convex waveguide lasers. *IEEE J. Quantum Electron.* QE-17: 1930.

45. Chao, C. P., S. Y. Hu, P. Floyd, K. K. Law, S. W. Corzine, J. L. Merz, A. C. Gossard, and L. A. Coldren. 1991. Fabrication of low-threshold InGaAs GaAs ridge waveguide lasers by using in situ monitored reactive ion etching. *IEEE Photon. Tech. Lett.* 3:585.

46. Shieh, C. L. and D. E. Ackley. 1994. Top emitting VCSEL with etch stop layer. U.S. Patent No. 5,293,392.

47. Hirao, M., S. Tsuji, K. Mizuishi, A. Doi, and M. Nakamura. 1980. Long wavelength InGaAsP/InP lasers for optical fiber communication systems. *J. Opt. Commun.* 1:10.

48. Mito, I., M. Kitamura, K. Kobayashi, S. Murata, M. Seki, Y. Odagiri, H. Nishimoto, M. Yamaguchi, and K. Kobayashi. 1983. InGaAsP double-channel-planar-buried-heterostructure laser diode (DC-PBH LD) with effective current confinement. *IEEE J. Lightwave Tech.* LT-1:195.

49. Dutta, N. K., S. G. Napholtz, R. Yen, T. Wessel, T. M. Shen, and N. A. Olsson. 1985. Long wavelength InGaAsP ($\lambda \sim$ 1.3 μm) modified multiquantum well laser. *Appl. Phys. Lett.* 46:1036.

50. Mito, I., M. Kitamura, K. Kaede, Y. Odagiri, M. Seki, M. Sugimoto, and K. Kobayashi. 1982. InGaAsP planar buried heterostructure laser diode (PBH-LD) with very low threshold current. *Electron. Lett.* 18:2.

51. Tsang, W. T. and R. A. Logan. 1979. GaAs-Al$_x$Ga$_{1-x}$As strip buried heterostructure lasers. *IEEE J. Quantum Electron* QE-15:451.

52. Nelson, R. J., P. D. Wright, P. A. Barnes, R. L. Brown, T. Cella, and R. G. Sobers. 1980. High-output power InGaAsP (1 = 1.3 μm) strip-buried heterostructure lasers. *Appl. Phys. Lett.* 36:358.

53. Ishikawa, H., H. Imai, T. Tanahashi, Y. Nishitani, M. Takusagawa, and K. Takahei. 1981. V-grooved substrate buried heterostructure InGaAsP/InP laser. *Electron. Lett.* 17:465.

54. Tamari, N. and H. Shtrikman. 1982. High-T_0 low-threshold crescent InGaAsP mesa-substrate buried-heterojunction lasers. *Electron. Lett.* 18:177.

55. Murotani, T., E. Oomura, H. Higuchi, H. Namizaki, and W. Susaki. 1980. InGaAsP/InP buried crescent laser emitting at 1.3 μm with very low threshold current. *Electron. Lett.* 16:566.

56. Ogura, M. 1995. In situ etching and regrowth process for edge- and surface-emitting laser diodes with an AlGaAs/GaAs buried heterostructure. *J. Vacuum Sci. Tech. B* 13:1529.

57. Rochus, S., M. Hauser, T. Röhr, H. Kratzer, G. Böhm, W. Klein, G. Tränkle, and G. Weimann. 1985. Submilliamp vertical-cavity surface-emitting lasers with buried lateral-current confinement. *IEEE Photon. Tech. Lett.* 7:968.

58. Itaya, Y., T. Matsuoka, Y. Nakano, Y. Suzuki, K. Kuroiwa, and T. Ikegami. 1982. New 1.5 μm wavelength GaInAsP/InP distributed feedback laser. *Electron. Lett.* 18:1006.

59. Zory, P. and L. D. Comerford. 1975. Grating-coupled double-heterostructure AlGaAs diode laser. *IEEE J. Quantum Electron.* QE-11:451.

60. Aiki, K., M. Nakamura, and J. Umeda. 1976. Lasing characteristics of distributed-feedback GaAs-GaAlAs diode lasers with separate optical and carrier confinement. *IEEE J. Quantum Electron.* QE-12:597.

61. Utaka, K., S. Akiba, K. Sakai, and Y. Matsushima. 1984. $\lambda/4$-shifted InGaAsP/InP DFB lasers by simultaneous holographic exposure of positive and negative photoresists. *Electron. Lett.* 20:1008.

62. Numai, T., M. Yamaguchi, I. Mito, and K. Kobayashi. 1987. A new grating fabrication method for phase-shifted DFB LDs. *Jpn. J. Appl. Phys.* 26:L1910.

63. Suzuki, A. and K. Tada. 1980. Fabrication of chirped gratings on GaAs optical waveguides. *Thin Solid Films* 72:419.

64. Okai, M., S. Tsuji, N. Chinone, and T. Harada. 1989. Novel method to fabricate corrugation for a $\lambda/4$-shifted distributed feedback laser using a grating photomask. *Appl. Phys. Lett.* 55:415.

65. Sekartedjo, K., N. Eda, K. Furuya, Y. Suematsu, F. Koyana, and T. Tanbun-Ek. 1984. 1.5 μm phase-shifted DFB lasers for single-mode operation. *Electron. Lett.* 20:80.

66. Reinhart, F. K., R. A. Logan, and C. V. Shank. 1975. GaAs-Al$_x$Ga$_{1-x}$As injection lasers with distributed Bragg reflectors. *Appl. Phys. Lett.* 27:45.

67. Abe, Y., K. Kishino, T. Tanbun-Kk, S. Arai, F. Koyana, K. Matsumoto, T. Watanabe, and Y. Suematsu. 1982. Room-temperature cw operation of

1.60 μm GaInAsP/InP buried-heterostructure integrated laser with Butt-jointed built-in distributed-Bragg-reflection waveguide. *Electron. Lett.* 18: 410.

68. Mikami, O., T. Saitoh, and H. Nakagome. 1982. 1.5 μm GaInAsP/InP distributed Bragg reflector lasers with built-in optical waveguide. *Electron. Lett.* 18:458.

69. Hasnain, G., K. Tai, L. Yang, Y. H. Wang, R. J. Fischer, J. D. Wynn, B. Weir, N. K. Dutta, and A. Y. Cho. 1991. Performance of gain-guided surface emitting lasers with semiconductor distributed Bragg reflectors. *IEEE J. Quantum Electron.* QE-27:1377.

70. Morgan, R. A., G. D. Guth, M. W. Focht, M. T. Asom, K. Kojima, L. E. Rogers, and S. E. Callis. 1993. Transverse mode control of vertical cavity top-surface emitting lasers. *IEEE Photon. Tech. Lett.* 5:374.

71. Lee, Y. H., B. Tell, K. Brown-Goebeler, and J. L. Jewell. 1990. Top-surface-emitting GaAs four-quantum-well lasers emitting at 0.85 μm. *Electron. Lett.* 26:710.

72. Tell, B., Y. H. Lee, K. F. Browngoebeler, J. L. Jewell, R. E. Leibenguth, M. T. Asom, G. Livescu, L. Luther, and V. D. Mattera. 1990. High-power cw vertical-cavity top surface-emitting gas quantum well lasers. *Appl. Phys. Lett.* 57:1855.

73. Hasnain, G., K. C. Tai, L. Yang, Y. H. Wang, R. J. Fischer, J. D. Wynn, B. Weir, N. K. Dutta, and A. Y. Cho. 1991. Performance of gain-guided surface emitting lasers with semiconductor distributed Bragg reflectors. *IEEE J. Quantum Electron.* QE-27:1377.

74. Ding, G., S. W. Corzine, M. R. T. Tan, S. Y. Wang, K. Hahn, K. L. Lear, and J. S. Harris, Jr. 1996 (October). Dynamic behavior of VCSELs under high speed modulation. OSA annual meeting, WR6, Rochester, N.Y.

75. Law, J., and G. P. Agrawal. 1996 (October). Effect of carrier diffusion on modulation and noise characteristics of vertical-cavity surface-emitting lasers. OSA annual meeting, WII8, Rochester, N.Y.

76. Schwartz, D. B., C. K. Y. Chun, B. M. Foley, D. H. Hartman, M. Lebby, H. C. Lee, C. L. Shieh, S. M. Kuo, S. G. Shook, and B. Webb. 1995. A low cost, high performance optical interconnect. *Proc. 45th Electron. Compon. Tech. Conf.,* 376.

77. Nordin, R. A., D. B. Buchholz, R. F. Huisman, N. R. Basavanhally, and A. F. J. Levi. 1993. High performance optical data link array technology. *Proc. 43rd Electron. Compon. Tech. Conf.,* 795.

78. Shieh, C. L., M. S. Lebby, and J. Lungo. 1995. Method of making a VCSEL. U.S. Patent No. 5,468,656.

79. Scherer, A., H. G. Craighead, M. L. Roukes, and J. P. Harbison. 1988. Electrical damage induced by ion beam etching of GaAs. *J. Vacuum Sci. Tech.* B6:227.

80. Lebby, M. S., C. L. Shieh, and H. C. Lee. 1994. High efficiency VCSEL and method of fabrication. U.S. Patent No. 5,359,618.
81. Huffaker, D. L., D. G. Deppe, and K. Kumar. 1994. Native-oxide ring contact for low threshold vertical-cavity lasers. *Appl. Phys. Lett.* 65:97.
82. Huffaker, D. L., D. G. Deppe, and T. J. Rogers. 1994. Transverse mode behavior in native-oxide-defined low threshold vertical-cavity lasers. *Appl. Phys. Lett.* 65:1611.
83. Lear, K. L., K. D. Choquette, R. P. Schneider, and S. P. Kilcoyne. 1995. Modal analysis of a small surface emitting laser with a selectively oxidized waveguide. *Appl. Phys. Lett.* 66:2616.
84. Dallesasse, J. M., N. Holonyak, A. R. Sugg, T. A. Richard, and N. Elzein. 1990. Hydrolyzation oxidation of Al_xGa_{1-x}As-AlAs-GaAs quantum well heterostructures and superlattices. *Appl. Phys. Lett.* 57:2844.
85. Sugg, A. R., E. I. Chen, T. A. Richard, N. Holonyak, and K. C. Hsieh. 1993. Native oxide-embedded Al_yGa_{1-y}As-GaAs-In_xGa_{1-x}As quantum well heterostructure lasers. *Appl. Phys. Lett.* 62:1259.
86. Sugg, A. R., E. I. Chen, T. A. Richard, N. Holonyak, and K. C. Hsieh. 1993. Photopumped room-temperature continuous operation of native-oxide-Al_yGa_{1-y}As-GaAs-In_xGa_{1-x}As quantum-well-heterostructure lasers. *J. Appl. Phys.* 74:797.
87. Maranowski, S. A., A. R. Sugg, E. I. Chen, and N. Holonyak. 1993. Native oxide top-confined and botton-confined narrow stripe p-n Al_yGa_{1-y}As-GaAs-In_xGa_{1-x}As quantum well heterostructure laser. *Appl. Phys. Lett.* 63:1660.
88. Choquette, K. D., R. P. Schneider, K. L. Lear, and K. M. Geib. 1994. Low threshold voltage vertical-cavity lasers fabricated by selective oxidation. *Electron. Lett.* 30:2043.
89. Huffaker, D. L., J. Shin, and D. G. Deppe. 1994. Low threshold halfwave vertical-cavity lasers. *Electron. Lett.* 30:1946.
90. Yang, G. M., M. H. MacDougal, and P. D. Dapkus. 1995. Ultralow threshold current vertical-cavity surface-emitting lasers obtained with selective oxidation. *Electron. Lett.* 31:886.
91. Weigl, B., M. Grabherr, R. Michalzik, G. Reiner, and K. J. Ebeling. 1996. High-power single-mode selectivity oxidized vertical-cavity surface-emitting lasers. *IEEE Photon. Tech. Lett.* 8:971.
92. M. H. MacDougal, P. Daniel Dapkus, V. Pudikov, H. M. Zhao, and G. M. Yang. 1995. Ultralow threshold current vertical-cavity surface-emitting lasers with AlAs oxide-GaAs distributed Bragg reflectors. *IEEE Photon. Tech. Lett.* 7:229.
93. M. H. MacDougal, G. M. Yang, A. E. Bond, C. K. Lin, D. Tishinin, and P. D. Dapkus. 1996. Electrically-pumped vertical-cavity lasers with Al_xO_y-GaAs reflectors. *IEEE Photon. Tech. Lett.* 8:310.

94. G. G. Hadley, K. D. Choquette, and K. L. Lear. 1996. Understanding waveguiding in vertical-cavity surface-emitting lasers. *CLEO Tech. Digest Ser.* 9:425.

95. Y. A. Wu, C. J. Chang-Hasnain, and R. Nabiev. 1993. Single mode emission from a passive-antiguide-region vertical-cavity surface-emitting laser. *Electron. Lett.* 29:1861.

96. Y. A. Wu, G. S. Li, R. F. Nabiev, K. D. Choquette, C. Caneau, and C. J. Chang-Hasnain. 1995. Single-mode, passive antiguide vertical cavity surface emitting laser. *IEEE J. Select. Top. Quantum Electron.* 1:629.

97. K. L. Lear, R. P. Schneider, K. D. Choquette, S. P. Kilcoyne, J. J. Figiel, and J. C. Zolper. 1994. Vertical cavity surface emitting lasers with 21% efficiency by metalorganic vapor phase epitaxy. *IEEE Photon. Tech. Lett.* 6:1053.

98. T. Yoshikawa, H. Kosaka, K. Kurihara, M. Kajita, Y. Sugimoto, and K. Kasahara. 1995. Complete polarization control of 8 × 8 vertical-cavity surface-emitting laser matrix arrays. *Appl. Phys. Lett.* 66:908.

99. K. D. Choquette and R. E. Leibenguth. 1994. Control of vertical-cavity laser polarization with anisotropic transverse cavity geometries. *IEEE Photon. Tech. Lett.* 6:40.

100. M. Shimizu, T. Mukaihara, T. Baba, F. Koyama, and K. Iga. 1991. A method of polarization stabilization in surface emitting lasers. *Jpn. J. Appl. Phys.* 30:L1015.

101. T. Mukaihara, F. Koyama, and K. Iga. 1992. Polarization control of surface emitting lasers by anisotropic biaxial strain. *Jpn. J. Appl. Phys.* 31:1389.

102. Ser, J. H., Y. G. Ju, J. H. Shin, and Y. H. Lee. 1995. Polarization stabilization of vertical-cavity top-surface-emitting lasers by inscription of fine metal-interlaced gratings. *Appl. Phys. Lett.* 66:2769.

103. Baba, T., Y. Yogo, K. Suzuki, F. Koyama, and K. Iga. 1993. Near room temperature continuous wave lasing characteristics of GaInAsP/InP surface emitting laser. *Electron. Lett.* 29:913.

104. Dudley, J. J., M. Ishikawa, B. I. Miller, D. I. Babic, R. Mirin, W. B. Jiang, J. E. Bowers, and E. L. Hu. 1992. 144°C operation of 1.3 μm InGaAsP vertical cavity lasers on GaAs substrates. *Appl. Phys. Lett.* 61:3095.

105. Ram, R. J., L. Yang, K. Nauka, Y. M. Houng, M. Ludowise, D. E. Mars, J. J. Dudley, and S. Y. Wang. 1993. Analysis of water fusing for 1.3 μm vertical cavity surface emitting lasers. *Appl. Phys. Lett.* 62:2474.

106. Dudley, J. J., D. I. Babic, R. Mirin, L. Yang, B. I. Miller, R. J. Ram, T. Reynolds, E. L. Hu, and J. E. Bowers. 1994. Low threshold, wafer fused long wavelength vertical cavity lasers. *Appl. Phys. Lett.* 64:1463.

107. Babic, D. I., K. Streubel, R. P. Mirin, N. M. Margalit, J. E. Bowers, E. L. Hu, D. E. Mars, L. Yang, and K. Carey. 1995. Room-temperature continuous-wave operation of 1.54-μm vertical-cavity lasers. *IEEE Photon. Tech. Lett.* 7:1225.

108. Margalit, N. M., D. I. Babic, K. Streubel, R. P. Mirin, R. L. Naone, J. E. Bowers, and E. L. Hu. 1996. Submilliamp long wavelength vertical cavity lasers. *Electron. Lett.* 32:1675.

109. Ettenberg, M., H. F. Lockwood, and H. S. Sommers, Jr. 1972. *J. Appl. Phys.* 43:5047.

110. Imai, H., M. Morimoto, H. Sudo, T. Fujiwara, and M. Takusagawa. 1978. Catastrophic degradation of GaAlAs DH laser diodes. *Appl. Phys. Lett.* 33:1011.

111. Gfeller, F. R., and D. J. Webb. 1990. Degradation and lifetime studies of high-power single-quantum-well AlGaAs ridge lasers. *J. Appl. Phys.* 68: 14.

112. Imafuji, O., T. Takayama, H. Sugiura, M. Yuri, H. Naito, M. Kume, and K. Itoh. 1993. 600 mW CW single-mode GaAlAs triple-quantum well laser with a new index guided structure. *IEEE J. Quantum Electron.* 29:1889.

113. Lau, K. Y., P. L. Derry, and A. Yariv. 1988. Ultimate limit in low threshold quantum well GaAlAs semiconductor lasers. *Appl. Phys. Lett.* 52:88.

114. Orenstein, M., A. C. Von Lehmen, C. Chang-Hasnain, N. G. Stoffel, J. P. Harbison, and L. T. Florez. 1991. Matrix addressable vertical cavity surface emitting laser array. *Electron. Lett.* 27:437.

115. Vakhshoori, D., J. D. Wynn, G. J. Zydik, and R. E. Leibenguth. 1993. 8×18 top emitting independently addressable surface emitting laser arrays with uniform threshold current and low threshold voltage. *Appl. Phys. Lett.* 62: 1718.

116. Uchiyama, S. and K. Iga. 1985. Two-dimensional array of GaInAsP/InP surface-emitting lasers. *Electron. Lett.* 21:162.

117. Von Lehmen, A., C. Chang-Hasnain, J. Wullert, L. Carrion, N. Stoffel, L. T. Florez, and J. Harbison. 1991. Independently addressable InGaAs/GaAs vertical cavity surface emitting laser arrays. *Electron. Lett.* 27:583.

118. Khoe, G. K. and G. Kuyt. 1978. Realistic efficiency of coupling light from GaAs laser diode into parabolic-index optical fibers. *Electron. Lett.* 14: 667.

119. Reith, L. A., P. W. Shumate, and Y. Koga. 1986. Laser coupling to single-mode fibre using graded-index lenses and compact disk 1.3 μm laser package. *Electron. Lett.* 22:836.

120. Mathyssek, K., J. Wittmann, and R. Keil. 1985. Fabrication and investigation of drawn fiber tapers with spherical microlenses. *J. Opt. Commun.* 6:142.

121. Cheng, W. H., and J. H. Bechtel. 1993. High-speed fibre optic links using 780 nm compact disc lasers. *Electron. Lett.* 29:2055.

122. Soderstrom, R. L., S. J. Baumgartner, B. L. Beukema, T. R. Block, and D. L. Karst. 1993 (June). CD lasers optical data links for workstations and midrange computers. ECTC'93, 505, Orlando.

123. Armiento, C. A., M. Tabasky, C. Jagannath, T. W. Fitzgerald, C. L. Shieh, V. Barry, M. Rothman, A. Negri, P. O. Haugsjaa, and H. F. Lockwood. 1991. Passive coupling of InGaAsP/InP laser array and singlemode fibres using silicon waferboard. *Electron. Lett.* 27:1109.

124. Tan, Q. and Y. C. Lee. 1996. Soldering technology for optoelectronic packaging. *Proc. 46th Electron. Compon. Tech. Conf.*, 26.

125. Zeeb, E., B. Möller, G. Reiner, M. Ries, T. Hackbarth, and K. J. Ebeling. 1995. Planar proton implanted VCSEL's and fiber-coupled 2-D VCSEL arrays. *IEEE J. Select. Top. Quantum Electron.* 1:616.

126. Wong, Y. M., D. J. Muehlner, C. C. Faudskar, D. B. Buchholz, M. Fishteyn, J. L. Brandner, W. J. Parzygnat, R. A. Morgan, T. Mullally, R. E. Leibenguth, G. D. Guth, M. W. Focht, K. G. Glogovsky, J. L. Zilko, J. V. Gates, P. J. Anthony, B. H. Tyrone, Jr., T. J. Ireland, D. H. Lewis, Jr., D. F. Smith, S. F. Nati, D. K. Lewis, D. L. Rogers, H. A. Aispain, S. M. Gowda, S. G. Walker, Y. H. Kwark, R. J. S. Bates, D. M. Kuchta, and J. D. Crow. 1995. Technology development of a high-density 32-channel 16-Gb/s optical data link for optical interconnection applications for the optoelectronic technology consortium (OETC). *J. Lightwave Tech.* 13:995.

127. Kosaka, H., A. K. Dutta, K. Kurihara, Y. Sugimoto, and K. Kasahara. 1995. Gigabit-rate optical-signal transmission using vertical-cavity surface-emitting lasers with large-core plastic-cladding fibers. *IEEE Photon. Tech. Lett.* 7:926.

128. Matsuda, K., T. Yoshida, Y. Kobayashi, and T. Chino. 1996. A surface-emitting laser array with backside guiding holes for passive alignment to parallel optical fibers. *IEEE Photon. Tech. Lett.* 8:494.

129. Hahn, K. H. 1995 (May). POLO — Parallel optical links for gigabyte data communication. *Proc. 45th Electron. Compon. Tech. Conf.*, 368, Las Vegas, Nev.

130. Rowlette, J. R., Jr., M. K. Kallen, J. Stack, and W. Lewis. 1996. Laser micromachining of polymer waveguides for low cost passive alignment to VCSELs. *Proc. LEOS '96* 2:169.

131. Ou, S. S., M. Jansen, J. J. Yang, J. J. Mawst, and T. J. Roth. 1991. High power cw operation of InGaAs/GaAs surface-emitting lasers with 45° intracavity micromirrors. *Appl. Phys. Lett.* 59:2085.

132. Fallahi, M., M. Dion, F. Chatenoud, I. M. Templeton, and R. Barber. 1993. High power emission from strained DQW circular-grating surface-emitting DBR lasers. *Electron. Lett.* 29:2117.

133. Fallahi, M., F. Chatenoud, M. Dion, I. Templeton, R. Barber, and J. Thompson. 1995. Circular-grating surface-emitting distributed Bragg reflector lasers on an InGaAs-GaAs structure for 0.98-μm applications. *IEEE J. Select. Top. Quantum Electron.* 1:382.

134. Zah, C. E., M. R. Amersfoort, B. Pathak, F. Favire, P. S. D. Lin, A. Rajhel, N. C. Andreadakis, R. Bhat, C. Caneau, and M. A. Koza. 1996. *IEEE Photon. Tech. Lett.* 8:864.
135. Guenter, J. K., R. A. Hawthorne, III, D. N. Granville, M. K. Hibbs-Brenner, and R. A. Morgan. 1996. Reliability of proton-implanted VCSELs for data communications. *Proc. SPIE* 2683:102.
136. Ladany, I., Ettenberg, M., Lockwood, H. F., and Kressel, H. 1977. Al_2O_3 half-wave films for long-life cw lasers. *Appl. Phys. Lett.* 30:87.

Chapter 17 | Receiver, Laser Driver, and Phase-Locked Loop Design Issues

Dave Siljenberg

IBM Corporation, Rochester, Minnesota 55901

17.1. Introduction

This chapter introduces receiver, laser driver, and phase-locked loop (PLL) designs used in high-speed fiber optic communications. The goal is not to explain the details of their design but to explore the real-world environment within which these designs must perform. I will also outline the strategies the designer must use to build a robust design.

17.2. Fiber Optic Receivers

17.2.1. DESCRIPTION

Fiber optic receivers must convert the optical signal into an electrical signal and amplify it enough to be treated as a digital signal. Although there are other topologies, I will use the transimpedance amplifier configuration to explain various design trade-offs. The transimpedance receiver has good sensitivity and dynamic range, making it one of the most popular configurations.

17.2.2. TRANSIMPEDANCE RECEIVER

A block diagram of a typical receiver is shown in Fig. 17.1. The transimpedance amplifier consists of an inverting amplifier of gain, -A, with a

655

HANDBOOK OF FIBER OPTIC
DATA COMMUNICATION
$35.00

Fig. 17.1   Fiber optic preamplifier block diagram.

feedback resistor, R. The input impedance, Zin, is R/A. The input impedance must be kept low to maximize bandwidth. The high-frequency photodiode currents will either go into the amplifier or the parasitic capacitance of the photodiode and wiring. Lower Zin increases the current into the amplifier. This says that R should be kept low and A kept high. There is also an output amplifier shown with a gain of G. It is used to provide additional gain and drive signals off chip at low impedance.

17.2.3.   THERMAL NOISE

The dominant thermal noise comes from the feedback resistor. It generates a noise current i_n:

$$i_n = \frac{4kT}{R} \text{ BW}, \tag{17.1}$$

where k is Boltzmann's constant, T is temperature in degrees Kelvin, R is the feedback resistance, and BW is the noise bandwidth. This equation indicates that noise is reduced when the resistor is increased. This contradicts the previous bandwidth discussion. In addition, an increased gain, A, reduces dynamic range by reducing the maximum input. These trade-offs are summarized in Table 17.1.

Table 17.1   Receiver Trade-Offs

Parameter to be increased	R	A
Bandwidth	Decrease	Increase
Sensitivity	Increase	Neutral
Maximum input	Decrease	Decrease

In general, the designer must find the trade-off that achieves the design objective. Moui [1] presents a good discussion of receiver design including open-loop and transimpedance amplifiers. Meyer *et al.* [2] present a concise description of transimpedance design.

17.2.4.   PACKAGING

Because the photodiode lead has the smallest signals, it must be packaged with the most care. This is done to reduce parasitic capacitance and coupled noise. The packaging can fit into three categories: discrete, hybrid, and integrated.

In the discrete package, the photodiode and transimpedance amplifier are in separate packages and mounted on a common board or substrate. This is the lowest cost approach. However, it has the largest parasitics and coupled noise susceptibility. Generally, some shielding is needed to yield acceptable noise susceptibility.

The hybrid mounts the photodiode and transimpedance amplifier in die form on a common substrate. This improves parasitics and coupled noise immunity at increased cost.

The ultimate packaging is integrated. With an integrated circuit (IC) process that is compatible with a photodetector and amplifier design, one can achieve the best parasitics and coupled noise suseptability at the low cost of one IC. An example is a GaAs IC reported by Rogers [3]. The photodetector is responsive to 780-nm wavelengths and the transimpedance amplifier is implemented in metal semiconductor field-effect transistors.

17.2.5.   NOISE TRADE-OFFS

Although analysis of the thermal noise contribution is important, it is quite common that coupled noise sources will dominate in the circuit's real environment. Three sources of coupled noise will be discussed.

17.2.5.1.   Power Supply Rejection

For optical links in a data communications environment it is assumed that the receivers are imbedded in large digital systems and computers. The power supplies for those systems have noise in excess of 100 mV due to logic switching. Because digital clock rates are moving beyond 200 MHz, the noise can extend to 1 GHz. Care must be taken in the receiver design to suppress this noise and to prevent it from becoming larger than the thermal noise contribution.

An example calculation illustrates the point. If the design is intended to receive light at -20 dBm, the photodiode will have 10 μW average power or 20 μW peak-to-peak power impinging on it. A common photodiode efficiency is 0.5 mA/mW, in which case the photodiode current is 10 μA peak to peak. If the transimpedance gain and feedback resistor are -20 and 2 $K\Omega$, respectively, and the output stage has a gain of 2, we can calculate the effect of a typical power supply rejection ratio of -30 dB. The transimpedance stage will convert the 10-μA current to 20 mV peak to peak. The output stage will produce 40 mV peak to peak. One hundred millivolts of power supply noise will produce 3.2 mV of noise at the output. This is acceptable because the signal-to-noise ratio is greater than 10. Note, however, that the PSRR can be no worse than -30 dB over the entire bandwidth. This becomes more difficult at high frequencies.

17.2.5.2.   Coupled Noise

Another often overlooked noise source is on the fiber optic transceiver itself. Most transceivers have large logic swings in close proximity to the photodiode lead. Even worse is the laser drive signal. As will be discussed, the laser signal can be as large as 25 mA peak to peak with 300 ps rise times. The typical photodiode currents at -20 dBm can be 10 μA. The laser is often physically adjacent to the photodiode. It is very easy for the laser driver current to induce signals in the receiver. The discrete package option is most susceptible, whereas the integrated is the least susceptible.

17.2.5.3.   Electrostatic Discharge and Electromagnetic Interference Rejection

The receiver must also be immune to ESD and EMI. There are few modeling tools that the designer can use to predict ESD or EMI susceptibility. Therefore, a designer must test the design to ensure a rugged design. Again, the discrete package option is most susceptible, whereas the integrated is the least susceptible.

17.3.   Laser Drivers

This section covers the critical issues involved in modulating a serial bit stream on semiconductor lasers.

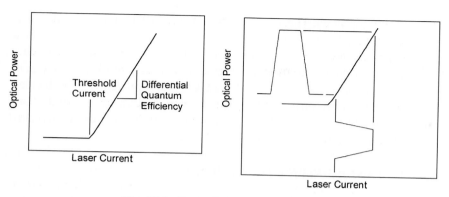

Fig. 17.2   Laser diode power vs current.

17.3.1. DESCRIPTION

The lasers described all have the same optical power vs current characteristic as shown in Fig. 17.2. This includes longwave and shortwave lasers. The laser has a characteristic threshold current, the current at which lasing begins, and the differential quantum efficiency, the slope of the optical power vs current curve. The laser driver must convert the bit stream into the correct current levels. This means driving the correct DC bias current and AC modulation current to produce an undistorted optical output pulse as shown in the right side of Fig. 17.2.

17.3.1.1. Laser Topologies

Lasers come in several topologies as shown in Fig. 17.3. If the case is to be grounded, the first laser requires current source from a positive supply. The second laser requires a current sink from a negative supply. The third topology offers more latitude because the case can be grounded regardless

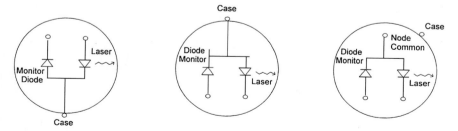

Fig. 17.3   Laser diode topologies.

of the driver topology. These choices will be the first step in defining a laser driver topology. All these topologies are shown with monitor diodes. These diodes are separate devices mounted in the package with the laser. They receive light from the back facet of the laser. The current from them can be used in a feedback loop to control the DC optical power. Newer devices, called vertical cavity surface-emitting lasers (VCSELs), are available. They have no back facet and therefore no monitor diode. They need to be biased with open-loop current sources.

17.3.2. LASER DRIVER TOPOLOGIES

An open collector laser driver is shown in Fig. 17.4. This topology makes use of high-speed NPN transistors found in many technologies. If the laser and driver are packaged with too much inductance, L1, fast rise times will cause ringing in the waveform. This is due to the high impedance of the

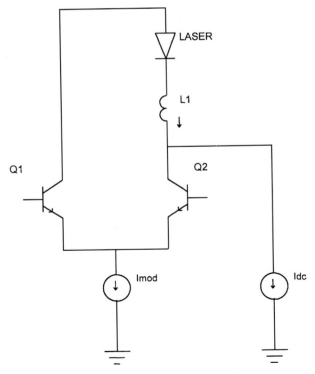

Fig. 17.4   Open collector laser driver.

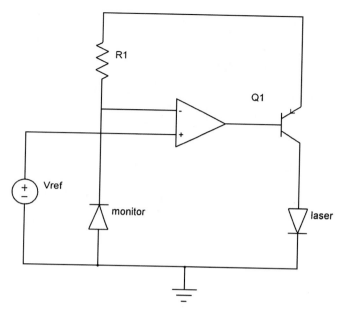

Fig. 17.5 DC laser bias circuit.

NPN collector and the low impedance of the laser. The laser impedance can be as low as 4Ω for shortwave lasers found in CD players and as high as 50Ω for VCSEL lasers.

17.3.2.1. DC Bias

A basic closed-loop DC bias circuit is shown in Fig. 17.5. Current from the monitor diode flows through R_1, generating a voltage at the input of the op amp. The op amp and transistor Q_1 control the laser current so that the voltage on the negative input of the op amp equals Vref.

17.3.2.2. AC Drive

Figure 17.6 shows a 50Ω driver that reduces ringing mentioned in the first driver. It also shows the driver AC-coupled to the laser. This is needed for a common cathode laser and a positive power supply. If the laser impedance is too low a resistor may needed in series with C_1. Figure 17.7 shows an emitter follower based design. This design can provide more drive current but suffers from emitter follower ringing or oscillation that is impedance sensitive.

Fig. 17.6   Terminated collector laser driver.

Fig. 17.7   Emitter follower laser driver.

17.3.2.3. Packaging

Similar to the receiver, the packaging can fit into two categories: discrete and hybrid. Integrated is not widely used because the semiconductor requirements for lasers are incompatible with transistors.

In the discrete package, the laser and laser driver are in separate packages and mounted on a common board or substrate. This is the lowest cost approach. However, it has the largest parasitics and electromagnetic radiation.

The hybrid mounts the laser and laser driver in die form on a common substrate. This improves parasitics and electromagnetic radiation at increased cost.

17.3.2.4. Electromagnetic Interference

Despite the fact that fiber optic transmission inherently produces no electromagnetic radiation, laser drivers do. This is compounded by the high speeds generated by serializing a parallel data stream. For example, if there is an 8-bit parallel data bus at 100 MHz it will have rise times on the order of 2.5 ns. If the bus is transported a distance on parallel cable significant shielding will be required to bring the radiation under acceptable limits. If this bus is serialized the rise times will become eight times shorter. If a code is used the edges will be faster. This means the laser driver will be producing 300 ps edges at 10–25 mA for a typical laser. The fiber will not radiate but the wiring from the laser driver to the laser must minimize loop area and may need shielding. Figure 17.8 shows a representation of a 1 Gb/s laser driver that is in production. The laser (in a TO style package) is within 25 mm of the laser driver IC. Despite this layout the assembly needs full metal enclosure to meet class B limits.

17.3.2.5. Termination and Layout Issues

Care must be used in laying out the connection between the laser driver and the laser. Parasitic capacitance and inductance of the laser and driver package must be minimized. If the path delay is a significant fraction of the rise time, transmission line techniques must be used. This may be difficult because many lasers have only a few ohms impedance. Laser drive return currents must be carefully returned to the driver. Failure to do these things will result in signal degradation and larger EMI radiation.

Fig. 17.8   1 Gb/s optical link card.

17.4.   Clock Recovery and PLLs

17.4.1.   DESCRIPTION

Two timing functions are required in a serial data link: a transmit synthesizer and a clock recovery circuit. Because these functions run at bit rates that present very demanding design challenges, the transmit synthesizer (illustrated in Fig. 17.9) must generate a bit rate clock from the byte rate or reference clock. The most common approach is to use a PLL or a delay-locked loop (DLL).

Fig. 17.9   Synthesizer block diagram.

There are many good sources for PLL design [4, 5]. The particular PLL shown is based on a charge pump and loop filter [6]. This design is shown because of its popularity and ease of integration. As shown Refs. [4]–[6], the PLL can be modeled with linear small signal equations. When this is done the PLL has a simplified closed-loop transfer function of the form:

$$H(s) = K_v \frac{I_p}{N} \frac{sRC_1 + 1}{s^2 2\pi C_1 + s K_v \dfrac{I_p}{N} RC_1 + K_v \dfrac{I_p}{N}}, \qquad (17.2)$$

where K_v is the voltage-controlled oscillator (VCO) gain in Hz/V, I_p is the peak current out of the charge pump, R and C_1 are the loop filter components, and N is the divide ratio or number of VCO cycles per reference clock cycle. This function describes the low-pass transfer function of the PLL. This means that phase shifts or jitter that are slower than the loop bandwidth will be accurately transferred by the PLL. Phase shifts or jitter that are faster than the loop bandwidth will be filtered out.

It needs to be pointed out that this is a simplified transfer function. The capacitor C_2 filters out the charge pump pulses and accounts for parasitic capacitance. C_2 makes the exact transfer function third order. However, a second-order function is a good approximation because C_2 must be kept small in order to minimize excess phase shift that can harm stability. This transfer function is plotted in Fig. 17.10, where K_v is 160 MHz/V, I_p is 600 μA, R is 430 Ω, C_1 is 10 nF, and N is 10.

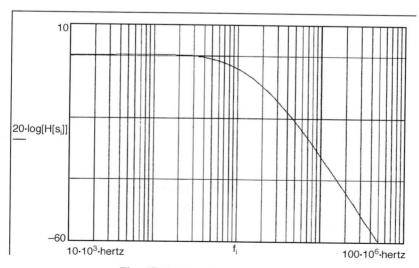

Fig. 17.10 PLL frequency response.

17.4.1.1.   Reference Clock Jitter

Because the reference comes from a clock on a large digital system, it will contain jitter from many digital sources in the system. The reference clock jitter has many frequency components due to system events happening at different rates. DRAM's get refreshed at a slow rate. System clocks generate noise in excess of 100 MHz. These noise sources get coupled to the reference clock through the power supply, adjacent wires, and ground bounce on logic ICs.

The effect of this jitter on the transmit synthesizer can be illustrated by examining reference clock phase shift occurring at low and high frequencies. If the reference clock shifts at low frequencies relative to the PLL bandwidth, the synthesizer will track it and propagate the jitter to the serial bit stream. If the reference clock shifts at high frequencies relative to the PLL bandwidth, the synthesizer will not be able to keep up and will filter the jitter out. At intermediate frequencies the synthesizer PLL will deliver a fraction of the jitter to the serial bit stream. The total jitter in a reference clock is a mixture of frequencies. Therefore, the synthesizer PLL will reduce but not eliminate the jitter in the reference clock. In order to predict how much jitter is propagated the designer must know the PLL bandwidth and estimate the spectral content of the reference clock jitter.

From the previous discussion, one might conclude that the synthesizer PLL bandwidth must be kept as low as possible. In the next section I will discuss why that is not always a good idea.

17.4.1.2.   Power Supply Rejection

All practical VCOs will vary in frequency due to noise on the power supply. This can be modeled as a power supply coefficient K_v. We can illustrate the jitter this induces by again examining noise at low and high frequencies. If the supply voltage varies at a slow rate relative to the PLL bandwidth, any phase shift the VCO makes will be compared to the reference clock by the phase detector. This will generate an equal and opposite control voltage change canceling out the jitter. If the supply voltage varies at a high rate relative to the VCO frequency, not the PLL bandwidth, the VCO does not get time to phase shift before the next VCO cycle. At intermediate frequencies the PLL will generate jitter in response to the power supply noise. In contrast to the previous discussion, this indicates that the PLL bandwidth should be as large as possible to cancel out the power supply-induced jitter. In practice the bandwidth will be a compromise between these two extremes.

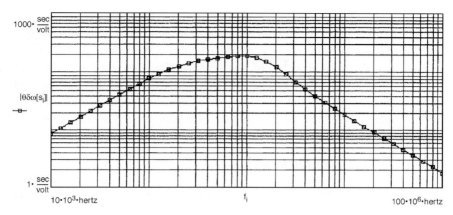

Fig. 17.11   PLL power supply rejection.

The power supply rejection can be modeled using the following equation:

$$P(s) = \frac{K_p}{s + \dfrac{K_v}{2\pi} \dfrac{I_p}{N} Z(s)}, \qquad (17.3)$$

where K_p is the VCO power supply coefficient in Hz/V, K_v is the VCO gain in Hz/V, I_p is the peak current out of the charge pump, $Z(s)$ is the impedance of the loop filter, and N is the divide ratio or number of VCO cycles per reference clock cycle. This equation is plotted in Fig. 17.11. The VCO gain is 160 MHz/V. The VCO supply coefficient is 11 MHz/V, which is 5%/V of the 220 MHz VCO center frequency. N is 10, I_p is 600 μA, and the loop filter values are as follows: $R = 430 \ \Omega$, $C_I = 10$ nF, and $C_2 = 100$ pF.

Note that the curve shows power supply-induced jitter of 200%/V at 900 kHz. This indicates that if the system has 0.1 V of power supply noise at 900 kHz, the PLL will induce 20% jitter in the serial bit stream! This is usually not acceptable. The designer must consider power supply filtering or designing a VCO with less than 5%/V power supply coefficient. This illustrates the design trade-off described previously. In this example, the designer must understand the reference clock jitter and power supply noise and design an optimal trade-off.

17.4.1.3.   Clock Recovery

The analysis of the transmit synthesizer reveals some basic PLL design issues. The design of the clock recovery circuit (CRC) is a more difficult task.

It must derive a bit rate clock based on the incoming data stream edges. The CRC must contend with an absence of transitions due to long strings of 1's and 0's. The CRC must also tolerate jitter in the incoming bit stream caused by noise and intersymbol interference. The following sections will discuss the trade-offs between various clock recovery methods and present some examples of CRC designs. CRC designs can be separated into analog and digital architectures. The analog designs include PLLs and DLLs. The digital designs consist of oversampled architectures.

17.4.1.4. Analog PLL Clock Recovery

The basic analog PLL clock/data recovery block diagram is shown in Fig. 17.12.

The PLL shown in Fig. 17.12 can be analyzed with the same linearized equations discussed for the synthesizer. This PLL will generate jitter due to power supply noise, as discussed previously. It will also respond to incoming jitter in a similar fashion as the synthesizer. The incoming jitter for the synthesizer was in the reference clock. The incoming jitter to the CRC comes from the serial data stream. The data stream has jitter from many sources. Major sources of generated jitter include the synthesizer, laser noise, fiber dispersion, and receiver noise.

The CRC PLL has a variable loop gain. In the synthesizer the divider ration N was fixed. That ratio determines the number of VCO cycles between every correction from the phase detector. The larger N is, the more infrequent are the phase detector correction pulses, lowering the loop gain. In the CRC the phase detector cannot generate a correction pulse without a data transition. In the CRC N is data dependent. Therefore, loop gain is data dependent. For $101010 \cdots$ NRZ data the gain is high and for long

Fig. 17.12   PLL-based clock/data recovery.

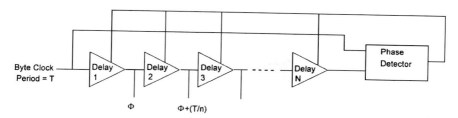

Fig. 17.13   Delay-locked loop.

strings of 1's and 0's the gain is low. The CRC PLL must be designed for both extremes of gain.

17.4.1.5.   DLLs

A typical DLL is shown in Fig. 17.13. By making phase-shifted copies of the byte rate clock we can generate bit rate timings. If we compare this to VCO-based PLLs we see that none of the DLL circuits run at the bit rate. Therefore, a slower technology can be used. However, the DLL needs more circuits. This makes it attractive to complementary metal oxide semiconductor developers.

Several advantages of the DLL over the PLL are reported [7]. These include (i) delay is not accumulated [8], (ii) the DLL is described by a single linear equation, and (iii) the loop locks quickly. The DLL lends itself to synthesizers but is difficult to use for clock recovery because an incoming data edge can be locked to any one of the phases.

17.4.1.6.   VCOs

The multivibrator VCO is shown in Fig. 17.14. It has been used in a large number of designs. J. T. Wu [9] reports a VCO with a large tuning range. Its major drawback is power supply rejection. Although the design looks differential its response to power supply noise is single ended. This can be understood by understanding the switch point. If Q_1 is off, current source I_1 discharges the capacitor pulling the emitter of Q_1 low. When the emitter of Q_1 is low enough Q_1 turns on and Q_2 turns off. Any power supply noise at that time will move the base voltage of Q_1, which moves the turn-on time of Q_1.

Another major topology is the ring oscillator. It is more compatible with digital IC processing because it does not need capacitors or analog current sources [10]. It can also be made fully differential for power supply

670   **Dave Siljenberg**

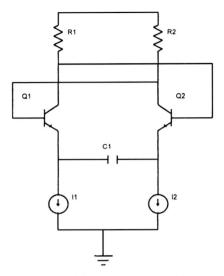

Fig. 17.14   Multivibrator VCO.

rejection [11]. The ring's main limitation is limited tuning range and difficult power supply and temperature compensation.

17.4.1.7.   Phase Detectors

A two-latch synthesizer phase detector is described in Ref [12]. This design provides phase and frequency detection because for every VCO transition there is a reference transition.

   The phase detector for the CRC is complicated by the absence of transitions in the serial data. The phase detector described by Deremer *et al.* [13] produces an increase and decrease pulse for every data edge. These pulses are averaged by the loop filter. A phase detector gain can be calculated for linear modeling. The design is shown in Fig. 17.15.

   A single-latch phase detector is shown in Fig. 17.16. If the data rise before the rising clock edge, the clock is late. A zero is latched and Q bar is high, charging up the loop filter. If the data rise after the rising clock edge, the loop filter is discharged. Unlike the two-latch design, this method is always correcting even if there are no data transitions. The CRC phase is constantly alternating above and below phase lock. To minimize these phase excursions the loop filter must be larger than the linear design. It is impossible to use a linearized gain factor for analysis. Designers must use transient simulators to optimize this design.

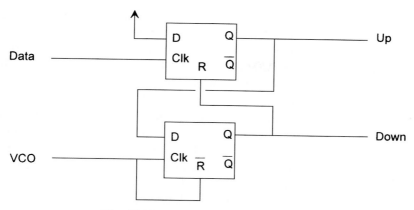

Fig. 17.15 Two-latch data phase detector.

The two-latch linear design provides the best trade-off between lock time and jitter because the loop filter can be smaller. However, skews in the latches or offsets in the charge pump create a static phase error. This happens because the PLL will settle into lock when the net charge in the loop filter is zero. If there are skews or offsets, there must be a phase difference between the VCO and data in the opposite direction to compensate. Great care must be taken to reduce these skews to an acceptable level.

The single-latch design will lock slower because of the large loop filter needed to keep the phase jitter to acceptable levels. However, offsets cause no phase error because the loop is only satisfied when the clock and data edges are lined up, even with charge pump skew. The logic must only be matched up to the input of the latch. In the linear design all the logic and the charge pump must be balanced to produce equal up and down pulses when the clock is in phase with the data.

17.4.1.8. Digital Data Recovery

Digital data recovery is typified by oversampled techniques described in Refs. [7] and [14]. These techniques involve sampling the incoming data stream with several phases of byte rate clocks. In 3× oversampling the phases would be spaced at one-third bit intervals. In addition, these clocks

Fig. 17.16 Single-latch data phase detector.

run at the byte rate. For example, for $3\times$ oversampling and 8-bit bytes there are 24 sampling clock phases. This generates multiple copies of the data stream. Digital circuits examine the sampled streams to find the data transitions. Once the transitions are determined, the correct data can be found by selecting data from the correct phase clock. For example in the previous example there are clock phases 0–23. If a transition is found between phases 3 and 4 the correct data can be found with phases 5, 8, 11, etc. Note also that these clock phases are synchronous with the local clock and not the remote clock that sent the data stream. Therefore, the decision regarding where to sample is constantly moving because these two clocks are within a crystal tolerance of the same frequency, typically 100 parts per million. Chen [15] describes the trade-off between jitter tolerance (defined below) and oversampling in a digital data recovery scheme.

17.4.1.9.   Digital vs Analog Trade-Offs

For a given technology, the analog techniques can produce the highest data rates. However, the analog techniques are more difficult to integrate with adapter logic. Therefore, as technologies advance the digital techniques replace analog at a given data rate.

17.4.2.   JITTER

Having discussed the jitter mechanisms in analog PLLs we can cover parameters necessary for defining the jitter in a communications system. These concepts are outlined in a Fibre Channel technical report [16].

17.4.2.1.   Jitter Generation

Jitter generation is the jitter contained in the bit stream source. The jitter comes from the reference clock jitter and power supply noise as mentioned in earlier sections. Thermal noise and cross-talk in the transmit VCO also adds jitter.

17.4.2.2.   Jitter Tolerance

Jitter tolerance is a measure of the amount of jitter the receiver can tolerate with no errors. It includes the jitter generation and jitter added by the media including intersymbol interference.

17.5. Conclusion

In this chapter I have summarized some of the most popular serial link circuits. Design trade-offs and real-life application concerns have been highlighted. As computer systems become faster and more widespread, the demand for high-speed serial optical connection will grow. The speed, power, and price of those connections will improve. This chapter provides designers with some information and techniques to use in those designs.

References

1. Moui, T. V. 1984 (June). Receiver design for high-speed optical-fiber systems. *J. Lightwave Tech.* **2**(3):243–267.
2. Meyer, R. G. *et al.* 1986 (August). A wide-band low-noise monolithic trans-impedance amplifier. *IEEE J. Solid-State Circ.* **21**(4):530–533.
3. Rogers, D. L. 1986 (November). Monolithic integration of a 3 Ghz MESFET detector/preamplifier using a refactory-gate ion-implanted MESFET process. *IEEE Electron. Dev. Lett.* **7**(11):600–602.
4. Gardner, F. M. 1979. *Phase lock techniques,* 2nd ed.
5. Wolaver, D. H. 1991. *Phase-locked loop circuit design.* New York: Prentice Hall.
6. Gardner, F. M. 1980 (November). Charge-pump phase-lock loops. *IEEE Trans. Commun.* **28**(11):321–332.
7. Marbot, R. *et al.* 1993. Integration of multiple point to point serial links in the gigabits per second range. Hot Links Symposium.
8. Lee, T. H., K. S. Donnelly, J. T. C. Ho, J. Zerbe, M. G. Johnson, and T. Ishikawa. 1994 (December). A 2.5 V CMOS delay-locked loop for a 10 Mbit, 500 megabyte/s DRAM. *IEEE J. Solid-State Circ.* **29**(12):1491–1496.
9. Wu, J. T. 1990 (February). A bipolar 1 Ghz multi-decade monolithic variable-frequency oscillator. *ISSCC Dig. Tech. Papers,* 106–107.
10. Razavi, B. 1996 (March). A study of phase noise in CMOS oscillators. *IEEE J. Solid-State Circ.* **31**(3): 331–343.
11. Walker, R. *et al.* 1992 (February). A 2-chip 1.5 Gb/s bus-oriented serial link interface. *ISSCC Dig. Tech. Papers,* 226–227.
12. Soyuer, M., and R. G. Meyer. 1990 (August). Frequency limitations of a conventional phase-frequency detector. *IEEE J. Solid-State Circ.* **25**(4):1019–1022.
13. Deremer, R. L., L. W. Freitag, and D. W. Siljenberg. 1990 (September). High speed phase detector. *IBM Tech. Disclosure Bull.* **33**(4):259–260.

14. Yang, C. K., and M. A. Horowitz. 1996. A 0.8 um CMOS 2.5 GB/s over sampled receiver for serial links. IEEE 1996 ISSCC Proceedings, 200–201.
15. Chen, D.-L. 1996 (August). A power and area efficient CMOS clock/data recovery circuit for high-speed serial interfaces. *IEEE J. Solid-State Circ.* **31**(8):1170–1176.
16. "Fibre Channel—Jitter Working Group," draft proposed X3 Technical Report, X3 Technical Committee of ANSI.

Chapter 18 | Packaging Assembly Techniques

Glenn Raskin

Motorola, Incorporated, Chandler, Arizona 85248

18.1. Packaging Assembly — Overview

Packaging can be thought of on many levels and serves many different functions. The basic function of packaging is the transfer of energy from one level to another. Whether that energy is electrical, thermal, or optical defines the level or type of packaging function. Traditionally, one considers the first level of packaging as that which deals with interconnection of semiconductor or optoelectronic devices to some kind of leadframe. The second level of packaging concerns itself with interconnection of the aforementioned leadframe to a circuit board, on which other packaged devices are also mounted.

18.1.1. *REQUIREMENTS FOR OPTOELECTRONIC PACKAGING*

The requirements for the packaging drive the technologies that are used and manufacturing techniques that are incorporated. In the area of optoelectronics one is primarily concerned with the transfer of optical and electrical energy from board level to the device. For semiconductor devices the focus is typically limited to the electrical domain. Other areas of concern and interest in packaging focus on thermal transfer, mechanical strength, and electromagnetic and other considerations. The discussions in Sections 18.2, 18.3, and 18.4 apply to optoelectronic as well as microelectronic assemblies.

675

HANDBOOK OF FIBER OPTIC
DATA COMMUNICATION
$35.00

Table 18.1   **Considerations for Packaging of Devices**

Electrical performance

Optical performance

Thermal performance

Reliability

Cost

Manufacturability

Testability

Device protection

Size

Weight

Mechanical integrity

Before focusing on the specific details, it is useful to look at the over-all picture. Table 18.1 attempts to capture the key considerations one must make in packaging assembly techniques. In any given application one must consider the factors in Table 18.1 and design a package that meets all of the requirements. It may be necessary to make compromises and trade-offs in each of the areas when making packaging decisions. For optoelectronics the requirements that drive device packaging revolve around some of the key characteristics of the system itself. If one is trying to construct a single-mode package the alignment tolerances to a single-mode fiber are much greater than those of a multimode system. These tighter alignments may require certain packaging techniques that limit the user's choice of tech-nologies. Another key input required for optoelectronic packaging is the amount of acceptable attenuation, or coupling loss, that is allowed. If the power budget allocated for the packaging is not great, this makes efficient coupling extremely critical.

Another key consideration in optoelectronic packaging is the thermal and reliability requirements. A laser's performance and expected lifetime are much more sensitive to its operating temperature compared to classical integrated circuits. This strong dependence of lifetime and reliability on temperature has limited the applications of laser devices and driven the development of more exotic thermal packages for optoelectronics.

Finally, one of the most important considerations, and the one typically given the least ink in text, is that of cost and manufacturability. In the end, whatever technology and performance is used the product must be cost-effective to the end user. If the required technology to meet the performance needs of the product is costly, often another avenue must be taken. Because cost is a prime driver in the industry, we often must compromise in other areas of performance to meet the cost goals of the product itself.

18.2. First-Level Interconnects

First-level interconnects are those that connect the device to the leadframe. There are a variety of interconnect technologies used in manufacturing. In this text I will review the three most common types: wire bonding, tape automated bonding (TAB), and flip chip.

18.2.1. WIRE BOND

The most common of all first-level interconnects in the microelectronic as well as the optoelectronic industry is known as wire bonding. Wire bonding is done using a number of techniques and in a variety of packages. The flowchart in Fig. 18.1 describes a typical process flow for wire bonding of devices.

I will discuss in-depth the wire bond technologies used in the industry and also review die attach and reliability concerns. The three most common wire bonding processes are ultrasonic, thermosonic, and thermocompression.

Fig. 18.1 Typical wire bond assembly flowchart.

18.2.1.1. Thermocompression and Thermosonic Wire Bond

Thermocompression bonding was the earliest type of wire bonding, demonstrated by AT&T in 1957 [1]. Thermocompression bonding is typically done using ultrapure (>99%) gold wire. A wire bonder is composed of a stage that holds the package and the bonding head, which feeds and places the wires. The bonding head forces the gold wire into contact with the die or substrate metallurgy that forms a bond through heating. Typical bond head temperatures between 300°C and 400°C are required to form the bond.

Thermosonic bonding is very similar to thermocompression bonding except that the wedge bond is performed using ultrasonic excitation to assist in bond formation. This allows for the use of lower temperatures than in thermocompression bonding.

Thermocompression and thermosonic bonding typically form a ball bond when contacting the device and a wedge bond onto the substrate. A schematic of ball and wedge thermosonic bonds is shown in Fig. 18.2. The process flow for both thermocompression and thermosonic wire bonding is the same. In either case we begin with a capillary holding a gold wire, typically from 0.001 to 0.05 in. in diameter, near its end. The wire that extends beyond the capillary is heated using either a hydrogen flame or through capacitive discharge. This forms a ball on the end of the wire two or three times the diameter of the wire itself. This ball is then compressed and the wire is heated to form the bond on the die (the substrate is also typically heated to a background temperature). The wire is then drawn through and looped over the die edge to the substrate. A wedge bond is then

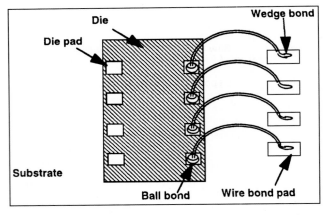

Fig. 18.2   Schematic showing ball and wedge wire bonds.

formed on the substrate, the wire is severed, and then the wire is fed through the capillary in preparation of the next bond [2].

18.2.1.2. Ultrasonic Wire Bond

Ultrasonic bonding differs from thermocompression bonding in that it forms only wedge bonds. Industry-standard ultrasonic bonding utilizes aluminum wire and can be accomplished on very narrow pad pitches. With ultrasonic bonding the wire is typically fed through a Sono-trode or stylus. The stylus forces the wire against the bond pad and applies a burst of ultrasonic energy to the interface. This process forms a cold wedge between the wire and metal on the bond surface. The stylus then loops up above the die surface and extends to the next bond point for the next wire bond.

18.2.1.3. Die Attach for Wire Bond

In order to wire bond to a device, the bottom of the device must be attached to a substrate of some kind, as shown in Fig. 18.2. The process of die attach can be accomplished utilizing a number of different materials from solders to organic adhesives to filled glass compounds [3]. Metallurgical attachment of the die to the substrate can be accomplished by metallizing the backside of the wafers. The die is then attached to the substrate using solder alloys. Solders offer low thermal and electrical resistance and are typically alloys of gold or lead.

Organic adhesives, most often epoxies, are the most common form of die attach. The adhesives are filled with precious metals to aid in thermal and electrical transfer. Epoxies provide reasonable performance and have been developed for highly manufacturable and relatively low-cost implementations. Glass adhesives, filled with silver, are a recent development that have not achieved wide acceptance in manufacturing areas.

18.2.1.4. Wire Bond Reliability

Wire bonding is a common process that has achieved very high yields and reliability levels for most applications. Many of the manufacturing defects have been eliminated as wire bond technology has reached its current maturity level. The most common reliability issues encountered in wire bonding today are due to gold aluminum mixing, die attach voids, and stress-induced creep. The mixing of gold and aluminum can lead to Kirkendahl voiding in extreme cases. This can be avoided by limiting the amount of time a unit is exposed to extreme (>400°C) temperature, by limiting the amount of impurities in the gold, and by forming the correct alloys in the bond. Stress-induced creep can occur in metallic joints due to low-melting eutectics (e.g., thallium) that weaken the grain boundaries. Once again, purity in the gold wires is key to avoiding reliability problems [4].

Reliability issues are also a concern in the die attach process. If excessive voids or bubbles are present in die attach material failure can occur with repeated thermal cycling. Failures in this case can be loss of bond line, leading to poor thermal conduction from the die to the substrate. Stresses can be so severe in die attach defects that cracking of the die itself occurs.

18.2.2. TAB

Tape automated bonding was developed in the 1960s by General Electric and is in large use today for devices that require low profiles (e.g., display line drivers) or very high lead counts. TAB differs from wire bonding in that the leadframe in this embodiment is directly attached to the device. The flowchart in Fig. 18.3 describes a typical process flow for assembling TAB devices. As can be seen in Fig. 18.3, TAB interconnection requires additional wafer processing to add a bump structure. Bump processing as well as the technology of interconnection will be discussed in the text.

18.2.2.1. TAB Bump Structures

Tab bump structures typically contain an adhesion layer, a barrier metal, and a plated bump. An adhesion layer helps promote adhesion between the underlying wafer metallization (typically an aluminum alloy) and the bump metal (typically gold). The composition of the layer can be copper, titanium, or chromium among others. The barrier metal serves the purpose of separating the bump metallization from the underlying metal. As previously discussed, one must be concerned with gold aluminum alloys. The barrier metal prevents intermixing of the aluminum and gold metallurgy.

Fig. 18.3 TAB flowchart.

Fig. 18.4   Cross section of a device after completion of TAB bump processing.

Barrier metals are typically sputtered across the whole wafer and are composed of any number of alloys including copper, palladium, platinum, or nickel [5]. After sputtering the wafers use a mask-defined photoresist to determine where the gold bumps will be located. The bumps are plated in an electrolytic bath. Following the plating process the photoresist and unwanted adhesion/barrier metallization are striped off the wafer.

The cross-section schematic shown in Fig. 18.4 shows a device after completion of wafer processing.

18.2.2.2.   TAB Tape Leadframes

The TAB leadframe is typically made of copper, surface plated with either gold or tin, and comes in numerous formats (Fig. 18.5a). The leads are typically supported on a polyimide film that can be chemically etched, mechanically punched, or laser etched to provide the necessary windows where the leads need to be exposed. Schematics of a TAB tape layout and cross sections are shown in Figs. 18.5b and 18.5c.

High-performance tapes have also recently been implemented that use two metal layers. The additional metal layer can be constructed to help in power/ground distribution, thereby improving the electrical and thermal performance of the leadframe.

18.2.2.3.   TAB Bonding

Attachment of the TAB leadframe to the die is commonly referred to as the inner lead bond (ILB). The leadframe is attached to the substrate in the area known as the outer lead bond. The bond is usually a single metallurgy

a

b

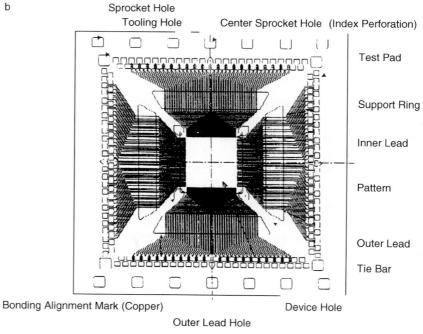

Fig. 18.5 (a) Examples of TAB tape formats. (b) TAB tape layout.

Fig. 18.5 *Continued* (c) Cross section of a TAB tape. (a) through (c) are courtesy of Shindo Co., Ltd. [6].

gold bond or eutectic gold tin bond. The ILB process is performed using either thermocompression or thermosonic single-point bonding. The most common process used today is gang thermocompression bonding. The advantage of a gang process, as opposed to wire bond or a single-point TAB attach, is in throughput. All the interconnections in a gang process are made simultaneously, in a single operation, allowing for excellent throughput. The disadvantages of gang bonding include the requirement of high levels of planarity, alignment, and uniformity across the interface. Single-point ultrasonic TAB bonding is similar to wire bonding.

Gang thermocompression bonding requires a heated tool, better known as a thermode, to join the die to the leadframe. The thermode, whether a solid block or multiple pieces, is heated to a temperature slightly above that required to make the metallurgical bond (approximately 300–400°C for tin–gold and 400–500°C for gold–gold bonding). The compression force works to guarantee good thermal contact between the thermode, leads, and the bumps. The force applied typically ranges from 10 to 50 g per lead. The dynamics of the process itself are very different when comparing a eutectic bond to a gold–gold thermocompression bond. Tin–gold eutectic bonding forms a low-melting-point liquidous alloy at the lead bump interface. This alloy allows for a lower stress, lower temperature bond compared to the single metallurgy bond.

Akin to wire bonding, thermosonic bonding techniques can also allow for lower bonding temperatures by using ultrasonic energy to aid in the attachment of the leadframe to the die. Thermosonic bonding is done in a single-point configuration and therefore does not have the throughput advantages of thermocompression gang bonding.

18.2.2.4. TAB Tape Reliability

TAB interconnections have shown themselves to be extremely reliable in a number of applications. As with wire bonding, care must be taken in the manufacturing process to guarantee good reliability. Key areas in ensuring reliability are in the integrity of the barrier/adhesion metals. The actual bond reliability relies on a high level of purity in the metallurgical components and especially in surface quality. As with most assembly processes the time the device is exposed to elevated temperatures must be limited. This is especially true for many optoelectronic devices in which low-temperature films are often used to limit the stress and interdiffusion of the laser die.

18.2.3. FLIP CHIP

Flip-chip interconnection relies on direct attachment of the device to the substrate. Wire bond and TAB interconnects rely on peripheral attachment. In other words, the interconnection is made on the outside edges of the die. Flip chip, on the other hand, can be both a peripheral and an area array attach. By using the area array approach a much higher density of interconnections can be achieved. This is shown in Fig. 18.6, in which the possible number of pads for typical wire bond pitches of 5.6 and 4.8 mils is compared to that of a typical flip-chip array pitch of 10 mils over a range of die sizes. For designs in which the number of interconnections are large, this can offer significant electrical and cost advantages. Another advantage of flip-chip attachment is that it can provide perhaps the lowest inductance interconnect between the die and the substrate. Low inductance is achieved by minimizing the length of the interconnect itself. Typical bump heights are on the order of 0.004 in. compared to 0.100 in. for wire bonds and 0.050–0.100 in. for TAB leads.

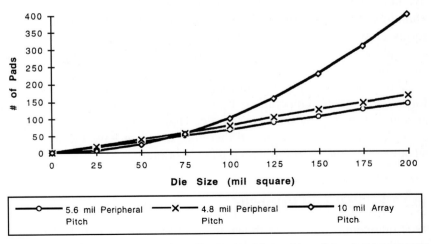

Fig. 18.6 Comparison of the number of interconnections with peripheral and array technologies.

Fig. 18.7   Flow diagram for C4/plated bump processes.

There are three major methods for flip-chip attach; C4 (controlled collapse chip connection), plated solder, and z-axis films. The most common embodiment is C4, which has a history closely tied with developments at IBM in the early 1960s [4]. In this embodiment an evaporated tin lead solder bump is used as the interconnect media between the die and the substrate. The plated solder solution is similar to the C4 process but relies on a plated bump rather than on an evaporative deposition. A typical flowchart that details the C4/plated bump processes is shown in Fig. 18.7. I will review in detail the process required to make a bump for flip-chip attach and the assembly processes required for attachment to the substrate.

The z-axis interconnects are a relatively new development that rely on adhesive films filled with conductive particles. In compression the conductive particles form a low-resistance path making contact to the substrate. Although promising, the use of conductive adhesives is currently very limited in the industry.

18.2.3.1.   Flip-Chip Bump Processes

The bump process metallurgy and structure can be readily compared to the bump process used in the TAB system. Whereas the TAB system is usually reliant on a gold bump system, flip chip relies on tin lead solder metallurgy.

C4 bump processing requires numerous steps to achieve the structure shown in Fig. 18.8. Wafers with vias in the top passivation over the pads are clamped to metal masks. After clamping the mask directly onto the wafer the ball-limiting metallurgy (BLM) is evaporated. Typically, this is a film of chromium, copper, and gold deposited in a single evaporator. The sandwich of metals that make up the BLM serves the purpose of adhesion and barrier layers as well as oxidation protection prior to solder deposition. The solder is then deposited typically in a separate evaporator. The metal mask is removed and the wafer is then reflowed. Reflow of the bumps is accomplished in an inert environment (usually H_2 at $350°C$) that transforms the bumps into the spherical shape as shown in Fig. 18.8. The spherical

Fig. 18.8   C4 bump cross section.

shape is formed during reflow due to the surface tension of the solder bump. This surface tension is an important aspect in solder-attach processes because it provides for what is commonly known as self-alignment. Upon reflow the liquidous metal will move to the lowest energy state it can achieve. This low-energy state centers the solder ball on the BLM [7].

Solder-plated bumps are manufactured in a very similar fashion to TAB-plated bumps. The only difference in the process is that solder is plated onto the wafers rather than onto gold. Numerous solder alloys have been considered. The most common for a solder bump is 95% lead/5% tin. This provides a high melting point (approximately 315°C) that can withstand the temperature that the package is exposed to during its attachment to the lower temperature solders (such as eutectic 63% tin/37% lead with a melting point of 183°C).

18.2.3.2.   Flip-Chip Substrates

Solder flip-chip attach processes are very much dependent on the substrates to which the die is being coupled. Ceramic substrates are usually metallized with copper or nickel films. Organic substrates, such as printed circuit boards, use lower melting point solders that allow for joining at lower temperatures. The footprint of the substrate is a mirror image of the die. This is shown schematically in Fig. 18.9 [8,9].

18.2.3.3.   Flip-Chip Attach

Attachment of the solder-bumped die to the substrate is a fairly straightforward process. Flux is applied to the substrates to enhance attachment and the die is placed fairly accurately onto the substrate, as shown in Fig. 18.9.

Fig. 18.9   Flip-chip die attached to a substrate.

The assembly is then reflowed and the connection of the die to the substrate is made in one process. In this way, hundreds of interconnections can be made quite rapidly. The placement of the die onto the substrate must be good enough so that some of the bumps are on the pad, but it does not need to be precise. Upon reflow, the self-alignment of the solder helps to correct for small misalignments. After attachment the residual flux is cleaned using any of a number of solvents or water. At this point the package may be capped and assembly completed. In some applications (high bump counts, large die, and large thermal expansion mismatches between the substrate and die) an underfill is applied between the die face and the substrate. The reasons why underfill is applied will be covered in more detail in the following section. The underfill itself is an epoxy that is dispensed around the die edge. Surface tension pulls the underfill into the area between the die and substrate. Following this process an underfill cure is required.

18.2.3.4.   Flip-Chip Reliability

Because the die are being attached directly to the substrates, the amount of thermal expansion mismatch between the two is critical. The stress caused by the mismatch is mostly absorbed in the bumps themselves. Thermal expansion mismatch can lead to excessive stress on the bumps and reliability failures if not properly managed. For alumina ceramic substrates, the coefficient of thermal expansion (CTE) is approximately $6 \times 10^{-6}/°C$, matching silicon and gallium arsenide fairly well. For organic substrates the CTE is much higher (approximately 12 to $16 \times 10^{-6}/°C$) and the mismatch stresses to the die are higher [10]. These mismatches cause a number of considerations in the application of flip-chip attachments. The first consideration is to use only materials that are well matched (such as alumina and silicon). The second consideration is that one may limit the size of the field of bumps that is allowed for flip-chip die. By limiting the bump field's size, the amount of stress on the outer bumps, where the extreme of the expansion mismatch is observed, is lessened. The third way to reduce the stress of thermal mismatch is to use an underfill epoxy between the die and the substrate. By acting as a compliant interstitial member the epoxy reduces the effective stress that is seen by the bumps themselves.

Long-term effects of solder creep and fatigue are well known and can be avoided through careful design and the matching of materials as discussed previously. If one uses these guidelines, the reliability of solder-bumped flip-chip assemblies is excellent.

18.3.   Package Types

There are numerous kinds of packages and formats available in the industry. The form factor of a package and its material makeup are driven by the application in which it will be used. Packages provide mechanical and environmental protection to the device while providing the electrical interconnections between the device and the circuit board. Packages in the optoelectronic industry can be divided into three categories; metal, plastic, and ceramic.

18.3.1.   METAL PACKAGES

Metal packages were the first packages used for the earliest transistors. As the microelectronics industry moved toward higher levels of integration, other packaging technologies were developed. The ceramic and plastic technologies were driven by the higher integration levels and the need for greater interconnection density between the device and the package. Metal packages in the form of transistor outline (better known as TO cans) are used widely in the optoelectronic industry today as the basis for numerous devices. These packages provide a cost-efficient method to package devices within an optoelectronic package. Schematic drawings of a TO can are shown in Fig. 18.10 [11].

TO cans utilize wire bonds (as discussed in Section 18.2.A) to interconnect the device to the package. As can be seen from Fig. 18.10b, the interconnection is made to the base of the package and to a post connected to the leads. TO cans are often integrated into optoelectronic modules in which other components (integrated circuits, discretes, etc.) are present.

Fig. 18.10 (a) TO can outline. (b) TO cross section with interconnection.

This provides the end user with a compact subassembly that serves a number of uses in optoelectronic products.

18.3.2. PLASTIC PACKAGES

In a plastic package the die is typically attached to a leadframe with any of the interconnection technologies discussed in Section 2. The leadframe and device are then encapsulated using a molding process. The molded plastic then serves as the package of the device. This embodiment is very popular in the microelectronics industry because it provides for a low-cost, highly automated assembly. In optoelectronics, molded packages are less common due to the thermal and performance limitations of the devices.

The first plastic packages used for integrated circuits were introduced in the 1960s and are known as flat pacs. These packages were used to replace metal cans in which higher lead counts and surface mounting was desired. Closely following the flat pack was the introduction of the dual in-line package (DIP). Both the flat pack and the DIP are cavity packages that are wire bonded and then the plastic is molded around the body and leads of the package. Following the introduction of the flat pac and the DIP packages, a number of other outlines were developed. Examples of flat pac and the DIP outlines are shown in Figs. 18.11a and 18.11b [12].

The key elements in a plastic package are the leadframe and the mold compound. Lead frames are typically made of copper with small amounts

Fig. 18.11   (a) Cross section of a flat pac. (b) Cross section of a DIP.

of iron, tin, phosphorous, or other elements to form an acceptable alloy. The design of the leadframe is optimized for manufacturing. The leadframe must accommodate the flag (which is where the die is attached), the fingers that radiate around the die (where wire bonds will be attached), and the exterior pins that are attached to the circuit board. Lead frames are typically mechanically stamped or chemically etched from copper sheets (though other alloys such as alloy 42, a mix of nickel and iron, are also widely used). After formation the flag and bond fingers are plated with silver or gold, often utilizing a nickel barrier layer. The leadframes are formed in strips that allow for automated equipment to handle a number of packages simultaneously.

The mold materials are usually thermoset compounds. The most common of these compounds are epoxy–Novolac-based molding materials. Compounds themselves are based on many components such as the resin, curing agent, mold release, accelerator, flame retardant, filler, and colorant. The mold compound must adhere well to both the leadframe and the surface of the device. Transfer molding typically preheats the resin pellet and then forces the compound into the mold cavity that holds the leadframe. The material actually flows as a viscous liquid when the viscosity drops enough due to the heat and force applied. As the resin flows through the small gates of the mold cavity it is heated through friction. This heating allows the material to cure and compact in a relatively short time. Curing of the plastic compound occurs in the mold, and in some packages it also occurs in a separate process known as postmold cure.

Some of the key concerns with molding around a device include how the mold compound affects the wire bonds that are present. Wire sweep is one of the major causes of defects in molded plastic packages. While the mold compound is filling the cavity the flow front of the material can displace the wires so that shorting or even wire fracture can occur. Careful design of the leadframe, cavity, and wire bond placement can minimize this effect. Other concerns with plastic package manufacture center around residual stresses of the molded body, voids in the mold compound, mold/device interface or mold/flag interface, and moisture-related issues. The susceptibility of plastic packages to moisture absorption and adsorption is one of the major limiters to its applications.

18.3.3. CERAMIC PACKAGES

Ceramic packages are used extensively in the microelectronic and opto-electronic industries. They come in a variety of forms, such as DIPs, flat packs, chip carriers, and pin-grid arrays (PGAs). Many of the formats discussed in the arena of plastic packages are offered in ceramic versions and vice versa. Ceramics cover a broad area of materials that offer a variety of material characteristics. This allows the flexibility of choosing a ceramic material that can best meet one's application. Ceramics are available that have thermal expansion coefficients that match a spectra of materials, from semiconductors ($3 \times 10^{-6}/°C$) to metals ($17 \times 10^{-6}/°C$). Ceramics thermal conductivities range from the insulator to conductor regime (up to 220 W/m $\cdot$ K) and dielectric constants from 4 to 10,000 [13]. The other major advantages of ceramics are its dimensional stability and its resistance to high-temperature processes. These characteristics allow ceramics to be used for highly integrated, densely routed packages and in ones in which hermetic sealing is a requirement. In these respects, ceramic packaging services an area of applications not covered by either of the materials discussed previously. Ceramic packages are made from a slurry of alumina, glass, and an organic binder or by utilizing dry processes and sintering of preforms.

For higher density of routings, ceramic packages utilize films of dielectrics and conductors on top of the alumina base layers. The two major technologies used in this area are thin film and thick film. Thin-film technology allows for higher routing density by utilizing lithographic techniques to define the conductors. Thick-film ceramic substrates are often used where seperate power and ground planes are desired. Whether using thick or thin film, the ability to have multilayer packages is a significant advantage in the areas of not only routing density but also thermal and electrical distribution.

Ceramic packages are often used for "high-reliability" applications. This is based on the ability of ceramic packages to be hermetic. By hermetically sealing the lids of the package and controlling the environment of the die cavity, ceramic packages can be made to be more resistant to caustic environments than can plastic packages. There are specific considerations that one must give to the reliability of ceramic packages. The pitfalls that arise with these package types are usually related to corrosion problems and mechanical cracking in the packages. Corrosion can be avoided by careful avoidance or removal of ionic contamination. Lid sealing of the package is also critical and must be monitored continuously in a manufacturing environment. Cracking must be managed through good manufacturing handling and careful treatment of the different materials used in the package, so as not to create a high-stress situation.

18.4.   Package to Board Attach

Attachment of the packaged device to the board is also a key consideration in the manufacturing environment. This second level of interconnection is often done to cards, boards, flexible substrates, and back planes. The key considerations at this level of attachment are its ability to be reworked, its density, and, of course, its performance. The two major types of second-level interconnection can be classified as pinned and surface-mount technologies (SMT).

18.4.1.   PINNED PACKAGES

Pinned packages are a common interconnection on nearly all boards in the industry today. The most common pinned packages are the DIP and PGA formats discussed earlier. Density of a pinned package is strongly dependent on the format of the package as shown in Fig. 18.12. The density relationship between peripheral and array layouts shown in Fig. 18.12 indicates that for large pin counts area arrays allow for higher packing densities.

Pins on a ceramic package are attached using a number of different methods. The two most common methods require that the pins be inserted through the package or brazed onto the surface. Inserted pins travel through the body of the package and therefore restrict trace routing and the placement of the device to where there are no pins. In this way we have a peripheral pin layout. Brazed-on pins allow the package designer more flexibility because the device or routing metals can be directly above the pins.

Fig. 18.12 Graph of the possible numbers of pins vs package size for a variety of pin pitches.

Therefore, brazed-on pins are often used for full-array pinned packages. Of course, in the industry through-hole pins are used in partial array configurations where a few rows of peripheral pins are used and the active routing is placed either between the pins or in areas where there are no pins present. Schematics of through-hole and brazed-on pins are illustrated in Figs. 18.13a and 18.13b.

A pinned package is attached to the board through a socket (which allows for easy replacement of the component) or, more permanently, solder

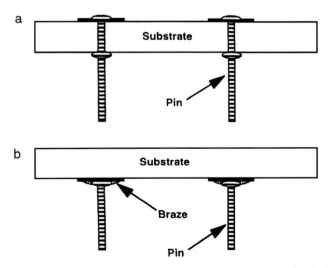

Fig. 18.13 (a) Schematic of a pin through a substrate. (b) Schematic of a pin brazed onto the surface of a substrate.

attached to the board. The pins on the package are attached to the board, which has a grid of plated-through holes filled with solder. The package is then wave-soldered into place. Wave soldering allows the solder deposition and joint formation to occur in one continuous process. In wave soldering the board and the components are exposed to flux, molten solder (typically eutectic lead tin at 230–260°C) for 3–10 s and then cleaned. In this fashion multiple components can be attached to a board at one time.

One of the major advantages of the wave-solder process is that the packages are exposed to relatively low temperatures (120–160°C) for single-sided boards. The reason for this is that the exposure of the board to the wave-solder temperature occurs on the underside of the board. This limits the amount of thermal stress that is applied to the package.

18.4.2. SURFACE-MOUNT PACKAGES

Surface-mount packages cover a broad number of formats as discussed previously (such as PLCC, SOJ, QFPs, TAB, and BGAs) and can provide a number of formats for the leads (gull-wing, J-lead, leadless, and butt joints). The use of SMT in the industry is growing due to reduced board space, cost, and the promise of higher performance solutions. Interconnection of SMT packages is considered to be more difficult and has been the focus of a great deal of industry research during the past decade. The focus of this development has been in allowing for smaller pitches (and sometimes blind mating) and in making the surface-mount joints less susceptible to thermal-cycle fatigue failures. The problem of thermal-cycle fatigue is analogous to the problems discussed in Section 18.2.C.4 because these joints are typically composed of solder.

Joining an SMT to a board can be accomplished through the same wave-soldering process described in Section 18.4.A. This process is in common use for SMT packages with lead pitches above 1.27 mm but is not considered adequate for packages with finer pitches due to solder-shorting issues. The most common processes used for SMT attach are known as vapor phase reflow (VPR) and infrared reflow (IR). These processes are often preceded by an application of solder paste to the mother board and treatment of the paste prior to the introduction of the packaged components. For VPR, the board and mounted components are loaded into a chamber containing vapor and boiling liquidous perfluorocarbon compounds. Perfluorocarbons are used because they have narrow and predictable boiling temperature ranges and are fairly inert. The assembly is rapidly brought to the desired temperature (175–250°C) and then reflowed [14].

One of the major advantages in VPR is that the processing time needed is very rapid, typically less than 5 min. This advantage is also the source of some of the significant problems with VPR. The rapid temperature rise can cause thermal shock of the components, problems with solder uniformity during reflow, and component movement due to solder wetting.

IR utilizes IR lamps providing radioactive energy to the surface components. The wavelength used by the IR lamps is usually between 1 and 5 μm. This technique is good for surface heating but causes problems with differential heating of the board surface. Differential heating is caused by different absorption of the IR energy by the components on the board. Often, the bodies of the package are heated to higher temperatures more readily than the solder lands. This problem can exasperate the temperature shock problems and lead to substantial temperature gradients across an assembled board. To overcome these shortcomings, IR is typically coupled with connective heat transfer to provide for a more stable process [15].

Other second-level interconnection technologies are also being investigated. These technologies include laser reflow, pressure interconnections, and conductive adhesives. These techniques have not found wide acceptance to date but continue to be investigated.

18.5. Optical Interconnect

The previous discussions incorporated manufacturing techniques that are common to both the microelectronics and optoelectronics industries. In this section I will focus on some of the manufacturing technologies specific to optoelectronic devices. The specific challenges encountered with packaging optoelectronic devices revolve around the transfer of optical power between the device and the outside world.

18.5.1. COUPLING

In the regime of assembly techniques, we are concerned with how efficiently we can couple light into (for detectors) or out of our devices [for light-emitting diodes (LEDs) and lasers]. This coupling can be in the free space regime (e.g., CD-ROMs, bar code scanners, or imaging) or be concerned with the transfer of data over a fiber of any of a number of compositions. The assembly technologies used are very dependent on the application and what level of coupling is required. For coupling of devices to a single-mode glass fiber (in which the core diameter is on the order of 6–8 μm) alignment

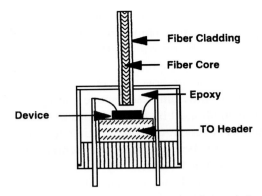

Fig. 18.14   Schematic of a device that is directly coupled to a fiber.

tolerances are quite tight. Coupling of devices to multimode glass fibers (in which the core diameter is on the order of 50–150 μm) is less critical, and coupling to large-diameter plastic fibers (>1 mm) requires even less attention to coupling accuracy.

The manufacturing techniques that are required to meet the application's coupling needs are quite varied. The simplest method of coupling a device to a fiber is known as direct or "butt" coupling. Figure 18.14 shows a typical direct-coupled configuration [16]. This direct coupling is often used for plastic fiber and some multimode applications but is not adequate for many multimode and nearly all single-mode applications.

In addition to butt coupling, light interconnection to fibers can also be accomplished by using waveguides, lenses, and prisms. This is especially useful when manufacturing tolerances need to be enhanced, as in the case of the smaller core fibers.

18.5.2.   WAVEGUIDES

Waveguides come in a number of configurations. These configurations include those made of polymers, ceramics, glass, and semiconductors; however, a waveguide can be formed from essentially any family of materials whose refractive indexes are different. In some cases, a waveguide allows for passive coupling of light to or from the device to the fiber. In other cases, waveguides are used to transfer light or to alleviate difficult tolerances.

The most commonly used waveguide is the silica fiber itself. Many of the processing techniques used for waveguide manufacture marry polymer or wafer technologies to those used in the manufacture of fiber and connectors.

Attachment of the waveguide to the device can be directly coupled and often acts as an intermediate between the device itself and the fiber.

18.5.3. LENSES

Lenses are commonly used in the optoelectronic industry. Lenses serve to focus and shape the light emitted from a laser or LED into a fiber. Nearly all single-mode fiber optic products use some kind of lens to help focus and shape the beam profile for coupling to a fiber. In single-mode fiber applications one must contend with small core diameters (6–8 μm) and relatively low acceptance angles (numerical aperture = 0.15, $\alpha = 8.6°$). The lenses and fiber must be placed at the focal distance from the light source to achieve the smallest spot sizes and most efficient coupling. In many cases this involves accurate placement of a semispherical lens on the surface of the optoelectronic device itself. This lens is secured through the addition of an adhesive. The fiber itself can also be secured at the focal distance from the lens by the same adhesive material. Figure 18.15 shows the use of a microlens in the well of a Burrus diode. This is often used in concert with active alignment to achieve optimal coupling. Although rigorous, this technique has provided the best solution for single-mode optical coupling in the industry.

18.5.4. PRISMS

Prisms are also used in the interconnection of optical energy. In the case of prisms, one can bend light as well as selectively filter out unwanted energies. Prisms offer interesting alternatives in the selection of which

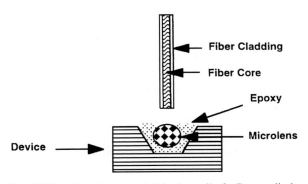

Fig. 18.15   Microlens epoxied in the well of a Burrus diode.

radiation energies are desired. Bending light can be quite useful in coupling from fibers to devices when the end of the fiber is not in the same plane as that of the active devices.

References

1. Anderson, O. L. 1975 (November). Bell Laboratories Record.
2. Steidel, C. A. 1982. Assembly techniques and packaging. In *VLSI technology,* ed. S. M. Sze, 551. New York: McGraw-Hill.
3. Shukla, R., and N. Mencinger. 1985 (July). A critical review of VLSI die-attachment in high reliability application. *Solid State Tech.,* 67.
4. Koopman, N., T. Reiley, and P. Totta. 1989. Chip-to-package interconnections. In *Microelectronics packaging handbook,* eds. R. Tummala and E. Rymaszewski. New York: Van Nostrand Reinhold.
5. Kawanobe, T., K. Miyamoto, and M. Hirano. 1983 (May). Tape automated bonding process for high lead count LSI. *33rd Electron. Comp. Conf. Proc.,* 221–226.
6. Shindo Company, LTD. Shindo TAB brochure.
7. Miller, L. F. 1969 (May). Controlled collapse reflow chip joining. *IBM J. Res. Dev.* 13:239–250.
8. Fired, L. J., J. Havas, J. S. Lechaton, J. S. Logan, G. Paal, and P. Totta. 1982 (May). A VLSI bipolar metallization design with three-level wiring and area array solder connections. *IBM J. Res. Dev.* 26:362–371.
9. Ohshima, M., A. Kenmotsu, and I. Ishi. 1982 (November). Optimization of micro solder reflow bonding for the LSI flip chip. Second Internation Electronics Packaging Conference.
10. Greer, S. E. 1978 (May). Low expansivity organic substrate for flip-chip bonding. *28th Electron. Comp. Conf. Proc.,* 166–171.
11. Motorola Inc. 1983. Optoelectronics Device Data.
12. Robock, P., and L. Nguyen. 1989. Plastic packaging. In *Microelectronics packaging handbook,* eds. R. Tummala and E. Rymaszewski. New York: Van Nostrand Reinhold.
13. Tummala, R. 1989. Ceramic packaging. In *Microelectronics packaging handbook,* eds. R. Tummala and E. Rymaszewski. New York: Van Nostrand Reinhold.
14. Hall, W. J. 1982. Vapor phase processing considerations for surface mounted devices. *Proc. Int. Electron. Pack. Soc. Conf.,* 41–46.
15. Dow, S. 1986 (October). Infrared reflow soldering: Another approach. *Surface Mount Tech.,* 28–29.
16. Midwinter, J. 1977. *Optical fibers for transmission,* 81. New York: Wiley.

Chapter 19 | Alignment Metrology and Techniques

Darrin P. Clement

Maponics, East Thetford, Vermont 05043

Ronald C. Lasky

Consultant, Medway, Massachusetts 02053

Daniel Baldwin

Manufacturing Research Center, School of Mechanical Engineering,
Georgia Institute of Technology, Atlanta, Georgia 30332

19.1. Introduction

This chapter will provide the reader with some benchmarks to use when determining characteristics of coupling in actual components. The theory behind such data will only be presented in the references. Alignment will be discussed in both optical and mechanical contexts. We will assume that the reader has a familiarity with the concepts herein and that this chapter will be used primarily as a reference.

We will then discuss the current techniques for achieving reliable mating between optical components for data communication. Active manual, active automated, and passive automated processes will be explored with an emphasis on expected performance and relative costs. It should be noted that throughout the chapter, we will not discuss telecommunications applications except as they may apply to data communication in the present or near future.

19.2. Interface Definition and Importance

An optical "interface" includes several interconnection elements. The cases considered here involve the mating of an optical source to an optical fiber and/or the coupling between a fiber and a photodetector.

699

HANDBOOK OF FIBER OPTIC
DATA COMMUNICATION
$35.00

A "source" may be either a laser diode (LD), a light-emitting diode (LED), or a vertical cavity surface-emitting laser (VCSEL). A lens is typically situated between the source and the fiber waveguide to increase coupling efficiency.

A packaging structure for datacom contains all of these elements and provides the required degree of precise optomechanical alignment among all elements. This structure is an optical subassembly (OSA), and we distinguish those for transmitter and receiver ends of the link as TOSAs and ROSAs, respectively. With the optoelectronic elements packaged in this way, the problem becomes an optomechanical one of aligning the OSA to fibers.

When using an LED, the fiber into which light is coupled would be multimode, whereas single-mode fibers require LD sources. In either case, the fibers used must be ferruled in order to provide durability and accuracy in optomechanical alignment. These cylindrically ferruled fibers are joined to the OSA module by "plugging" them into a bore (an alignment sleeve) located in the OSA. The ability for the ferrule to stay within the bore is characterized through spring and retention forces. Typically accepted values are 7.8–11.8 N for the ferrule spring force and 2.9–5.9 N for the bore retention force [1, 2].

A popular design scheme for datacom is the "connectorized" module. Connectorized in this context is the provision of an optomechanical construction at the module that can provide secure alignment and retention of the fiber while also allowing simple manual detachment and reconnection.

Generally, the trend in datacom hardware is to employ "duplex" connectors/modules, meaning that the connector hardware accommodates a pair of fibers, one each for the sending and receiving paths. Thus, at the transceiver module, the TOSA and ROSA are situated alongside each other and both fibers can be connected or detached in the same manual operation. The fiber distributed data interface, Fibre Channel Standard (a derivative of SC) and the Enterprise System Connection series of duplex connectors/modules are examples that embody this design [3]. Ribbon connectors are used for array devices whereby multiple fibers (usually multimode) are coupled into a linear (one-dimensional) array of lasers or detectors. These are all discussed in Chapter 1. To date, several new connector types are under development for use with VCSEL array transmitters.

19.3. Light Coupling

19.3.1. MIN/MAX COUPLED POWER LIMITATIONS

Because we are concerned with the pluggability of fiber optic connectors into OSAs, both mechanical and optical characteristics must be considered when determining the amount of coupling deviation acceptable. In order to achieve successful data transmission, criteria must be established that put a lower limit on what is considered "enough" coupled power. The coupled power range (CPR) is a measure of how much variation in light transfer is allowed when mating any given fiber with any given transmitter optical subassembly to ensure adequate system performance and yet not violate laser safety requirements.

Definition.   Coupled power range is the allowable difference between the minimum and maximum allowed power.

In order to account for all the losses along a system's length, a concept referred to as the "link budget" must be understood and has been explained in Chapter 7.

Whenever a receiver is said to have a sensitivity value, it must also be understood that this is with respect to an attainable bit error rate (BER). Data communication often requires a BER of 10^{-15}, which means that only 1 in 10^{15} bits can be erroneous. Telecommunication requirements are more relaxed and can be as low as BER $= 10^{-9}$. Some data links have specifications midway between the two values (e.g., 10^{-12}). See Ref. [4] for a trade-off analysis.

For example, a series of components (fiber connectors, fiber length, couplers, etc.) could result in approximately 13-dB total link loss. The minimum light that a typical photodetector can receive for a successful BER of 10^{-15} is approximately -23 dBm. This leaves a minimum launching power of -10 dBm. Thus, the TOSA and connector must achieve at least -10 dBm of coupling every time (e.g., within a 3 σ limit, or less than 27 faulty connections per 10,000). These numbers will vary depending on particular system implementations.

Maximum coupled power is a strong concern in laser systems (compared to LED systems) simply because of the higher available intensities. Some local area networks are user accessible, meaning that nontechnicians are able to connect/disconnect optical fiber cables between computers.

This access provides robustness but also presents potential safety hazards to the user. Additionally, there is concern that technicians may be exposed to focused/amplified light when inspecting an optical link for problems. International laser safety standards place a limit of less than 0.600 mW (−2.2 dBm) of optical power in the link [5], whereas the American National Standards Institute standard requires less than 100 s of exposure to 1 mW of light at $\lambda = 1.3$ or 1.55 μm [6]. See Chapters 7 and 8 for more information regarding planning a link and safety considerations.

This situation is unique to datacom because the telecom industry typically has user-inaccessible optical fibers. Thus, if the telecommunication power received at a detector is too weak, the laser output can simply be increased with much less concern for safety. Their limit of maximum coupled power is more likely to be dictated by the characteristics of the components. Recent developments have provided a different perspective on laser safety concerns. The alignment of source to fiber need not be controlled so as to keep the launched power below a certain value. Automatic power controls and open fiber controls, which electronically adjust the power while monitoring the transmission, are permissible under International Electrotechnical Commission (IEC) rules.

Another reason to have limits on output power, even with LED sources, is receiver saturation. Receiver sensitivity can be very high, approaching −40 dBm or better, but the dynamic range of the receiver may be limited. If the power at the detector is too high and saturation occurs, response speed can be degraded to a point at which link performance may be compromised.

19.3.2. MULTIMODE VS SINGLE-MODE CONNECTIONS

We should at this point contrast the cases of single- and multimode. In doing so, we will necessarily have to compare the situation of laser sources with that of LED sources. Although a laser can be used with either fiber type, LEDs are for all practical purposes suitable only for multimode fiber.

Let us assume two hardware sets, one for single mode (SM) and one for multimode (MM), that have the same absolute tolerances on dimensions, angles, and distances. For example, a loss of 1 dB in a SM system could occur with a transverse displacement of the fiber core by approximately 1 μm. It takes approximately 15 μm of similar multimode displacement (50-μm diameter core) to give rise to the same fractional loss of power [7]. The analysis of coupling geometry and efficiency is more difficult for the multimode case; however, connector hardware parts that fail optomechanical tolerances for single mode are often serviceable for multimode applications. See Refs. [8–14] for details.

Multimode fiber used in conjunction with LED sources can be very low cost. Currently, there are a number of commercial offerings of fiber optic transceivers designed for use with multimode fiber that provide distance–bandwidth products in the hundreds of Mb/km. Requirements for precision in the critical dimensions of the multimode connector interface are readily achievable with good- to high-quality machining. In certain applications, even a plastic fiber with core diameter of 100 μm may suffice.

We must bear in mind that the discussion has emphasized the need for short-distance data transmission at high data rates. For much longer distances, single-mode fiber should be used to ensure high bit rates in excess of 1 Gb/s. For this reason, the long-term strategy may be to use single-mode fiber "to the curb" supplanted by multimode fiber "to the home or office." Although MM fiber is no cheaper than SM fiber, it has been possible to reduce component cost with multimode solutions.

However, many companies are now marketing an "array" type of transmitter/receiver set. In these parallel transmission systems, an array of VCSELs are coupled into a ribbon fiber. The highly directional VCSELs coupled into multimode fibers provide relaxed mechanical tolerances in order to achieve consistent coupling.

19.4. Elements of Coupled Power

The practical coupling considerations for datacom connections are outlined in this section. Coupling to the TOSA and ROSA requires different considerations. The large active area of a typical photodetector and the larger MM fiber core diameter result in the fiber/ROSA interface having much less stringent optomechanical constraints than the TOSA/fiber interface. Thus, the TOSA/fiber interface will be emphasized in this chapter.

Poor control of the many tolerances involved can cause two basic problems in connector/module mating. The first problem may be observed when one plugs the same connector into the same module and obtains extensive variations in the launched power. This variation is referred to as plug repeatability (PR). Two decibels is typically the most extensive acceptable PR variation.

Definition.   Plug repeatability is the variation in coupled power for multiple connections between the same components.

The second variation is referred to as cross-plug range (XPR) and is measured by subtracting the lowest power reading from the highest power reading in a large sample of different connectors mated to the same module. A large cross-plug range is of concern because it is a significant source of the variation of the power launched to any given link. In many cases, improving the "peak" performance is less important than ensuring a small change in performance when interchanging connections.

Definition.   Cross-plug range is the difference between the measured lowest power and highest power for matings between multiple connectors and the same transceiver.

In order to reduce XPR between components, we must establish the sources of coupling variations. Once these are defined, it becomes an engineering problem to optimize component performance.

19.4.1. OPTICAL ALIGNMENT

An optical parameter that plays an important role in coupling is the beam itself. Most beams can be characterized as approximately Gaussian: for example, a SM fiber's fundamental mode, an LD's far-field (although elliptical), the distribution of modes in a MM fiber, and the beam from an LED. Values for each of these parameters vary widely and often must be determined through direct measurement [15]. Rather than use separate terms for each (e.g., mode field diameter, spot size, numerical aperture, etc.), we will simply use source beam spot (SBS) and fiber beam spot (FBS), although the reader should be aware that different definitions of "width" are used: $1/e$, $1/e^2$, full width at half maximum, etc. When you are given a value for a beam spot, make sure you know which definition has been used. Even when all mechanical alignments are perfect, a difference in beam spots between components will reduce the coupling efficiency. Although the FBS is well measured and controlled, the SBS is usually poorly known or controlled.

19.4.2. PLANAR ALIGNMENT

Physical misalignments are the most obvious source of coupling variation when mating two optical components such as a ferrule and an OSA bore. We can define the interface plane as the plane normal to the optical axis. Longitudinal displacements occur along the (optical) z-axis and are not

(a) Ceramic Solid Bore

(b) Metal Split-Sleeve Bore

Fig. 19.1 Solid and split-sleeve bores [17].

significant [11, 16]. We do consider some lateral misalignment in the x and y directions, or r (where $r^2 = x^2 + y^2$).

Ferruled fibers are inserted into bores in the OSAs that house the transmitter or receiver. Two style of bores are available: solid and split sleeve. A solid bore has complete rotational symmetry in the sense that it is a "perfect" cylinder as shown in Fig. 19.1a. It is typically constructed of a rigid, ceramic material. The split-sleeve bore, in contrast, is separated lengthwise along the cylinder wall, as shown in Fig. 19.1b. Its composition is of a more flexible material. The split-sleeve bore is designed so that its effective diameter will enlarge slightly to accommodate a larger ferrule.

The solid bore inner diameter (BD) and ferrule diameter (FD) also determine coupling performance. The difference in these two diameters creates an offset in solid bore TOSAs that can vary from insertion to insertion in the same module/connector pair. Hence, these two elements are critical in plug repeatability for solid bore TOSAs. Typical values for the BD are 2.5010–2.5025 mm. For the ferrule dimensions, some specifications require a diameter of 2.4992–2.5000 mm, within a 3-standard deviation criterion. See Fig. 19.2 as a reference. These parameters give the largest ferrule-to-bore difference possible of 3.3 μm. The minimum difference is 1 μm, and assuming normal distributions the average would be 2.15 μm.

Beam centrality (BC) is the position of the laser beam axis with respect to the mechanical axis of the bore. It is desirable that the two be coaxial (making BC = 0) so that the laser beam center is exactly in the center of the bore. The ferrule/core eccentricity (FCE) is the corresponding centrality misalignment element in the connector. The FCE is typically controlled to

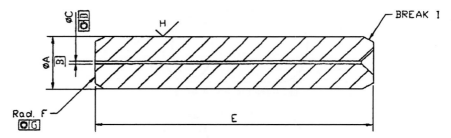

Fig. 19.2   Example schematic of a ferrule.

<1.6 μm. It is non-zero when an imperfect ferruling process results in a fiber core that is nonconcentric with the ferrule.

These parameters are shown in Fig. 19.3. Note that to ease interpretation, it is assumed for now that the ferrule fits "perfectly" within the bore so that the ferrule and bore centers coincide. (The degree of "coincidence" is explained below.) Therefore, we can define an angle, ψ, between the directions of BC and FCE with respect to some fixed coordinate system. (ψ itself does not constitute a new "element.") For the BC and FCE given, the ψ value will be randomly distributed throughout 360°C. For $\psi = 0°$, the misalignments will have the least effect, whereas $\psi = 180°$ will place the fiber core the farthest from the light spot. The scales of the BC and FCE have been grossly enlarged in Fig. 19.3 for clarity.

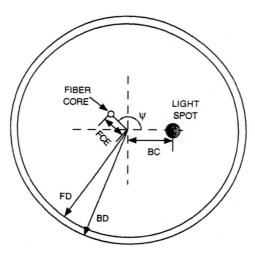

Fig. 19.3   BC and FCE (exaggerated scale) [17]. *Note that the ferrule and bore centers coincide.

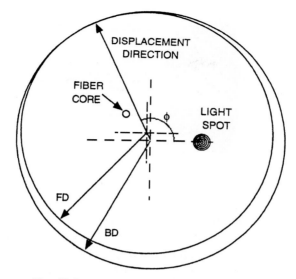

Fig. 19.4 BD and FD (exaggerated scale) [17].

These same elements are shown in Fig. 19.4 for an arbitrary direction, ϕ, of displacement. If $FD - BD = 0$, then ϕ is meaningless and we revert back to Fig. 19.3. Depending on the ψ orientation between BC and FCE, ϕ can either improve coupling or degrade it. Because ψ and ϕ are random variables, the best we could hope for is to establish a range of possible distances between the fiber core and the light spot.

The random fluctuation of all these dimensions can be reduced during manufacturing through a process known as tuning, whereby the eccentricity is intentionally aligned in a preferred direction [1].

It should be observed, however, that split-sleeve TOSAs will not be sensitive to these elements; by design, the split-sleeve walls flex to achieve a snug fit for a wider range of FD.

19.4.3. ANGULAR ALIGNMENT

The ferrule/bore diameter difference can lead to a source of misalignment known as the angular offset, or tilt. However, due to the small lateral displacement of $BD - FD = 3.3 \, \mu m$, and the long length of the bore of approximately 4 mm, this tilt will be small. The largest tilt due to the ferrule/bore difference would be $0.05°$. This can be shown to contribute insignificantly to the CPR [16].

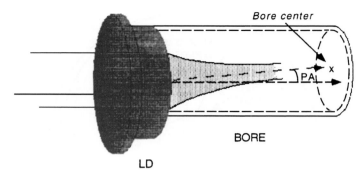

Fig. 19.5   Laser PA [17].

There are, however, other possibly significant sources of tilt. One potential for tilted beams can be found in the TOSA. Even if the beam center spot can be successfully imaged to the bore center (i.e., BC = 0), the laser may possess some tilt as shown in Fig. 19.5. This tilt is known as the pointing angle (PA). As explained previously, before the laser and bore are welded, the laser is repositioned for maximum coupling. If there is a significant PA to the laser light, the weld will still be optimum for a small BC, as shown, and the engineer will be oblivious to the fact that any tilt is present!

The pointing angle can be as large as 3° and must be taken as an unknown. It is hoped that as datacom applications become more common, the need for well-controlled tolerances will drive manufacturers to produce modules with much smaller laser tilts. A necessary precursor to this optimization is the development of an accurate measurement technique.

The fiber end-face angle (EFA) is due to the polishing process of the ferruled fiber. It is common to intentionally polish fibers at angles ranging from 5 to 10° in order to reduce back reflections from the fiber surface. The de facto industry standard is 8°.

However, even supposedly "flat" fibers may have a non-zero EFA. Fibers are typically arc-polished so that the hemispherical tip deforms slightly when two ferruled fibers are butt-coupled. This process, known as the physical contact polish, assumes that the polished apex coincides with the fiber core center. If the polishing tool is not precisely aligned, the end-face normal of the fiber will not be parallel to the fiber/ferrule axis. A non-zero fiber FCE further complicates this situation.

Refraction of the entering/emerging light will cause the effective light axis to be nonparallel, or tilted, with respect to the alignment axis. As seen

Fig. 19.6   Fiber EFA.

in Fig. 19.6, a beam emerging from an obliquely polished fiber is refracted according to Snell's Law:

$$n_1 \sin(\text{EFA}) = n_0 \sin(\text{EFA} + \theta). \qquad (19.1)$$

It can easily be seen that when a fiber end face is polished at an angle EFA, the beam emerges at an angle θ with respect to the fiber axis (which is an angle EFA $+ \theta$ with respect to the end-face normal). Using the small angle approximation, θ can be found:

$$n_1(\text{EFA}) \approx n_0 \,(\text{EFA} + \theta)$$

or

$$\theta \approx \left(\frac{n_1}{n_0} - 1\right)\text{EFA}. \qquad (19.2)$$

Using 1.47 for the fiber's index of refraction, and a value of 1 for air, the relation between θ and EFA is $\theta \approx (0.47)$ EFA. Note that for analytical purposes, θ can be seen as the fiber analog to the TOSA pointing angle.

Random sampling of nonrejected spherically polished fibers shows up to $0.3°$ of effective tilt, θ. The measurement of the end-face angle can typically be made using interferometric techniques so that samples with, for example, $>10.1°$ EFA can be rejected, but this will produce large numbers of unused cables, increasing manufacturing costs.

19.4.4. MODULE ALIGNMENT

The module itself can also contribute to coupling variations through the subassembly misalignment (SAM) and connector ferrule float (FF). SAM is the axial misalignment of the entire OSA/bore assembly from its desired true position within the transceiver module. Remember that we are discussing duplex interconnections so that both the transmitting and receiving

Fig. 19.7 Connector/module mating elements.

bore/ferrule matings must be aligned (Fig. 19.7). If one of the couples cannot be mated, then by design the other pair would be prevented from achieving successful plugging.

FF is the amount of lateral movement or float possessed by the connector ferrules. Too rigid a setting would impede ease of use and/or would make component damage more likely. Thus, spring forces built into the design allow a little "give." FF could compensate for a slight SAM. Of course, too large a FF could also prevent proper mating because of the "sloppiness" of the ferrule position. In split-sleeve OSAs, excessive SAM may cause a large off-axis force on the split sleeves that can cause plug repeatability problems or damage the split sleeve. It is important to note, however, that these two elements do not strongly affect the plug repeatability or cross-plug range for solid bore systems because the bore stiffness overwhelms them. A detailed account of the forces involving FF and SAM is given in Ref. [2].

Other characteristics are the module shroud dimension (SD) and the connector body dimension (CD). The clearance between the two is the variable

Fig. 19.8 Split-sleeve bore PR [17].

that affects plug repeatability and cross-plug range. Excessive clearance can create several mechanically stable plug modes that can affect plug repeatability in the same module/connector pair. Data indicate that this clearance has a much stronger effect on split-sleeve OSAs than on solid bores (Figs. 19.8 and 19.9).

19.4.5. PUTTING IT ALL TOGETHER

For reference, Table 19.1 provides typical values for the 12 elements.

A typical CPR might be from −2.2 to −10 dBm—a 7.8-dBm range. All of the variables that affect the CPR are usually analyzed with a Monte Carlo technique to ensure that the CPR can be achieved. The CPR can be viewed as the sum of variations due to aging effects on the electrical

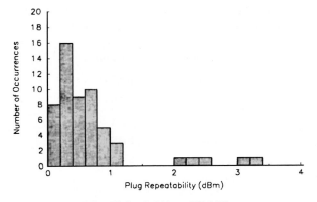

Fig. 19.9 Sold bore PR [17].

Table 19.1   Typical Values of the 12 Elements

Element	Typical value	CPR influence	
		Solid bore	Split sleeve
SBS	5–20 μm[a]	✓✓	✓✓
FBS	9.5 ± 0.5 μm for SM[a]	✓	✓
BC	3.3 ± 2.0 μm	✓✓	✓✓
FCE	<1.6 μm (0.7 μm)	✓	✓
BD	2.5010–2.5025 mm	✓	
FD	2.4992–2.5000 mm	✓	
PA	<3°	✓✓	✓✓
EFA	0.1–0.2° (barring rejection)	✓	✓
SAM	Spec. ± 0.34 mm		✓
FF	Spec. ± 0.34 mm		✓
SD	Spec. ± 0.125 mm		✓
CD	Spec. ± 0.125 mm		✓

Note. From Ref. [17]. ✓, determines the influence of a parameter's variability. If a parameter either has little variation or has little effect on the coupling efficiency to begin with, it will be ranked with fewer ✓. Given these measurements, it is desirable to predict how much variation to expect from a given product that could include any combination of values.
[a] Using $1/e^2$ width definition.

components and to changes in coupling from mechanical/optical variations in the connectors and TOSAs.

In general, all 12 elements affect XPR. However, if we limit ourselves to solid bore TOSAs, only 8 variables are operable. Four variables are related to the connector and 4 to the module. These variables are

Module	Connector
BD	FD
BC	FCE
PA	EFA
SBS	FBS

Each of these factors will produce variability in both PR and XPR.

The actual distance between the fiber core and the laser spot is essentially determined by BC, FCE, BD, and FD as shown in Fig. 19.10 (compare with

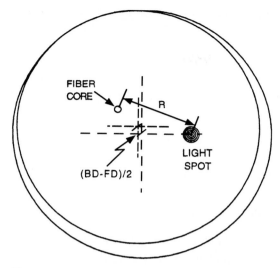

Fig. 19.10   Fiber core and laser spot separation [17].

Figs. 19.3 and 19.4). Knowing these quantities along with the directions of displacement ψ and ϕ, we can develop a relation for the distance R:

$$R^2 = \left[\text{BC} - \text{FCE} \times \cos \psi - \frac{1}{2}(\text{BD} - \text{FD}) \times \cos \phi \right]^2$$
$$+ \left[\text{FCE} \times \sin \psi + \frac{1}{2}(\text{BD} - \text{FD}) \times \sin \phi \right]^2 \qquad (19.3)$$

It is clear that the maximum value for R occurs when ψ and ϕ are $180°$ and the minimum occurs for both directions equal to zero. Thus,

$$R_{\text{max}} = \text{BC} + \text{FCE} + \frac{1}{2}(\text{BD} - \text{FD})$$

and

$$R_{\text{min}} = \text{BC} - \text{FCE} - \frac{1}{2}(\text{BD} - \text{FD}). \qquad (19.4)$$

Clearly, BC, FCE, and DB − FD can give rise to an R_{min} of zero. Figure 19.11 shows the theoretical results for coupling from LDs with SBS = 9.5 and 14 μm to a standard fiber with FBS = 9.5 μm for purely radial offset R. Actual coupling losses can be up to 3 dB greater than those shown in Fig. 19.11, but the shapes of the experimental curves are

Fig. 19.11   Coupling loss as a function of radial offset [17].

surprisingly close to those in Fig. 19.11. It is believed that the difference is due to lens aberrations in the TOSA [7, 18].

Different modules will necessarily have different PAs, even if BC is well controlled. Figure 19.12 shows the effect on coupling loss that these various pointing angles will create. Generally speaking, this variation can be responsible for up to 10-dB differences in loss between matings of different modules with a given connector. However, plug repeatability will not likely be affected because the PA for a given module can be assumed constant.

The fiber EFA has a much less severe influence on coupling efficiency when it is assumed that other misalignments are small. In fact, because it is common that EFA < 0.2°, one can use Fig. 19.12 to estimate its

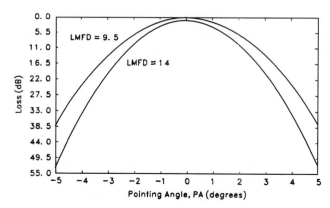

Fig. 19.12   Coupling loss as a function of laser pointing angle [17].

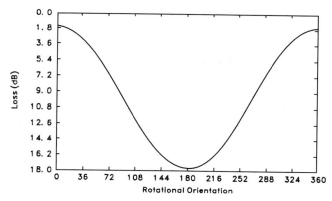

Fig. 19.13   Change in coupling loss during fiber rotation for a large end-face angle [17].

effect on overall coupling loss, provided one sets PA $= \theta = 0.47$ (EFA) from Eq. (19.2). The variability comes from different insertions of one connector/module pair. With 360° of possible rotation within the bore, a fiber with an EFA $= 0.2°$ will produce minimal coupling variation.

It should be noted, however, that many applications call for intentionally angled end faces to reduce back reflection into the source. The industry standard for the EFA in this case is 8°. Here, for the same parameter values as in the previous paragraph, different insertion orientations have dramatic effects, as shown in Fig. 19.13 [8].

It is well worth noting that in our discussions, we have primarily considered isolated misalignments: radial, axial, and angular. When all these are present, they can each become more significant to coupling. When the radial offset is significant, the PA has a more dramatic effect and vice versa. Nemoto and Makimoto [16] give analytic solutions when the three misalignment "categories" can be assumed to occur in the same plane. Out-of-plane misalignments are explored in Ref. [8].

19.5.   Alignment Techniques

Alignment technology is critical to the production of low-cost optodatacom components. It is considered by some to be the critical issue that may delay the proliferation of optical data communication. Currently, it can cost up to $100 to align and secure a laser diode in a SM optical subassembly. Developing technologies to reduce this cost by one or two orders of magnitude is paramount to the proliferation of optodatacom into the consumer markets.

19.5.1. ALIGNMENT REQUIREMENTS

Although SM alignment requirements are often stated as submicrometer, experience has shown that 1–3 μm control of the center of the laser spot to center of the SM fiber is usually adequate. MM technology is relaxed to approximately five times that of SM technology. There are active (light source turned on), manual alignment technologies for manufacturing available in the late 1990s to perform both SM and MM assembly. Unfortunately, these techniques are slow, serial, and hence expensive. Passive and active automated alignment techniques are on the horizon that may address this need.

19.5.2. CURRENT ALIGNMENT AND
SECURING TECHNOLOGY

As stated in previous chapters a transceiver module generally consists of a ROSA and a TOSA assembled on a substrate with associated drive and mux/demux circuitry. The critical alignments that we are discussing are needed to assemble the TOSA. The alignment tolerances in assembling the ROSA are considerably more relaxed as discussed previously. To review the assembly process and structure of the TOSA see Chapter 5.

In a typical TOSA the optical axis of the laser (or other source) is first aligned to that of a lens (Fig. 19.14). This alignment is currently performed with the light source active, using some type of robot. When the alignment is optimized the robot measures a maximum in light output. At this point the securing of the lens and laser positions is performed, usually using some type of welding or occasionally adhesive. The welding process, although fast and strong, can also result in a shifting as the weld cools and shrinks. The assembled laser and lens are then aligned to the center of the bore of the TOSA. This process is also performed actively, aligning the light from the laser/lens combination with the bore center by measuring light coupled into a ferruled fiber in the bore. Using a ferruled fiber in the bore contributes a 1- or 2-μm offset immediately because there is clearance between the ferrule and bore and the fiber is never perfectly centered in the ferrule. These errors were discussed previously in this chapter: FCE and BD − FD. After the laser/lens combination is aligned by the robot to the bore/ferrule, another weld is formed to secure these two components. Unfortunately, another "weld shift" is usually experienced. Hence, with the weld shift and bore to ferrule error, it is typically difficult to obtain better

TO can *Packaged diode laser mounted in a TO-5 header can includes a PIN photodiode to monitor the beam output and a heat sink to reduce the effects of temperature instabilities.*

Glass window
Cap
Heat sink
Laser chip
Monitor photodiode
Stem

OSA

LASER DIODE PKG
SELFOC LENS
FERRULE STOP

MODULE
RX IC
DESERIALIZER
ROSA
TOSA
TX IC
CLK REC CHIP
SERIALIZER

Common Solution Today:

Laser TO can → Optical Sub-Assembly (OSA) → Module
- Three levels of packaging
- Several active alignments

Fig. 19.14 A typical OSA.

than 3-μm alignment in a welded TOSA. These steps are time consuming and hence expensive, especially in SM technology.

19.5.3. ACTIVE AUTOMATED ALIGNMENT

Active automated alignment involves aligning a source to an optical detector (i.e., axis) while both source and detector are active. A primary advantage of active alignment techniques is that they produce known good assemblies with optimum or near-optimum alignment. This is in contrast to passive alignment techniques, which must be tested postmanufacture. Cost is believed to be the primary disadvantage with this alignment technique. The prevailing activity in optoelectronics packaging is to develop passive alignment technology for low-cost assembly. The most common method for active automated alignment involves mechanical alignment techniques. Recent developments in selective silicon etching and silicon micromachining have enabled development of combined active and passive alignment techniques, simplifying the assembly process.

There are numerous automated alignment techniques reported in the literature, with each having unique advantage. A representative set can be found in Refs. [19–23]. A typical assembly process involves first the temporary insertion of an optical fiber into a large area photodetector to determine the maximum optical signal strength. This signal strength is used to determine the maximum relative signal strength during active alignment. Then the optoelectronic package is assembled and the source and detector are activated. Next, after fiber is removed from the photodetector, it is inserted into the optoelectronic package and brought into the vicinity of the photodetector or source depending on the function of the package. Micromanipulation is then used to align the fiber to the photodetector or source. The end of the fiber is manipulated in five degrees of freedom (i.e., x, y, z, θ_x, and θ_y) while the signal strength on the detector is monitored. The photodetector signal strength is maximized to find the optimum alignment. At this point, the fiber is permanently affixed in the package to secure the aligned position. Typical fixturing processes include soldering, adhesive attach, and welding.

Active automated alignment in optoelectronic assemblies is typically accomplished using one of three techniques for maximizing the detector signal as the fiber is precision aligned relative to the detector. Near-optimum alignment is achieved using simple linear five-axis optimization in which the fiber is moved independently along each of five axes (i.e., x, y, z, θ_x, and θ_y) to identify a local maximum signal strength [22]. Each axis is traversed independently, whereas previously analyzed axes are fixed in their local maxima positions. The primary drawback with this technique is that it rarely produces optimum alignment and is prone to locating local maxima having signal strength considerably lower than the maximum. A second active alignment technique utilizes minimum path search techniques to determine the optimum alignment (Fig. 19.15). These search techniques are derived from fundamental research performed in robotic motion and automated assembly. Minimum path searches involve incrementally moving the fiber along each degree of freedom (i.e., δx, δy, δz, $\delta\theta_x$, and $\delta\theta_y$) and measuring the signal strength. The path motion space is then analyzed to identify the incremental motion yielding the maximum positive gradient

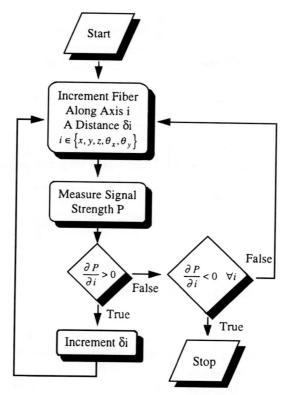

Fig. 19.15 Typical process flow for minimum path search technique for active automated alignment.

of the signal strength. The path variables are then incremented based on the local optima. The fiber is again incrementally moved along each degree of freedom and the path of maximum response is again identified. This process is repeated until all combinations of incremental moves generate negative signal strength gradients corresponding to the optimum alignment. Identification of localized maxima is a potential problem with this technique, although various techniques have been developed to minimize the problem. With this technique, the fiber is incrementally moved along a path of increasing signal strength until the maximum is reached.

The third technique for active automated alignment is direct feedback control of the micropositioning system during active alignment [24]. One technique used for feedback control utilizes a phase-sensitive detector in the optoelectronic assembly and drives the fiber alignment axis using a

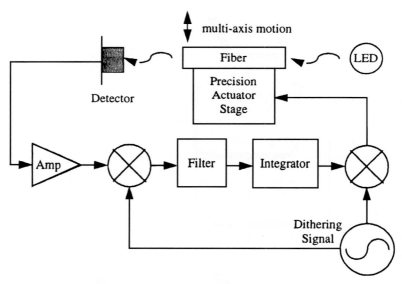

Fig. 19.16   Typical feedback control loop for active automated alignment (adapted from Ref. [24]).

sinusoidal dithering motion superimposed on a monotonic translation. The output of the phase-sensitive detector is directly proportional to the spatial offset error (used for feedback control) of the optical fiber from the optimum alignment position (Fig. 19.16). Moreover, the polarity of the phase-sensitive detector output corresponds to the positive and negative approach directions for the optimum alignment position. For zero-offset positioning of the fiber from optimum (within 0.1–1%), an integral feedback controller can be used. Additional options include proportional–integral and proportional–integral–derivative as feedback control systems. The primary advantage of this technique is its rapid response and settling time for locating an optimum alignment. For multiaxis alignment, this technique has a significant speed advantage in that it allows optimization of the various axes simultaneously. This is done by having the micropositioning system actuators for the different axes dithered at different frequencies. Different phase-sensitive detectors are then used to extract the spatial offset information pertaining only to the appropriate axis. The spatial offset errors from each phase-sensitive detector are then used as the feedback control error signal for actuating the independent axis transducers controlling the fiber position.

Assembly of optical fiber is typically achieved using a microactuator to mechanically align the fiber relative to a detector or source. The manual alignment technique is the most common method. Here the fiber is grasped with a microgripper attached to a precision multiaxis stage and translated/rotated using precision lead screw actuators. To minimize vibration and backlash, the stages typically ride on air bearings and are mounted to large granite slabs. Actuator drive systems incorporate submicrometer encoders and closed-loop position control to ensure precise alignment.

19.5.4. HYBRID ACTIVE/PASSIVE AUTOMATED ALIGNMENT

Due to the excessive cost associated with five degrees of freedom, precision actuators, and the relatively low-volume utilization of the precision assembly systems, recent efforts have focused on combined active and passive alignment assembly. Numerous techniques have been reported using various degrees of passive alignment [25–31]. Tewksbury et al. [32] also present a comprehensive review of optical interconnect technology. The primary advantage of combined active and passive assembly technology is significantly reduced cost. This is because combined active and passive alignment requires only one or two precision actuators compared with five for a full degree of freedom assembly. Numerous combined active and passive alignment techniques have been developed. We will focus on a few representative techniques here.

Due to the relatively high cost of optoelectronic component assembly and optical fiber alignment, there has been a heightened interest in fiber array assembly technology. Jackson et al. [33] present several approaches for aligning linear arrays of optical fibers (similar to ribbon cables used in electronics) to multichannel GaAs lasers and detector arrays. The assembly process is simplified using anisotropically etched precision silicon V grooves etched in a rectangular silicon block. The fibers are secured in the V grooves using high-temperature adhesives or solders forming a passively aligned subassembly. In the case of solder attach, the fibers must be metallized with a solder-wettable metallization system to ensure reliable adhesion and robust performance. This is particularly important for subsequent processing steps that can subject the assembly to numerous high-temperature thermal excursions. The fiber array and V-groove assembly significantly reduces assembly complexity by transforming the multifiber array into a single-fiber subassembly that can be aligned and assembled as a single unit. This enables multiple interconnects to be formed during a single assembly process and significantly reduces the cost per interconnect.

To further simplify assembly of the fiber array to the vertical GaAs laser array or the detector array, the fiber V-groove subassembly is beveled on one end (typically at a 35 or 45° angle relative to the horizontal). The beveled end acts as a mirror reflecting the vertically transmitted light from the lasers into the fibers or from the fibers into the detectors. In the case of the 45° bevel, the fiber end must be metallized to enhance reflectivity and reduce coupling losses; the 35° bevel totally reflects the transmitted light and does not require metallization. The alignment of the fiber array to the laser array and/or detector array is achieved using standard precision actuator systems. To simplify the alignment process, photolithographically defined ridges and contacts are produced on the laser array and/or detector. These act as alignment faducials for precision alignment of the fiber array. This assembly can have an angular registration better than $\pm 1°$ and positioning alignment within $\pm 3 \mu$m [33] (Fig. 19.17). As one would expect, these tolerances are well within the requirements for multimode assembly.

A second representative assembly processing using combined active and passive alignment is presented by MacDonald *et al.* [34]. The particular assembly integrates a laser, a back-facet monitor diode [positive–intrinsic–negative (PIN)], and coupling optics and is based on a silicon subassembly.

Fig. 19.17   Beveled fiber array coupling approach for a GaAs detector array (adapted from Ref. [33]).

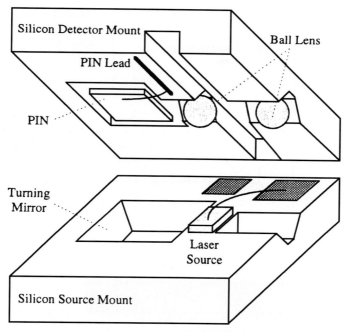

Fig. 19.18 Schematic drawing of laser source and back facet monitor subassembly (adapted from Ref. [34]).

As shown in Fig. 19.18, the silicon source submount houses the laser. Registration of the active area of horizontal (side) emitting laser to the source submount is achieved with a solder column bond site as shown in Fig. 19.19. A large pad of solder is sputtered onto the source submount bond site, and textured compression bond sites are formed on stand-off columns of the source submount using selective silicon etching. The laser is then tacked, junction side down (i.e., the metallized side down), to the texture bond site. Mechanical alignment of the laser is achieved using a precision actuator system capable of compression bonding. Permanent bonding of the laser is achieved by reflowing the solder, which balls due to surface tension effects, forming a permanent electrical and mechanical bond. The process typically results in a ± 2-μm registration of the laser active area to the source submount. The source and detector submounts also have selectively etched cavities and V grooves. One such element is the turning mirror in the source submount that reflects back transmitted light from the laser onto the surface-illuminated PIN on the detector submount. Key elements to the assembly are the front and back facet ball lenses, which

Fig. 19.19 Solder column bond site featuring textured compression tacking and bump bonding (adapted from Ref. [34]). (a) As plated. (b) After solder reflow. (c) Laser tacked into textured surface and solder bump bonded.

are placed in selectivity etched lens holders (V grooves intersecting at 90°) that precisely establish the position of the spheres relative to the surface of the source and detector submounts. The position of the two lenses is controlled by photolithography and etching to within ± 1 μm. When the two submounts are assembled, the surfaces of the two submounts are brought into contact, aligning the laser and lenses in the direction normal to the submount surface (the z direction). The axial degree of freedom along the optical signal path is establish by the internal references (V grooves and columns) defined by photolithography and selective etching. The lateral degree of freedom is established by sliding the two submounts relative to one another. Alignment is optimized along a single axis by actively monitoring the near-field intensity at the output of the front facet lens.

Therefore, in this assembly technique, active alignment is used on only a single axis. The remaining axes are aligned passively.

A third combined active and passive alignment assembly process is presented by Solgaard *et al.* [35]. Their innovative optoelectronic packaging technology uses silicon surface micromachined alignment mirrors for active alignment of an optical fiber to a laser or detector. Their alignment technique is based on silicon optical bench technology in which active and passive optoelectronic components are assembled on a silicon chip. Typically, accurate passive positioning of the components is achieved by solder bumps and mechanical stops that are defined using conventional photolithography. Although this technology has found wide application, it does not provide adequate positioning accuracy for high-performance applications such as CATV laser modules. The technology Solgaard *et al.* have developed incorporate movable, micromechanical structures that are fabricated directly on the silicon chip by polysilicon surface micromachining [34]. The independent optoelectronic components are assembled to the package using standard optical-bench technology. The inherent inaccuracy of the passive alignment of the optoelectronic components is corrected by movable aligning microstructures as shown in Fig. 19.20.

Two schemes are used for alignment. The first uses an external servo feedback control system to automatically align the micromechanical structure. After alignment, the servo is disconnected and the micromechanical structure is left in a permanent position. The second technique is to integrate the servo feedback control, detector, and source in the package so that accurate alignment is guaranteed throughout the life of the package.

Fig. 19.20   Fiber-coupled, semiconductor laser module with on-chip alignment mirror.

Fig. 19.21   Schematic of micromachined micromirror structure.

The alignment mirrors are micromachined polysilicon plated with gold to provide high reflectivity. As shown in Fig. 19.21, the lower portion of the micromachined alignment mirror is connected by polysilicon hinges to another polysilicon plate that can translate linearly (i.e., a slider plate) along polysilicon guides etched in the silicon carrier. The top of the mirror is also hinged to a back support, which is hinged to a second slider. Moving the sliders simultaneously translates the mirror linearly. Translating one slider relative to the other tilts the mirror to a controlled angle. The two degrees of freedom (translation and tilt) provides the necessary beam alignment to compensate for the transversal offset of the lens and laser assembled using the conventional silicon optical-bench technology.

Potential concerns with this alignment technology include possible flexing of the structure during motion, dimensional tolerance loss due to loose hinges, and the possibility of slider motion during shock and vibration. However, due to the extremely light masses, inertial shock forces and vibration acceleration do not translate the micromirrors. This was demonstrated for shock tests up to 500 G. The micromachined structure is held in place by the friction forces in the slides and hinges.

19.5.5.   PASSIVE AUTOMATED ALIGNMENT

Passive automated alignment involves aligning a light source to an optic axis automatically without turning the light source on. This approach has a disadvantage in that it is not possible at the time of manufacture to determine if the alignment is successful because the light source is not turned on. However, with state-of-the-art manufacturing techniques any manufacturing loss due to this phenomenon could be minimized.

One of the most promising technologies for passive automated alignment involves using the surface tension of solder. This technique is shown in

Fig. 19.22   Solder ball surface tension.

Fig. 19.22. This self-alignment phenomenon was first observed at IBM [36]. Because their integrated circuit technology uses solder balls to interconnect, it was observed that if one did not precisely place the integrated circuit (IC) on the package, then the surface tension of the melted solder balls would ensure the centering of the IC. To use this technique to assemble the light source to a lens or fiber, the following process could be used. The light source is placed with an accuracy of approximately 25 μm by a standard IC die placement machine. Upon reflow of the solder, the surface tension moves the light source into alignment with the pads (Fig. 19.23). Hence, the light source and fiber are aligned with approximately the same accuracy as the pads were to the optical axes. Lee *et al.* and Lin *et al.* [36–39] have shown that this technique is capable of accuracies down to micrometer levels.

With the advent of VCSELs it is likely that a light source may be passively mounted directly to a fiber. A possible process to perform this assembly might be as follows. Photolithographic processes are used to form the pads (likely copper) on both the VCSEL and the fiber (Fig. 19.24). Current art enables better than micrometer location of the pads with respect to the VCSEL optic axis, in the VCSEL process. Unfortunately, a new tool would likely need to be invented to form the pads on the fiber to within micrometer accuracy with respect to the optic axis of the fiber. This tool is shown in

Fig. 19.23   Solder reflow alignment.

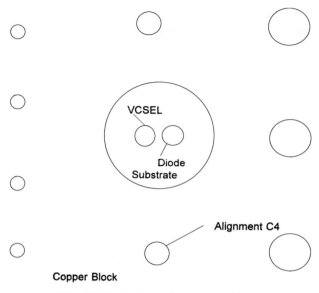

Fig. 19.24   Solder pad layout (top view).

Fig. 19.25. To form the pads accurately with respect to the optic axis of the fiber, the tool must focus on light emitted from the fiber and use this point as a reference to photolithographically image the pads in the correct spot. The lasers shown on this tool would image the pads in the photoresist on the fiber stub. To perform this operation the lasers would have to be moved to the proper location. This movement would require state-of-the-art motion systems and good vibration damping to achieve the desired micrometer tolerance control. An additive or subtractive could be used to form the metal pads.

Because the SBS of the VCSEL can be controlled by design, it is possible to select it to match that of a SM fiber. In this case a lens is not needed to achieve reasonable coupling. Hence, a structure such as that in Fig. 19.26 would be possible. The fiber is tilted to minimize reflected light into the laser. It is evident that the solder balls require a gradation in size. This size gradation is possible with solder jetting technology [40].

The amount of light coupled will be critical to the successful functioning of the link. The theoretical coupling of light as a function of the distance of the VCSEL from the fiber and its transverse offset is shown in Fig. 19.27. A mode diameter of 10 μm has been chosen for both the fiber and the VCSEL in this calculation. As seen in Fig. 19.28, the coupling effects from longitudinal (z) offset are far less dramatic. The coupling model of

Fig. 19.25 Proposed tool for formation of solder pads on fiber stub.

Nemoto and Makimoto [16] has been used. Our experience has shown that this model provides excellent relative results; however, absolute amounts of light coupled are typically 30–50% less. This discrepancy is likely due to abberration in the light source.

   Although somewhat more cumbersome, it is still possible to use this approach to align edge emitting light sources to lenses or fibers, albeit with less coupling due to the typical mismatch between the SBS and the FBS.

   To date, this process has still not been implemented into practice. Our belief is that the process could be implemented with initial costs of tens of dollars per SM alignment for process volumes of less the one hundred thousand. As volumes approach a million, cost per alignment should approach several dollars or less. These kinds of price reductions will be needed to enable optoelectronic communication to become a consumer product.

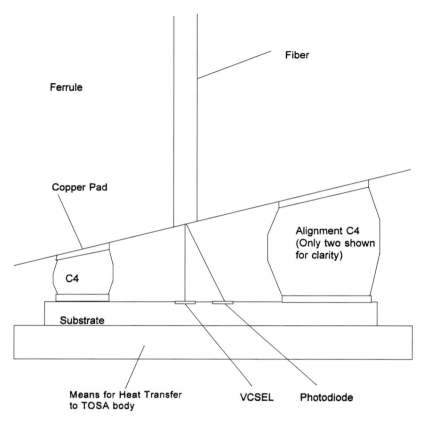

Fig. 19.26 Passive alignment with solder balls.

19.6. Conclusion

Analyzing coupling efficiencies and the effects of alignment technology is an inherently complex process. This chapter has outlined the relevant parameters and the principles needed. However, for detailed calculations, many references have been presented that explore theoretical and experimental techniques. Currently, all models show a slight deviation in absolute coupling efficiencies but the relative results are well established.

Various alignment methods have been presented but it is understood that this is an area that is still under development for optodatacom. Passive automated alignment of components is the ultimate goal of optodatacom manufacturing. Parallel components and VCSEL technology in particular

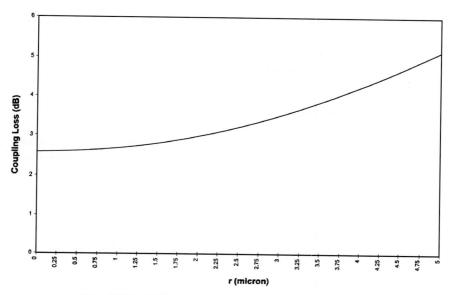

Fig. 19.27 VCSEL-to-fiber coupling for transverse offset.

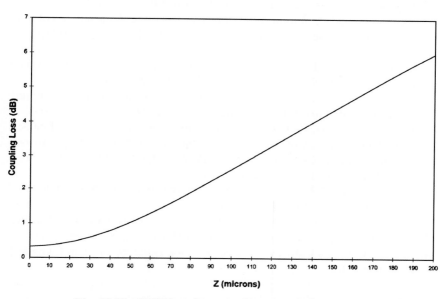

Fig. 19.28 VCSEL-to-fiber coupling for longitudinal offset.

present new opportunities to explore new manufacturing techniques and to
reduce the cost of components.

References

1. Vukovic, M. 1996 (July). Analyze connector parameters to realize high performance. *Lightwave,* 50–56.
2. Paz, O., H. B. Schwartz, D. E. Smith, and E. B. Flint. 1992 (December). Measurements and modeling of fiber optic connector pluggability. *IEEE Trans. Comp. Hybrids Manuf. Tech.* **15**(6):983–991.
3. Aulet, N. R., D. W. Boerstler, G. DeMario, F. D. Ferraiolo, C. E. Hayward, C. D. Heath, A. L. Huffman, W. R. Kelly, G. W. Peterson, and D. J. Stigliani, Jr. 1992 (July). IBM enterprise systems multimode fiber optic technology. *IBM J. Res. Dev.* **36**(4):553–576.
4. Saleh, B. E. A., and M. C. Teich. 1991. *Fundamentals of photonics.* New York: Wiley.
5. International Electrotechnical Commission (IEC). International standard 825-1 and 825-2, safety of laser products. Central Bureau of the IEC, 3 rue de Varembe, Geneva, Switzerland.
6. American National Standards Institute. 1988. American National Standard for the safe use of optical fiber communication systems utilizing laser diode and LED sources. American National Standards Institute No. Z136.2.
7. Kawano, K. and O. Mitomi. 1986 (January). Coupling characteristics of laser diode to multimode fiber using separate lens methods. *Appl. Opt.* **15**(1):136–141.
8. Clement, D. P., and U. Österberg. 1995. Laser diode to single-mode coupling using an "out-of-plane misalignment" model. *Opt. Eng.* **34**(1):63–74.
9. Hudson, M. C. 1974 (May). Calculation of the maximum optical coupling efficiency in multimode optical waveguides. *Appl. Opt.* **13**(5):1029–1033.
10. Kogelnik, H. 1964. Coupling and conversion coefficients for optical modes. Proceedings of the Symposium on Quasi-Optics, June 8–10. In *Microwave research institute symposia series,* Vol. XIV. Brooklyn, N.Y.: Polytechnic Press of the Polytechnic Institute of Brooklyn.
11. Marcuse, D. 1977 (May/June). Loss analysis of single-mode fiber splices. *Bell Syst. Tech. J.* **56**(5):703–718.
12. Miller, C. M. 1986. *Optical fiber splices and connectors.* New York: Dekker.
13. Neumann, E. G. 1988. *Single-mode cables.* Heidelberg: Springer-Verlag.
14. Snyder, A. W., and J. D. Love. 1983. *Optical waveguide theory,* Part 1. London: Chapman & Hall.
15. Anderson, W. T., and D. L. Philen. 1983 (March). Spot size measurements for single-mode cables—A comparison of four techniques. *J. Lightwave Tech.* **1**(1):20–26.

16. Nemoto, S., and T. Makimoto. 1974. Analysis of splice loss in single-mode fibres using a Gaussian field approximation. *Opt. Quantum Electron.* 11:447–457.

17. Lasky, R. C., U. Österberg, and D. J. Stigliani, eds. 1995. *Optoelectronics for datacommunication.* New York: Academic Press.

18. Karstensen, H., and K. Drögemuller. 1996 (May). Loss analysis of single-mode fiber couples with glass spheres or silicon plano-convex lenses. *J. Lightwave Tech.* **8**(5):739–747.

19. Bargar, D. S. 1988. An automated fiber alignment, fixing, and hermetic sealing system. *SPIE* 994:11–17.

20. Bristow, J., Y. Liu, T. Marta, K. Johnson, B. Hanzal, A. Peczalski, S. Bounnak, Y. S. Liu, and H. Cole. 1996. MCM board level optical interconnects using passive polymer waveguides with hybrid optical and electrical multichip module packaging. *SPIE* 2691:18–24.

21. Gabler, C., L. Li, S. Hackwood, and G. Beni. 1986. An optical alignment robot system. *SPIE* 703:8–28.

22. Goldman, L. S. 1972 (May 15–17). Proceedings of the 22nd Electronic Components Conference, 332–339, Washington, D.C.

23. Karioja, P., K. Tukkiniemi, V. Hikkinen, and I. Kaisto. 1993. Inexpensive packaging techniques of fiber pigtailed laser diodes. *SPIE* 1851:48–53.

24. Goodwin, J. 1986. Dynamic alignment of small optical components. *SPIE* 703:2–7.

25. Han, H., J. E. Schramm, J. Mathews, and R. A. Boudreau. *SPIE* 2691:118–123.

26. Kalman, R. F., E. R. Silva, and D. F. Knapp. 1996. Single-mode array optoelectronic packaging based on actively aligned optical waveguides. *SPIE* 2691:124–129.

27. Kilcolyne, M. K., J. W. Scott, and J. Plombon. 1993. Packaging of optical interconnect arrays for optical signal processing and computing. *SPIE* 1851:80–88.

28. Schmid, P., and H. Melchior. 1984. Coplaner flip-chip mounting technique for picosecond devices. *Rev. Sci. Instrum.* 55:1854–1858.

29. Spitzer, M. B., D. P. Vu, and R. P. Gale. 1996. Applications of circuit transfer technology to displays and optoelectronic devices. *SPIE* 2691:34–42.

30. Weber, R., F. Fidorra, M. Hamacher, H. Heidrich, and G. Jacumelt. 1993. Multi fiber/chip coupling and optoelectronic integrated circuit packaging based on flip chip techniques. *SPIE* 1894:149–152.

31. Yap, D., W. W. Ng, D. M. Bohmeyer, H. P. Hsu, H. W. Yen, M. J. Tabasky, A. J. Negri, J. Mehr, C. A. Armiento, and P. O. Haugsjaa. 1996. RF optoelectronic transmitter and receiver arrays on silicon waferboards. *SPIE* 2691:110–117.

32. Tewksbury, S. K., L. A. Hornak, H. E. Nariman, and S. M. Lansjoen. 1993. Opportunities and issues for optical interconnects in microelectronic systems. In *Optoelectronics: Technologies and applications,* eds. A. Selvarajan, K. Shenai, and V. K. Tripathi. Bellingham, Wash.: SPIE.

33. Jackson, P., A. J. Moll, E. B. Flint, and M. F. Cina. 1988. Optical fiber coupling approaches for multi-channel laser and detector arrays. *SPIE* 994:40–47.
34. MacDonald, W. M., R. E. Fanucci, and G. E. Blonder. 1993. Si-based laser sub-assembly for telecommunications. *SPIE* 1851:42–47.
35. Solgaard, O., M. Daneman, N. C. Tien, A. Friedberger, R. S. Muller, and K. Y. Lau. 1995. Optoelectronic packaging using silicon surface-micromachined alignment mirrors. *IEEE Photon. Tech. Lett.* 7:41–43.
36. Lin, W., S. K. Patra, and Y. C. Lee. 1995. Design of solder joints for self-aligned optoelectronic assemblies. *IEEE Trans. Comp. Packaging Manuf. Tech. Part B,* 18:543–551.
37. McCroarty, J., B. Yost, P. Borgesen, and C. Y. Li. *Mater. Res. Soc. Symp. Proc.* 264:423–435.
38. Patra, S. K., J. Ma, V. Ozguz, and S. H. Lee. 1994. *SPIE* 2153:118–131.
39. Patra, S. K., and Y. C. Lee. 1991. *Electron. Packaging* 113:337–342.
40. Snyder, M. D., R. Lasky. 1995. Proceedings of the MRS, Spring 1995 meeting. San Francisco.

Appendix A: Measurement Conversion Tables

English-to-Metric Conversion Table

English unit	Multiplied by	Equals metric unit
Inches (in.)	2.54	Centimeters (cm)
Inches (in.)	25.4	Millimeters (mm)
Feet (ft)	0.305	Meters (m)
Miles (mi)	1.61	Kilometers (km)
Fahrenheit (F)	$(°F - 32) \times 0.556$	Celsius (C)
Pounds (lb)	4.45	Newtons (N)

Metric-to-English Conversion Table

Metric unit	Multiplied by	Equals English unit
Centimeters (cm)	0.39	Inches (in.)
Millimeters (mm)	0.039	Inches (in.)
Meters (m)	3.28	Feet (ft)
Kilometers (km)	0.621	Miles (mi)
Celsius (C)	$(°C \times 1.8) + 32$	Fahrenheit (F)
Newtons (N)	0.225	Pounds (lb)

735

HANDBOOK OF FIBER OPTIC
DATA COMMUNICATION
$35.00

Absolute Temperature Conversion

Kelvin (K) = Celsius + 273.15

Celsius = Kelvin − 273.15

Area Conversion

1 square meter = 10.76 square feet = 1550 square centimeters

1 square kilometer = 0.3861 square miles

Metric Prefixes

Yotta = 10^{24}

Zetta = 10^{21}

Exa = 10^{18}

Peta = 10^{15}

Tera = 10^{12}

Giga = 10^{9}

Mega = 10^{6}

Kilo = 10^{3}

Hecto = 10^{2}

Deca = 10^{1}

Deci = 10^{-1}

Centi = 10^{-2}

Milli = 10^{-3}

Micro = 10^{-6}

Nano = 10^{-9}

Pico = 10^{-12}

Femto = 10^{-15}

Atto = 10^{-18}

Zepto = 10^{-21}

Yotto = 10^{-24}

Appendix B: Physical Constants

Speed of light $= c = 2.99\ 792\ 458 \times 10^8$ m/s

Boltzmann constant $= k = 1.3801 \times 10^{-23}$ J/K $= 8.620 \times 10^{-5}$ eV/K

Planck's constant $= h = 6.6262 \times 10^{-34}$ J/S

Stephan–Boltzmann constant $= \sigma = 5.6697 \times 10^{-8}$ W/m^2/K^4

Charge of an electron $= 1.6 \times 10^{-19}$ C

Permittivity of free space $= 8.849 \times 10^{-12}$ F/m

Permeability of free space $= 1.257 \times 10^{-6}$ H/m

Impedance of free space $= 120\ \pi$ ohms $= 377$ ohms

Electron volt $= 1.602 \times 10^{-19}$ J

HANDBOOK OF FIBER OPTIC
DATA COMMUNICATION
$35.00

Appendix C: Index of Professional Organizations

AIP, American Institute of Physics
ANSI, American National Standards Institute
APS, American Physical Society
ASTM, American Society for Test and Measurement
CCITT, International Telecommunications Standards Body (see ITU)
CDRH, U.S. Center for Devices and Radiological Health
DARPA, Defense Advanced Research Projects Association (also referred to as ARPA)
EIA, Electronics Industry Association
FCA, Fibre Channel Association
FDA, U.S. Food and Drug Administration
HSPN, High Speed Plastic Network Consortium
IEC, International Electrotechnical Commission
IEEE, Institute of Electrical and Electronics Engineers
IETF, Internet Engineering Task Force
IrDA, Infrared Datacom Association
ISO, International Standards Organization
ITU, International Telecommunications Union (formerly CCITT)
NBS, National Bureau of Standards
NFPA, National Fire Protection Association
NIST, National Institute of Standards and Technology
OIDA, Optoelectronics Industry Development Association

HANDBOOK OF FIBER OPTIC
DATA COMMUNICATION
$35.00

OSA, Optical Society of America
OSHA, U.S. Occupational Health and Safety Administration
PTT, Postal, Telephone, and Telegraph Authority
SI, System International (International System of Units)
SIA, Semiconductor Industry Association
SPIE, Society of Photooptic Instrumentation Engineers
TIA, Telecommunications Industry Association
UL, Underwriters Laboratories

Appendix D: OSI Model

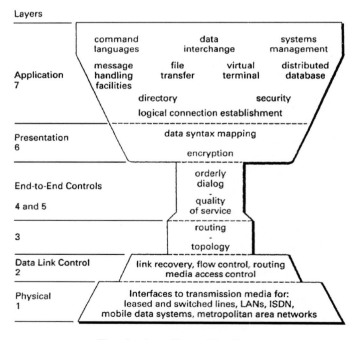

The wineglass of layered functions.

HANDBOOK OF FIBER OPTIC
DATA COMMUNICATION
$35.00

Appendix E: Network Standards and Documents

The IEEE defines a common path control (802.1) and data link layer (802.2) for all the following LAN standards; although FDDI is handled by ANSI, it is also intended to fall under the logical link control of IEEE 802.2 (the relevant ISO standard is 8802-2).

IEEE LAN Standards

> 802.3-code sense multiple access/collision detection (CSMA/CD) also known as Ethernet. A variant of this is Fast Ethernet (100BaseX). 802.3z is the emerging gigabit Ethernet standard. IEEE 802.3ae is the proposed 10 Gbit/s Ethernet standard.
>
> 802.4-Token bus (TB)
>
> 802.5-Token ring (TR)
>
> 802.6-Metropolitan area network (MAN) [also sometimes called switched multimegabit data service (SMDS) to which it is related]
>
> 802.9-Integrated services digital network (ISDN) to LAN interconnect
>
> 802.11-Wireless services up to 5 Mb/s
>
> 802.12-100VG AnyLAN standard
>
> 802.14-100BaseX (version of Fast Ethernet)

HANDBOOK OF FIBER OPTIC
DATA COMMUNICATION
$35.00

ANSI Standards

Fast Ethernet-ANSI X3.166

FDDI-ANSI X3.263

Fiber distributed data interface (FDDI): ANSI X3T9.5 (the relevant ISO standards are IS 9314/1 2 and DIS 9314/3)

Physical layer (PHY)

Physical media dependent (PMD)

Media access control (MAC)

Station management (SMT)

Because the FDDI specification is defined at the physical and data link layers, additional specifications have been approved by ANSI subcommittees to allow for FDDI over single-mode fiber (SMF-PMD), FDDI over copper wire or CDDI, and FDDI over low-cost optics (LC FDDI). A time-division multiplexing approach known as FDDI-II has also been considered.

Serial byte command code set architecture (SBCON): ANSI standard X3T11/95-469 (rev.2.2., 1996); follows IBM's enterprise systems connectivity (ESCON) standard as defined in IBM documents SA23-0394 and SA22-7202 (IBM Corporation, Mechanicsburg, PA)

Fibre channel standard (FCS)

ANSI X3.230-1994 rev. 4.3, physical and signaling protocol (FC-PH)

ANSI X3.272-199x rev. 4.5 (June 1995) Fibre Channel arbitrated loop (FC-AL)

High-performance parallel interface (HIPPI)

ANSI X3.183-mechanical, electrical, and signaling protocol (PH)

ANSI X3.210-framing protocol (FP)

ANSI X3.218-encapsulation of ISO 8802-2 (IEEE 802.2) logical link protocol (LE)

ANSI X3.222-physical switch control (SC)

Serial HIPPI has not been sanctioned as a standard, although various products are available.

HIPPI 6400 is currently under development.

Synchronous optical network (SONET): originally proposed by

Bellcore and later standardized by ANSI and ITU (formerly CCITT) as ITU-T recommendations G.707, G.708, and G.709.

Other standards

Asynchronous transfer mode (ATM): controlled by the ATM Forum

Appendix F: Data Network Rates

The "fundamental rate" of 64 kb/s derives from taking a 4-kHz voice signal (telecom), sampling into 8-bit wide bytes (32 kb/s), and doubling to allow for a full-duplex channel (64 kb/s). In other words, this is the minimum data rate required to reproduce a two-way voice conversation over a telephone line. All of the subsequent data rates are standardized as multiples of this basic rate.

DS0	64 kb/s	
T1 = DS1	1.544 Mb/s	24 × DS0
DS1C	3.152 Mb/s	48 × DS0
T2 = DS2	6.312 Mb/s	96 × DS0
T3 = DS3	44.736 Mb/s	672 × DS0
DS4	274.176 Mb/s	4032 × DS0

Note: framing bit overhead accounts for the bit rates not being exact multiples of DS0).

STS/OC is the SONET physical layer ANSI standard. STS refers to the electrical signals and OC (optical carrier) to the optical equivalents. Synchronous digital hierarchy (SDH) is the worldwide standard defined by CCITT (now known as ITU, the International Telecommunications Union); formerly known as synchronous transport mode (STM). Sometimes the

744

HANDBOOK OF FIBER OPTIC
DATA COMMUNICATION
$35.00

notation STS-*XC* is used, where *X* is the number (1, 3, etc.) and *C* denotes the frame is concatinated from smaller frames. For example, three STS-1 frames at 51.84 Mb/s each can be combined to form one STS-3C frame at 155.52 Mb/s. Outside the United States, SDH may be called plesiochronous digital hierarchy (PDH). Note that OC and STS channels are normalized to a data rate of 51.84 Mbit/s, while the equivalent SDH specifications are normalized to 155.52 Mbit/s.

STS-1 and OC-1	51.840 Mb/s	
STS-3 and OC-3	155.52 Mb/s	same as STM-1
STS-9 and OC-9	466.56 Mb/s	
STS-12 and OC-12	622.08 Mb/s	same as STM-4
STS-18 and OC-18	933.12 Mb/s	
STS-24 and OC-24	1244.16 Mb/s	same as STM-8
STS-36 and OC-36	1866.24 Mb/s	
STS-48 and OC-48	2488.32 Mb/s	same as STM-16
STS-192 and OC-192	9953.28 Mb/s	same as STM-64
STS-256 and OC-256	13271.04 Mb/s	same as STM-86
STS-768 and OC-768	39813.12 Mb/s	same as STM-256
STS-3072 and OC-3072	159252.48 Mb/s	same as STM-1024
STS-12288 and OC-12288	639009.92 Mb/s	same as STM-4096

Higher speed services aggregate low speed channels by time division multiplexing; for example, OC-192 can be implemented as four OC-48 data streams. Note that although STS (synchronous transport signal) is analogous to STM (synchronous transport mode) there are some important differences. The first recognized STM is STM-1, which is eqivalent to STS-3. Similarly, not all STS rates have a corresponding STM rate. The frames for STS and STM are both set up in a matrix (270 columns of 9 bytes each) but in STM frames the regenerator section overhead (RSOH) is located in the first 9 bytes of the top 3 rows. The fourth STM row of 9 bytes is occupied by the administrative unit (AU) pointer, which operates in a manner similar to the H1 and H2 bytes of the SONET line overhead (LOH). The 9 bytes of STM frame in rows 5 through 9 is the multiplex section overhead (MSOH) and is similar to the SONET LOH. In SONET we defined virtual tributaries (VT), while SDH defines virtual containers (VC), but they basically work the same way. A VT or VC holds individual

E1 or other circuit data. VCs are contained in tributary unit groups (TUG) instead of SONET's VT groups (VTG). VCs are defined as follows:

DS-2	VC-2
E3/DS-3	VC-3
DS-1	VC-11
E1	VC-12

Also note that while there are many references to 10 Gbit/s networking in new standards, the exact data rates may vary. As this book goes to press, proposed standards for 10 Gigabit Ethernet and Fibre Channel are not yet finalized. The Ethernet standard has proposed two data rates, approximately 9.953 Gbit/s (compatible with OC-192) and 10.3125 Gbit/s. Fibre Channel has proposed a data rate of approximately 10.7 Gbit/s. There is also ongoing discussion regarding standards compatible with 40 Gbit/s data rates (OC-768).

The approach used in Europe and elsewhere:

E0	64 Kb/s	
E1	2.048 Mb/s	
E2	8.448 Mb/s	4 E1s
E3	34.364 Mb/s	16 E1s
E4	139.264 Mb/s	64 E1s

Thus, we have the following equivalences for SONET and SDH hierarchies:

SONET signal	*SONET capacity*	*SDH signal*	*SDH capacity*
STS-1, OC-1	28 DS1s or 1 DS3	STM-0	21 E1s
STS-3, OC-3	84 DS1s or 3 DS3s	STM-1	63 E1s or 1 E4
STS-12, OC-12	336 DS1s or 12 DS3s	STM-4	252 E1s or 4 E4s
STS-48, OC-48	1344 DS1s or 48 DS3s	STM-16	1008 E1s or 16 E4s
STS-192, OC-192	5376 DS1s or 192 DS3s	STM-64	4032 E1s or 64 E4s
STS-768, OC-768	21504 DS1s or 768 DS3s	STM-256	16128 E1s or 256 E4s

For completeness, the SONET interface classifications for different applications are summarized in the table below. Recently, a new 300-meter Very Short Reach (VSR) interface based on parallel optics for OC-192 data rates has been defined as well (see Optical Internetworking Forum document OIF2000.044.4 or contact the OIF for details). Industry standard network protocols are summarized in the tables below:

	Short reach	Intermediate reach	Long reach		Very long reach				
Distance (km)	<2	15	15	40	60	60	120	160	160
Wavelength (nm)	1,310	1,310	1,550	1,310	1,550	1,550	1,310	1,550	1,550
OC-1	SR	IR-1	IR-2	LR-1	LR-2	LR-3	VR-1	VR-2	VR-3
OC-3	SR	IR-1	IR-2	LR-1	LR-2	LR-3	VR-1	VR-2	VR-3
OC-12	SR	IR-1	IR-2	LR-1	LR-2	LR-3	VR-1	VR-2	VR-3
OC-48	SR	IR-1	IR-2	LR-1	LR-2	LR-3	VR-1	VR-2	VR-3
OC-192	SR	IR-1	IR-2	LR-1	LR-2	LR-3	VR-1	VR-2	VR-3

Common Fiber Optic Attachment Options
(without repeaters or channel extenders)

Channel	Fiber	Connector	Bit rate	Fiber bandwidth	Maximum distance	Link loss
ESCON (SBCON)	SM	SC duplex	200 Mb/s	N/A	20 km	14 dB
	MM 62.5 micron	ESCON duplex or MT-RJ	200 Mb/s	500 MHz-km 800 MHz-km	2 km 3 km	8 dB
	MM 50.0 micron	ESCON duplex or MT-RJ	200 Mb/s	800 MHz-km	2 km	8 dB
Sysplex Timer ETR/CLO	MM 62.5 micron	ESCON duplex or MTRJ	8 Mb/s	500 MHz-km or more	3 km	8 dB
	MM 50.0 micron	ESCON Duplex or MTRJ	8 Mb/s	500 MHz-km or more	2 km	8 dB
FICON/ Fibre Channel LX	SM	SC duplex or LC duplex	1.06 Gb/s	N/A	10 km	7 dB

(continued)

Common Fiber Optic Attachment Options
(without repeaters or channel extenders) (continued)

Channel	Fiber	Connector	Bit rate	Fiber bandwidth	Maximum distance	Link loss
FICON LX	MM w/MCP 62.5 micron	SC duplex or LC duplex	1.06 Gb/s	500 MHz-km	550 meters	5 dB
	MM w/MCP 50.0 micron	SC duplex or LC duplex	1.06 Gb/s	400 MHz-km	550 meters	5 dB
FICON/ Fibre Channel SX	MM 50.0 * micron	SC duplex or LC duplex	1.06 Gb/s	500 MHz-km	500 meters	3.85 dB
	MM 62.5 * micron	SC duplex or LC duplex	1.06 Gb/s	160 MHz-km	250 meters	2.76 dB
	MM 62.5 * micron	SC duplex or LC duplex	1.06 Gb/s	200 MHz-km	300 meters	3 dB
	MM 50.0 * micron	SC duplex or LC duplex	2.1 Gb/s	500 MHz-km	300 meters	
	MM 62.5 * micron	SC duplex or LC duplex	2.1 Gb/s	160 MHz-km	120 meters	
	MM 62.5 * micron	SC duplex or LC duplex	2.1 Gb/s	200 MHz-km	150 meters	
Parallel Sysplex coupling links-HiPer-Links	SM	SC duplex or LC duplex	1.06 Gb/s compati-bility mode	N/A	10 km (20 km on special request)	7 dB
	SM	LC duplex	2.1 Gb/s peer mode	N/A	10 km (20 km on special request)	7 dB
	MM w/MCP 50.0 micron	SC duplex or LC duplex	1.06 Gb/s	500 MHz-km	550 meters	5 dB
	MM 50.0 * Discontinued May 98	SC duplex	531 Mb/s	500 MHz-km*	1 km	8 dB *
OC-3/ ATM	SM	SC duplex	155 Mb/s	N/A	20 km	15 dB
155	MM 50 micron	SC duplex	155 Mb/s	500 MHz-km	2 km	

(continued)

Common Fiber Optic Attachment Options
(without repeaters or channel extenders) (continued)

Channel	Fiber	Connector	Bit rate	Fiber bandwidth	Maximum distance	Link loss
	MM 62.5 micron	SC duplex	155 Mb/s	500 MHz-km	2 Km	11 dB
	MM * 50 micron	SC duplex	155 Mb/s	500 MHz-km	1 km	
	MM * 62.5 micron	SC duplex	155 Mb/s	160 MHz-km	1 km	
OC-12/ ATM 622	MM 50 micron	SC duplex	622 Mb/s	500 MHz-km	500 m	
	MM 62.5 micron	SC duplex	622 Mb/s	500 MHz-km	500 m	
	MM * 50 micron	SC duplex	622 Mb/s	500 MHz-km	300 m	
	MM * 62.5 micron	SC duplex	622 Mb/s	160 MHz-km	300 m	
FDDI	MM 62.5 micron	Media Access Connector (MAC) or SC duplex	125 Mbit/s overhead reduces to 100Mb/s	500 MHz-km	2 Km	9 dB
	MM 50 micron	MAC or SC duplex	125 Mbit/s overhead reduces to 100 Mb/s	500 MHz-km	2 km	9 dB
Token Ring *	MM 62.5 micron	SC duplex	16 Mbit/s	160 MHz-km	2 km	
	MM 50 micron	SC duplex	16 Mbit/s	500 MHz-km	1 km	
Ethernet	MM * 50 micron 10Base-F	SC duplex	10 Mbit/s	500 MHz-km	1 km	
	MM * 62.5 micron 10Base-F	SC duplex	10 Mbit/s	160 MHz-km	2 km	
	MM * 50 micron 100Base-SX	SC duplex	100 Mbit/s	500 MHz-km	300 m	
	MM * 62.5 micron 100Base-SX	SC duplex	100 Mbit/s	160 MHz-km	300 m	
Fast Ethernet	MM 50 micron 100Base-F	SC duplex	100 Mbit/s	500 MHz-km	2 km	

(continued)

Common Fiber Optic Attachment Options
(without repeaters or channel extenders) (continued)

Channel	Fiber	Connector	Bit rate	Fiber bandwidth	Maximum distance	Link loss
	MM 62.5 micron 100Base-F	SC duplex	100 Mbit/s	500 MHz-km	2 km	
Gigabit Ethernet IEEE 802.3z	SM 1000BaseLX	SC duplex	1.25 Gb/s	N/A	5 km	4.6 dB
	MM * 62.5 micron 1000BaseSX	SC duplex	1.25 Gb/s	160 MHz-km 200 MHz-km	220 meters 275 meters	2.6 dB *
	MM w/MCP 62.5 micron 1000BaseLX	SC duplex	1.25 Gb/s	500 MHz-km	550 meters	2.4 dB
	MM * 50.0 micron 1000BaseSX	SC duplex	1.25 Gb/s	500 MHz-km *	550 meters	3.6 dB *
	MM w/MCP 50.0 micron 1000BaseLX	SC duplex	1.25 Gb/s	500 MHz-km	550 meters	2.4 dB

Notes:

* Indicates channels which use short wavelength (850 nm) optics; all link budgets and fiber bandwidths should be measured at this wavelength.
* SBCON is the non-IBM trademarked name of the ANSI industry standard for ESCON.
* All industry standard links (ESCON/SBCON, ATM, FDDI, Gigabit Ethernet) follow published industry standards. Minimum fiber bandwidth requirement to achieve the distances listed is applicable for multimode (MM) fiber only. There is no minimum bandwidth requirement for single mode (SM) fiber.
* Bit rates given below may not correspond to effective channel data rate in a given application due to protocol overheads and other factors.
* SC duplex connectors are keyed per the ANSI Fiber Channel Standard specifications.
* MCP denotes mode conditioning patch cable, which is required to operate some links over MM fiber.
* As light signals traverse a fiber optic cable, the signal loses some of its strength (decibels (dB) is the metric used to measure light loss). The significant factors that contribute to light loss are: the length of the fiber, the number of splices, and the number of connections. All links are rated for a maximum light loss budget (i.e., the sum of the applicable light loss budget factors must be less than the maximum light loss budget) and a maximum distance (i.e., exceeding the maximum distance will cause undetectable data integrity exposures). Another factor that limits distance is jitter, but this is typically not a problem at these distances.
* Unless noted, all links are long wavelength (1300 nm) and the link loss budgets and fiber bandwidths should be measured at this wavelength. For planning purposes, the following worse case values can be used to estimate the total fiber link loss. Contact the fiber vendor or use measured values when available for a particular link configuration:

    Link loss at 1300 nm   = 0.50 dB/km
    Link loss per splice   = 0.15 dB/splice (not dependent on wavelength)
    Link loss per connection   = 0.50 dB/connection (not dependent on wavelength)

* HiPerLinks are also known as Coupling Facility (CF) or Inter System Channel (ISC) links.
* All links may be extended using channel extenders, repeaters, or wavelength multiplexers. Wavelength multiplexing links typically measure link loss at 1550 nm wavelength, typical loss is 0.3 dB/km.

Common Ethernet Standards

Standard		IEEE	Data rate (Mbit/s)	Medium	Max. distance (m) 1/2 duplex	Full duplex
Ethernet	10BaseT	802.3i	10	2 pair 100 ohm Cat 3 UTP copper	100	100
	10BaseFL	802.3j	10	Optical fiber pair	2,000	< 2,000
	10BaseFB	802.3j	10	Optical fiber pair	2,000	
	10BaseFP	802.3j	10	Optical fiber pair	1,000	
Fast	100BaseTX	802.3u	100	2 pair 100 ohm Cat 5 UTP copper	100	100
Ethernet	100BaseFX	802.3u	100	Optical fiber pair	412	2,000
	100BaseT4	802.3u	100	4 pair 100 ohm Cat 3 UTP copper	100	
	100BaseT2	802.3y	100	2 pair 100 ohm Cat 3 UTP copper	100	100
Gigabit Ethernet	1000BaseLX	802.3z	1,000	62.5 micron multimode fiber	316	550
	1000BaseLX	802.3z	1,000	50 micron multimode fiber	316	550
	1000BaseLX	802.3z	1,000	Singlemode fiber	316	5,000
	1000BaseSX	802.3z	1,000	62.5 micron multimode fiber	275	275
	1000BaseLX	802.3z	1,000	50 micron multimode fiber	316	550
	1000BaseCX	802.3z	1,000	Shielded balanced copper jumpers	25	25
	1000BaseT	802.3ab	1,000	4 pair 100 ohn Cat 4 UPT copper	100	100

Proposed 10 Gigabit Ethernet Standard (IEEE 802.3ae) Draft Objectives (final draft target 2002)

Fiber type	distance	Physical layer
SM	2 km	1300 nm serial
SM	10 km	1300 nm serial
SM	40 km	1550 nm serial
MM (all types of fiber, including new fibers such as 50 micron/ 2000 MHz-km)	65 meters (very short reach, VSR)	**850 nm serial ** (preferred)** 1300 nm WWDM 850 nm CWDM
MM installed (62.5 micron, 160/500 MHz-km at 850/1300 nm)	300 meters (very short reach, VSR)	**1300 nm WWDM * (preferred)** 850 nm CWDM 850 nm serial

* Does not require new fiber installation to reach 300 meters distance, and is the preferred, minimal solution for this service class.
** Reaches 65 meters on installed fiber, and up to 300 meters over new MM fiber.
Note: previous guidelines calling for 100 meters over installed MM fiber and 300 meters over MM fiber (including new fiber types) were revised December 2000.

Transmit and Receive Levels of Common Fiber Optic Protocols

Protocol type	I/O Spec.
ESCON/SBCON MM and ETR/CLO MM	TX: −15 to −20.5 RX: −14 to −29
ESCON/SBCON SM	TX: −3 to −8 RX: −3 to −28
FICON LX SM (MM via MCP)	TX: −4 to −8.5 RX: −3 to −22
FICON SX MM	TX: −4 to −9.5 RX: −3 to −17
ATM 155 MM (OC-3)	TX: −14 to −19 RX: −14 to −30
ATM 155 SM (OC-3)	TX: −8 to −15 RX: −8 to −32.5
FDDI MM	TX: −14 to −19 RX: −14 to −31.8

(*continued*)

Transmit and Receive Levels of Common Fiber Optic
Protocols (continued)

Protocol type	I/O Spec.
Gigabit Ethernet LX SM (MM via MCP)	TX: -14 to -20 RX: -17 to -31
Gigabit Ethernet SX MM (850 nm)	TX: -4 to -10 RX: -17 to -31
HiPerLinks (ISC coupling links, IBM Parallel Sysplex, 1.06 Gbit/s compatibility mode)	TX: -3 to -11 RX: -3 to -20
HiPerLinks (ISC coupling links, IBM Parallel Sysplex, 2.1 Gbit/s peer mode)	TX: -3 to -9 RX: -3 to -20

Appendix G: Other Datacom Developments

Because the field of fiber optic data communication is expanding so rapidly, inevitably new technologies and applications will emerge while this book is being prepared for print. In addition, there will be some recent developments that have not been incorporated into the previous chapters. This appendix is an attempt to address these changes by including a partial list of related datacom devices and standards for reference purposes and to include brief descriptions of several emerging datacom technologies that the reader may encounter in the literature (product names and terminology used in this appendix may be copyrighted by the companies that developed them).

Free-Space Optical Links

Recent advances have made free-space optical data links practical in conditions when the weather is not a factor in attenuating the signal. Although bit error rates are typically on the order of 10^{-6} to 10^{-9} at 155 Mb/s, this is adequate for some applications such as voice and video transmission; recent experiments have shown that error rates as low as 10^{-12} can be obtained at 200 Mb/s using ESCON protocols on free-space optical links. Some examples of this technology and its applications include the following:

AstroTerra Corp. currently offers the TerraLink system, a tripod-mounted line-of-sight communication link with an 8-in. telescope aperture capable of 155 Mb/s transmission over 8 km in clear weather.

HANDBOOK OF FIBER OPTIC
DATA COMMUNICATION
$35.00

ThermoTrex Corp. in cooperation with SDL Inc. and funded by the Ballistic Missile Defense Organization, has demonstrated 1.2 Gb/s free-space communication over 150 km. This technology may form the basis of an aircraft-to-aircraft communications system and is scheduled to fly aboard the Space Technology Research Vehicle in 1998 to demonstrate communications among satellites.

Portable free-space laser communications are being developed by companies such as Leica Technologies, which has demonstrated a prototype binocular communications system for military applications at 100 kb/s over 3–5 km.

Optical Connectors

Older connector types, including the optical SMA connector, DIN, and LightRay MPX (a 2- to 12-element multifiber connector incompatible with the MTP/MPO) still resist being replaced and may be found in some installations. Many other legacy connectors continue to be used in large quantities, including the SC and ST connectors (recall that the ST was developed and trademarked by AT&T over 15 years ago). As new connector options emerge, many types of parallel or multifiber connectors have been proposed, most of which are mutually incompatible. For example, the MTP connector and its variations (MPO or MPX) do not mate with the SMC connector used on the Paroli parallel optical transceivers, yet both are being deployed as part of HIPPI 6400 installations. Recently, there has been increased interest in HIPPI 6400 for technical computing applications. Formerly known as SuperHIPPI and now known officially as Gigabyte System Network (GSN, a trademark of the High Performance Networking Forum), the physical layer of this protocol is now available as a draft standard (ANSI NC ITS T11.1 PH draft 2.6, dated December 2000, ISO/IEC reference number 11518-10) or online at www.hippi.org. This link provides a two-way, 12-channel-wide parallel interface running at 6400 Mbit/s (an increase from the standard HIPPI link rate of 800 Mbit/s). The link layer uses a fixed size 32-byte packet, 4B/5B encoding, and credit-based flow control, while the physical layer options include parallel copper (to 40 meters) or parallel optics (several hundred meters to 1 km).

Optical backplane connectors are likewise not yet standardized, and may include variants such as the MAC-II (an improved version of the original 12-fiber AT&T MAC connector) or the mini-MAC (a Bellcore approved version based on the MT ferrule). There have been various proposals for stacking multiple MT ferrules in a single connector; for example, a stack of 6 MTs forming a 72-fiber connector has been developed by LoDan West Corporation, and multichannel optical transceivers capable of utilizing this capacity in a full duplex 320 Gbit/s link were demonstrated at a recent Optical Fiber Conference (OFC). Research efforts continue to push the limits of stacked one-dimensional and two-dimensional connectors; an example is the 144-fiber "super-MT" developed by the POINT consortium between Amp, Allied Signal, Honeywell, and University of Columbia. Other examples include the 4×12 element Diamond MF multiple fiber connector.

Plastic Optical Fiber

At the Optical Fiber Conference 1997 in Dallas, Texas, plastic fiber applications were demonstrated for 155-Mb ATM and 100 Mb/s Ethernet LANs. This technology was developed by the High Speed Plastic Network Consortium (HSPN), a DARPA-funded venture consisting of Boeing, CEL, Boston Optical Fiber, Honeywell, Packard Hughes, and Lucent. Graded-index plastic fiber made from materials such as PMMA now offers bandwidth >3 Gb/s to 100 m and may be an alternative media for premises wiring. HSPN has submitted a proposal to the Optoelectronics Micro Networks OMNET program to continue development of plastic fibers.

Although some manufacturers have presented data indicating that plastic fiber will be unable to meet industry standard TIA/EIA-568A at 100 m, a number of companies are already developing plastic fiber applications. Some companies currently marketing plastic fiber or components including the following:

Amp Inc. offers plastic fiber connectors and components for airplanes and automobiles.

CEL (California Eastern Laboratories), a distributor for NEC Japan, offers the HiSpot transceiver with a visible red LED over step-index plastic fiber at 155 Mb/s using an F07 connector; applications include Fast Ethernet, ATM, and IEEE 1394 (Firewire).

Spec Trans Specialty Optics (distributor for Toray Industries of Japan) offers 200-μm step-index fiber; they have proposed to the ATM Forum a standard for 200-μm hard clad silica at 155 Mb/s, 100 m and plastic fiber for 50 Mb/s, 50 m.

Transceivers

Although discussion over the next generation of Fibre Channel continues, some companies are now offering an alternative, the gigabit interface connector (GBIC). This is a gigabit module that offers the convenience of being able to unplug a copper interface and replace it with an optical interface at a later time for increased distance and bandwidth. It features a parallel rather than a serial interface; the copper and optical modules plug into the same electrical connection point, which includes a pair of rails to guide the modules during plugging. Found mainly in workstation applications, the GBIC is available from several sources including IBM Rochester, Minnesota.

Low-cost bidirectional transceivers for fiber in the loop and similar applications (fiber to the curb or home) are also being developed. For example, a recent device reported by Amp in collaboration with Lasertron, Digital Optics Corp., GTE Labs, BroadBand Technologies, and the University of Colorado offers a connectorized, single-mode bidirectional transceiver with a source, modulator, and detector integrated into a single package. Each module includes an InP diode laser, silicon CMOS modulator/demodulator electronics, a PIN photodiode, an integrated beam splitter, and a built-in SC for ferrule-based fiber attachment. The beam splitter is used for bidirectional transmission at 1.55-μm wavelength and reception at 1.3-μm wavelength. Data rates up to 1.2 Gb/s is at 1.3 μm and a burst mode, 50 Mb/s receiver at 1.55 μm, are available in a package measuring $85 \times 17 \times 10$ mm.

There are a number of trends in the future development of datacom transceiver technology. Among these is the migration to lower power supply voltages, following the trend of digital logic circuits. Most digital logic has migrated from 5 to 3.3 V and is on a path toward 2.5-V operation; optical transceivers are emerging that will be able to operate at 1 Gb/s speeds using a single 3.3-V power supply, although many sources will remain at 5 V for some time, especially VCSELs.

Although the telecommunications market continues to have distinct requirements from the datacommunications market, there is certainly some merging of requirements between these two areas as well. The telecom market is currently driving development of analog optical links for cable television applications, including bidirectional wavelength multiplexed links that transmit in one direction at 1300 nm and in the other at 1550 nm. This is achieved by incorporating some form of optical beam splitter in the transceiver package. Many of these devices incorporate optical fiber pigtails rather than connectorized transceiver assemblies. Packaging is

another evolving area; existing datacom industry standards, such as the
1 × 9 pin serial transceiver or 2 × 9 pin transceiver with integrated clock
recovery, may give way to surface-mount technology in the near future.
In the telecom industry, the standard optical transceiver package has been
the dual in-line pin (DIP) package (14 pins and two rows); recent developments in the so-called mini-DIP (8 pins and two rows) may develop into a
new packaging standard for the industry. Finally, integrated optoelectronic
integrated circuits are advancing to the point where they may replace the
standard T0 can and ball lens assembly for optical device packaging and
alignment in the next 5 years.

VCSELs hold a great deal of promise as low-cost optical sources for data
communications, especially for optical array interconnections. Research
continues on VCSELs that can operate at long wavelengths (1300 nm);
although some devices have been demonstrated under laboratory conditions, the technology is probably several years away from commercial
applications.

Parallel optical transceivers will face new challenges in the area of laser
safety. Although many parallel transceivers will be able to meet U.S. class
1 safety standards, it is more difficult to meet the international class 1
limits specified by IEC 825. One possible alternative is international certification as a class 3A product, which would require a warning label on
the transceiver; it is unclear whether the IEC will allow short-wavelength
lasers to hold this classification or whether the market will accept such
products instead of class 1 sources.

Acronyms

1M	a laser safety classification
2R	retime/reshape
3R	retime/reshape/regenerate
AAL	ATM adaptation layer
AC	alternating current
ACK	acknowledgment
ACTS	advanced communication technologies and services
A/D	analog to digital
AFNOR	Assoc. Française de NORmalisation (ISO member)
AIP	American Institute of Physics
ANS	American National Standards
ANSI	American National Standards Institute
AOI	add or invert
AON	all-optical network
APC	auto power control
APD	avalanche photodiode
API	application programmer interface
APS	American Physical Society
AR	antireflective
ARP	address resolution protocol
ARPA	Advanced Research Project Agency
ASCI	Advanced Strategic Computing Initiative
ASCII	American Standard Code for Information Interchange (developed by ANSI)
ASE	amplified spontaneous emission
ASIC	application specific integrated circuit
ASTM	American Society for Test and Measurement
ATM	asynchronous transfer mode
ATMARP	ATM address resolution protocol

AUA	another useless acronym (from Preface to the Second Edition)
AWG	array waveguide grating
BD	solid bore inner diameter
BER	bit error rate
BGA	ball grid array
BGP	border gateway protocol
BH	buried heterostructure
BiCMOS	bi-junction transistor/complementary metal oxide semiconductor
BIP	bit interleaved parity
BIS	Bureau of Indian Standards
B-ISDN	broadband integrated services digital network
bit/s	bits per second
BLIP	background limited in performance
BLM	ball-limiting metallurgy
BLSR	bidirectional line switched ring
BNR	Bell Northern Research
BOL	beginning of lifetime
BPS	bytes per second
BSI	British Standards Institute
BSY	control code for busy, as defined by the ANSI Fibre Channel Standard
BTW	behind the wall
BUS	broadcast and unknown server
BWPSR	bi-directional path-switched rings
C4	controlled collapse chip connection
CAD	computer-aided design
CAM	computer-aided manufacturing
CATV	community antenna television (cable TV)
C-band	conventional wavelength band used in wavelength multiplexing (1520–1570 nm)
CBD	connector body dimension
CBGA	ceramic ball grid array
CBR	constant bit rate
CCD	charge coupled devices

CCGA	ceramic column grid array
CCITT	Consultative Committee for International Telephone and Telegraph (International Telecommunications Standard Body)-now ITU
CCW	channel control words
CD	compact disc
CDL	converged data link
CDMA	code-division multiple access
CDR	clock and data recovery
CDRH	U.S. Center for Devices and Radiological Health
CE mark	Communaute Europeenne mark
CECC	CENELEC Electronic Components Committee
CEL	California Eastern Laboratories
CENELEC	Comité Européen de Normalisation ELECtrotechnique or 'European committee for electrotechnical standardization'
CFR	Center for Devices and Radiological Health
CICSI	Building Premises Wiring Standard Body
CIP	carrier-induced phase modulation
CLEC	competitive local exchange carrier (local telephone service)
CLEI	common language equipment identification (Bellcore)
CLI	common line interface
CLO	control link oscillator
CLP	cell loss priority bit
CML	current-mode logic
CMOS	complementary metal oxide semiconductor
CMU	Carnegie Mellon University
CNS	Chinese National Standards
COB	chip on board
COGS	cost of goods sold
COSE	Committee on Optical Science & Engineering
CPR	coupled power range
CPU	central processing unit
CRC	clock recovery circuit
CRC	cyclic redundancy check
CS	convergence sublayer
CSMA/CD	carrier sense multiple access/with collision detection
CSP	channeled substrate planar

CTE	coefficient of thermal expansion
CVD	chemical vapor deposition
CW	continuous wave
CWDM	coarse wavelength-division multiplexing (aka. wide-WDM)
CYTOP	a transparent fluorpolymer
DA	destination address
DARPA	Defence Advanced Research Projects Association (see also ARPA)
DAS	dual-attach station
DASD	direct access storage device
DBR	diffraction Bragg reflector
DBR	distributed Bragg reflector
DC	direct current
DCD	duty cycle distortion, also data carrier detect (modems)
DCF	dispersion compressing fiber
DCN	data communications network
DDM	directional division multiplexing
DDOA	dysprosium doped optical amplifier
DE	diethyl (as in diethylzinc-DEZn)
DEC	Digital Equipment Corporation
DEMUX	demultiplexer
DFB	distributed feedback
DH	double heterojunction/double heterostructure
DHHS	Department of Health Services in the IEC
DIN	Deutsches Institut für Normung, (the German Institute for Standardization) or Data Input
DIP	dual in-line package
DJ	derministic jitter
DLL	delay locked loop
DM	dimethyl (as in dimethylzinc-DMZn)
DMD	differential mode delay
DMM	digital multimeter
DP	data processing
DRAM	dynamic random access memory
DS	Dansk Standardiseringsrad
DSF	dispersion shifted fiber
DSL	digital subscriber line

DSO	digital sampling oscilloscope
DSP	digital signal processing
DTPC	design for total product cost
DUT	device under test
DVD	digital video disk
DWDM	dense wavelength division multiplexers
EBP	end of bad packet (delimiter for Infiniband data packets)
EC	European Community
ECF	echo frame
ECH	frames using the ECHO protocol
ECHO	European Commission Host Organization
ECL	emitter coupled logic
EDFA	erbium-doped fiber amplifier
EELD	edge-emitting semiconductor laser diode (also ELED)
EFA	end face angle
EGP	end of good packet (delimiter for Infiniband data packets)
EH	hybrid mode where electric field is largest in transverse direction
EIA	environmental impact assessments
EIA standard	Electronics Industry Association
EIA/TIA	Electronics Industry Association/Telecommunications Industry Association
ELAN	emulate local area network
ELED	edge-emitting semiconductor laser diode (also EELD)
ELO	epitaxial liftoff
EMBH	etched-mesa buried heterostructure
EMC	electromagnetic compatibility
EMI	electromagnetic interference
ENIAC	Electronic Numeric Integrator and Computer
EOF	end of frame
EOL	end of lifetime
EOS	everything over SONET
ESCON	Enterprise system (IBM) connection
ESD	electrostatic discharge
ESPRIT	European Strategic Program for Research in Information Technology
ETDM	electronic time division multiplexing

ETR	external timing reference
ETSI	European Telecommunication Standards Institute
FBG	fiber Bragg grating
FBS	fiber beam spot
FC	fibre channel
FC	frame control
FC	connector-threaded fasteners
FC	flip chip, a manufacturing chip bonding procedure
FCA	Fibre Channel Association
FC-AL	Fibre Channel arbitrated loop
FCC	Federal Communications Commission
FCE	ferrule/core eccentricity
FCS	Fibre Channel Standard
FCSI	Fibre Channel Standard Initiative
FCV	Fibre Channel Converted/FICON conversion vehicle
FD	ferrule diameter
FDA	Food and Drug Administration
FDDI	fiber distributed data interface
FDMA	frequency division multiple access
FET	field effect transistor
FF	ferrule float
FICON	Fibre Channel Connection or Fiber Connection
FIFO	first in, first out
FIG	fiber image guides
FIT	failures in time
FJ	Fiber-Jack
FLP	fast link pulse
flops	floating point operation performed per second
FO	fiber optic
FOCIS	fiber optic connector intermatability standard
FOTP	fiber optic test procedures
F_Port	fabric port, as defined by the ANSI Fibre Channel Standard
FP	Fabry-Perot
FPA	Fabry Perot amplifier
FPGA	field-programmable gate array
FQC	Fiber Quick Connect (feature in the IBM Global Services structured cable offering)

FRPE	flame-retardant polyethylene
FSAN	full service access network
FSR	free spectral range
FTC	fiber quick connect
FTS	fiber transport services
FTTC	fiber-to-the-curve
FTTH	fiber-to-the-home
FTTO	fiber-to-the-office
FWHM	full width at half maximum
FWM	four wave mixing
GBE or GbE	Gigabit Ethernet
GBIC	gigabit interface connector
GDPS	geographically dispersed Parallel Sysplex
GFC	generic flow control
GHz	gigahertz
GI	graded index
GLC	Gigalink Card
GOSS	State Committee of the Russian Federation for Standardization Metrology and Certification
GP	gross profit
GRIN	graded index
GRP	glass reinforced plastic
GSN	gigabit serial network (also known as HIPPI 6400)
GVD	group velocity dispersion
HAN	home area network
HC	horizontal cross-connect
HE	hybrid mode where magnetic field is largest in transverse direction
HEC	header error correction
HFET	heterostructure field effect transistor
HIBITS	high bitrate ATM termination and switching (a part of the EC RACE program)
HIPPI	high-performance parallel interface
HMC	hardware management console
HOCF	hard polymer clad (glass) fiber
HOEIC	hybrid optoelectronic integrated circuit
HPC	high-performance computing

HPCF	hard polymer clad fiber
HR	highly reflective
HSPN	High Speed Plastic Network Consortium
IB	InfiniBand
IBM	International Business Machines
IC	integrated circuit
IC	intermediate cross-connect
ICCC	International Conference on Computer Communications
ID	identification
IEC	International Electrotechnical Commission
IEEE	Institute of Electrical and Electronics Engineers
IETF	Internet Engineering Task Force
IFB	imaging fiber bundles
IHS	Information Handling Services
ILB	inner lead bond
ILEC	incumbent local exchange carrier (local telephone service)
ILMI	interim local management interface
InATMARP	inverse ATM address resolution protocol
I/O	input/output
IODC	Institute for Optical Data Communication
IP	internet protocol
IPI	intelligent physical protocol
IPT	integrated photonic transport (inband)
IR	infrared reflow
IrDA	Infrared Datacom Association
IRED	infrared emitting diode
ISC	intersystem channel gigabit links
ISI	intersymbol interference
ISO	International Organization for Standardization
ISP	internet service provider
IT	information technology
ITO	indium tin oxide
ITS	Institute for Telecommunication Sciences
ITU	International Telecommunications Union
IVD	inside vapor deposition
IXC	interexchange carrier (long distance telephone service)

JEDEC	Japanese-based telecom/datacom standards
JIS	Japanese Industrial Standards
JSA	Japanese Standards Association
KGD	known good die
LAN	local area network
LANE	local area network emulation
L-band	long wavelength band used in wavelength multiplexing (1560–1610 nm)
LC	late counter
LC	"Lucent Connector"
LCOS	liquid crystal on silicon
LCT	link confidence test
LD	laser diode
LEAF	large effective area fiber
LEC	LAN emulation client
LECID	LEC(LAN emulation client) identifier
LED	light-emitting diode
LES	LAN emulation server
LIA	Laser Institute of America
LIS	logical IP subnet
LLC	logic link control
LOH	line overhead
LOS	loss of signal
LP	linear polarized mode
LPE	liquid phase epitaxy
LSZH	low smoke zero halogen cable
LUNI	LANE user-network interface
LVD	low voltage directive
LVDS	low voltage differential signal
LX	long wavelength (1300 nm) transmitter
MAC	media access control
MAN	metropolitan area network
MBE	molecular beam epitaxy
MBGA	metal ball grid array

MC	main cross-connect
MCM	multichip module
MCP	mode conditioning patch cables
MCVD	modified chemical vapor deposition
MDF	main distribution facility
MDI	medium-dependent interfaces
MEM	microelectromechanical
MESFET	metal semiconductor field effect transistor
METON	metropolitan optical network
MFD	modal field diameter
MIB	management information base
MIC	media-interface connector
MII	media-independent interface
MIMD	multiple instruction multiple data stream
MJS	methodology for jitter specification
MM	multimode
MMF	multimode fiber
MOCVD	metal organic chemical vapor deposition
MONET	multiple wavelength optical network
MOS	metal oxide semiconductor
MOSIS	MOS implementation system
MPE	maximum permissible exposure
MPLmS	multi-protocol lambda switching
MPLS	multi-protocol label switching
MPO	trade name for multifiber optical connector using a 12 fiber interfaces; also known as MPX
MPX	trade name for multifiber optical connector using a 12 fiber interface; also known as MPO
MQW	multiquantum well
MS	multi-standard
MSA	multi-source agreements
MSI	medium scale integration
MSM	metal-semiconductor-metal
MT	multi-fiber termination
MTP	multi-fiber terminated push-on connector
MT	RT multi-fiber RJ-45 latched connector
MT-RJ	multi-termination RJ-45 latch (a type of optical connector)
MU	multi-termination unibody
MUX	multiplexer
MZ	Mach-Zehnder

NA	numerical aperture
NAND	not and (logic gate)
NAS	network attached storage
NBS	National Bureau of Standards (now NIST)
NC	narrowcast
NC&M	network control and management
NEBS	network equipment building system
NEC	National Electric Code
NEP	noise equivalent power
NEXT	near end crosstalk
NFPA	National Fire Protection Agency
NGI	next generation Internet
NGIO	next generation input/output
NIC	network interface card
NIF	neighbor information frame
NIST	National Institute of Standards & Technology
NIU	network interface units
NLOG	non-linear optical gate
NMOS	negative-channel metal oxide semiconductor
NMS	network management system
NOR	not or (logic gate)
N_Port	nodal port, as defined by the ANSI Fibre Channel Standard
NPN	negative-positive-negative
NRC	National Research Council
NRZ	nonreturn to zero
NSPE	National Society of Professional Engineers
NTIA	National Telecommunication and Information Administration
NTT	Nippon Telegraph and Telephone Corporation
NUMA	non-uniform memory architecture
NZDSF	non-zero dispersion shifted fiber
OA	optical amplifiers
OADM	optical add-drop multiplexer
OAMP	operations, administration, maintenance, and provisioning (sometimes OAM&P)
OCI	optical channel interface
OCI	optical chip interconnect
OCLD	optical channel laser detector
OCM	optical channel manager

ODC	optical data center
ODSI	Optical Domain Service Interconnect coalition
OE	optoelectronic
O/E/O	optical/electrical/optical
OFB	ordered fiber bundles
OFC	optical fiber control system
OFN	optical fiber nonplenum/nonriser/nonconductive
OFNP	optical fiber nonconductive plenum listing
OFNR	optical fiber nonconductive riser listing
OFSTP	optical fiber system test procedure
OIDA	Optoelectron Industry Development Association
OIF	Optical Internetworking Forum
OIIC	optically interconnected integrated circuits
OLIVES	optical interconnections for VLSI and electronic systems (a part of the EC ESPRIT program)
OLS	optical label switching
OLTS	optical loss test set
OMA	optical modulation amplitude
OMB	office of management and budget
OMC	operational management committee
OMNET	optical micro-network
OMX	optical multiplexing modules
ON	Osterreisches Normungistitut (Austrian Standards Institute)
ONTC	Optical Network Technology Consortium
OPFET	optical field effect transistor
OSA	open system adapter
OSA	optical subassembly, also called the coupling unit
OSA	Optical Society of America
OSHA	Occupational Safety & Health Administration
OSI	open systems interconnection
OSIG	optical signal
OSNR	optical signal to noise ratio
OSPF	open shortest path first
OTDM	optical time division multiplexer
OTDR	optical time domain reflectometer
OTF	optical transfer function
OTN	optical transport network
OUT	output
OUTB	average output

OVD	outside vapor deposition
OXBS	optical crossbar switches
OXC	optical cross connects
PA	preamble
PA	pointing angle
PANDA	polarization maintaining and absorption reducing fiber
PAROLI	parallel optical link
PBGA	plastic ball grid array
PBX	private branch exchange
PC	personal computer
PC	polycarbonate-a thermoplastic compound
PCA	product cost apportionment
PCB	printed circuit board
PCI	peripheral component interconnect
PCM	physical connection management
PCVD	plasma-assisted chemical vapor deposition
PD	photodiode
PDF	probability density function
PDG	polarization dependent gain
PDL	polarization dependent loss
PECL	postamplifier emitter coupled logic unit
PECVD	plasma-enhanced chemical vapor deposition
PF	perfluorinated
PGA	pin-grid arrays
PHY	physical layer of OSI model
PIN	positive-intrinsic-negative
PLCC	plastic leaded chip carrier
PLL	phase-lock loop
PLOU	physical layer overhead unit
PLS	primary link station
PMD	physical medium-dependent sublayer
PMD	polarization mode dispersion
PMF	parameter management frame
PMF	polarization maintaining fiber
PMI	physical media-independent sublayer
PMMA	polymethylmethacrylate-a thermoplastic compound
PMT	photomultiplier tube
POF	plastic optical fiber

POI	parallel optical interconnects
POINT	polymer optical interconnect technology
PON	passive optical network
POP	post office protocol
POTS	plain old telephone system
PPL	phase locked loop
PPRC	peer to peer remote copy
PR	plug repeatability
PRBS	pseudo-random binary sequence
PSTN	packet switched telephone network
PT	payload type
PTT	local telecom authority
PVC	polyvinyl chloride (a plastic)
PVL	parallel vixel link
QCSE	quantum confined Stark effect
QE	quantum efficiency
QFP	quad flat pack
QOS	quality of service
QW	quantum well
RACE	Research in Advanced Communications for Europe
RAID	redundant array of inexpensive disks
RARP	reverse address resolution protocol
RC	resistor capacitor
RCDD	registered communication distribution designers
RCE	resonant-cavity enhanced
RCLED	resonant-cavity light-emitting diode
RECAP	resonant-cavity enhanced photodetector
RF	radio frequency
RFI	radio frequency interference
RHEED	reflection high-energy electron diffraction
RIE	reactive ion etching
RIN	reflection induced intensity noise
RISC	reduced instruction set computing
RJ	random jitter
RJT	control code for reject, as defined by the ANSI Fibre Channel Standard
RMAC	repeater media access control

RMS	root mean square
ROSA	receiver optical subassembly
RSOH	regenerator section ovehead
Rx	receiver
SA	source address
SA	Standards Association of Australia
SAGCM	separate absorption, grading, charge sheet and multiplication structure
SAGM	separate absorption grading and multiplication
SAM	separate absoprtion and multiplication layers
SAM	sub-assembly misalignment
SAN	storage area network
SAR	segmentation and reassembly sublayer
SAS	single attach station
SASO	Saudi Arabian Standards Organization
S-band	short wavelength band used in wavelength multiplexing (1450–1510 nm)
SBCCS	single byte command code set
SBCON	single byte command code sets connection architecture
SBS	source beam spot
SBS	stimulated Brillouin scattering
SC connector	subscriber connector (spring latch fasteners)
SC-DC	subscriber connector, dual connect ferrule
SCI	scalable coherent interface
SC-QC	subscriber connector, quad connect ferrule
SCSI	small computer system interface
SD	shroud dimension
SD	start delimiter
SDH	synchronous digital hierarchy
SDM	space division multiplexing
SDO	standards developing organizations
SEAL	simple and efficient adaptation layer
SEED	self electrooptic effect devices
SETI	search for extraterrestrial intelligence
SFF	small form factor
SFP	small form factor pluggable
SFS	Suomen Standardisoimislitto Informaatiopalvelu (Finland Standards Information)

SI	System International (International System of Units)
SIA	Semiconductor Industry Alliance
SIC	Standard Industrial Classification
SIF	status information frame
SIMD	single instruction-stream, multiple data-stream
SiOB	silicon optical bench (non-hermetic)
SIPAC	Siemens Packaging System
SIS	Standardiseringkommisiionen I Sverige (Swedish Standards Commission)
SJ	sinusoidal jitter
SL	superlattice
SLA	secure level agreement
SL-APDs	superlattice avalanche photodiodes
SM	single mode
SMC	surface mount component
SMD	surface mount device
SMDS	switched megabit data services
SMF	single-mode fiber
SMP	shared memory processor
SMT	station management frames
SMT	surface mount technique
SMU	Sanwa multi-termination unibody
SNMP	simple network management protocol
SNR	signal-to-noise ratio
SNZ	Standards Association of New Zealand
SOF	start of frame
SOHO	small office/home office
SOJ	small outline packages with J leads
SONET	Synchronous Optical NETwork
SP	shelf processor
SPC	statistical process control
SPE	synchronous payload envelope
SPIBOC	standardised packaging and interconnect for inter and intra board optical connections (a part of the EC ESPRIT program)
SPIE	Society of Photooptic Instrumentation Engineers
SQNR	signal-to-quantum-noise ratio
SQW	single quantum well
SRAM	static random access memory

SRM	sub-rate multiplexing
SRS	stimulated Raman scattering
SSA	serial storage architecture
SSA	storage system architecture
S-SEED	symmetric self electrooptic effect devices
SSI	small scale integration
SSP	storage service provider
ST connector	subscriber termination
STP	shielded twisted pair
STS	synchronous transport signal
SWDM	sparse WDM
SWF	surface wave filter
SX	short wavelength (850 nm.) transmitter
TAB	tape access bonding
TAB	tape automated bonding
TAXI	transparent asynchronous transceiver/receiver interface
TBGA	Tape ball grid array
TC	transmission convergence sublayer
TCM	time compression multiplexing
TCP/IP	transmission control protocol/internet protocol
TDFAs	thulium doped fiber amplifiers (to be used in the wavelength range of 1450-1510 nm.)
TDM	time division multiplexing
TDMA	time-division multiple access
TE mode	transverse electric (electric field normal to direction of propogation)
TE	triethyl (as in triethylgallium-TEGa)
TERKS	trade name of a data base used by Telcordia (formerly Bell Labs) to track inventory and depreciation on telecommunications hardware
THC	through-hole component
THT	token holding timer
THz	tera hertz
TIA	Telecommunications Industry Association
TIA	trans-impedance amplifier (also TZA)
TM mode	transverse magnetic (magnetic field normal to direction of propogation)
TM	trimethyl (as in trymethylgallium-TMGa)

TO	transistor outline
TOSA	transmitter optical subassembly
TP	twisted pair
TRT	token rotation timer
TRX	transceiver
TTI	time to instalation
TTL	transistor-transistor logic
TTRT	target token rotation time
TUG	tributary unit groups
TWA	traveling wave amplifier
Tx	transmitter
TZA	trans-impedance amplifier (also TIA)
UDP	user datagram protocol
UHV	ultra high vacuum
UI	unit interval
UL	Underwriters Laboratories
ULP	upper level protocol
UNI	user-network interface
USNC	United States National Committee
UTP	unshielded twisted pair
UV	ultra-violet
VAD	vapor axial deposition
VBR	variable bit rate
VC	virtual circuit
VCC	virtual channel connection
VCI	virtual channel identifier
VCO	voltage-controlled oscillator
VCSELs	vertical cavity surface-emitting lasers
VDE	Verband Deutscher Electrotechniker (Association of German Electrical Engineers)
VDI	Vereins Deutscher Ingenierure (German Standards)
VF-45	trade name of a small form factor optical connector developed by 3M Corporation with an RJ-45 latch
VLIW	very long instruction word
VLSI	very large-scale integration
VPI	virtual path identifier
VPN	virtual private network

VPR	vapor phase reflow
VT	virtual tributary
VTG	virtual tributary group
WAN	wide area network
WDM	wavelength-divison multiplexing
WWSM	wide spectrum WDM
XDF	extended distance feature
XGP	Multi-Gigabit Ethernet pluggable
XPR	cross plug range
XRC	extended remote copy
ZBLAN	fluorozirconate

Some useful acronym searches on the World Wide Web
The One Look Dictionary www.onelook.com
Stammtisch Beau Fleuve Acronyms http://www.plexoft.com/SBF/(Best accessed through The One Look Dictionary)
Acronym Finder http://www.acronymfinder.com/

Glossary

absorption	the loss of light when passing through any material, due to conversion to other energy forms, such as heat.
acoustooptic tunable filter	a light filter tuned by acoustic (sound) waves. This is accomplished using polarizers and an acoustic diffraction grating, which creates a resonant structure that rotates the polarization.
active area	the area in a semiconductor that either absorbs or emits radiation. Also known as effective sensing area.
aggregate capacity	a measure of the total information handling capability of a smart pixel array. It combines the individual channel data rate with the total numer of channels in the array to produce an aggregate information-carrying capacity.
alignment	the connection of optical components to maximize signal transmitted.
alignment sleeve	part of an ESCON connector into which the ferrules are inserted, assuring accurate alignment.
annealing	the process of heating and slowly cooling a material. This makes glass and metal stabilize their optical, thermal, and electric properties, and can reverse lattice damage from doping to semiconductors.
architecture	overall structure of a computer system, including the relationship between internal and external components.
asynchronous	a form of data transmission where the time that each character, or block of characters, starts is arbitrary. Asynchronous data has a start bit and a stop bit, because there is no regular time interval between transmissions, and no common clock reference across the system.
attenuation	the decrease in signal strength caused by absorption and scattering. The power or amplitude loss is often measured in dB.

Auger nonradiative recombination	when an electron and hole recombine, and then pass excess energy and momentum into another electron or hole. This process does not generate any additional radiation.
avalanche photodiodes	photodiodes where the primary carriers, electron/holes pairs created by the incident photon, are accelerated through a voltage, and then collide with neutral atoms, creating more secondary carriers. These secondary carriers may accelerate and create new carriers. This process, known as photomultiplication, can multiply the signal by up to two orders of magnitude.
backbone network	a primary conduit for traffic that is often both coming from, and going to, other networks.
backplane	circuit board with sockets to connect other cards, especially communication channels.
back-reflections	reflections back to the laser from devices in its path.
band gap	the energy difference between the top of the valence band and the bottom of the conduction band of a solid. The size of the band gap will determine whether a photon will eject a valence electron from a semiconductor.
bandwidth	the range of frequencies over which a fiber-optic medium or device can transmit data. This range, expressed in hertz, is the difference between the highest and the lowest frequencies for optical filter elements. For MM fibers, the range is expressed as a product of the bandwidth and distance, in MHz-km.
baud	the signaling speed, as measured by the maximum number of times per second that the state of the signal can change. Often a "signal event" is simply the transmission of a bit. Baud rate is measured in $\sec^{-1}$.
beacon process	a token ring process which signals all remaining stations that a significant problem has occurred, and provides restorative support. Like a string of signal fires, the beacon signal is passed from neighbor to neighbor on the ring.

beating	when superimposing waves of different frequencies, beating occurs when the maximum amplitude of these waves match up. Coherent beating occurs when the data signal beats with itself out of phase, and can cause cross-talk.
biasing	applying a voltage or a current across a junction detector. This changes the mode of the detector, which can affect properties such as the noise and speed. (*See* photoconductive and photovoltaic mode)
bias voltage	the voltage applied when biasing. (*See* biasing)
bit	either 0 or 1, the smallest unit of digital communications.
bit error rate	the probability of a transmitted bit error. It is calculated by the ratio of incorrectly transmitted bits to total transmitted bits.
bit synchronization	when a receiver is delivering retimed serial data at the required BER.
Bragg reflector	see distributed Bragg reflector.
breakdown voltage	the bias voltage in an avalanche photodiode where the current gain begins to approach infinity. In other words, the bias voltage when an avalanche photodiode is no longer responding to incident photons.
Brillouin scattering	the scattering of photons (light) by acoustic phonons (sound waves). A special case of Raman scattering.
broadcast-and-select network	a network that has several nodes connected in a star topology. (*See* topology)
bulkhead splice, splice bushing	a unit that allows two cables with unlike connectors to mate.
burn-in	the powering of a product before field operation, to test it and stabilize its characteristics.
brust mode switching networks	an approach that minimizes latency by transmitting large amounts of data in a short time over the network.
bus	(*See* optical bus)
bus topology	(*See* topology)
butt coupling	a mechanical splice between a fiber and a device. This technique is done with the signal on, and the fiber secured into position when measured signal is maximized. (*See* splice)

butterfly package	(*See* dual-in-line pin package)
byte	eight bits, numbered 0 to 7. (*See* octet)
cable jacket	the outer material that surrounds and protects the optical fibers in a cable.
cable plant	passive communications elements, such as fiber, connectors and splices, located between the transmitter and receiver.
calibration	the comparison to a standard or a specification. Also to set a device to match a specification.
channel	a single communications path or the signal sent over that path.
chatter	any transient response greater than 0.5 volts, ingoing between the signal detect negative level to assert its level, and of any duration that can be sensed by the channel logic.
"chicken and egg" syndrome	when it is hard to find someone to take the first risk on a new technology, which delays its implementation. For example: when customers don't want to commit to using an emerging technology without some assurance that it will be fully supported by manufacturing, and suppliers don't want to commit to putting emerging technology into production without having a customer who is committed to using it.
chirp	a pulsed signal whose frequency lowers during the pulse.
cholesteric liquid crystals	(*See* liquid crystals)
chopping	when an optical signal is varied through use of a mechanical or electrical "chopper" to block the signal at a certain time. (*See* lock-in amplifier)
chromatic dispersion	(*See* dispersion)
cladding	the material of low refractive index used to cover an optical fiber, which reflects escaping light back into the core, as well as strengthens the fiber.
cleaving	to split with a sharp instrument along the natural division in the crystal lattice.

client

a computer or program that can download data or request a service over the network from the server. (*See* server)

clock generation/ multiplication

synchronizing a terminal's internal clock to the received bit stream.

clock recovery

reconstructing the order of operations after serial/ parallel conversion.

coherent beating

(*See* beating)

correlators

a device that detects signal from noise by computing correlation functions, which is similar to transforms.

costs of goods sold

the sum of parts cost, manufacturing cost, serviceability and warranty cost, obsolescence and scrap and a portion of the supply chain cost. (*See* parts cost, manufacturing cost, supply chain cost)

coupled power range

the allowable difference between the minimum and maximum allowed power.

coupling

connecting two fibers or the connector between two fibers.

cross-plug range

the difference between the measured lowest power and highest power for matings between multiple connectors and the same transceiver

cross-talk

the leaking of or interference between signal in two nearby pixels, wires, or fibers.

cutoff wavelength

the wavelength for which the normalized power becomes linear to within 0.1 dB.

daisy chain

a bus wiring scheme where devices are connected to each other in sequence (device A is wired to device B is wired to device C, etc.), like a chain of daisies. (*See* topology)

dark current

the current measured from a detector when no signal is present.

data dependent jitter

(*See* jitter)

data rate

short for data transfer rate, the speed at which devices transmit digital information. Units include bits per second, but are more likely to be in the range of megabits per second (Mbit/s), or even gigabits per second (Gbit/s) in the field of fiber optics.

de-adjustment	the variation of the coupled power after several matings with the same connector.
decoding	(*See* encoding/decoding)
detectivity	the reciprocal of the noise equivalent power (NEP). This can be a more intuitive figure of merit, because it is larger for more sensitive detectors.
deterministic jitter (DJ)	(*See* jitter)
development expense	the total development investment in people, test equipment, prototypes, etc.
device	any machine or component that attaches to a computer.
dielectric mirror	a multilayer mirror that is an alternative to a distributed Bragg reflector (DBR) in semiconductor laser manufacture.
diffraction	when a wave passes through an edge or an opening, secondary wave patterns are formed which interfere with the primary wave. This can be used to create diffraction gratings, which work similarly to prisms.
differential mode delay	delay variation caused by differences in group velocity among the different propagation modes of an optical fiber.
directional	division multiplexing. (*See* multiplexer)
dispersion	the distortion of a pulse due to different propogation speeds. Can be chromatic, which is caused by the wavelength dependence of the index of refraction, or intermodal, caused by the different paths traveled by the different modes.
dispersion-flattened fiber	specialty obtical fiber whose cross-section is designed to minimize dispersion over a broad range of input wavelengths.
dispersive self-phase modulation	nonlinear property of an optical fiber, in which the light passing through the fiber induces a change in the refractive index of the core, and chromatic dispersion causes either an up-chirp or down-chirp in the optical frequency.
distortion	the change in a signal's waveform shape.
distributed Bragg reflector (DBR)	a mirror used in the manufacture of semiconductor lasers and photodetectors. It is made from multiple

layers of semiconductors that have a band gap in the wavelength of interest.

distribution panel — a central panel from which a signal is routed to points of use. (*See* patch panel)

divergence — the bending of light rays away from each other, for example the spreading of a laser beam with increased distance.

doping — adding an impurity to a semiconductor.

double-clad optical amplifier — an optical fiber amplifier in which the pump light is delivered via an outer coating on the same fiber that performs the amplification.

down-chirp — a linear decrease in frequency over time.

dual homing architecture — a type of network with two hubs that fail over to each other in the event of a disaster at either location. (*See* topology)

dual hubbed rings — variation on a hubbed ring architecture in which all traffic flows through not one but two hub sites in a network. (*See* topology)

dual-in line pin package, butterfly package — a cavity package for a semiconductor laser that is wire bonded and then has the plastic molded around the body and leads of the package.

duplex — a communications line that lets you send and receive data at the same time.

duty cycle — the pulse duration times the pulse repetition frequency. In other words, the percentage of time an intermittent signal is on.

duty cycle distortion (DCD) — the ratio of the average pulse width of a bit to the mean of twice the unit interval. (*See* unit interval)

dynamic range — the ratio of the largest detectable signal to the smallest detectable signal, such as the receiver saturation charge to the detection limit (also known as linear dynamic range).

effective sensing area — (*See* active area)

electrooptic transducer — a device that converts an electric signal to an optical signal and vice versa. An example of this is the photocell and laser in a transceiver. (*See* transducer)

emulation — the use of program to simulate another program or a piece of hardware.

encoding/ decoding	encoding is the process of putting information into a digital format that can be transmitted using communications channels. Decoding is reversing the process at the end of transmission.
end node	a node that does not provide routing, only end user applications.
epitaxy	method of growing crystal layer on a substrate with an identical lattice, which maintains the continuous crystal structure. Styles include molecular beam epitaxy, vapor phase epitaxy, etc.
etalon	an optical device with two reflective mirrors facing each other to form a cavity.
etched mesa	a flat raised area on an electronic device, created during the photolithography process.
extinction ratio	the ratio of the power level of the logic '1' signal to the power of the logic '0' signal. It indicates how well available laser power is being converted to modulation or signal power.
eye-diagram	an overlay of many transmitted responses on an oscilloscope, to determine the overall quality of the transmitter or receiver. The relative separation between the two logic levels is seen by the opening of the eye. Rise and fall times can be measured off the diagram. Jitter can be determined by constructing a histogram of the crossing point.
fabric, switched network	a FCS network in which all of the station management functions are controlled at the switching point, rather than by each node. This approach removes the need for complex switching algorithms at each node. The telephone system, where the dialer supplies the phone number, can be used as an analogy.
Fabry-Perot resonance	resonance that occurs when waves constructively interfere. Fabry-Perot resonance occurs in a cavity surrounded by mirrors. Semiconductor lasers, such as VCSELs, take advantage of Fabry-Perot resonance to stimulate emission in their active region.
facet passivation	growing a thin oxide layer over the semiconductor (for datacom the EELD) facet, to limit environmental exposure and natural oxidation.

Fast Ethernet	A LAN running at speeds up to 100 MBd. (*See* LAN)
Fermi level	the maximum energy of the electrons in a solid, which determines the availability of free electrons. If the Fermi level is in the conduction (top) band, the material is a conductor (metal). If the Fermi level is in the valence (lower) band, the material is an insulator. If the Fermi level is between the conduction and the valence band, the material is a semiconductor.
ferrule	a cylindrical tube containing the fiber end that fits within high tolerance into the flange of the transceiver port.
fiber optic link	the transmitter, receiver, and fiber optic cable used to transmit data.
fiber ribbon	multifiber cables and connectors.
Fibre Channel Connection (FICON)	an I/O interface standard that mainframe computers use to connect to storage devices. This standard is eight times faster than the previous fiber optics standard, ESCON, due to a combination of new architecture and faster link rates. (*See* architecture)
field installable connectors	optical connectors that can be installed on-site at a customer location, as opposed to factory installed connectors that can only be attached at an authorized manufacturing center.
flicker noise	noise with a $1/f$ spectrum, which occurs when materials are inhomogeneous.
flip chip mounting	a semiconductor substrate in which all of the terminals are grown on one side of the substrate. It is then flipped over for bonding onto a matching substrate.
flip flop	a device that has two output states, and is switched by means of an external signal.
flow shop	a manufacturing area that makes only one product.
footprint	the amount of desk or floor space used by a component.
frame	1. In the SONET transmission format, a 125 microsecond frame contains 6480 bit periods, or 810 octets (bytes), that contain layers of information. These layers include the overhead, or instructions between computers, and the voice or audio channel. SONET can be used to support telephony.

2. A technique used by web pages to divide the screen into multiple windows.

frame errors
errors from missing or corrupted frames, as defined by the ANSI Fibre Channel Standard.

frequency agile
capable of being easily adjusted over a range of operating frequencies.

frequency chirping
inducing either an up-chirp or down-chirp in an optical signal. (*See* up-chirp and down-chirp)

frozen process
a process where, if there are major changes of equipment, process parameters, agents or even parts, a partial or total requalification of the process must be performed.

fusion splice
(*See* splice)

gain
amplification. In a photodetector, the number of electron–hole pairs generated per incident photon.

GGP fiber
a specialty optical fiber, proprietary to 3M Corporation, which is more resistant to mechanical fractures when bent; used in optical connectors such as the VF-45.

Gigabit Ethernet (GBE)
a standard for high-speed Ethernet, which can be used in backbone environments to interconnect multiple lower speed internets. (*See* backbone network)

Gigalink card
a laser-based transceiver card that runs at approximately 1 Gb/s and works as a transponder.

graded index fiber
a fiber with a refractive index that varies with radial distance from the center.

gross profit
revenue minus cost of goods sold (COGS).

group velocity
the transmission velocity of a wave packet, which is made of many photons with different frequencies and phase velocities.

hermetic seal
a seal that air and fluids cannot pass through.

heterostructure
a semiconductor layered structure, with lattice matched crystals grown over each other.

Hill gratings
the first types of in-fiber Bragg diffraction gratings, named after the researcher who discovered them.

homologation
confirming that a product follows the rules of each country in which it is used.

hubbed ring   optical network architecture in which all data traffic flows through a single common location or hub. (*See* topology)

hybrid integration   a technique of developing smart pixels where optical devices are grown separately from the silicon electronic circuitry, and then are bonded together. (*See* smart pixel array)

image halftoning   an image compression technique whereby a continuous-tone, gray-scale image is printed or displayed using only binary-valued pixels.

Infiniband   an architecture standard for a high-speed link between servers and network devices. It will initially run at 0.3 GB/sec, but eventually scale as high as 6.0 GB/sec. It is expected to replace peripheral component interconnect (PCI). (*See* architecture)

intelligent optical network   optical network that also controls higher level switching and routing functions above the physical layer.

intermodal dispersion   (*See* dispersion)

intersymbol interference   the distortion by a limited bandwidth medium on a sequence of symbols, which causes adjacent symbols to interfere with each other.

intrinsic   without impurities.

jitter   The error in ideal timing of a threshold crossing event. The CCITT defines jitter as short-term variations of the significant instants (rising or falling edges) of a digital signal from their ideal position in time. Jitter can be both deterministic and random. Low-frequency jitter can be tracked by the clock recovery circuit, and does not directly affect the timing allocations within a bit cell. Other jitter will affect the timing. Data dependent jitter, a type of deterministic jitter (DJ), includes intersymbol interference (ISI). DJ may also include duty cycle distortion (DCD), sinusoidal jitter, and other non-Gaussian jitter. Random jitter (RJ) can be defined as the peak-to-peak value of the bit error rate (BER) of 10^{-12}, or approximately 14 times the standard deviation of the Gaussian jitter distribution. (*See* intersymbol interference, duty cycle distortion, bit error rate)

job shop — a manufacturing area that deals with a variety of products, each treated as a custom product.

jumper cable — an optical cable that provides a physical attachment between two devices or between a device and a distribution panel. In this way they are different from trunk cables.

junction capacitance — the capacitance formed at the pn junction of a photodiode.

lambertian — scattering that obeys Lambert's cosine law—the flux per unit solid angle leaving a surface in any direction is proportional to the cosine of the angle between that direction and the normal to the surface. Matt (not shiny) surfaces tend to be lambertian scatterers.

latency — the delay in time between sending a signal from one end of connection to the receipt of it at the other end.

lattice-matched — when growing a crystal layer on a substrate, the junction is lattice-matched if new material's crystal structure fits with the substrate's structure.

launch reference cable — a known good test cable used in loss testing.

layer — a program that interacts only with the programs around it. When a communications program is designed in layers, such as OSI, each layer takes care of a specific function, all of which have to occur in a certain order for communication to work. (*See* layered architecture)

layered architecture — a modular way of designing computer hardware or software to allow changes in one layer not to affect the others.

legacy product — a product an organization has already invested in, or has currently installed. New technology must be compatible with legacy products.

line state | a continuous stream of a certain symbol(s) sent by the transmitter that, upon receipt by another station, uniquely identifies the state of the communication line. For example Q for quiet or H for halt.

linearity range | the range of incident radiant flux over which the signal output is a linear function of the input.

link | the fiber optic connection between two stations, including the transmitter, receiver, and cable, as well as any other items in the system, such as repeaters.

link-level errors | errors detected at lower level of granularity than frames, as defined by the ANSI Fibre Channel Standard.

link budget | range of acceptable link losses.

link loss analysis | a calculation of all the losses (attenuation) on a link.

liquid crystals, nematic, smectic, and cholesteric | a material that has some crystalline properties and some liquid properties. In optics these materials usually have elongated molecules that are rod shaped. If they are oriented randomly they have different optical properties than when they are aligned. Nematic liquid crystals tend to have the rods oriented parallel but their positions are random. Cholesteric liquid crystals are a subcategory of nematic, in which the molecular orientation undergoes a helical rotation about the central axis. Smectic liquid crystals have a permanent dipole moment that can be switched by an externally applied electric field. This gives them many photonics applications, such as smart pixel arrays.

local area network (LAN) | a network limited to about 1 km radius.

lock-in-amplifier | a device used to limit noise by encoding input data with a known modulation, made by chopping the signal. The amplifier "knows" from this reference signal when the signal to be detected is on and when it is off. This allows the detector to low-pass filter the data, which narrows the bandwidth of the detector, making it more precise. (*See* chopper)

long wavelength	approximate 1300 nm (1270–1320) or approximately 1550 nm.
loopback testing	looping a signal back across a section of the network to see if it works properly. If a transceiver passes a unit loopback test, but fails a network loopback test, the problem is in the cables, not the transceiver.
loss	attenuation of optical signal.
machine capability	the statistical safety for processes performed by a tool or a machine.
macro-bending losses	losses due to nanometer size deviations in the fiber.
micro-bending losses	losses due to visible bends in the fiber.
manufacturing cost	the cost to manage procurement of parts, inventory, system assembly and test, and shipping costs.
margin	the amount of loss, beyond the link budget amount, that can be tolerated in a link.
master/slave	an architecture where one device (the master) controls other devices (the slaves).
mechanical splice	(*See* splice)
meshed rings	network topology in which any node may be connected to any other node. (*See* topology)
metrology	the science of measurement.
metropolitan area network (MAN)	interconnected LANs with a radius of less than 80 km (50 miles).
mini-zip	a type of zipcord fiber cable with a smaller outer diameter than standard zipcord.
mode mixing	the changing of the modal power distribution following a splice.
mode partition noise	within a laser diode, the power distribution between different longitudinal modes will vary between pulses. Each mode is delayed by a different amount due to the chromatic dispersion and group velocity dispersion in the fiber, which causes pulse distortion. (*See* distortion, dispersion, and group velocity)
mode scrambled launch	a type of optical coupling into a multimode fiber or waveguide that attempts to uniformly excite all modes in the target waveguide or fiber.

modulate/ demodulate	modulation is when one wave (the carrier) is changed by another wave (the signal). Demodulation is restoring the initial wave.
monolithic arrays	a technique of simultaneous fabrication of electronic and optical circuits on the same substrate, to produce high-speed smart pixels. (*See* smart pixel array)
Monte Carlo simulation	any type of statistical simulation which accounts for the probability of various events. This technique is useful when dealing with non-Gaussian probability distributions. An example is the probability of absorption and scattering of photons traveling through a medium.
multimode fiber	a type of fiber in which light can travel in several independent paths.
multiplexer	a device that combines several signals over the same line. Wavelength division multiplexing (WDM) sends several signals at different wavelengths. Directional division multiplexing (DDM) combines the laser diode and photodiode using a coupler. Time compression multiplexing is similar to DDM, when a diode is in the transmission mode and the other in the receiving mode ("ping-pong transmission"). Space division multiplexing (SDM) requires two fibers—one for upstream transmission, and one for downstream transmission. Time-division multiplexing (TDM) transmits more than one signal at the same time by varying the pulse duration, pulse amplitude, pulse position, and pulse code, to create a composite pulse train.
narrowcast (NC)	to direct a program to a specific, well-defined audience.
nematic liquid crystals	(*See* liquid crystals)
neural networks	a system of programs and data structures that simulates the brain. They use a large number of simple processors in parallel, each with local memory. The neural network is "trained" by feeding it data and rules about relationships between the data.

noise equivalent power (NEP)	the flux in watts necessary to give an output signal equal to the root mean square noise output of a detector.
noise floor	the amount of noise self-generated by a device.
numerical aperture (NA)	The NA defines the light-gathering ability of a fiber, or an optical system. The numerical aperture is equal to the sine of the maximum acceptance angle of a fiber.
Nyquist frequency	the highest frequency that can be reproduced when a signal is digitized at a given sample rate. In theory, the Nyquist frequency is half of the sampling rate.
octet	eight bits in a row. Also known as a byte.
ongoing inventory management cost	the cost to maintain and manage an inventory, as compared with procuring parts for inventory stocking.
opened rings	network topology in which a ring has been opened at one point and turned into a linear network with add/drop of channels at any points. (*See* topology)
optical bus	a facility for transferring data between several fiber-connected devices located between two end points, when only one device can transmit at a time.
optical bypass	an optical switch that diverts traffic around a given location.
optical interface	where the optical fiber meets the optical transceiver.
optical internetworking	allows IP switching layer to operate at the same line rate as a DWDM network, typically using an OC-48c connection.
optical power	the time rate of flow of radiant energy of a signal, expressed in watts.
optical seam	part of a network that does not allow passthrough of traffic meant for other nodes.
overshoot	waveform excursions above the normal level.
packet switching	protocols where data is encoded into packets, which travel independently to a destination where they are decoded.
paradigm	a model, example, or pattern.
parallel optical links	links that transform parallel electrical bit streams directly into parallel optical bit streams. Possible

mechanisms include transceivers made from VCSELs and array detectors sending data through fiber ribbon. (*See* VCSEL and fiber ribbon)

parts cost
: the sum of all bill of material components used in a design.

passivation layer
: an oxide layer coating a semiconductor, to reduce contamination by making the surface less reactive.

patch panel
: a hardware unit that is used as a switchboard, to connect within a LAN and outside for connection to the internet or a WAN.

phase-locked loop
: a circuit containing an oscillator whose output phase locks onto and tracks the phase of a reference signal. The circuit detects any phase difference between the two signals and generates a correction voltage that is applied to the oscillator to adjust its phase. This circuit can be used to generate and multiply the clock signal. (*See* clock generation/multiplication)

phase velocity
: the speed of a wave, as determined by a surface of constant phase.

phasor formalism
: a polar method of displaying complex (real and imaginary) quantities.

photoconductive
: the mode of a reverse-biased detector. This reduces the capacitance of the detector, and thus increases the speed of response of the diode. It is the preferred mode for pulsed signals.

photoconductor
: a non-junction type semiconductor detector where incident photons produce free charge carriers, which change the electrical conductivity of the material. Lead sulfide and lead selenide are examples of this type of detector, as well as most MSMs.

photolithography
: a process for imprinting a circuit on a semiconductor by photographing the image onto a photosensitive substrate, and etching away the background.

photomicrograph
: a photograph of an object that is magnified more than 10 times its size.

photonics
: the study of photon devices and systems.

photoresist implant mask
: used in photolithography to cover parts of a photosensitive medium when imprinting a circuit on a semiconductor

photovoltaic	the mode of a junction unbiased detector. Because 1/f noise increases with bias, this type of operation has better NEP at low frequencies.
pigtail	a short fiber permanently fixed on a component, used for connecting the component to the fiber optic system.
PIN (or p-i-n) photodiode	a diode made by sandwiching p-type (doped with impurities so the majority of carriers as holes), i-intrinsic (undoped), and n-type (doped with impurities so the majority of carriers as electrons) semiconductor layers. Photons absorbed in the intrinsic region create electron-hole pairs that are then separated by an electric field, thus generating an electric current in a load circuit.
plenum-rated cable	cable that can be used in a duct work system (plenum), which has smoke retardant properties.
plug repeatability	the variation in coupled power for multiple connections between the same components.
point-to-point transmission	carrying a signal between two end points without branching to other points.
polarized connector	a connector that can only plug in one position, so that it is aligned properly.
polarized light	light whose waves vibrate along a single plane, rather than randomly. There are several polarization modes such as TE, where the electric field is in the direction of propagation, and TM, where the magnetic field is in the direction of propagation.
preamplifier	a low-noise amplifier designed to be located very close to the source of weak signals. Often the first stage of amplification.
private mode	when a port or repeater receives only packets addressed to the attached node. Also known as normal mode.
process window	a defined variation of process parameters that characterize equipment and production used in series production.
processor	the part of a computer that interprets and executes instructions. Sometimes used to mean microprocessor or central processing unit (CPU), depending on context.

product cost apportionment (PCA) or supply chain costs	shipping and distribution expenses for raw material and assemblies from suppliers and delivery of some final product to the end customer.
promiscuous mode	when a port or repeater receives all packets, not only those addressed to the attached node.
protocol	the procedure used to control the orderly exchange of information between stations on a data link, network, or system. There are several standards, in code set; such as ASCII, transmission mode; asynchronous or synchronous, and non-data exchanges; such as contract, control, failure detection, etc.
pump laser diode	a short wavelength (typically 900 nm range) laser used to provide a pump input or gain in an optical fiber amplifier.
quantum confined Stark effect	a mechanism for changing the optical absorption of a quantum well by applying an electric field. Because of this effect quantum wells are used in optical modulators.
quantum efficiency (QE)	the ratio of the number of basic signal elements produced by detector (usually photoelectrons) to the number of incident photons.
quantum well	a heterostructure with sufficiently thin layers that quantum effects begin to affect the movement of electrons. This can increase the strength of electrooptical interactions by confining the carriers to small regions.
raised floor	a floor made of panels that can be removed for easy access to the wiring and plumbing below.
Raman scattering	scattering off phonons (quanta of vibration). A special case is Brillouin scattering, which is when an acoustic phonon (sound wave) is involved.
random jitter (RJ)	(*See* jitter)
refractive index	the ratio of the speed of light in a vacuum to the speed of light in a material at a given wavelength.
regeneration	insures that there is sufficient optical power for a signal to reach its destination.

repeater	a device placed in a data link to amplify and reshape the signal in mid-transmission, which increases the distance it can travel.
reshaping	removes pulse distortion caused by dispersion.
resonant cavity photodetectors	a photodetector made by placing a photodiode into a Fabry-Perot (FP) cavity to enhance the signal magnitude.
response time	the time it takes a detector's output to rise when subjected to a constant signal.
responsivity	the ratio of the detector output to the radiation input. It is usually expressed as a function of wavelength.
retiming	restores a timing reference to a signal to remove jitter and improve clock/data recovery.
ring topology	(*See* topology)
ringing	waveform overshoot and oscillations. (*See* overshoot)
roadmap	a long-range projection for the future of a type of product, for example, "The National Technology Roadmap for Semiconductors."
run length	the number of consecutive identical bits, such as the number of 1's or 0's in a row, in the transmitted signal. The pattern 010111100 has a run length of four.
running disparity	the difference between the number of 1's and 0's in a character, which is often tracked as a special parameter in network management software.
saturation	when the detector begins to form less signal output for the same increase of input flux.
scalability	the ability to add power and capability to an existing system without significant expense or overhead.
Schottky-barrier photodiodes	a variation on the PIN photodiode where the top layer of semiconductor material has been eliminated in favor of a reverse biased, metal-semiconductor-metal (MSM) contact. This results in faster operation, but lower signal. The advantage of this approach is improved quantum efficiency, because there is no recombination of carriers in the surface layer before they can diffuse to either the ohmic contacts or the depletion region. (*See* PIN photodiode)

scintillation	rapid changes in the irradiance of a laser beam.
serializer/ deserializer	a serializer is a device that converts parallel digital information into serial. A deserializer converts it back.
server	a central computer where data is deposited, and can be accessed over the network by other computers, known as clients.
serviceability	the ease in which a product can be serviced and inspected.
short wavelength	780–850 nm.
Shot noise	noise made by the random variations of in the number and speed of the electrons from an emitter.
shunt resistance	the resistance of a silicon photodiode when not biased.
signal-to-noise ratio	the ratio of the detector signal to the background noise.
simplex	a one-way communications line, which cannot both send and receive data.
single mode fiber	fiber where the light can only propagate through one path
skew	the tendency for parallel signals to reach an interface at different times. The skew for a copper cable is 2-nanosecond bit periods over 20 m, while for fiber ribbons it is under 10 picoseconds/m.
slope efficiency	the differential quantum efficiency of the laser combined with the losses of the optical coupling.
smart pixel array	an array of optical devices (either detectors, modulators, or transmitters) which are directly connected to logic circuits. By integrating both electronic processing and individual optical devices on a common chip, one may take advantage of the complexity of the electronic processing circuits and the speed of the optical devices.
smectic liquid crystals	(*See* liquid crystals)
space division multiplexing	(*See* multiplexer)
splice	to join together two pieces at their ends to form a single one. When applying to optical cable, to form a

permanent joint between two cables, or a cable and a port. There are two basic types of splice, mechanical splices and fusion splices. Mechanical splices place the two fiber ends in a receptacle that holds them close together, usually with epoxy. Fusion splices align the fibers and then heat them sufficiently to fuse the two ends together.

splice bushing — (*See* bulkhead splice)

splice loss — the loss in optical power due to splicing cables.

spoofing — sending acknowledgments of data transfer before data is actually received at its destination; usually done to artificially reduce latency in a network.

spun fiber — a manufacturing process in which the fiber is rotated during the drawing process in an effort to remove polarization dependence.

Stark effect — the splitting of spectral lines due to an incident electric field. The quantum confined Stark effect is a special case of that. It uses an electric field to wavelength modulate sensitivity of quantum well detectors. (*See* quantum confined Stark effect)

star topology — (*See* topology)

statistical process control — the use of statistical techniques to analyze, monitor, and control a process. Quality is often monitored in this way.

storage area network (SAN) — a high-speed network, or section of an enterprise network, of storage devices, for access by local area networks (LAN) and wide area networks (WAN).

strain relief — a design feature that relieves the pressure on a connector, which could otherwise cause it to crack or unplug.

striping — simultaneously allowing data transfer through multiple ports.

Strowger switch — developed in the late 1800s, this early telecommunications switch replaced human switchboard operators by automatically making phone connections using electrical signals on the phone line.

substrate — the base layer of support material on which crystals are grown. Products grown on substrates include

	semiconductor detectors and integrated circuits. A silicon wafer is an example of a substrate.
supply chain costs	(*See* product cost apportionment [PCA])
surface wave filter	a surface acoustic wave (SAW) phase filter often used to generate and multiply the clock signal. (*See* clock generation/multiplication)
switched base mode	operating mode of the IBM 2029 Fiber Saver DWDM device, in which unprotected channels are passed through a dual fiber optical switch that protects availability in the case of a fiber cut only, not in the case of equipment failure.
switched network	(*See* fabric)
synchronous	a system is synchronous if it can send and receive data at the same time, using a common timing signal. Because there is regular time interval between transmissions, there is no need for a start or stop bit on the message.
thermal noise	noise caused by randomness in carriers generation and recombination due to thermal excitation in a conductor; it results in fluctuations in the detector's internal resistance, or in any resistance in series with the detector. Also known as Johnson or Nyquist noise.
time	compression multiplexing. (*See* multiplexer)
time	division multiplexing. (*See* multiplexer)
timing reference, trigger	signals used by the oscilloscope to start the waveform sweep.
TO can	metal packaging in the form of transistor outline for semiconductor lasers.
token	a frame with control information, which grants a network device the right to transmit.
topology	the physical layout of a network. The three most common topologies are bus, star, and ring. In a bus topology all devices are connected to a central cable. (*See* optical bus and backbone) In a ring topology in which the terminals are connected serially point to point in an unbroken circle. In a star topology

all devices are connected to a central hub, which copies the signal to all the devices.

transceiver   a package combining a transmitter and a receiver.

transducer   a device that converts energy from one form to another. (*See* electrooptic transducer)

transimpedance amplifier   an operation amplifier and variable feedback resistance connected between the input and output of the amplifier. They are used to amplify low photodiode signals, and provide high dynamic range, good sensitivity, and bandwidth.

transponder   a receiver/transmitter that can reply to an incoming signal. Passive transponders allow devices to identify objects, such as credit card magnetic strips. Active transponders can change their output signal, such as radio transmitter receivers. Satellite systems use transponders to uplink signal from the earth, amplify it, convert it to a different frequency, and return it to the earth.

trigger   (*See* timing reference)

troubleshooting   a systematic method to find the reason for a problem.

trunk fiber   a fiber between two switching centers or distribution points.

two beam holographic exposure   a technique used to fabricate gratings using an expanded laser beam that is divided by a beamsplitter and then recombined, creating an interference pattern. This pattern is transferred photolithographically onto the surface of a semiconductor substrate.

ultrasonic   sound waves at a frequency too high for humans to hear. These waves can be used to excite metals used in wire bonding, or to clean items before placing them in vacuum.

undershoot   waveform excursions below the normal level.

unit interval (UI)   the shortest nominal time between signal transition. The reciprocal of baud, it has units of seconds.

up-chirp   a linear increase in frequency over time.

vertical cavity surface-emitting laser (VCSEL)   a semiconductor laser made from a bottom distributed Bragg reflector (DBR), an active region, and a top DBR.

virtual tributary	a SONET format with lower bandwidth requirements, to allow services like a DS1 or T1 signal to be carried on a SONET path without remultiplexing the voice channels.
wafer	a flat round piece of silicon that is used as a substrate on which to manufacture integrated circuits.
wavelength-division multiplexing	(*See* multiplexer)
wavelength-routed networks	a network with several routers which choose the light path of the signal in the network by its wavelength. This architecture is being developed for all-optical networking, and has WDM applications.
wide area network (WAN)	a network that is physically larger than a LAN, with more users.
wire bond	connecting wires to devices.
word	four contiguous bytes.
zipcord	a type of optical fiber consisting of two unibody cables connected in the middle by a thin, flexible outer coating.

Some useful dictionaries on the World Wide Web
The One Look Dictionary www.onelook.com
The Photonics Dictionary www.photonicsdictionary.com
Academic Press Dictionary of Science and Technology
www.harcourt.com/dictionary/browse/

Index